CLINICAL
ASPIRATION
CYTOLOGY

Edited by

Joseph A. Linsk, M.D.

Chief, Department of Oncology/Hematology
and Associate Pathologist
Atlantic City, Medical Center
Atlantic City, New Jersey

Sixten Franzen, M.D.

Emeritus Director
Cytology Laboratory
Karolinska Hospital
Radiumhemmet
Consultant in Cytology
Sophiahemmet
Stockholm, Sweden

with 12 contributors

CLINICAL ASPIRATION CYTOLOGY

Second Edition

PHILADELPHIA

J. B. LIPPINCOTT COMPANY

CAMBRIDGE   NEW YORK   ST. LOUIS   SAN FRANCISCO   LONDON   SINGAPORE   SYDNEY   TOKYO

Acquisitions Editor: Lisa McAllister
Sponsoring Editor: Delois Patterson

Compositor: Progressive Typographers
Printer/Binder: Halliday Lithograph

The authors and publisher have exerted every effort to ensure that drug selection and dosage set forth in this text are in accord with current recommendations and practice at the time of publication. However, in view of ongoing research, changes in government regulations, and the constant flow of information relating to drug therapy and drug reactions, the reader is urged to check the package insert for each drug for any change in indications and dosage and for added warnings and precautions. This is particularly important when the recommended agent is a new or infrequently employed drug.

Library of Congress Cataloging-in-Publication Data
Clinical aspiration cytology.

Includes bibliographies and index.
1. Diagnosis, Cytologic.   2. Biopsy, Needle.
I. Linsk, Joseph A.   II. Franzén, Sixten.
[DNLM: 1. Biopsy, Needle.   2. Cytodiagnosis—
methods.   WB 379 C641]
RB43.C54   1989   616.07′582   88-13310
ISBN 0-397-50826-3

To the memory of Nils Söderström

Contributors

GERT AUER, M.D.
Associate Professor of Pathology
Karolinska Hospital
Stockholm, Sweden

DAVID B. KAMINSKY, M.D., F.I.A.C.
Chairman, Department of Pathology
Eisenhower Memorial Center
Rancho Mirage, California

MEHRDAD NADJI, M.D., M.I.A.C.
Professor of Pathology
Director, Cytopathology
University of Miami School of Medicine
Miami, Florida

RAFFAEL PERRONE H. DONNORSO, M.D.
Professor and Director, Department of Cytology
Regina Elena Tumor Institute
University of Rome
Rome, Italy

TORSTEN LÖWHAGEN, M.D.
Attending Cytopathologist
Cytology Laboratory, Radiumhemmet
Karolinska Hospital
Stockholm, Sweden

ALBERT J. SALZMAN, M.D., F.A.C.R.
Associate Clinical Professor of Radiology
Hahnemann School of Medicine
Philadelphia, Pennsylvania
Chief Attending Radiologist
Atlantic City, New Jersey

MIGUEL A. SANCHEZ, M.D.
  Associate Director of Pathology and Laboratory Medicine
  Englewood Hospital
  Englewood, New Jersey
  Clinical Associate Professor of Pathology
  New York University School of Medicine
  New York, New York
  Visiting Assistant Professor of Pathology
  Albert Einstein College of Medicine
  New York, New York

BERND–UWE SEVIN, M.D., PH.D.
  Associate Professor, Department of Gynecologic Oncology
  University of Miami School of Medicine
  Miami, Florida

JAN–SILVESTER WILLEMS, M.D.
  Attending Cytopathologist
  Cytology Laboratory, Radiumhemmet
  Attending Pathologist, Department of Pathology
  Karolinska Hospital
  Stockholm, Sweden

LAMBERT SKOOG, M.D., Ph.D.
  Associate Professor
  Karolinska Institute
  Attending Cytopathologist
  Division of Clinical Cytology
  Karolinska Hospital
  Stockholm, Sweden

EDNEIA M. TANI, M.D., Ph.D.
  Associate Professor
  Karolinska Institute
  Attending Cytopathologist
  Division of Clinical Cytology
  Karolinska Hospital
  Stockholm, Sweden

ANDERS ZETTERBERG, M.D.
  Associate Professor of Pathology
  Karolinska Hospital
  Stockholm, Sweden

Preface

It is now over 50 years since the first substantive clinical paper on aspiration biopsy cytology (ABC) was published by Martin and Ellis and over 30 years since the procedure has been applied to a large volume of cases at the Karolinska Hospital in Stockholm. The number of publications, meetings, and tutorials dealing with this clinicocytologic tool has risen dramatically, particularly in the last 10 years. Not only have the basic cytopathologic descriptions been repeatedly presented, but many articles have appeared with cytologic descriptions of rare tumors, infections, and inflammations.

The acceptance of this method as a valid clinical tool in Scandinavia and other parts of Europe preceded by 20 years its current focal utilization in the Western Hemisphere. In spite of the increasing attention in both geographic areas, it is important to realize that the bulk of literature has appeared in specialized journals of cytology and in texts often not readily available at booksellers or most hospital libraries. Most of the work and teaching occur in tertiary academic centers and are only slowly filtering into local and regional hospitals, clinics, and offices.

One continuing source of cytologic material as well as focus of interest in the procedural aspects of ABC is the specialty of interventional radiology. Intrathoracic and nonpalpable intra-abdominal masses have been the main targets. Although these procedures are carried out in many community hospitals, it must be recognized that the vast bulk of targets suitable for aspiration, namely palpable lesions, are seen in the outpatient setting.

The technique and cytopathology of ABC have not yet become part of medical school curricula. With the exception of a few major medical centers, residency training programs in medicine, surgery, and pathology have generally not incorporated the findings of ABC into the training programs. ABC has not yet become an integral part of the average physician's diagnostic armamentarium. Finally, the method is used primarily at sophisticated medical centers and has not yet contributed to the medical needs of vast geographic areas of the world.

It is for these reasons that the contents of this new edition are directed to the clinical training of medical students, residents, and practicing physicians, as well as to pathologists. Basic cytomorphologic patterns as well as descriptions of unusual tumors and non-neoplastic disorders including immune deficiency disease are presented within a clinical context. The findings in normal tissue and technical aspects of aspirating including imaging procedures are included in each chapter, where appropriate.

The separate chapter on imaging techniques has been updated. The cytologic smears depicted are placed within a clinical context and are presented as they might appear during the course of routine aspirations on a medical or surgical service. That magnification which reveals the most important diagnostic features of each lesion is presented and will vary from case to case.

Finally, advances in the study of DNA content, electron microscopy, immunocytochemistry, and estrogen receptor determinations on cytologic material, are included in a separate chapter.

Joseph A. Linsk, M.D.
Sixten Franzen, M.D.

Preface to the First Edition

The place of aspiration biopsy cytology (ABC) as a diagnostic technique as well as a source of research data has now been secured. Pioneer papers from the 1930s, 1940s, and 1950s were lonely harbingers of the avalanche of interest and reports appearing in the past few years. The majority of well-developed institutional programs have appeared in European clinical centers, but tutorials, seminars, and training programs are now finding sponsorship in North America. Most recent reports are concerned with technical methods of reaching previously unaccessible sites, such as intrathoracic and retroperitoneal targets, and cytomorphologic studies of specific organs such as thyroid, breast, pancreas, and so on.

We attempt with this volume to deal holistically with the entire body, illustrating the diagnostic methods as well as the cytologic yield, expected and unexpected, obtained from all anatomic sites. The relationship of individual organs to other potential targets in this region (*e.g.,* salivary glands and lateral neck masses) is stressed, as well as the interrelatedness of clinical medicine and morphologic diagnosis. Particularly with respect to tumors, ABC is presented as a direct bridge between the lesion and final diagnosis. As a result, this volume will be valuable to clinicians as well as pathologists. It presents many of the problems of oncology from a unique viewpoint in that direct and immediate diagnosis becomes potentially available in the clinician's office, in oncologic, hematologic, radiotherapeutic, and surgical clinics, and at the bedside.

The text is arranged in the following manner. Each anatomic region — head and neck, thorax, abdomen, and pelvis — is dealt with as a unit to emphasize the clinical presentation of an often unknown pathologic process. Within the regions and as separate chapters, special organ and tissue cytology are considered in detail. Abundant photographs illustrating many varieties of all important, as well as rare, tumors and related lesions are distributed throughout the text. The illustrations do not all present classical morphologic patterns, not often found in practice. Legends form an important part of the text.

The material presented has been selected from large slide and case archives accrued and studied over a period of 30 years. Many of the cell patterns and clinical concepts described here, however, are not new. They appeared originally in the publications of important contributors to the field of ABC. This volume has in effect been erected on the shoulders of Drs. Martin and Ellis, Franzen, Lopes – Cardozo, Söderström, Zajicek, and others. It is designed to broaden the diagnostic skills of both the clinician and the histopathologist. The authors hope this goal has been fulfilled.

Joseph A. Linsk, M.D.
Sixten Franzen, M.D.

Acknowledgments

The preparation of this second edition was initiated after seminal discussions with Lisa Biello, Senior Editor at J. B. Lippincott, who encouraged an increase in textual material and acquiesced to a significant expansion of the critical color plates. Delois Patterson of the Medical Editorial Department was extremely helpful in providing material from the first edition for us and our contributing authors. She also made valuable suggestions concerning problems of cropping old material and introducing new.

Thanks to Debora Conrad for helping prepare slides as well as providing valuable secretarial aid and to Susan Kitz, Janet Sulpizi, and Bernadette Mastracola for secretarial help.

We are grateful to colleagues who have responded to inquiries or offered case material, microphotographs, and suggestions including Maged Khoory, M.D., C. K. Lin, M.D., Pierre Luigi Esposti, M.D., William J. Frable, M.D., and Jack Lubin, M.D.

In the critical area of microphotography we are particularly grateful to colleagues who have offered their time and counsel including Dorothy Rosenthal, Jefferson Davis Morgan, and Ahmad Dibadj.

Finally, to the contributors to this text who have sequestered time for preparation from extremely busy professional lives, we understand the efforts involved in meeting deadlines. We welcome Drs. David Kaminsky, Miguel Sanchez, Gert Auer, Anders Zetterberg, Lambert Skoog, and Edneia Tani who have contributed entirely new and exciting material. We are pleased that Dr. Raffael Donnorso could join with Dr. Kaminsky in preparation of the critical chapter on thoracic lesions. We are also entirely pleased with the efforts of Drs. Albert Salzman, Mehrdad Nadji, Bernd-Uwe Sevin, Jan-Silvester Willems, and Torsten Löwhagen in enlarging and enriching their chapters.

Contents

Acronyms

P	Papanicolaou
H & E	Hematoxylin and eosin
R	Romanowsky
MGG	May – Grünwald — Giemsa
LN	Lymph node
FNA	Fine needle aspiration
ABC	Aspiration biopsy cytology
BMD	Benign mammary dysplasia
FAN	Fibroadenoma
S – R cell	Sternberg – Reed cell
N/C ratio	Nuclear: cytoplasmic ratio
Lg bodies	Lymphoglandular bodies

List of Color Plates

CLINICAL
ASPIRATION
CYTOLOGY

Introduction

JOSEPH A. LINSK

SIXTEN FRANZEN

Aspiration biopsy cytology (ABC) is a branch of diagnostic cytology that interprets changes in cells extracted from within organs, tumors, and nonneoplastic abnormal tissues. ABC contrasts with exfoliative cytology, which studies cells shed or scraped from surface epithelia or mesothelia. The diagnostic criteria of both branches have many common features as well as many important differences.

Morphologic features common to aspirated and exfoliated cells include nucleolar and nuclear size, shape, number, texture, and tinctorial characteristics. Nuclear-cytoplasmic ratio and cytoplasmic shape, texture, and content are additional structural characteristics.

BRANCHES OF CYTOLOGY

Exfoliative

Although diagnosis of carcinoma from exfoliated cytologic material is aided by knowledge of clinical data, in practice a cancer diagnosis is most often rendered by cytotechnologists with confirmation by pathologists on morphologic criteria alone.[6,10,14]

The identification of appropriate criteria in individual cells is paramount. These cancer-specific criteria may appear after a series of graded changes, each of which can also be identified (i.e., normal, atypia, dysplasia, cancer). Subtle alteration in size, shape, texture, density, and tinctorial aspects of cell and nuclear membranes, nucleoli, cytoplasm, and chromatin may define a cancer cell. Cell groups and clusters have diagnostic significance in specimens scraped from surfaces (cervical scrape) and also in body fluids (adenomatous and papillary groupings). However, individual cell cytology remains the most significant visual data in arriving at a diagnosis.

On screening one or many slides prepared from an exfoliated specimen, few abnormal cells and cell clusters may be found. The diagnosis of dysplasia and cancer is painstaking and requires extensive training, but it is regularly and remarkably made on often sparse cellular evidence.

Aspiration

In contrast, fine needle ABC usually yields many diagnostic cells, often hundreds or thousands on a single slide. With such a profusion of material, group patterns become important; and individual cell cytology with its minute variations may require less emphasis. As with the normal-appearing lymphocyte in a well-differentiated lymphoma, individual cancer cells may appear relatively bland and should be considered within the context of the entire smear.

Occasionally, screening may be required, particularly if the aspirated target is necrotic or from an organ in which discrete metastatic cells may be extracted with normal cells. However, rapid identification often becomes feasible with experience. Cells are distributed in sheets, clusters, groups, follicles, papillary fronds, and singly. In some instances, extracted stroma provides additional diagnostic criteria (see the section on fibroadenoma in Chap. 6).

Cellular and stromal contents provide consistent diagnostic criteria that permit rules for instruction and learning, as in other morphologic disciplines. For pathologists, material for study is readily available at the surgical bench, where specimens can be freely aspirated, slides prepared, and cytologic diagnosis correlated with histopathology. Experience has shown that trained pathologists who attend appropriate seminars and study preparations from their own laboratories will rapidly increase their level of proficiency. Of course, difficulties and pitfalls exist, and caution is needed at all times.

Acceptance of ABC as an important clinical tool and pathologic diagnostic method has to some extent been curtailed because of tradition. The experience of one of us in trying to introduce ABC to a major tumor institute has been recounted.[8,9] Many of the current ABC enthusiasts have had and continue to have problems introducing and fostering the method among fellow pathologists and clinicians for economic, political, and egoistic reasons, among others. In the current regulatory and competitive climate to which a traditionally conservative profession has been subjected, tilting toward and yearning for the old ways are understandable.

While the strength of tradition as well as other factors has influenced general acceptance of the method, it is no longer necessary to justify it. In the past, time and space have been devoted to establishing its validity as a primary diagnostic tool. Much emphasis has been placed on false-positive and -negative results and on correlation with histopathologic findings. At present, the only nonnegotiable caveats are that the cytopathologist have requisite training and experience before assuming responsibility for tumor and other diagnoses and that negative reports become the responsibility of the clinician, who must then decide to accept the report, repeat the aspiration, or proceed to surgical biopsy.

MATERIALS

Syringes

Various-sized syringes and holders and syringe modifications[13] have been used successfully. The one-handed Franzen syringe and holders for disposable plastic syringes are illustrated in Figures 10-1 and 11-1. The syringe chamber should allow 10 ml of suction.

Needle Guide

The curved Franzen needle guide (see Figs. 10-1 and 11-1) is generally used. The proximal end is flared;

and the distal ring fits over the index finger, allowing fingertip palpation of the lesion. The mobile flat plate, attached midway, helps to stabilize the guide against the palm of the hand. The curved hollow rod is rigid.

Needles

Fine needles range from 25 to 20 gauge or 0.6 mm to 0.9 mm in outer diameter. Lengths vary from 1 cm (0.5 in) to 20 cm (9 in). The finer the needle, the less trauma it causes. Longer fine needles are flexible and require a needle guide or stylet for control.

Glass Slides

Good-quality glass slides are desirable; and the smear can be made by placing one glass slide on another, if done carefully. Coverslips, particularly the Bürcker variety, may give better-quality smears in some hands.

Stains

The two major stains used in virtually all cytologic laboratories are the Papanicolaou (P) and a Romanovsky (R). Some pathologists prefer hematoxylin and eosin (H & E) stain. P stain is applied to wet-fixed material. Fixation is obtained by immersing the slide in 95% ethanol or spraying with several varieties of fixative. As an inexpensive and suitable substitute, hair spray has been used. P stains nucleoli, nuclear chromatin, and nuclear membranes and defines parachromatin spaces with much greater clarity than R stains. Hyperchromatism can be gauged, and cytoplasm of squamous cells is more clearly identified.

P stain is particularly valuable in thick smears and when cells are obscured by clotted blood and stromal content. It is critical in evaluating potential squamous cytology and should be used routinely in smears from metastatic lymph nodes, skin lesions, skin nodules, salivary glands, lungs, and pelvic masses.

May–Grünwald–Giemsa (MGG), applied to air-dried material, is one of a group of R stains, all of which produce similar color changes in the cell components. A rapid stain (Diff-Quik*) is particularly suitable in the office or clinic to obtain diagnostic

---

* Harleco, Gibbstown, NJ.

results while the patient waits. R stains cytoplasm, inclusions, granules, and intercellular substance very well. Nuclear chromatin, however, is not clearly delineated, and hyperchromatism cannot be judged. Chromatin may indeed be quite bland in malignant cells (see adenoid cystic carcinoma, Chap. 5). Nucleoli, though sometimes poorly visible, may appear as large irregular blue structures. The R stain's major use is in the study of endocrine, adenomatous, and hematopoietic tumors. Poorly differentiated and primitive tumor cells have uniform, finely granular nuclear chromatin (see Figs. 13-18 and 14-14). The leukemic blast cell is a prototype of this.

Tumor cells generally do not have light microscopic cytoplasmic inclusions or structures. Breast carcinoma and hepatoma, however, may contain characteristic cytoplasmic inclusions (see Chaps. 6 and 8). Similarly, carcinoid, endocrine, and neuroendocrine tumors contain cytoplasmic granules evident with R stain (see Chap. 8).

In practice, air-dried smears are sometimes more easily handled and more constantly fixed. The problem of rapid drying of small quantities of material before fixation is important.

Controversies concerning which stain is preferable are irrelevant. The cytopathologist should use the stain that most easily allows him to make a correct diagnosis.[3,4,9] In a prospective study of P and Wright–Giemsa staining of fluid cytology, independent observers were in complete diagnostic agreement in 87% of the cases. Differences in the remaining 13% were not considered clinically significant. Smears considered malignant by one method were at least atypical or suspicious by the other.[2] The morphology of body fluids is similar to that of aspiration smears, and this study therefore is quite relevant.

The choice of stain is often a result of habit. Exfoliative smears are routinely stained by the P method. Students, pathology trainees, and cytotechnologists learn cytologic diagnosis of material stained exclusively by that method. Hematologists routinely use R stains for primary diagnoses. In addition, the cytopathologist may express an interest in different material (air-dried versus fixed smears) but find that the clinicians are less compliant than anticipated. Experience has shown that study of both P- and R-stained smears provides the best overall bases for accurate diagnosis. Each stain may, however, highlight any of the following specific morphologic features of aspiration smears: individual intact cells, others stripped of cytoplasm, a variety of cell group-

ings, cytoplasmic and intranuclear inclusions, stromal components with multiple tinctorial characteristics, background fluid, and numerous cell fragments, cell shadows, necrotic debris, and calcific formations (see below.)

Mucin (Plate XII C), acid mucopolysaccharide (Plate VI A), and colloid (Plate I B) provide characteristic staining qualities with some consistency with R stains, but all staining methods may contribute.

The following is a list of textbook contributors to the field of aspiration cytology with a designation of the types of stains illustrated.

Frable[3]	P, Diff-Quik
Zajicek[15]	P, MGG
Kaminsky[5]	P
Koss[7]	P, H & E
Orell[11]	H & E, MGG
Lopes–Cardoza[10]	MGG

The largest clinical experience with ABC (S.F.[10]) has been based on the use of MGG. Subdivisions of tumor diagnosis (see above), in addition to the simple identification of cancer, rely heavily on cytoplasmic and stromal changes not readily identifiable in P and H & E stains.

METHOD

The routine method of aspiration for palpable masses is described below. Special techniques and aids for aspiration of the brain, base of skull, intrathoracic structures, kidney, breast, thyroid gland, retroperitoneum, prostate, and pelvic organs are discussed in their respective sections.

To proceed, thoroughly palpate the target area and delineate the most suspicious, usually the most firm, portion. Fix the mass with the palpating hand (see Fig. 1-1). An alcohol pad is sufficient for local sterilization. Gloves are not required. Briskly insert the needle into the target area without anesthesia (Fig. 1-1). (Exceptions to this step and the need for local anesthesia are discussed in the appropriate chapters.) Apply a full 10 ml of suction to the syringe while moving the needle short distances in one channel (Fig. 1-2). This produces a greater yield than if the needle direction is changed. The operator can judge the extent of these movements after evaluating yields in successive patients. Discontinue the aspiration if blood appears in the syringe tip (see Fig. 4-2). Release

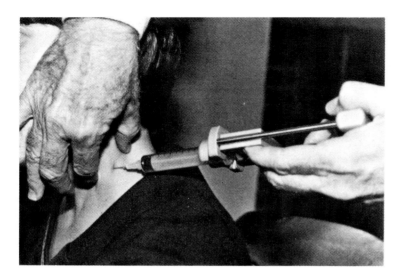

Fig. 1-1. The mass is transfixed and the needle is inserted briskly. (From Linsk JA, Franzen S: Fine Needle Aspiration for the Clinician, p. 19. Philadelphia, JB Lippincott, 1986)

suction before withdrawing the needle, and then apply pressure to the puncture site with the patient's or an aide's assistance. Rapidly separate the needle containing aspirated material from the syringe and draw air into the syringe (Fig. 1-3). Re-attach the needle to the syringe and express the material on a glass slide, generally forming a drop or droplets with small particles of tissue.

The Smear

Making the smear is critical because it determines the quality of the preparation the microscopist will examine. In practical terms, it may be the most important maneuver in the whole range of steps in the aspiration. The quality of referred slides can be frustrating, even though ample material has been extracted.

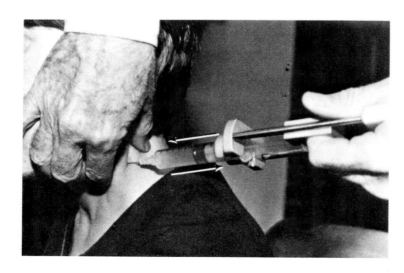

Fig. 1-2. With suction applied, the syringe and needle are oscillated (arrows) in one channel. (From Linsk JA, Franzen S: Fine Needle Aspiration for the Clinician, p. 20. Philadelphia, JB Lippincott, 1986)

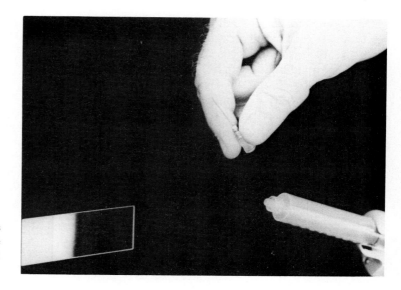

Fig. 1-3. The needle containing the aspirate is separated and air drawn into the syringe. The needle is reattached and the material expelled in droplets onto the slide.

Droplets deposited at one end of the slide may be collected under the smearing slide or coverslip and then pulled like a blood smear. Particles will gather at the end of the smear and may then be further gently crushed out in the manner of bone marrow particles (Fig. 1-4).

Often, however, the material is too solid to be pulled, and it is then gently crush-smeared with the smearing slide or an angled coverslip. The smearing slide is then lowered gently to draw the particles out into a thin preparation (Fig. 1-4).

Smearing must be gentle enough to avoid cell damage but selectively firm to convert tissue fragments into a cellular smear. If the volume of material is large, decanting the slide will reveal fragments. These are sequestered and also crush-smeared. If there is sufficient material, some of it can be trans-ferred to another slide. Smears are prepared from frothy, fatty, mucinous, and serous droplets even if particles are not identified, since cellular content may be surprisingly rich.

The reader is directed to the paper on smearing techniques authored by J.S. Abele and colleagues.[1]

NORMAL CELL CYTOLOGY

Knowledge of normal cell morphology and its non-malignant alterations is an important basis for all diagnostic cytology. It is particularly important in ABC, in which the aspirating needle will traverse normal tissue to reach the target and may extract mixtures of normal, inflammatory, necrotic, meta-plastic, and malignant cells.

Fig. 1-4. The aspirate is drawn out with a Bürcher coverslip (or another slide), and the particles are then crush—smeared and feathered out.

Nonmalignant cells fall into two groups. *Individual organs* such as salivary glands, breast, and liver yield identifiable single cells and cell groups with normal or varied but benign morphologic characteristics. Knowledge of the site of puncture will usually reduce any difficulty in identification. *General tissue cytology* consists of cells originating in capillary vessels, blood, lymph and fat, skeletal, joint, muscle, and cutaneous tissues. Many of these cells are considered in the following regional and individual organ chapters.

HISTOPATHOLOGY AND CLINICAL EVALUATION

Diagnosis by fine needle aspiration cytology is rooted in two distinct and equally important disciplines: histopathology and clinical evaluation. A background in tissue pathology aids in interpretation of intact cytologic structures (alveolar groups, cell balls, cell molding, and linear arrangements of columnar or cuboidal cells) as well as the nature of dispersed, usually poorly differentiated cells.

Histopathology will often follow cytology: at the time of definitive surgery when there is a need for tissue structure, as in lymphoma, because of inconclusive cytologic diagnosis, or because of lack of confidence in the method. It is not rare, however, in our experience, for aspiration cytology to reveal a diagnosis not evident on follow-up histopathologic examination. In addition, an apparent "false-positive" cytology based on negative histopathologic findings has been ultimately determined to be "false-negative" histopathology. A group of such cases is to be published (S.F.).

THE INTERVENTIONAL CYTOLOGIST

Erosion of the tradition of history-taking and physical examination because of emphasis on technological imaging and laboratory techniques has taken place in the training and practicing phases of modern medicine. This has been broadly addressed in our textbook, *Fine Needle Aspiration for the Clinician.*[9] Aspiration biopsy cytology has now extended the role of the cytologist well into this clinical arena. Although many cytopathologists are content, indeed prefer, to receive slides for morphologic diagnosis without seeing the patient, the addition of clinical data is deemed essential in rendering most diagnoses.

Cytologists and cytopathologists who secure their own specimens have become interventional cytologists, much as the radiologists carrying out invasive procedures are termed interventional radiologists. The advantages of hands-on cytology, assessing the mass, penetrating the target precisely to the satisfaction of the operator, and preparing smears in specific manners for specific stains, are unparalleled. With these advantages, however, come responsibilities.

Carrying out the role of the clinician requires the same clinical evaluation expected of a remote clinician referring slides. The cytopathologist will have to freshen his clinical skills. Although he will have the advantage of full understanding of pathologic processes and differential pathologic diagnoses, the act of seeing the patient includes a review of the history and a careful physical examination that will often extend beyond the site of the target lesion.

More commonly, pathologists are recipients of cytologic specimens. They should maintain a close working relationship with the clinician. Clinical history, description of the lesion, and clinical diagnosis are a mandatory part of the referral request.

Targets for aspiration often appear as masses of unknown origin. For this reason, *Clinical Aspiration Cytology* uses the regional approach to diagnosis. The regions covered are head and neck, thorax, abdomen, and pelvis. Clinical features and differential diagnosis of unidentifiable masses in, for example, the abdomen become part of the data leading to the final cytopathologic diagnosis. When the pathology is obviously organ specific, chapters are devoted to the varieties of pathologic changes. Organs covered by chapter are brain, skeleton, thyroid, salivary gland, breast, testis, prostate, and skin. Cytologic aspects of sarcomas and lymphomas, which may involve any structure, are also considered in separate chapters.

All negative smears are interpreted in the light of clinical findings. No negative report definitively rules out a disease process, since the needle may have missed the target.

In recent years, clinical procedures using the technology available in most radiology departments have extended the range of diagnostic cytologic specimens presented to the cytopathologist. These specimens include aspects of brain, base of skull, thyroid, lung, mediastinum, breast, liver, retroperitoneum, pelvic organs, prostate, and bone. It is valuable for the cytologist to maintain a cooperative exchange with radiologists, surgeons, and other clinicians carrying out these procedures.

ASPIRATION: WHAT IS FOUND?

All tissues and cystic fluids can be aspirated. The lack of adhesion of tumor plugs to the enveloping lymphatic walls and to the stroma of the tissue spaces invaded by the tumor allows one to withdraw large numbers of cells. This can be visualized by reviewing the tissue section in Figure 1-5 (see also Fig. 6-50). Tumor cells are extracted in numerous cohesive plugs (Fig. 1-6). Lack of cancer cell cohesion in poorly differentiated tumor varieties also encourages extraction and dispersal of cells on smears (Fig. 1-7). The connective tissue network woven *in situ* by collagen fibers is not easily extracted; however, benign cells are sometimes extracted in surprising numbers. Examples are fibroadenomas (see Chap. 6) and hyperplastic prostate (see Chap. 11). Diagnostic stromal and background material may also be obtained in fibroadenomas and salivary gland, testicular, and other tumors. Yield from lymphatic tissue, both benign and malignant, is generally large. Fat is easily

Fig. 1-6. Recurrent colon carcinoma, fixed abdominal mass. This is a densely cellular smear composed of discrete clusters of densely coherent, well-differentiated cells. (R stain, ×200)

extracted from lipomas and breast and subcutaneous tissue. It can be identified grossly on the smear before staining (see Fig. 1-8), and such observation may suggest a repeat aspiration in, for example, a breast mass. On the other hand, epithelial and nonepithelial cellular clumps and clusters have a fairly characteristic appearance on unstained slides resembling a cellular bone-marrow smear. Benign striated muscle (Fig. 1-9) and neural tissue (see Chap. 17) are sometimes obtained with a fine needle and have characteristic appearances.

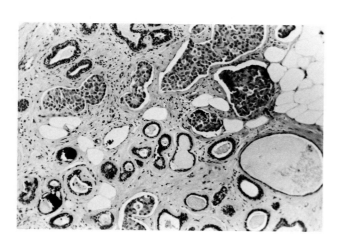

Fig. 1-5. Plugs of tumor cells within lymphatics. Tissue section. Such tumor tissue may provide an abundant tumor cell yield, leaving stroma *in situ*. (H & E stain, ×120)

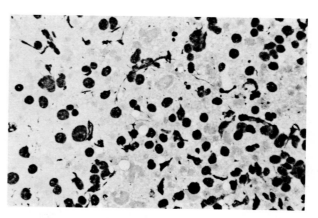

Fig. 1-7. Anaplastic prostate carcinoma. Shown are dispersed, poorly differentiated carcinoma cells. Nuclei are fragile with prominent nucleoli *(arrow)*. Cytoplasm is not evident. (R stain, ×400)

Fig. 1-8. Unstained smear. Lipoma. Glistening fatty droplets are evident grossly.

Inflammatory cells, particularly histiocytes, are easily identified but can be confused with tumor cells. Epithelioid cells, giant cells, and foreign material such as sutures and asbestos bodies (see Fig. 7-51) help in the diagnosis of granulomas and tumors.

Mucin, colloid, and acid mucopolysaccharide are also obtained from appropriate lesions in the gastrointestinal tract, breast, lung, salivary gland, and thyroid gland. They provide useful diagnostic information. Pigment (including melanin, hemosiderin, and bile) can be identified by clinical context as well as by special stains.

Finally, necrosis, cell shadows, proteinaceous material, and blood, old and new, provide diagnostic aid.

Multiple sites of cyst formation are easily aspirated. These sites include breast (fibrocystic disease), thyroid (degenerated colloid cyst and papillary carcinoma), parotid (Warthin's tumor and mixed tumor), branchial cysts, and cavitary squamous carcinoma.

Fig. 1-9. Striated muscle. Striations are evident in this high-power view. (R stain, ×330)

THE CYTOPATHOLOGIC DIAGNOSIS

Morphologic diagnosis begins with observation of the gross material on the glass slide. This is available to the pathologist only if he carries out the aspiration. The following observations may be made.

1. Clear cystic fluid will rarely yield diagnostic cellular content. Many cystic fluids are not submitted by clinicians, based on their experience of extremely low cytologic yield. Cystic fluids are discussed in Chapter 6. Smears, both air-dried and fixed, may, however, be prepared. Thyroid, salivary, and breast cystic fluids may yield cancer diagnoses; these are considered in Chapters 4, 5, and 6.

2. Turbid cystic fluid is more likely to yield cytologic clues of an inflammatory or neoplastic disorder, and slides should be prepared. All cystic fluid should be observed for tissue particles, which are then teased out and smeared. Expelling fluid into a watch glass (see Fig. 4-25) may facilitate this procedure. Turbid fluids may be centrifuged and the sediment smeared in selected cases.

3. Gross fat is easily identified. It forms droplets on glass (see Fig. 1-11). Smears prepared from a clinical lipoma may not necessarily require staining. Clinical judgment is needed. If there is suspicion of tumor or epithelial cells in fat (see Chap. 6), all slides should be stained. The gross appearance of fat may be deceptive.

4. Bloody (sanguinous) aspirates usually consist of abundant fluid material. If pure blood is drawn into the syringe, a vessel has been entered and the aspiration should be terminated. Mixed bloody yield appears as a pool, droplets, or a

drop of fluid. Blood clotting may occur rapidly, and what appear to be small tissue fragments may be aggregates of platelets. Multiple smears should be prepared. Tumor cells may be sequestered in a clotted portion.

5. Gelatinous semisolid material is the yield from a needle well placed within a lymph node or lymphoid mass (see Chap. 14). There is ample time before drying to prepare fixed and air-dried smears.

6. Fluid aspirates containing tissue particles or semisolid material with a granular structure are the usual yield from cellular benign and malignant tumors. Thousands of cells have been extracted to form cohesive mounds, and the cellularity of the aspiration is readily recognized in the unstained smear.

7. Sparsely cellular fibrotic tumors and nodules often yield small aspirates after radiation therapy. There may be a drop of tissue fluid that should be smeared and examined for occasional diagnostic cancer cells.

8. Noncellular targets punctured inadvertently or by design consist of tissue fluid, red blood cells, some white blood cells, and sometimes stromal strands.

After observation of the gross specimen on a glass slide, when this is possible, accurate diagnosis is heavily dependent on the availability of optimum smears and staining.

Before scanning a smear under low power, it is useful to glance directly at the glass slide. It is easy to distinguish smears composed of dispersed cells from those in which the cells form cohesive clusters (see Figs. 1-6 and 1-7). Lymphoid aspirates, acute inflammatory exudate (pus), and poorly differentiated tumor aspirates form smooth homogeneous carpets of cells (Fig. 1-10). Adenocarcinomas, including breast, lung, abdomen, and prostate aspirates, will often yield dappled smears with punctate stained cell groups. Thick, semisolid aspirates extracted from glandular targets such as salivary gland or prostate may smear out as overlapping masses (see Fig. 11-19). Mixed tumor of salivary gland in particular is readily recognized by the dense tissue yield that stains pink with R stain (Color Plate VI A). Necrosis appears finely granular and can often be recognized before microscopic examination.

Observing and evaluating the gross appearance of the stained smear is partly gamesmanship, but it does add another dimension to the whole process of diagnosis. It suggests possibilities and stimulates thought even before microscopic review.

In approaching the diagnostic moment with the slide on the stage of the light microscope, the cytopathologist should maintain a flexible and imaginative stance. Ideally, he should have the clinical data freshly in mind or immediately available. As with all morphologic disciplines, the material is readily recognized as a familiar malignant or specific benign disorder (e.g., a pyogenic abscess), as clearly negative for evidence of tumor or a specific benign disorder (e.g., breast fat), or as suspicious, atypical, or unfamiliar.

We have noted that some cytopathologists early in their experience with aspiration smears will insist on a surgical biopsy (which may be a craniotomy, thoracotomy, or laparotomy) because of unfamiliarity with an air-dried preparation or an uncommon tumor such as thymoma, carcinoid, or ependymoma. A willingness to accept aspiration specimens must be accompanied by a willingness to refer unfamiliar smears before subjecting the patient to possible unnecessary surgery.

The large bulk of tumor specimens submitted for cytodiagnosis does not require special stains, DNA analyses, enzyme specificity, or electron microscopy, all of which are discussed in Chapter 18. They consist of squamous carcinoma, ductal carci-

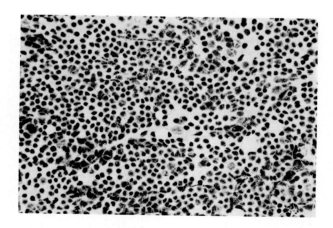

Fig. 1-10. Non-Hodgkin's lymphoma, supraclavicular node. The smear contains a carpet of many thousands of poorly differentiated lymphoid cells. Uniformity is less evident at higher powers. (R stain, ×330)

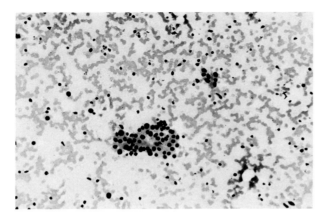

Fig. 1-11. Hashimoto's thyroiditis. There is a discrete cluster of enlarged epithelial cells with nuclear variation and abundant cytoplasm. Red blood cells and discrete lymphocytes are scattered in the background. (R stain, ×200)

noma, carcinomas of the lung, and adenocarcinomas arising in the abdominal viscera. The cell yield is usually abundant. Screening of a slide is not necessary, and the diagnosis is immediately evident to an experienced cytopathologist or cytotechnologist.

The low-power examination is necessary to assess cellularity and to locate the tumor groups that may be scattered within a fluid, bloody, or stromal background (Fig. 1-11). A cardinal rule, which may be violated under certain circumstances, states that diagnosis should not be attempted on insufficient cellular material. This applies particularly to slides submitted without clinical data. When the clinical picture is pathognomonic (Fig. 1-12) and aspiration is carried out for documentation prior to chemotherapy, radiation, or surgery, observation of an isolated but frankly malignant cluster may be adequate. An early, often minute, skin recurrence may yield a few cells but sufficient to confirm a diagnosis (Fig. 1-13).

Organization of the cell groups is critical in diagnosis. Microscopy reveals papillary groups (Fig. 1-14), follicular or alveolar groups, overlapping cell balls (Fig. 1-15), and loose clusters of poorly differentiated cells (Fig. 1-16). Benign epithelium with clearly defined cell borders and normal polarity is easily recognized. Liver (Fig. 1-17), benign prostatic hypertrophy (see Figs. 11-20 and 11-21), thyroid nodules (Fig. 1-18), and mesothelium (Fig. 1-19 and see Fig. 8-39) all may yield readily identifiable structures and sheets. A pitfall early in diagnostic experience is reading smears under high power, including oil emersion. Benign cells may appear enlarged with prominent nucleoli and finely granular chromatin (Figs. 1-20 and 1-21). In perspective, it is evident that these cells are part of a well-recognized benign pattern.

For the pathologist approaching cytodiagnosis of aspiration smears, it is important that the criteria of exfoliative cytology are not used uncritically. Individual cell cytology relying on nuclear hyperchromatism

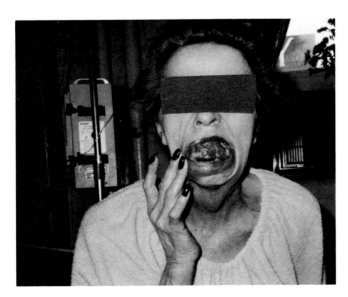

Fig. 1-12. Fungating carcinoma of the tongue. On surgical biopsy, only necrosis was obtained. Squamous carcinoma was diagnosed by fine needle aspiration. (From Linsk JA, Franzen S: Fine Needle Aspiration for the Clinician, p. 39. Philadelphia, JB Lippincott, 1986)

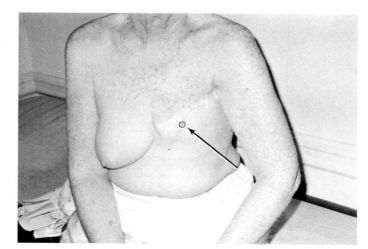

Fig. 1-13. Biopsies of small deposits in the skin *(arrow)* are easily done by tangential aspiration with a small needle. (From Linsk JA, Franzen S: Fine Needle Aspiration for the Clinician, p. 123. Philadelphia, JB Lippincott, 1986)

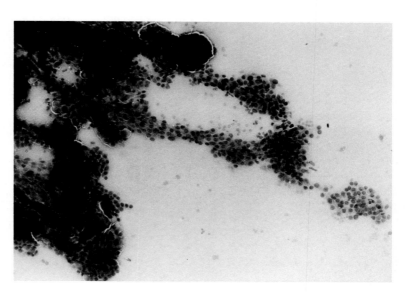

Fig. 1-14. Papillary carcinoma of thyroid. There are abundant papillary tongues of small tumor cells. (R stain, ×200)

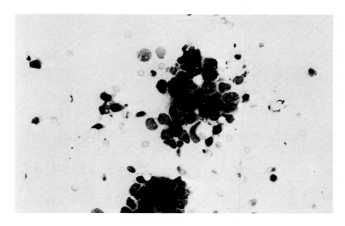

Fig. 1-15. Breast carcinoma. Cells are in three-dimensional clusters (cell balls). (R stain, ×330)

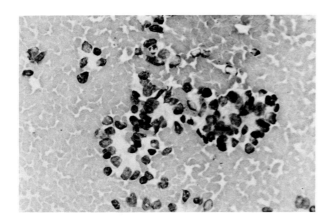

Fig. 1-16. Adenocarcinoma, abdominal mass. There are loose clusters that form acinar structures. The background is unclotted blood. (R stain, ×330)

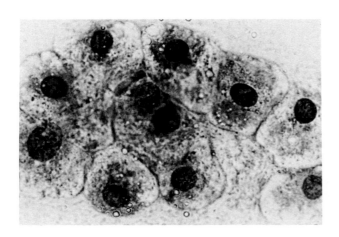

Fig. 1-17. Fragment of a liver morula. There are clearly defined cell borders highlighted by cloudy cytoplasm surrounded by a clear zone. (MGG stain, ×800)

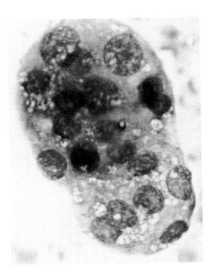

Fig. 1-18. Microfollicle from multinodular goiter. (MGG stain, ×800)

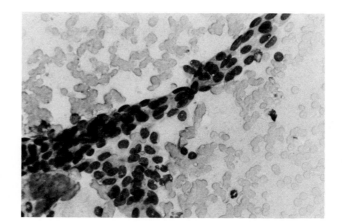

Fig. 1-19. Mesothelium. Some groups of mesothelial cells are usually present when the needle must pass through the pleural or peritoneal space. They are often surprisingly numerous and are easy to recognize when they form coherent sheets. (MGG stain, ×200)

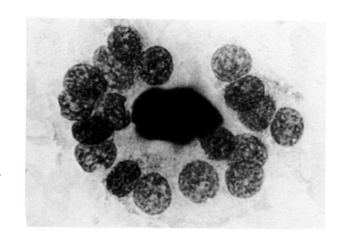

Fig. 1-20. Thyroid aspirate from case of multinodular goiter. The follicular wall fragment is an apparently bowl-shaped structure with some colloid left in the center. (MGG stain, ×1000)

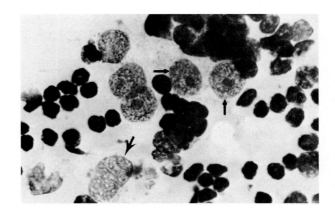

Fig. 1-21. Thymus. Needle aspirate from a surgical specimen of a normal thymus in a 3-year-old girl. The aspiration cytology of the thymus, very similar to that of lymph nodes, is identified by the thymus epithelium cells with very conspicuous nucleoli *(arrows)*. Note nucleus of a thymic macrophage *(barbed arrow)*. (MGG stain, ×140)

and thickened and irregular nuclear membranes is of limited use in diagnosis of most aspirated smears.

In summary reliance on the gestalt of the smear rather than individual cell cytology is important. This includes the cellularity, the arrangement of cell groups, cohesion and dispersion, background fluid and stroma, intranuclear and intracytoplasmic inclusions, and finally the classic cytologic hallmarks of malignancy or benignity, if present.

The identification of cancer on aspiration smears has become a routine skill in many cytopathologic laboratories. Extension of the method to the diagnosis of tumor subtypes and specific benign disorders is now an important dimension because of the proliferation of therapeutic options.

THE PATHOLOGIST'S ROLE

The pathologist has responsibilities beyond the distinction of benign and malignant smears. He must review the clinical data. He then judges the adequacy of the smear and decides whether to assay a diagnosis or request additional material. If smears have been obtained by stereotactic instrumentation or by transthoracic, transabdominal, transcranial, transvaginal, or transrectal routes, it is not always a simple request. Therefore, scrutiny of the smear combined with clinical data should be diligent. On the other hand, the pathologist dare not render a diagnosis that leads to major intervention without complete assurance of its correctness.

His responsibilities include diagnosis of malignancy, diagnosis of tumor type, identification of specific benign lesions (benign tumor or inflammation), and determination that the smear is negative for any specific malignant or benign lesion.

False-Negative Smear

False-negative smears are a result of the following:

1. Missing the target. This factor is reduced when the cytopathologist carries out his own examination and aspiration.
2. Interpretive error. Failure to identify cancer is similar to failure in histopathologic diagnosis.

It is the responsibility of the pathologist to communicate with the clinician the need for a repeat aspiration. The negative smear that is ultimately deter-mined to be false is not a clinical hazard as long as the clinician clearly recognizes that the clinical findings take precedence in the decision process. Reliance on a negative smear is the responsibility of the clinician. A negative diagnosis is not accompanied by a warranty that cancer is absent. The main serious effects of a false-negative finding are premature surgical intervention and loss of confidence in the whole method. Since missing the target is the main source of false-negative diagnoses, it is the cytopathologist's responsibility to inform the clinician of the need for a re-biopsy.

The Indecisive Smear

Clinicians do not readily accept the pronouncement that the smear is "probably cancer," "almost certainly cancer," "probably benign," and so forth. The histopathologic diagnoses with which they are familiar are almost invariably categorical, albeit in error at times. The indecisive smear does not necessarily mandate surgical intervention. The report may be accompanied by a number of recommendations as follows:

1. Repeat the aspiration
2. Follow the patient depending on clinical correlation
3. Obtain surgical biopsy

Probably more important for maintenance of continuity within the program is for the pathologist to discuss the case directly with the clinician.

ABC must be considered, on the one hand, as a branch of the discipline of pathology, with heavy reliance on the principles of cytologic diagnosis accrued over the past four decades of research and study of exfoliative cytologic specimens. On the other hand, ABC relates to the clinical spectrum of disease. This is seen both in the varied techniques used to extract material from patients and in the dependence on clinical, particularly oncologic knowledge and evaluation. In many disorders, ABC becomes an efficient bridge between the clinical observations and the definitive morphologic diagnosis.

Caution and conservatism cannot be overemphasized; and training, experience, and apprenticeship are essential. Nevertheless, ABC can render a specific diagnosis of malignant or benign disease, thus allowing for definitive surgery, radiation, or chemotherapy without further confirmation. Mastectomy, pneumonectomy, orchiectomy, pelvic and brain irradiation, and other irreversible procedures are currently being carried out on the basis of a fine needle aspirate.

It is one thesis of *Clinical Aspiration Cytology* that ABC is a valuable method that has previously been underused, particularly in North America. We hope that this description of the method and the cytology applied to disorders of the entire body will help correct this deficiency.

REFERENCES

1. Abele JS, Miller TR, King EB et al: Smearing technique for the concentration of particles for fine needle aspiration biopsy. Diag Cytopathol 1:59, 1985
2. Clare N, Rone R: Detection of malignancy in body fluid. Lab Med 17:147, 1981
3. Frable WJ: Thin-Needle Aspiration Biopsy. Philadelphia, WB Saunders, 1983
4. Frable WJ: Comment of letter by N. Gubin. Acta Cytol 29:649, 1985
5. Kaminsky DB: Aspiration Biopsy for the Community Hospital. New York, Masson, 1981
6. Koss LG: Diagnostic Cytology and Its Histopathologic Bases, 3rd ed. Philadelphia, JB Lippincott, 1979
7. Koss LG, Woyke S, Olszewski W: Aspiration Biopsy — Cytologic Interpretation and Histologic Bases. New York, Igaku – Shoin, 1984
8. Linsk JA: Aspiration cytology in Sweden: Aspiration biopsy — The Karolinska Group. Diag Cytopathol 1:332, 1985
9. Linsk JA, Franzen S: Fine Needle Aspiration for the Clinician. Philadelphia, JB Lippincott, 1986
10. Lopes – Cardoza P: Atlas of Clinical Cytology. Philadelphia, JB Lippincott, 1976
11. Orell SR, Sterrett GF, Walters MN–I et al: Manual and Atlas of Fine Needle Aspiration Cytology. Melbourne, Churchill Livingstone, 1986
12. Pak HK, Yokota S, Teplitz RI et al: Rapid staining technique employed in fine needle aspirations of the lung. Acta Cytol 25:178, 1981
13. Palmieri B: A new device for thin needle and core biopsies (letter). Acta Cytol 30:82, 1986
14. Saccomanno G: Diagnostic Pulmonary Cytology. Chicago, American Society of Clinical Pathology, 1978
15. Zajicek J: Aspiration Cytology, Parts 1 and 2. Basel, S Karger, 1975 and 1979

CHAPTER 2

Imaging Techniques in Aspiration Biopsy

ALBERT J. SALZMAN

Fine needle aspiration (FNA) of nonpalpable lesions requires, for the most part, some form of localization with one or more of the imaging techniques currently available in most modern radiology departments. The simplest, least costly, and most rapid imaging techniques should be employed whenever possible. There will be times, however, when unusual circumstances may require more sophisticated imaging devices to provide guidance for the aspiration biopsy. This chapter will stress the use of imaging systems readily available in most community hospitals. Procedures employing computed tomography (CT) scans or angiography will be discussed in those instances where their use is required. Table 2-1 summarizes available methods and their application.

The radiologist should be an integral part of the aspiration biopsy cytology (ABC) team. In many institutions the radiologist does the aspiration. Even where the cytologist or clinician performs the procedure, however, the radiologist should be consulted on the best topographic approach, the imaging guidance required, and the portion of the lesion that will yield the most diagnostic material. He should review all the diagnostic imaging material available. Aspiration should not be carried out without careful review of radiographs, nuclear scans, ultrasonograms, and CT scans. The various guidance control systems are reviewed below.

IMAGING PROCEDURES

Routine radiography should include films obtained at right angles to the lesion. This enables the clinician to judge the depth of the lesion and relate its position to various external body landmarks. Unidirectional or multidirectional tomography may be necessary in the chest. For lesions below the diaphragm, opacification studies may be required. These may consist of angiography, lymphangiography, endoscopic retrograde cholangiopancreatography (ERCP), and routine barium studies of the intestinal tract.[15,24,30,64,75]

The advent of image-amplified television monitoring has been a major factor in enhancing the use of FNA of most areas of the body, but especially of the chest.[10,62] It has allowed immediate needle localization with direct observation of the lesion to be biopsied and has made it possible to carry out the procedure in a lighted room. An image amplifier mounted on a C-arm allows the amplifier to be moved around the stationary patient. Biplane amplifiers allow simultaneous viewing of the lesion at right angles. Although these may be very convenient to use, most aspirations can be carried out with single-plane image amplifiers.

Ultrasonography has become extremely useful both in the evaluation of abdominal and retroperito-

Table 2-1. Localization Aids for Aspiration Biopsy

ANATOMIC AREA	IMAGING AID*
Brain	CT scans, ultrasound, angiography, plain films, nuclear scans
Breast	Xerography, mammography, ultrasound
Chest	Plain films, tomography, fluoroscopy, CT scans, ultrasound
Kidney	IVP, ultrasound, plain films, fluoroscopy, CT scans
Retroperitoneum (includes pancreas, lymph nodes)	Plain films, contrast GI studies, ultrasound, IVP, angiography, ERCP, CT scans, lymphangiography
Bone	Plain films, nuclear scans, CT scans
Thyroid gland	Nuclear scans, ultrasound

* IVP, intravenous pyelogram.

neal masses and as a guide to the actual aspiration biopsy.[4,28,31,32,34,60] Cystic and solid lesions may be distinguished, areas of necrosis visualized, and major vessels in the area of biopsy avoided. The localization procedure is simple and flexible and allows visualization in multiple body planes. Ultrasound is most useful in aspiration of lesions larger than 3 cm in diameter.[52] A number of guidance systems are available for use in conjunction with[32] or attached to[4,32,49] the imaging transducer after localization of the mass. We have been performing ultrasonic guidance with a freehand system and have not employed any of the available special transducers. Previously, A-mode scanning allowed visualization of the aspiration needle tip only within cystic lesions (Fig. 2-1), but not within solid masses. With the use of real-time sector and linear-array transducers, the needle may be visualized during passage and can be identified even within solid masses (Fig. 2-2). An additional advantage of ultrasonic guidance is the lack of exposure of the patient and operator to ionizing radiation. Because of the physical properties of acoustics, the use of ultrasound is limited in very obese patients or in the presence of large amounts of gas within the intestinal tract overlying the lesion. Aspiration of masses within the chest is limited by the presence of air in the lungs, but has been used successfully for aspiration of lesions in contact with the chest wall.[1,7] Ultrasound is not useful for study of lesions within bone.

Computed tomography has become increasingly important as a guidance system for FNA biopsy as more units are available for use and as reconstruction times become shorter. Several authors have described their experience with CT guidance in various

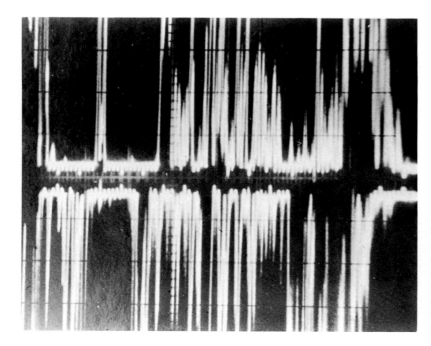

Fig. 2-1. A-Mode ultrasound in a patient with a right renal cyst *(upper tracing)*. The *left half* of the tracing shows the echo-free cyst. The *large central spike* represents the aspirating needle within the cyst. With a solid mass, extensive echoes within the mass would obscure the echoes from the aspirating needle.

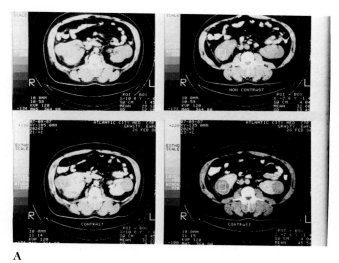
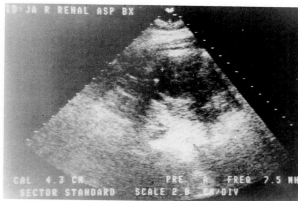

A   B

Fig. 2-2. A. Bilateral renal cell carcinomas. Pre- *(upper)* and post- *(lower)* contrast CT scans showing multiple bilateral, slightly enhancing low density renal masses. **B.** Fine needle aspiration of renal mass under ultrasonic control. This sector scan shows complex mass 4.3 cm below posterior surface *(lower cursor).* Tip of aspirating needle is seen as bright echo just below the cursor.

body areas during aspiration biopsy.[3,17,26,27,66,68] The craniocaudal level of a lesion is determined by placing a series of longitudinally oriented opaque catheters, differing in length by 1 cm, on the patient's skin (Fig. 2-3A), and then comparing the number of catheters visualized on the axial slice with a radiograph of the area obtained with the catheters in place (Fig. 2-3B). Alternatively, a digital radiograph may be obtained and used to indicate the precise craniocaudal level of the lesion to be biopsied (Fig. 2-3C). After placement of the needle, a confirmatory axial section can be obtained to check the position of the needle tip, which casts a black artifactual shadow on the image (see Figs. 2-16 and 2-18). If aspiration is to be carried out off the vertical axis in either the sagittal or coronal plane, to avoid the pleura or major vessels, the techniques described by van Sonnenberg[67] and Axel[3] may be used. A direct triangulation approach may be used in the sagittal plane by using direct computer-generated angle and distance from the lateral digital radiograph (Fig. 2-4) and in the coronal plane by computer measurements on the axial section showing the mass. A lateral digital radiograph may be obtained with the needle in place to localize the needle tip in off-axis punctures and an axial section obtained to confirm its position within the lesion.[69]

The drawbacks of CT localization include cost, radiation exposure, and increased length of the procedure. Radiation exposure may be reduced during the biopsy procedure by reducing the milliamperage for the confirmatory axial images, although these may have higher levels of "noise." Unlike ultrasound, CT scanning is not limited by the patient's size (as long as the patient fits within the gantry) or by air in the lungs or bowel. CT scanning may be used to evaluate skeletal lesions for biopsy.[21,23]

ASPIRATION OF SPECIFIC AREAS

Lung and Mediastinum

After reviewing all available radiographs, the approach to the lesion is selected. The depth of the lesion from the anterior to the posterior chest wall can usually be determined from the lateral radiograph after allowing 30% for magnification. If the lesion is not apparent on the lateral films, the depth is determined by tomograms obtained in the biopsy position (Fig. 2-5). A lesion larger than 7 cm may be biopsied at the patient's bedside after it has been suitably localized in relation to bony landmarks.[18] Generally,

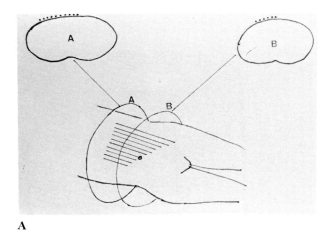

A

B

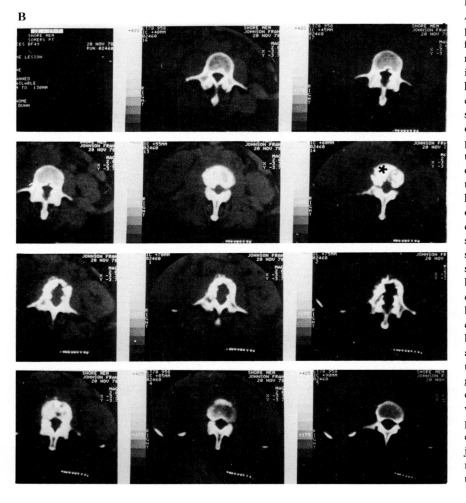

Fig. 2-3. Value of CT scanning. **A.** A series of radiopaque catheters placed on the patient's skin surface during the CT examination may be used for longitudinal localization. The catheters differ in length by 1 cm. At level *A* all the catheters are included in the CT section. At level *B* only six catheters will be seen; this can be compared to a scout radiograph to determine the longitudinal position of a lesion to be biopsied. **B.** A patient with solitary myeloma of a lumbar vertebra. The series of opaque markers are seen as *white* dots on the patient's posterior skin surface. The abnormal vertebra is seen in the two middle series of slices *(star);* by counting the number of catheters seen, the longitudinal position of this vertebra can be determined for biopsy. **C.** A digital radiograph of the area to be biopsied may be obtained in anteroposterior or lateral projection. By generating a series of longitudinal lines on this image by computer, the exact CT section on which the lesion to be biopsied appears can be easily determined by comparing the number of the projected line with the corresponding number appearing on the CT section.

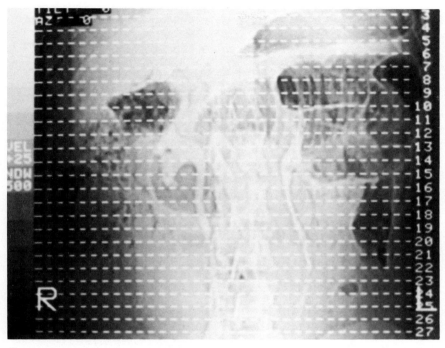

C

smaller lesions, or those in which bedside aspiration has been unsuccessful, will require aspiration under image-amplified fluoroscopy.

To proceed, choose the shortest approach and mark the point of entry on the skin. A 25-gauge needle is useful for this purpose. Insert the aspiration needle just above the rib and pass it to the appropriate depth into the lesion. The position of the needle may be verified in a number of ways. If biplane or C-arm fluoroscopy is available, verification is simple because the position of the needle can be seen in multiple planes without moving the patient. However, with single-plane equipment, the following techniques may be used: (1) the patient may be rotated into oblique or lateral positions under fluoroscopy; (2) the hub of the needle may be moved through a short excursion and the lesion observed for similar movement; (3) a supplementary level chest radiograph may be obtained with the needle in place to verify its position; (4) the lesion and needle can be observed for synchronous movement during respiration; or (5) circular movements of the fluoroscopy screen around the needle hub may be made, ensuring that the needle tip and lesion remain in juxtaposition. If the needle is not within the lesion, it will be seen to

shift away from the needle tip.[10,38] Techniques 2 and 4 will not be useful with hilar or fixed masses. After ensuring that the needle is satisfactorily positioned, carry out aspiration and examine the specimen.

Some problem areas should be noted in certain patients. Puncture of the superior vena cava or pulmonary arteries is not associated with any significant morbidity, since these are low-pressure systems.[18,67] Avoid puncturing the aorta, although inadvertent fine needle puncture has not resulted in any reported morbidity. The ascending aorta lies to the right of the midline and anteriorly, and the descending aorta lies to the left and posteriorly. Approach to lesions adjacent to the aorta may require some ingenuity but can usually be accomplished.

There are certain exceptions to choosing the shortest route to a lesion. Lesions located anteriorly beneath the medial end of the first rib and clavicle may require a posterior approach, while subscapular lesions should be approached anteriorly. Pleural lesions and those at the costovertebral angles may present special problems, since they are often difficult to visualize.[36] Haaga and colleagues have found CT guidance helpful when these lesions have not been successfully aspirated under fluoroscopic guid-

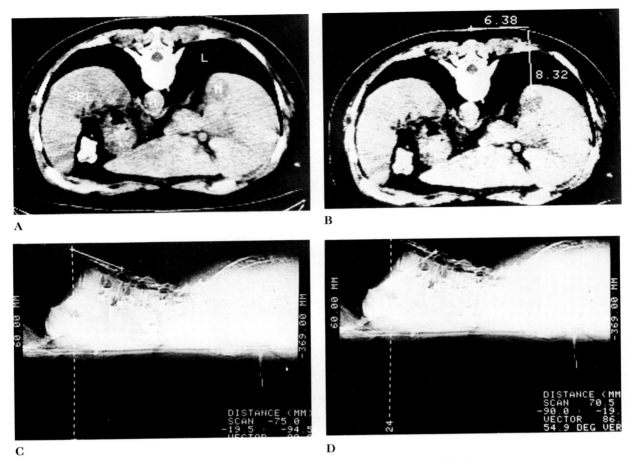

Fig. 2-4. A. Prone CT scan of liver shows low density mass in posterior segment of right lobe of liver. Lung *(L)* is along posterior surface of mass. **B.** Cursor program marks distance of mass from posterior skin surface (8.32 cm) and distance from midline to entry point on skin (6.38 cm). **C.** A lateral scout view shows level of CT section *(broken vertical line)* and distance (75.0 mm) to entry point to avoid pleural space. **D.** Lateral scout view measures depth (70.5 mm) along oblique path to liver mass from entry point established in **C.** Aspiration showed posthepatic hepatoma.

ance.[27] Ultrasonography may be useful for puncture of subpleural lesions.[1,7,33,71]

With cavitary lesions, aspiration should be carried out from the periphery of the lesion (Fig. 2-6). Similarly, large masses are likely to have necrotic centers; and aspiration is best carried out from the more peripheral portion of the tumor.[10,36] Some deeply situated masses may require insertion of an 18-gauge guide needle to the pleural surface, through which the aspirating needle is passed[10]; or a 23-gauge needle may be used to localize the lesion and

used as a guide over which a 19-gauge needle may be passed after cutting off the hub of the smaller needle.[68]

CT may be useful in puncture of necrotic masses,[18,22] for small apical and peripheral masses,[20] as an adjunct to fluoroscopy,[8] with lesions obscured by bone,[20] for mediastinal masses,[69] and for lesions adjacent to major vessels.[8,69] Separation of tumor masses from adjacent major vessels is greatly aided by axial sections obtained after the intravenous bolus injection of iodinated contrast agents (Fig. 2-7).

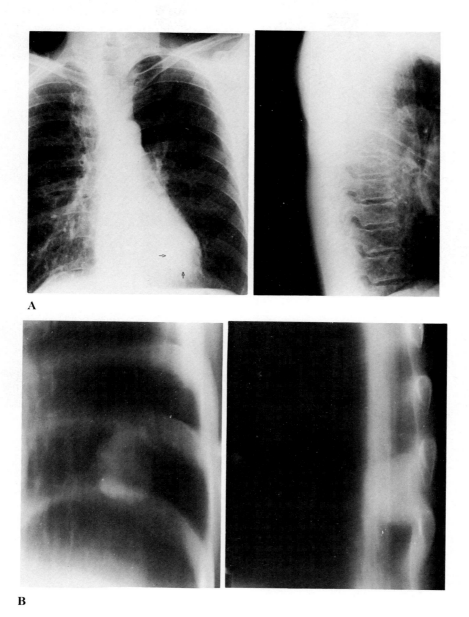

Fig. 2-5. Value of tomography. **A.** Posteroanterior and lateral chest films. On the posteroanterior film a mass in the left lower lobe *(arrows)* is just barely visible; it cannot be seen at all on the lateral film. **B.** Tomograms of the left lower lobe clearly demonstrate the neoplasm of the left lower lobe and show its position against the posterior chest wall. This mass is in an ideal location for aspiration biopsy.

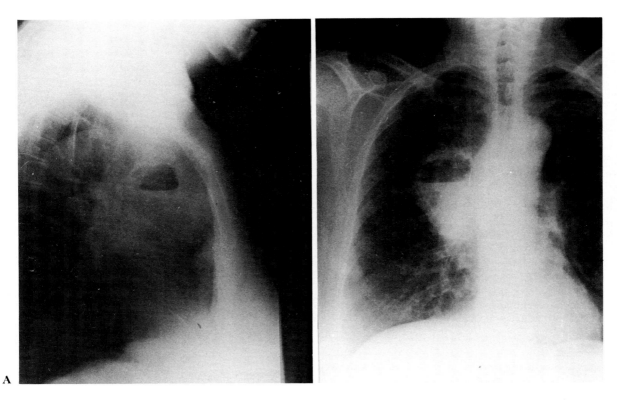

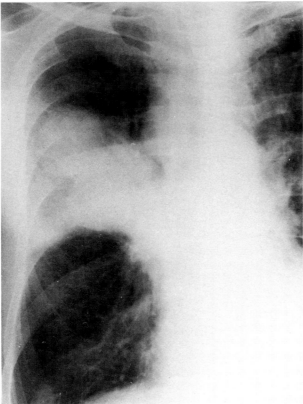

A

B

Fig. 2-6. **A.** Cavitary carcinoma of the right upper lobe. Biopsy is best obtained from the peripheral portion of the tumor. **B.** Larger masses may not show radiographic evidence of a necrotic cavity. Nevertheless, biopsy is best obtained from the more peripheral area of the tumor, where there is less likelihood of necrosis.

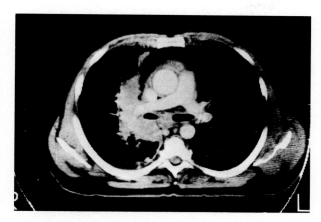

Fig. 2-7. Central carcinoma of the right lung with mediastinal invasion. CT scan of the chest at the level of the right pulmonary artery after intravenous bolus injection of contrast separates major vessels from tumor and provides a map for safest approach to avoid puncture of major vessels.

The most serious complication of lung aspiration is pneumothorax, especially in older or emphysematous patients.[10,18,27] Tao and colleagues consider advanced emphysema and pulmonary hypertension contraindications to needle biopsy of the lung.[62] Pneumothorax will be avoided to some extent if the needle is not handled while the patient is breathing.[30] Simple pneumothorax (Fig. 2-8A) may occur in up to 35% of patients and is of no clinical significance. Chest radiographs should be obtained both immediately and 2 to 3 hours after aspiration to evaluate the presence of an increasing pneumothorax.[18] Tension pneumothorax (Fig. 2-8B) requires immediate insertion of a catheter or chest tube.

The hazards of lung puncture are further considered in Chapter 7.

A 9F Teflon catheter with an attached Heimlich valve* may be inserted percutaneously in the radiology department.[56] To proceed, clean the skin and insert the catheter in the midclavicular line, in the second anterior intercostal space at the level of Louis's angle. After the catheter enters the pleural space, the obturator needle may be removed to check egress of air. Advance the catheter until the last side hole has entered the pleural space, with the catheter tip directed to the apex of the pleural cavity under fluoroscopy. Evacuate air by syringe and connect the

tube to the Heimlich valve. Obtain chest films at 1, 24, and 48 hours. If the pneumothorax continues to enlarge, chest-tube underwater drainage will be required; otherwise, the catheter may be removed at 48 hours if the lung remains expanded.

Patients may occasionally experience hemoptysis after aspiration, or the postbiopsy chest film may show evidence of parenchymal pulmonary hemorrhage. In the patient with a normal clotting mechanism, these are not serious complications; they are self-limited, and the patient only needs reassurance.

Bone

The most appropriate site for biopsy of skeletal lesions is decided on after complete review of all radiographs and nuclear scans. A superficial prominence of a non-weight-bearing bone is ideal for biopsy if such an area appears to be involved.[11] Fine needle aspiration generally will be successful only with destructive lesions, preferably those associated with destruction of cortical bone (Fig. 2-9). For a discussion of other bone biopsy needles and indications for their use, see the article by de Santos and colleagues.[14] Positive biopsies may be anticipated in 74% to 85% of lesions with positive radiographs, but in only 33% if the nuclear medicine scan alone is positive.[11,14]

Single-plane, image-amplified fluoroscopy is usually satisfactory for biopsy of lesions in the peripheral skeleton,[5,14] although Murphy[48] favors the use of biplane fluoroscopy. For vertebral lesions, biplane or C-arm image amplification is needed; or, if not available, single-plane fluoroscopy can be supplemented by lateral radiographs.[11,12,35] Although somewhat more time-consuming, the biopsy can be carried out entirely with frontal and lateral radiographs of the spine.[9,63,65] Lesions involving flat bones such as ribs, scapula, and ilium are best approached by passing the needle along the long axis of the involved bone.[14] CT guidance may be especially useful for lesions involving the thorax, pelvis, and spine[23]; to localize soft tissue masses adjacent to bone[21,23]; and to establish the area of maximum cortical thinning in a lytic lesion.[21] Zegel[74] has used isotopic localization for biopsy of rib and sternal abnormalities.

Before a detailed discussion of biopsy of lesions of the spine, a brief review of pertinent anatomy is in order. In the cervical region the carotid arteries lie anterolaterally. The aorta descends anterior and to the left of T4 – L3. In the thoracic region the esopha-

* Cook Catalogue No. TPT-1

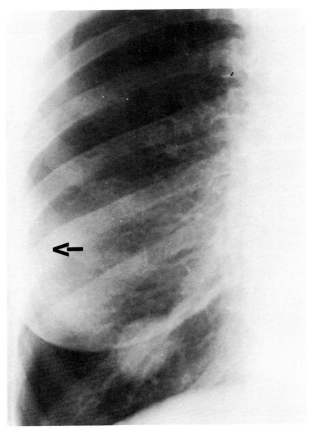
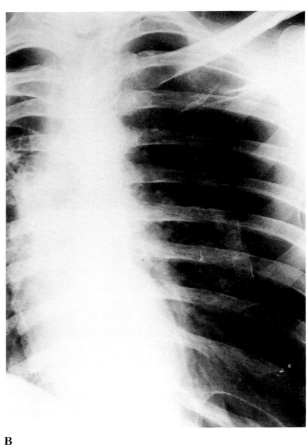

A   B

Fig. 2-8. **A.** Small asymptomatic pneumothorax after aspiration of a right lower lobe mass. This requires no treatment, and the patient must only be observed to ensure that the pneumothorax is not enlarging *(arrow).* **B.** A tension pneumothorax is apparent with depression of the ipsilateral diaphragm and contralateral displacement of the mediastinal structures. This requires immediate therapy and can be treated with a percutaneous catheter insertion (see text).

gus is to the right and anterior to T4–T10. The posteromedial limits of the lungs lie approximately 4 cm lateral to the midline of the thoracic spine. In the lumbar region the inferior vena cava lies to the right and anterior to T11–L3. The right renal artery and left renal vein pass anterior to L2. As a general rule, if a vertebral lesion in the thoracic or lumbar area can be approached from either side, the right side is preferred in order to avoid inadvertent puncture of the aorta. For lesions involving the neural arches of any portion of the spine, a simple posterior approach can be used. Vertebral body lesions are best approached

by a more individualized route, depending on the level of the spine involved.

In the cervical region the upper three vertebral bodies are best approached by an anterior, transpharyngeal route, while the lower four cervical vertebrae are best approached by a direct lateral route with the patient placed in the appropriate decubitus position.[50,59] In order to avoid the carotid arteries, the point of entry should remain posterior to the posterior margin of the sternomastoid muscle. Once the appropriate level is determined either fluoroscopically or radiographically, the needle is passed directly

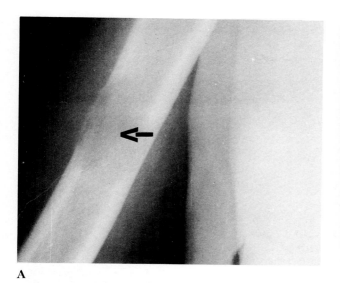

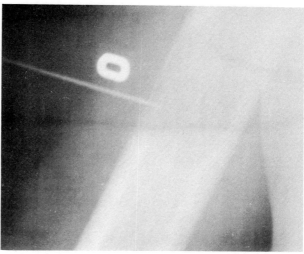

A   B

Fig. 2-9. A. Metastatic lung carcinoma to the mid-shaft of the humerus. Note that the tumor has destroyed the overlying cortex. This site of cortical disruption represents an ideal site for thin needle aspiration, since the tumor may be aspirated without having to pierce any cortical bone *(arrow)*. **B.** A fluoroscopic spot film obtained during the procedure shows the lead marker *(O)* placed on the skin surface over the lesion. The tip of the aspirating needle can be seen within the area of metastatic carcinoma.

to the involved vertebra and its position is again verified by right-angle views. The cervical vertebrae should be no deeper than 3.4 cm from the skin surface; and if the needle has passed more deeply than this, it will probably be anterior to the vertebral column.[50] If FNA is to be carried out, it is completed after verification of needle position. If an osteoblastic lesion is to be biopsied by a trephine type of needle, the thin needle may be used as a guide for the biopsy needle (Fig. 2-10).

A posterolateral approach is used for vertebral body biopsy in both the thoracic and lumbar regions. Although thoracic biopsies were considered too risky by early physicians, it is now felt that such biopsies can be carried out under direct fluoroscopic control without undue hazard.[11,35,65] With the patient prone, identify the appropriate level with a lead marker. Again, if the lesion to be biopsied is not lateralized, then a right-sided approach is preferable. In the thoracic area the point of entry is just above the neck of the rib caudal to the involved vertebra and no more than 4 cm from the midline to avoid lung puncture.[11] Valls and colleagues suggest starting 6.5 cm from the midline at an entry angle of 35° for biopsy of lumbar

vertebral lesions (Fig. 2-11).[65] Once the needle is advanced to bone, verify its position by observation at right angles with any convenient combination of fluoroscopy or radiography. After establishing proper position of the needle, carry out aspiration in much the same manner as with cervical spine lesions.

Abdomen

A number of guidance techniques are available to aid in aspiration of intra-abdominal and retroperitoneal masses. At present, ultrasound[4,31,49] and CT[52] scans are most frequently employed; but there are still certain areas in which other techniques have some merit. A brief discussion of the various guidance systems follows, after which some special situations will be discussed.

Plain films of the abdomen will rarely provide enough information to biopsy any but the largest of masses. Methods of employing various contrast agents have been described. Masses within the pancreas may be localized with barium in the upper gastrointestinal (GI) tract (Fig. 2-12), by angiography (Fig. 2-13), and after ERCP (Fig. 2-14).[30,64] Masses in

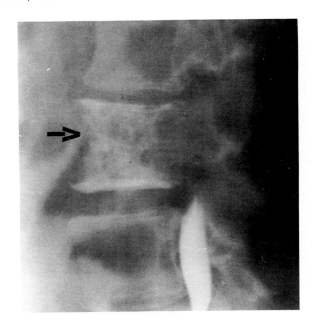

Fig. 2-10. Unusual, partially osteoblastic solitary plasmacytoma of L3 associated with extradural block *(arrow)*. With lesions showing this much reactive bone formation, fine needle aspiration is difficult. In such a situation, the fine needle may be inserted as a guide for a trephine type of needle biopsy.

the pancreatic head or those arising from the biliary system may be aspirated after percutaneous transhepatic cholangiography (Fig. 2-15).[15] After bipedal lymphangiography, contrast within the lymph nodes can be used for guidance of aspiration biopsy (Fig. 2-16).[16,75] Intravenous (IV) urography is employed for localization of renal masses during fluoroscopically guided aspiration biopsy. Urography may also be used in aspiration biopsy of metastatic retroperitoneal lymph node involvement. In this situation, the area of ureteral narrowing produced by the enlarged nodes is used as a guide during fluoroscopically controlled aspiration.[19]

Pancreas

Pancreatic masses will be most commonly localized under ultrasonic guidance.[28,31,42,58,60] CT scanning is equally noninvasive but is more time-consuming as

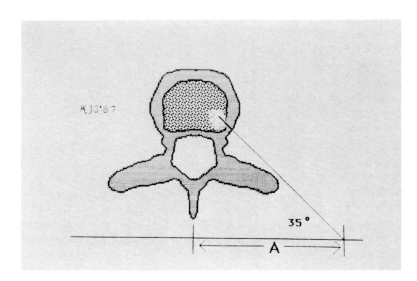

Fig. 2-11. Diagramatic representation of posterolateral approach used for biopsy of vertebral body lesions in the thoracic and lumbar regions. The distance *A* from the tip of the spinous process to the point of entry is 6.5 cm for lumbar spine aspiration and is used with an angle of entry of 35 degrees. In the thoracic region, *A* should be less than 4 cm in order to avoid puncture of the underlying lung. The angle of entry is correspondingly increased to take into account the shorter distance to the midline.

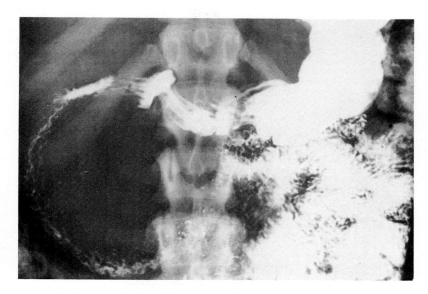

Fig. 2-12. A large mass within the head of the pancreas is widening and compressing the duodenal loop. A mass of this magnitude is easily aspirated under fluoroscopic guidance, with barium used to outline the mass's effect on the duodenum.

well as expensive, and it is associated with radiation exposure. CT will be used after an unsuccessful attempt at aspiration with ultrasonography or initially if the lesion is not visible by ultrasound, if the lesion is less than 3 cm in size,[52] or if the tumor involves the tail of the pancreas.[59] Angiography[64] and ERCP[30] may be more specific diagnostic procedures in pancreatic masses, but both suffer the disadvantage of being invasive. If these procedures are performed for diagnostic purposes, however, aspiration immediately after the diagnostic study should be considered.

After localization of the mass, mark a point on the skin for entry to provide the shortest route to the lesion. Although major vessels should be avoided, no special precaution is needed to avoid passage of the needle through either the liver or bowel.[60] Aspiration may be carried out by direct insertion of a 23-gauge needle into the mass. However, a second needle may help in one of two ways. First, a 19-gauge needle can be introduced through the skin to act as a stabilizing guide for the 23-gauge aspiration needle.[60] Second, a two-needle system, described by Ferrucci and Wittenberg as a "tandem needle" technique, can be used with CT localization.[16,17] The first needle is introduced into the mass, and its position is confirmed by a CT section. The second needle is then introduced parallel to the first, with the first needle acting as a directional guide. The first needle is left in place within the mass and, if additional aspirations are re-

quired, will act as a guide. The position of the needle within the mass can be confirmed on an axial CT section if a characteristic dark artifactual shadow is seen at the tip of the needle (Fig. 2-17).

Lymph Nodes

After bipedal lymphangiography, films are reviewed and an appropriate node is chosen for biopsy (see Fig. 2-16). The node is localized fluoroscopically, and a point over the selected node is marked on the skin. Aspiration of involved nodes is carried out under fluoroscopic guidance with either single-plane or bi-plane equipment. With larger masses, the depth of the lesion may be further evaluated by ultrasonography. The aspiration needle should be directed toward the area of defect in a node involved by metastasis. With lymphomas, where involvement of the node is usually more generalized, any portion of the node may be aspirated. Verify needle position by observing simultaneous movement of the node and needle. Initial introduction of an 18-gauge needle through the skin acts as a guide for the aspirating needle. The presence of lymphangiographic oil in the aspirated material verifies that the node has been aspirated. Generally, positive diagnoses will be made more frequently from nodes involved by metastatic disease than from those involved by lymphoma.[19,24,75]

(Text continues on p. 32)

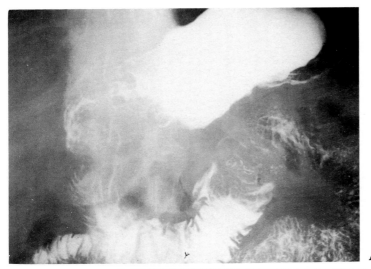

A

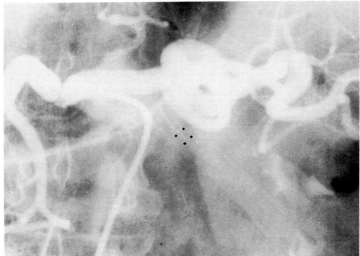

B

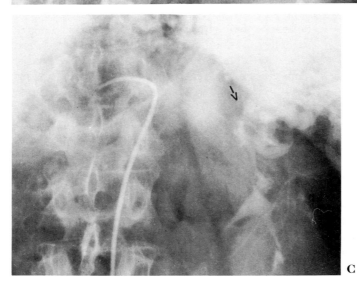

C

Fig. 2-13. **A.** Carcinoma of the body of the pancreas invading the third portion of the duodenum. Radiographically, this lesion would be difficult to distinguish from a primary duodenal neoplasm. **B.** Mid-arterial phase of a celiac arteriogram showing encasement of a pancreatic arterial branch by carcinoma of the body of the pancreas *(four squares)*. Here the biopsy could be performed immediately following the diagnostic arteriogram, while the catheter was still in place. **C.** Although biopsy may be guided by the arterial changes, the obstruction of the splenic vein is more easily seen *(arrow)* and more often present, since the veins are more easily compressed than the stiffer-walled arteries.

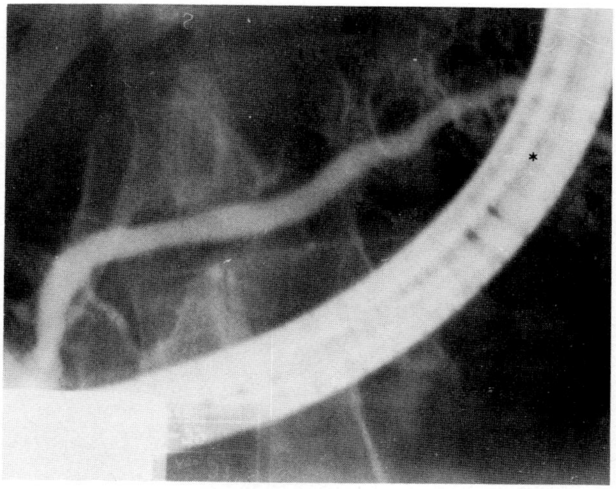

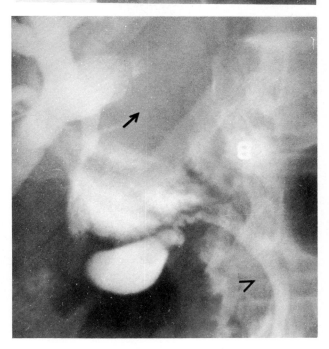

Fig. 2-14. ERCP film showing localized deflection of pancreatic duct *(asterisk)* as it crosses the endoscope by a mass at the juncture of the body and tail of the pancreas. Biopsy with fine needle aspiration can now be carried out while the patient is still in the fluoroscopic suite.

Fig. 2-15. Percutaneous transhepatic cholangiogram demonstrating obstruction of the common hepatic duct *(arrow)* by a primary bile duct carcinoma. Fine needle aspiration was obtained under fluoroscopic guidance from the area between the point of obstruction *(arrow)* and the common bile duct *(arrowhead)* opacified distal to the obstruction.

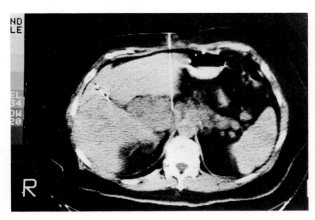

Fig. 2-16. Anterior transhepatic approach to mass in the pancreatic head. When the needle tip is in the plane of the axial section showing the mass, a characteristic black linear artifact is seen at the tip of the aspirating needle as seen in this example.

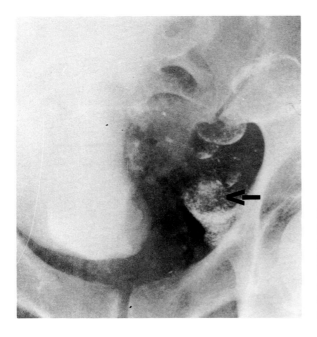

Fig. 2-17. Bipedal lymphangiogram in a patient with metastases to the left pelvic lymph nodes from carcinoma of the prostate. The bladder is opacified from simultaneously performed intravenous (IV) urograms. In performing fine needle biopsy from nodes involved by metastatic disease, carry the aspiration out from the filling defects within the involved nodes and not from the opacified portions of the node (*arrow*).

Liver

Diffuse liver disorders are biopsied by random aspiration. With focal lesions, ultrasonic guidance allows for simple and accurate localization.[2,44,57,76] As with other intra-abdominal aspirations, it may be helpful in patients with heavy abdominal musculature to introduce an 18-gauge needle through the abdominal wall, through which the aspirating needle may be introduced with successful aspirates in 83% to 94%. Although ultrasound is the most frequently used localization technique, CT is useful in lesions located in unusual locations, such as near the dome of the liver where it may be advisable not to cross the pleural space (see Fig. 2-4).

Kidney and Adrenal Gland

After evaluation of the kidneys by urography, aspiration of mass lesions is carried out through a posterior approach with the patient prone. Guidance is provided by fluoroscopy after opacification of the renal collecting systems by IV contrast material. Ultrasound, however, will usually provide more convenient and precise guidance without the use of contrast material. Ultrasonography will also allow the clinician to evaluate whether a lesion is cystic or solid. If a cyst is found, then direct aspiration of the cyst contents can be carried out with an 18-gauge needle rather than the usual 23-gauge needle used to aspirate solid masses.[31] Generally, the region of the renal

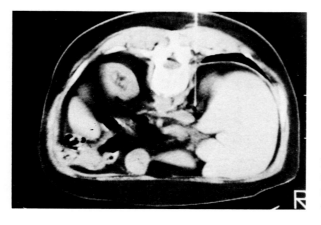

Fig. 2-18. Posterior approach to right adrenal mass shows the needle tip within a 15-mm right adrenal adenoma. Note the black linear artifact at the tip of the aspirating needle, confirming its position within the mass.

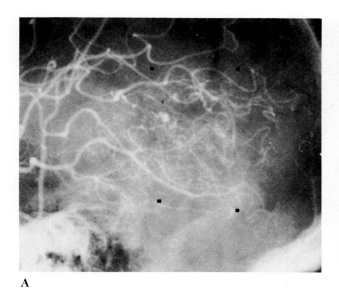 

A   B

Fig. 2-19. **A.** A carotid angiogram in a patient with a large vascular glioblastoma of the parietal lobe *(four squares)*. **B.** Lateral skull films obtained at the time of the angiogram, with an opaque *R* placed over the skull corresponding to the level of the tumor. A small bur hole placed at this point was used to carry out fine needle aspiration.

pelvis should be avoided to decrease the risk of puncturing a major renal artery or vein, which could result in formation of an arteriovenous fistula. Zajicek, however, carried out FNA of renal pelvic lesions and did not report any development of arteriovenous fistulae.[73]

Adrenal masses may be aspirated by ultrasonic guidance when the lesion is visible by this technique.[43] CT localization will generally be required in small adrenal tumors.[40] Pagani[51] has reported success employing CT guidance to aspirate normal-size adrenal glands in evaluating patients with small-cell carcinoma of the lung. We have employed a posterior paraspinal approach to the adrenal (Fig. 2-18), but various authors have described the use of anterior,[40] and lateral[43] approaches. If a pheochromocytoma is suspected either clinically or by vanillylmandelic acid (VMA) levels, hypertensive crises may be precipitated by FNA. In such cases, treatment with phentolamine, nitroprusside, or propanalol may be required.[6]

Retroperitoneum

Retroperitoneal masses are generally approached through the anterior abdominal wall in the same fashion that pancreatic lesions are aspirated.[24,25,31]

The comments in the section on the pancreas apply to other retroperitoneal lesions. Haaga has suggested administering a bolus of contrast material before CT localization in order to evaluate the relation of the retroperitoneal mass to the inferior vena cava and aorta and thus avoid puncturing these structures when masses adjacent to either of these major vessels are aspirated.[25] The same information may be obtained by ultrasound, but the presence of significant bowel gas may render ultrasonic guidance ineffective. Opacification of the displaced or compressed ureter during IV urography may be used as a guide in aspirating retroperitoneal masses.[19]

Brain

Previous reports of aspiration biopsy lesions have employed larger bore needles.[41,45,46,47,71] Moran[46,47] describes the use of a stabilization device.* With this device, but employing large-bore needles, he has achieved 97% accuracy in diagnosing masses larger than 1 cm. We have employed one of two guidance techniques. In the first system, the lesion is localized by biplane angiograms that relate the site of the neo-

* Available from Parol Co., 7532 Warner Avenue, St. Louis, MO 63117.

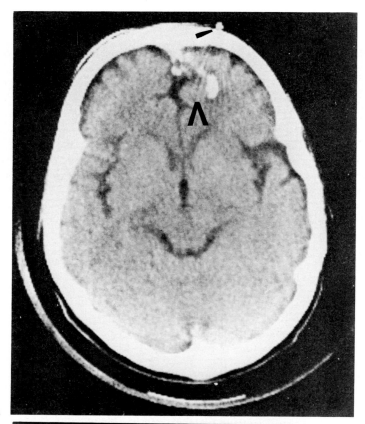

A

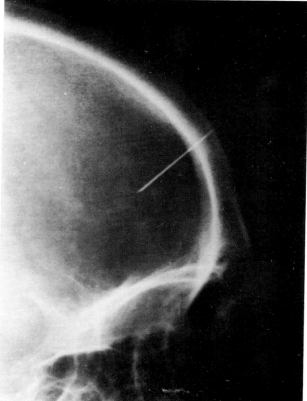

B

Fig. 2-20. A. Nonenhanced CT scan section showing a calcified oligodendroglioma in the left frontal lobe (viewer's right; *arrowhead*). A small opaque marker placed on the skin surface is seen as a *white dot* anteriorly *(line marker)*. This is used as the entrance point for the fine needle aspiration carried out through a small bur hole. B. Localization film showing the aspiration needle in place. The calcifications are faintly visible adjacent to the needle tip.

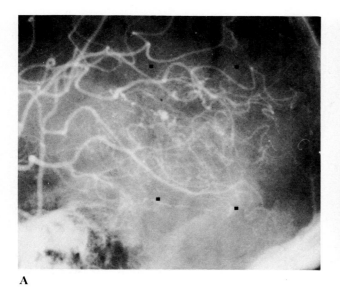
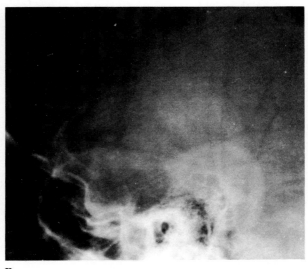

A   B

Fig. 2-19. **A.** A carotid angiogram in a patient with a large vascular glioblastoma of the parietal lobe *(four squares)*. **B.** Lateral skull films obtained at the time of the angiogram, with an opaque *R* placed over the skull corresponding to the level of the tumor. A small bur hole placed at this point was used to carry out fine needle aspiration.

pelvis should be avoided to decrease the risk of puncturing a major renal artery or vein, which could result in formation of an arteriovenous fistula. Zajicek, however, carried out FNA of renal pelvic lesions and did not report any development of arteriovenous fistulae.[73]

Adrenal masses may be aspirated by ultrasonic guidance when the lesion is visible by this technique.[43] CT localization will generally be required in small adrenal tumors.[40] Pagani[51] has reported success employing CT guidance to aspirate normal-size adrenal glands in evaluating patients with small-cell carcinoma of the lung. We have employed a posterior paraspinal approach to the adrenal (Fig. 2-18), but various authors have described the use of anterior,[40] and lateral[43] approaches. If a pheochromocytoma is suspected either clinically or by vanillylmandelic acid (VMA) levels, hypertensive crises may be precipitated by FNA. In such cases, treatment with phentolamine, nitroprusside, or propanalol may be required.[6]

Retroperitoneum

Retroperitoneal masses are generally approached through the anterior abdominal wall in the same fashion that pancreatic lesions are aspirated.[24,25,31]

The comments in the section on the pancreas apply to other retroperitoneal lesions. Haaga has suggested administering a bolus of contrast material before CT localization in order to evaluate the relation of the retroperitoneal mass to the inferior vena cava and aorta and thus avoid puncturing these structures when masses adjacent to either of these major vessels are aspirated.[25] The same information may be obtained by ultrasound, but the presence of significant bowel gas may render ultrasonic guidance ineffective. Opacification of the displaced or compressed ureter during IV urography may be used as a guide in aspirating retroperitoneal masses.[19]

Brain

Previous reports of aspiration biopsy lesions have employed larger bore needles.[41,45,46,47,71] Moran[46,47] describes the use of a stabilization device.* With this device, but employing large-bore needles, he has achieved 97% accuracy in diagnosing masses larger than 1 cm. We have employed one of two guidance techniques. In the first system, the lesion is localized by biplane angiograms that relate the site of the neo-

* Available from Parol Co., 7532 Warner Avenue, St. Louis, MO 63117.

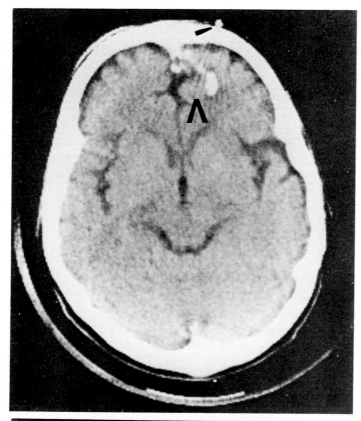

A

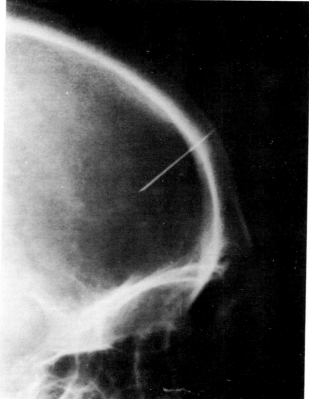

B

Fig. 2-20. A. Nonenhanced CT scan section showing a calcified oligodendroglioma in the left frontal lobe (viewer's right; *arrowhead*). A small opaque marker placed on the skin surface is seen as a *white dot* anteriorly *(line marker)*. This is used as the entrance point for the fine needle aspiration carried out through a small bur hole. B. Localization film showing the aspiration needle in place. The calcifications are faintly visible adjacent to the needle tip.

plasm to external landmarks on the patient's skull (Fig. 2-19). A burr hole is made at the chosen site, and the aspiration needle is then introduced in the radiology suite. Frontal and lateral radiographs of the skull are obtained for verification of needle position, and aspiration is completed.

The second technique uses the CT scan. After a determination of which CT section most clearly depicts the lesion, this section is rescanned with an opaque marker placed in the appropriate position on the patient's skull (Fig. 2-20). Once the position of the external marker is established in relation to the lesion, a burr hole is made at the site of the external marker. The depth of the lesion can be determined from the CT section, and aspiration is then carried out. If necessary, the position of the needle in relation to the mass can be established by repeating the CT section with the needle in place within the tumor.

Breast

Although most breast masses are readily palpated for aspiration biopsy, nonpalpable lesions require mammography. These minimal cancers are manifest either as mass lesions or as areas of irregular small calcifications with or without an associated mass. To facilitate aspiration, introduce the aspiration needle into the area of the breast in which the suspicious changes have been identified on the mammogram. The needle should be introduced in a position similar to that used for obtaining the mammogram, since there is considerable mobility of most breast masses with change in position. Mammograms are then obtained in craniocaudal and mediolateral projection with the needle in place to ensure that the needle is in the suspicious area. The needle may be repositioned, if necessary, and the position re-evaluated by mammography. Stereotactic localization and use of a screw-controlled needle[59] may allow more rapid and accurate localization.

Thyroid Gland

Nuclear imaging of the thyroid gland, using either ^{123}I or ^{99m}Tc, will evaluate the functional status of thyroid nodules. Although most nodules that show no function will be benign (cyst or adenoma), approximately 15% will prove to be malignant. Occasionally nuclear scans will suggest the presence of a nonpalpable nodule. To aspirate these, establish their rela-

tion to superficial landmarks by scanning the thyroid gland with ^{60}Co markers placed on the skin surface.

Ultrasonography of the thyroid gland is useful in two situations: to establish whether a thyroid gland nodule is cystic or solid and to evaluate the palpable nodule that is not seen on the nuclear scan. This may occur if the nodule has function equal to the normal thyroid gland or if the nodule is hidden by the activity of normal thyroid tissue anterior or posterior to it. Because ultrasonography does not irradiate the thyroid gland, it is useful in following the size of thyroid nodules that are not removed surgically.

Miscellaneous

CT may be employed for guidance in FNA of nonpalpable tumors of the head and neck. Ljung[39] has described her experience employing five basic anatomic approaches: transnasal, infratemporal, posterolateral, suboccipital, and regional direct.

Ultrasonic criteria for malignant prostatic masses have been accurate in only 50% of cases.[29] However, intrarectal sonography may be used to diagnose and serve as a guide for FNA of nonpalpable prostatic nodules. The needle is introduced through the perineum either freehand[54,55] or with use of a special guide plate.[29]

Similarly, ultrasonography has been employed to guide FNA of gynecologic tumors. This may be performed by way of a transperineal approach with a multichannel guide and intrarectal transducer.[37] Alternatively, standard pelvic ultrasound techniques and a suprapubic approach may be employed.[13]

REFERENCES

1. Afschrift M, Nachtegaele P, Voet D et al: Puncture of thoracic lesions under sonographic guidance. Thorax 37:503, 1982
2. Alspaugh J, Bernardino ME, Sewell CW et al: CT directed hepatic biopsies: Increased diagnostic accuracy with low patient risk. J Comput Assist Tomog 7:1012, 1983
3. Axel L: Simple method for performing oblique CT-guided needle biopsies. Am J Roentgen 143:341, 1984
4. Buonocore E, Skipper GJ: Steerable real-time sonographically guided needle biopsy. Am J Roentgen 136:381, 1981
5. Carrera GF, Gonyo JE, Barthelemy CR: Fluoro-

scopically guided percutaneous bone biopsy. JAMA 246:884, 1981

6. Casola G, Nicolet V, vanSonnenberg E et al: Unsuspected pheochromocytoma: Risk of blood-pressure alterations during percutaneous adrenal biopsy. Radiology 159:733, 1986

7. Cinti D, Hawkins HB: Aspiration biopsy of peripheral pulmonary masses using real-time sonographic guidance. Am J Roentgen 142:1115, 1984

8. Cohan RH, Newman GE, Braun SD et al: CT assistance for fluoroscopically guided transthoracic needle aspiration biopsy. J Comput Assist Tomogr 8:1093, 1984

9. Coley BL, Sharp GS, Ellis EB: Diagnosis of bone tumors by aspiration. Am J Surg 12:215, 1931

10. Dahlgren S, Nordenstron B: Equipment and biopsy technique In: Dahlgren S, Nordenstron B (eds): Transthoracic Needle Biopsy. Stockholm, Almqvist and Wiksell, 1966

11. Debnam JW, Staple TW: Needle biopsy of bone. Radiol Clin North Am 12:157, 1975

12. Debnam JW, Staple TW: Trephine bone biopsy by radiologists. Radiology 116:607, 1975

13. deCrespigny LC, Robinson HP, Davoren RAM et al: Ultrasound-guided puncture for gynaecological and pelvic lesions. Aust NZ J Obstet Gynaecol 25:227, 1985

14. DeSantos LA, Lukeman JM, Wallace S et al: Percutaneous needle biopsy of bone in the cancer patient. AM J Roentgenol 130:641, 1978

15. Evander A, Ihse I, Lunderquist A et al: Percutaneous cytodiagnosis of carcinoma of the pancreas and bile ducts. Am J Surg 188:90, 1978

16. Ferrucci JT, Wittenberg J, Mueller PR et al: Diagnosis of abdominal malignancy by radiologic fine-needle aspiration biopsy. Am J Roentgen 134:323, 1980

17. Ferruci JT, Wittenberg J: CT biopsy of abdominal tumors: Aids for tumor localization. Radiology 129:739, 1978

18. Fink I, Gamsu G, Harter LP: CT-guided aspiration biopsy of the thorax. J Comput Assist Tomogr 6:958, 1982

19. Freiman DB, Ring EJ, Oleaga JA et al: Thin needle biopsy in the diagnosis of ureteral obstruction with malignancy. Cancer 42:714, 1978

20. Gatenby RA, Mulhern CB, Broder GJ et al: Computed-tomographic-guided biopsy of small apical and peripheral upper-lobe lung masses. Radiology 150:591, 1984

21. Gatenby RA, Mulhern CB, Moldofsky PJ: Computed tomography guided thin needle biopsy of small lytic bone lesions. Skeletal Radiol 11:289, 1984

22. Gobien RP, Stanley JH, Vujic I et al: Thoracic biopsy: CT guidance of thin-needle aspiration. Am J Roentgen 142:827, 1984

23. Golimbu C, Firooznia H, Rafii M: Use of CT-guided percutaneous bone biopsy in staging of genitourinary tumors. Urology 22:322, 1983

24. Gothlin JJ: Postlymphangiographic percutaneous fine needle biopsy of lymph nodes guided by fluoroscopy. Radiology 120:205, 1976

25. Haaga JR: New techniques for CT guided biopsies. Am J Roentgen 133:633, 1979

26. Haaga JR, Alfidi RJ: Precise biopsy localization by computed tomography. Radiology 118:603, 1976

27. Haaga JR, Reich NE, Harilla TR et al: Interventional CT scanning. Radiol Clin North Am 15:449, 1977

28. Hancke S, Holm HH, Koch F: Ultrasonographically guided percutaneous fine needle biopsy of the pancreas. Surg Gynecol Obstet 140:361, 1975

29. Hastak S, Gammelgaard J, Holm HH: Ultrasonically guided transperineal biopsy in the diagnosis of prostatic cancer. J Urol 128:69, 1982

30. Ho C, McLoughlin MJ, McHattie JD et al: Percutaneous needle aspiration biopsy of the pancreas following endoscopic retrograde cholangiopancreatography. Radiology 125:351, 1977

31. Holm HH, Pedersen JF, Kristensen JK et al: Ultrasonically guided percutaneous puncture. Radiol Clin North Am 13:493, 1975

32. Holm HH, Torp-Pedersen S, Larsen T et al: Percutaneous fine needle biopsy. Clin Gastroent 14:423, 1985

33. Izumi S, Tamaki S, Natori H et al: Ultrasonically guided aspiratio needle biopsy in disease of the chest. Am Rev Respir Dis 125:460, 1982

34. Juul N, Torp-Pedersen S, Holm HH: Ultrasonographically guided fine needle aspiration biopsy of retroperitoneal mass lesions. Brit J Rad 57:43, 1984

35. Lalli AF: Roentgen guided aspiration biopsies of skeletal lesions. J Canad Assoc Radiol 21:71, 1970

36. Lalli AF, McCormack LJ, Zelch M et al: Aspiration biopsies of chest lesions. Radiology 127:35, 1978

37. Larsen T, Torp-Pedersen S, Bostofte E et al: Transperineal fine needle biopsy of gynecological tumors guided by transrectal ultrasound: A new method. Gyn Oncol 22:281, 1985

38. Linsk JA, Salzman AJ: Diagnosis of intrathoracic tumors by thin needle cytologic aspiration. Am J Med Sci 263:181, 1972

39. Ljung BME, Larsson SG, Hanafee W: Computed tomography-guided aspiration cytologic examination in head and neck lesions. Arch Otolaryngol 110:604, 1984

40. Luning M, Neuser D, Kursawe R et al: CT guided percutaneous fine needle biopsy in the diagnosis of small adrenal tumours. Europ J Radiol 3:358, 1983

41. Marshall LF, Jennett B, Langitt TW: Needle biopsy for the diagnosis of malignant glioma. JAMA 28:1417, 1974

42. Martinez A, Velasco M, Caceres J: Fine-needle biopsy of the pancreas using real-time ultrasonography. Gastroint Radiol 9:231, 1984

43. Montali G, Solbiati L, Bossi MC et al: Sonographically guided fine-needle aspiration biopsy of adrenal masses. Am J Roentgen 143:1081, 1984

44. Montali G, Solbiati L, Croce F et al: Fine-needle aspiration biospy of liver focal lesions ultrasonically guided with a real-time probe. Report on 126 cases. Brit J Radiol 55:717, 1982

45. Moran CJ, Nadich TP, Gado MH et al: Central nervous system lesions biopsied or treated by CT guided needle placement. Radiology 131:681, 1979

46. Moran CJ, Naidich TP, Marchoski JA: CT-guided needle placement in the central nervous system: Results in 146 consecutive patients. Am J Roentgen 143:861, 1984

47. Moran CJ, Naidich TP, Marchosky JA et al: A simple stabilization device for intracranial aspiration procedures guided by computed tomography. Radiology 144:183, 1982

48. Murphy WA: Radiologically guided percutaneous musculoskeletal biopsy. Orth Cl NA 14:233, 1983

49. Otto RC: Results of 100 fine needle punctures under real-time sonographic control. J Belge Radiol 65:193, 1982

50. Ottolenghi CE, Schajowicz G, Fermin AD: Aspiration biopsy of the cervical spine. J Bone Joint Surg 46:715, 1964

51. Pagani JJ: Normal adrenal glands in small cell lung carcinoma: CT-guided biopsy. Am J Roentgen 140:949, 1983

52. Palaez JC, Hill MC, Dach JL et al: Abdominal aspiration biopsies. Sonographic vs. computed tomographic guidance. JAMA 250:2663, 1983

53. Rabinov K, Goldman H, Robash H: The role of aspiration biopsy of focal lesions in lung and bone by simple needle and fluoroscopy. Am J Roentgen 101:932, 1967

54. Rifkin M, Kurtq AB, Goldberg BB: Prostate biopsy utilizing transrectal ultrasound guidance: Diagnosis of nonpalpable cancers. J Ultrasound Med 2:165, 1983

55. Rifkin M, Kurtz AB, Goldberg BB: Sonographically guided transperineal prostatic biopsy: Preliminary experience with a longitudinal linear-array transducer. Am J Roentgen 140:745, 1983

56. Sargent EN, Turner AF: Emergency treatment of pneumothorax. A simple catheter technique for use in the radiology department. Am J Roentgen 109:531, 1970

57. Schwerk WB, Durr H, Schmitz-Moormann P: Ultrasound guided fine-needle biopsies in pancreatic and hepatic neoplasms. Gastroint Radiol 8:219, 1983

58. Schwerk WB, Schmitz-Moormann P: Cytohistologic diagnoses and echo pattern of lesions. Cancer 48:1469, 1981

59. Silver CE, Koss LG, Brauer RJ et al: Needle aspiration cytology of tumors at various body sites. Current Problems in Surgery 22:1, 1985

60. Smith EH, Bartrum RJ, Young CC et al: Percutaneous aspiration biopsy of the pancreas under ultrasonic guidance. N Engl J Med 292:825, 1975

61. Stevens GM, Weigen JF, Lillington GA: Needle aspiration biopsy of localized pulmonary lesions with amplified fluoroscopic guidance. Am J Roentgen 103:561, 1968

62. Tao LC, Pearson FG, Delarue NC et al: Percutaneous fine needle aspiration. Cancer 45:1480, 1980

63. Tatsuta M, Yamamoto R, Yamamura H et al: Cytologic examination and CEA measurement in aspirated pancreatic material collected by percutaneous fine-needle aspiration biopsy under ultrasonic guidance for the diagnosis of pancreatic carcinoma. Cancer 52:693, 1983

64. Tylen U, Arnesso B, Lindberg LG et al: Percutaneous biopsy of carcinoma of the pancreas guided by angiography. Surg Gynecol Obstet 142:737, 1976

65. Valls J, Ottolenghi CE, Shajowicz F: Aspiration biopsy in diagnosis of lesions of vertebral bodies. JAMA 136:376, 1948

66. vanSonnenberg E, Wittenberg J, Ferruci JT et al: Triangulation method for percutaneous needle guidance. Am J Roentgen 137:757, 1981

67. vanSonnenberg E, Lin AS, Deutsch AL et al: Percutaneous biopsy of difficult mediastinal, hilar, and pulmonary lesions by computed-tomographic guidance and a modified coaxial technique. Radiology 148:300, 1983

68. Vogelzang RL, Matalon TA, Neiman HL et al: Lateral scout radiograph in CT-guided aspiration biopsy. Am J Roentgen 140:164, 1983

69. Williams RA, Haaga JR, Karagiannis E: CT guided paravertebral biopsy of the mediastinum. J Comp Assist Tomogr 8:575, 1984
70. Yang P-C, Luh K-T, Sheu J-C et al: Peripheral pulmonary lesions: Ultrasonography and ultrasonically guided aspiration biopsy. Radiology 155:451, 1985
71. Yeates A, Enzmann DR, Britt RH et al: Simplified and accurate CT-guided needle biopsy of central nervous system lesions. J Neurosurg 57:390, 1982
72. Zajicek J: Aspiration biopsy cytology: 1. Cytology of supradiaphragmatic organs. In Wied GL (ed): Monographs in Clinical Cytology, Part 1. Basel; S. Karger, 1974
73. Zajicek J: Aspiration biopsy cytology: 2. Cytology of supradiaphragmatic organs. In Wied GL (ed): Monographs in Clinical Cytology, Part 2. Basel S. Karger, 1974
74. Zegel HG, Turner M, Velchik MG et al: Percutaneous osseous needle aspiration biopsy with nuclear medicine guidance. Clinical Nuclear Med 9:89, 1984
75. Zornoza J, Wallace S, Goldstein OHM et al: Transperitoneal percutaneous retroperitoneal lymph node aspiration biopsy. Radiology 125:111, 1977
76. Zornoza J, Wallace S, Ordonez N et al: Fine needle aspiration biopsy of the liver. Am J Roentgen 134:331, 1980

CHAPTER 3

Head, Neck, and Regional Nodal Sites

JOSEPH A. LINSK

SIXTEN FRANZEN

Lesions of the head and neck are readily accessible to the fine needle, and numerous publications describe the efficacy of aspiration biopsy cytology (ABC) in this region.* Cytologic smears from primary lesions may contain cells derived from epithelial, lymphoid, plasmacytic, melanotic, osseous, and supporting soft tissues.[42] Similarly, metastatic masses may yield a variety of cell patterns. Therefore, clinical data, including the patient's age and sex and the clinical presentation of the lesion, are helpful and generally necessary, particularly in cytologic diagnosis of referred smears. Inspection and palpation of the primary mass, satellite lesions, and regional nodes should encompass the anatomic sites listed below.

ABC Targets of the Head and Neck Region
Scalp
Peri-orbital region and lacrimal glands
Ears, including pinna
Pre-auricular and postauricular regions
Nose
Cheek and lips
Angles of jaw
Jaw
High jugulodigastric areas

Cervical regions — anterior, posterior, lateral
Notch of the carotid
Suprasternal and supraclavicular areas
Intra-oral mucosa, tonsils, and tongue

METHOD

Most of the targets listed above are surface lesions that are easily palpated, aspirated, smeared, and stained as described in Chapter 1. Tight skin overlying lesions of portions of the nose and the ear is best punctured obliquely with a 25-gauge needle (see Chap. 13). Important rules for examination of the head and neck have been described by Conley.[9]

A light source is necessary for careful examination of the intra-oral cavities. Carry out palpation with a gloved finger before the puncture. This includes particularly the base of the tongue (Fig. 3-1). An anesthetic spray is useful. Depress or deflect the tongue with a tongue stick or a finger. Aspiration can be carried out with ease after these preparations. Transpharyngeal aspiration of the retropharyngeal space and cervical vertebral bodies may require the use of fluoroscopy (see Fig. 16-2). Transnasal puncture of the skull base yielding pituitary tumors, chordoma, or metastatic carcinoma is shown in Figure 3-2 (also see Chap. 17).

* See References 11, 15, 16, 17, 21, 22, 25, 38, 45, 50, 51, 56, 58, 61, 66, 69, 70.

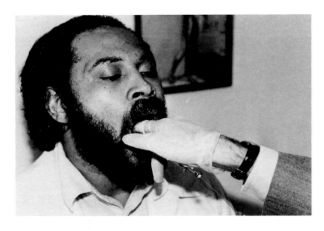

Fig. 3-1. Gloved finger palpates posterior tongue with prostate needle guide allowing direct aspiration (see Chap. 11).

The advent of computed tomography–guided aspiration has extended the reach of the needle to lesions in the skull base, skull fossae, sinus structures, and larynx.[38]

PRIMARY AND METASTATIC ASPIRATION TARGETS

Scalp, Ears, Face, Nose, and Orbit

Scalp

The most common tumorous scalp lesion that may become a target for fine needle aspiration (FNA) is the sebaceous or epidermoid cyst. It is usually recognized clinically, but when hard may simulate a tumor. The aspirate consists of semisolid, yellow-tan, granular material with a sour odor. In the cytologic smear, the background is amorphous and contains inflammatory cells, squames, and well-differentiated squamous cells (Fig. 3-3). These structures are benign. Primary squamous carcinomas of epidermoid cysts are quite rare or even nonexistent.[40,64] Aspiration of carcinoma cells indicates that the "clinical cyst" is a metastatic deposit. Intact sebaceous cells are not usually part of the content of these cysts, and when found on aspiration are considered products of a sebaceous-lined cyst (steatocystoma multiplex; Fig. 3-4).

The primary malignant lesions of the scalp are squamous and basal cell carcinoma and melanomas (see Chap. 13). Basal and squamous lesions primarily occur on the bald scalp from solar exposure.[40] Aspira-

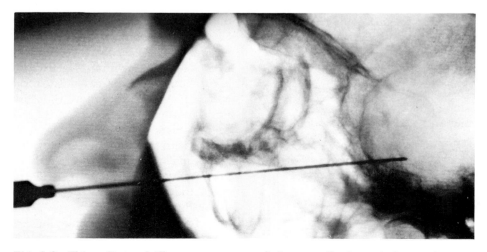

Fig. 3-2. This radiograph illustrates passage of a long needle through the nasal cavity to penetrate an enlarged sella.

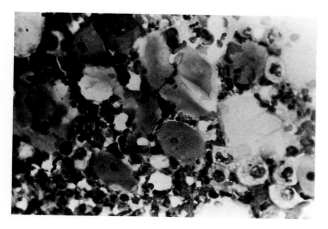

Fig. 3-3. Epidermoid cyst. The smears are predominantly composed of squames and squamous cells with pyknotic nuclei. In addition, there are inflammatory cells and small histiocytes, but few lipid storage cells. (Romanowsky [R] stain, ×132)

tion biopsy of the local lesion can be carried out to demonstrate that a doubtful lesion is indeed a cancer requiring surgery or radiation.

Metastatic scalp lesions are dermal or subcutaneous nodules. The usual metastases arise from primary tumors in breast, lung, and kidney. However, a variety of primary lesions may have scalp deposits.[65] Aspiration may identify the primary tumor and shorten diagnostic studies (see later discussion). Often, however, the cell types are nonspecific adenocarcinoma or squamous carcinoma; and search for a primary cancer is required.

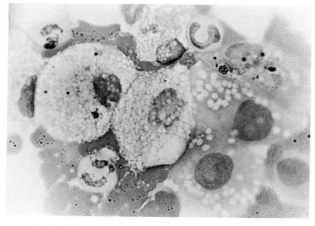

Fig. 3-4. Sebaceous-lined cyst (steatocystoma multiplex). The smear is rich in fat, storage (sebaceous) cells, and inflammatory cells. (R stain, ×330)

Ears

The usual significant ear lesions are basal cell epitheliomas and squamous carcinomas. The clinician is not infrequently concerned with diagnosis of a small nodular lesion of the ear, which will either be surgically biopsied or be observed in follow-up examination. Fine needle aspiration utilizing a 25-gauge needle will usually produce sufficient cells to settle the diagnosis.

Facial Lesions

Facial lesions multiply, particularly in the elderly. Fine needle aspiration is useful in distinguishing basal cell carcinoma from inflammatory lesions. Palpable nodules composed of lymphoid tissue and occurring in the cheeks and lips may be benign or malignant. Malignant lymphoid facial nodules usually occur in the context of the known lymphoma. Other facial lesions including melanoma are considered in Chapter 13.

Nose and Sinuses

Two unusual tumors may present as targets for FNA yielding distinctive cytology. Esthesioneuroblastoma is a tumor presenting as a granular polypoid mass arising from the upper third of the nasal septum. It is within reach of the fine needle. Cytology is that of neuroblastoma described in Chapter 15 and by Jobst and colleagues.[31] It may metastasize to cervical lymph nodes and must be considered in differential diagnosis of cervical adenopathy, particularly in patients with nasal obstruction or epistaxis, or both.[19]

Primary meningioma of the nasal cavity and paranasal sinuses has also been reported.[32] Such lesions are diagnosable by FNA, and the cytology is the same as that of intracranial meningioma as described in Chapter 17 and of extracranial meningioma (see Fig. 5-28).

Advanced tumors of the sinuses may produce facial bulges that become targets for aspiration. The bony wall is often thinned or destroyed, allowing fine needle penetration. Squamous carcinoma complicated by necrosis and secondary infection may be aspirated (Fig. 3-5).

Sarcomas of all varieties, but particularly rhabdomyosarcoma in children, may become targets in the head and neck region. Sarcomas are considered in Chapter 15.

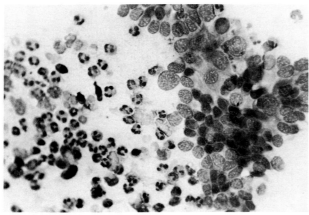

Fig. 3-5. Advanced tumor invading cheek. Large clusters of fairly intact nonkeratinized tumor cells together with acute inflammatory exudate. (R stain, ×132)

Destructive lesions of skull, facial bones, and mandible are all suitable targets when fluoroscopy is used. Types of cytology are considered in Chapters 9 and 16.

The Orbits and Eyes

The orbits, eyelids, lacrimal glands, and eyes are all suitable targets for aspiration.[11,12,13,58] Orbital lymphomatous deposits occur in disseminated non-Hodgkin's lymphoma[33] and in that context are easily diagnosed by FNA. Primary orbital lymphomas are discussed in Chapter 14. Not all lymphoid deposits here pursue a malignant course.

The pathologic lacrimal gland may be both visible and palpable. It is an easy target for FNA and may be the site of a variety of pathologic disorders, including adenoid cystic carcinoma, acinic cell carcinoma,[13] lymphoma, and sarcoidosis. All are potentially diagnosable by ABC. Intra-orbital recurrence postexenteration of melanoma, adenoid cystic carcinoma of lacrimal gland tumors, and other primaries can be aspirated with a 25-gauge needle.

Most eyelid tumors are basal cell epitheliomas, considered in Chapter 13. The cytology of a meibomian gland carcinoma has been described. The cytologic picture consists of pleomorphic reticular nuclei with abundant vacuolated cytoplasm containing globules of neutral fat (positive oil–red–O stain).[12]

Direct puncture of a metastatic mass palpable within the orbital brim is easily carried out. The tech-

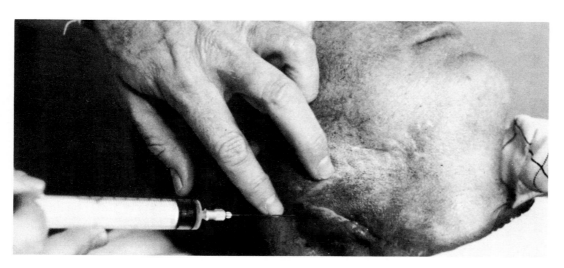

Fig. 3-6. Aspiration of recurrent tumor in scarred, postirradiation area. Use of a plastic syringe without a holder.

nique for intra-orbital aspiration of melanoma was described by Czerniak and colleagues.[11] The cytology of melanoma is described in Chapter 13.

The technique of fine needle sampling without aspiration is described by Zajdela and colleagues[68] in the cytologic diagnosis of orbital and peri-orbital palpable tumors.

Oral and Intra-Oral Primary Lesions

Lips, tongue, tonsils, floor of the mouth, palate, buccal mucosa, and jaw can be readily inspected, palpated, and aspirated. Intra-oral bulges secondary to obstructing and advanced sinus lesions are also suitable targets, yielding smears discussed above under facial bulges. Squamous carcinoma is the major tumor found, and its cytology is described later. Nonsquamous lesions include salivay gland tumors, lymphomas, and inflammatory masses. Salivary gland tumors are discussed in Chapter 5.

In suspected recurrent disease (*i.e.*, after surgery or radiation), both surface and intra-oral aspiration are often diagnostic without repeat surgical biopsy. This is useful because taut tissue at the site of radical surgery or radiation is not conveniently excised. Tumors appear within scars, nodes, muscle bundles, dermal and subcutaneous sites, and intra-orally.[21] Figure 3-6 illustrates aspiration of a scarred postradiation region with a 10-ml plastic syringe. Peristomal recurrence after laryngectomy for glottic and subglottic carcinoma provides a smooth, taut, indurated target. The cytologic yield may consist of scant squamous cells intermixed with inflammation and cellular debris.

The Jaw

Lytic lesions of the jaw are readily accessible to FNA with radiographic control. A variety of lesions diagnosed cytologically have been reported by Ibrahim and colleagues.[30] Giant cell granulomas produce spindle cells and multinucleated giant cells. Hemosiderin is present in the brown tumor of hyperparathyroidism that also yields multinucleated giant cells. Odontogenic cysts contain anucleated squames and superficial squamous cells. Ameloblastoma yields both basoloid cells in sheets and keratinized squamous cells. The cytology of eosinophilic granuloma (Fig. 3-7) is discussed in Chapter 16. A variety of metastatic as well as inflammatory lesions, including myeloma and actinomycosis, may also be identified.

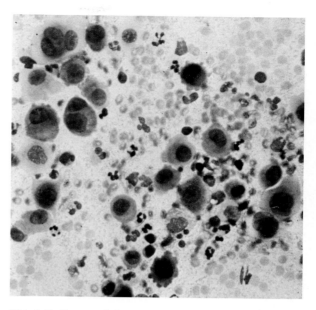

Fig. 3-7. Eosinophilic granuloma, skull lesion. There are scattered histiocytic cells, some double nucleated with characteristic violet-gray cytoplasm. Eosinophils, polymorphonuclear leukocytes, and RBC are in background. (R stain, ×800)

Direct or lymphatic extension of head and neck tumors can produce distructive lesions of the mandible and maxilla.[3] Radiation necrosis of the mandible can be distinguished from destructive metastases by FNA. This is illustrated in Figure 3-8 in a patient who received 6000 rads to a lesion on the floor of the mouth.

Metastatic Node

Lymphatic drainage in primary lesions is not always predictable and is often altered by prior surgery.[53] Such alterations, as well as the possibility of an unsuspected pathologic lesion, must be kept in mind and should encourage aspiration rather than presumption.

Metastases in nonpalpable and nonenlarged lymph nodes are often detected in the surgical pathology laboratory during examination of a tumor specimen with accompanying nodes. Such nodes rarely become targets for aspiration. Enlargement of a node by metastatic deposits indicates fairly diffuse involvement or replacement by tumor (Fig. 3-9). As a result, random puncture of an enlarged metastatic

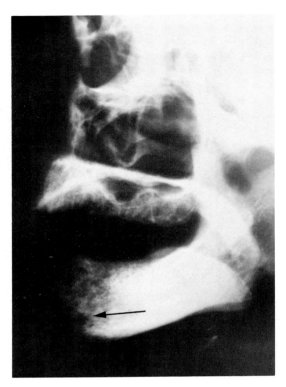

Fig. 3-8. Destruction of irradiated mandible *(arrow)* cytology differentiates recurrent tumor from radiation necrosis.

node will almost invariably obtain tumor cells, most often without a lymphoid component. *Puncture of a clinically enlarged and pathologic node that fails to produce foreign tumor cells is fairly good evidence against metastases and should lead to consideration of inflammation or lymphoma.*

Failure to obtain or identify cancer cells in metastatic nodes occurred in 17 patients (6.6%) of a series of 275 studied at Karolinska.[16] In a recent series of 350 node aspirations, most of which were cervical in location, 209 were categorized as malignant and 30 as suspicious. Virtually all were proved malignant on history. Of 102 cytologically benign cases, 69 resolved without surgery. In the remaining 33, 24 were found to be reactive, 7 to be lymphoma, one to be undifferentiated carcinoma, and one to be malignant fibrous histiocytoma. Thus there were nine that were cytologically false-negative.

In contrast, in a series of 925 surgically biopsied patients, benign nodes were found in 63% of cervical, 60% of axillary, and 71% of inguinal node biopsies.[37]

The large percentages of benign diagnoses could undoubtedly have been reduced by pre-operative FNA. It must be emphasized, however, that negative cytologic findings from palpable lesions are always subordinate to clinical findings.

A false positive report of metastatic carcinoma in a lymph node may occur with carotid body tumors and neck cysts (see later discussion) and calcifying epithelioma of Malherbe (see Chap. 13).

Relatively small firm nodes in regional sites and draining tumor beds may be suitable targets for a 25-gauge needle and yield surprise positive smears. Do not hesitate to puncture on slight clinical suspicion.

The ease of FNA renders clinical judgment as a safe guide to therapy obsolete. Neither incisional nor excisional biopsy is recommended because of cell spillage, but either may be necessary where a primary tumor has not been identified and aspiration is inconclusive.

Studies in which end results of radiotherapy are based on clinical diagnosis of palpable nodes[43] should be reconsidered in the light of the high percentage of benign nodes at surgical biopsy.[37] Problems in diagnosis are largely resolved by FNA.

The identification of metastatic cancer in cervical nodes is followed by an intensive search for a primary tumor both above and below the clavicles. In a series of reports, the primary tumor was not found in over 50% of patients.[8,14,16,41] Identification of cell type in a cervical node aspirate will often narrow the search for a primary tumor. Cytologic types and potential primary sites are discussed later. Of 522 metastatic cervical nodes of primary tumor in the head and neck region, 456 originated from oral, pharyngolaryngeal, nasal, and paranasal cavities or from the skin. Sixty-six originated from thyroid, salivary, and lacrimal glands.[16]

Cervical Node Metastases

Squamous Carcinoma

From Scalp Lesions. Occipital and posterior cervical nodes are the regional nodes for benign or malignant scalp lesions. In the absence of scalp lesions, nodes in these areas may suggest *de novo* disease (lymphoma). However, failure to detect or to recall a prior scalp lesion may lead to misdiagnosis, as illustrated in the following case report:

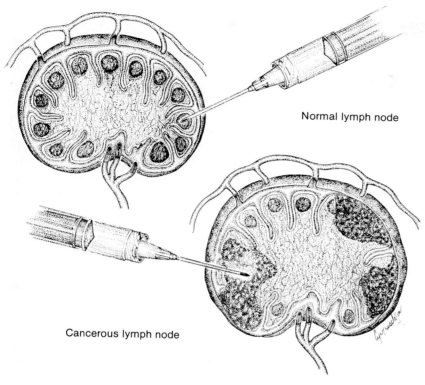

Normal lymph node

Cancerous lymph node

Fig. 3-9. The metastatic node. These drawings illustrate the heterogenous cell yield that may be obtained as a needle is oscillated through follicular center and medullary cords of a nonmalignant hyperplastic node. Metastatic deposits may segmentally occupy a slightly or moderately enlarged node. Though aspiration will obtain cancer as well as lymphocytes, pure cancer is obtained when nodes are totally replaced and enlarged.

■ *Case 3-1*

A bald radiologist aged 67 years, with chronic lymphocytic leukemia (CLL) for 22 years, presented intermittent generalized lymphadenopathy depending on his therapy. Three years earlier, a squamous carcinoma of the scalp had been eradicated with radiation. Rather large, fixed occipital nodes were interpreted by a hematologist as progressive lymphomatous nodes. He was not made aware of the prior scalp carcinoma. When the nodes failed to respond to chemotherapy, they were aspirated, yielding keratinizing squamous carcinoma (Fig. 3-10).

An increase of squamous head and neck carcinoma in well differentiated lymphocytic lymphoma was recently reported.[49]

From Tongue, Tonsils, and Intra-oral Mucosa. Anterior and lateral cervical nodes, including those in the submandibular region, are the prime sites of metastases from the rather easily identified intra-oral lesions, as illustrated in the following case report:

■ *Case 3-2*

A man aged 80 years presented with firm nodes in the right anterior cervical chain. Aspiration yielded moderately well-differentiated squamous carcinoma. On inspection, there was a small area of induration and ulceration along the right lateral margin of the tongue. Immediate aspiration without anesthesia with a 25-gauge needle demonstrated a primary squamous carcinoma, expediting diagnosis and treatment.

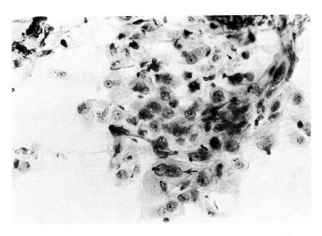

Fig. 3-10. Squamous carcinoma, occipital node. This is a large cluster of malignant squamous cells, some with keratinization and nucleoli (*arrow*). Cellular debris, squames, and a few inflammatory cells are present. (P stain, ×132)

From Nasopharyngeal Carcinoma. Nasopharyngeal carcinoma (see below) is notoriously occult on routine inspection and often invisible with more intensive examination. Opthalmoneurologic signs and symptoms arising from any cranial nerve together with cavernous sinus involvement are clues to the origin of palpable neck nodes.[26,55] In up to 90% of patients, clinically positive nodes may be the primary presenting abnormality.[44] The jugulodigastric node at the angle of the jaw and the high lateral jugular nodes are frequent sites of metastases from nasopharyngeal carcinoma.[44]

From Laryngeal, Hypopharyngeal, and Base of Tongue Primary Tumors. Since symptoms from primary lesions of the larynx, hypopharynx, and base of tongue are usually antecedent, cervical nodes will only occasionally be the presenting abnormality.[54] However, in up to 45% of patients with laryngeal and hypopharyngeal carcinoma, metastases will recur in the skin or nodes and become targets for aspiration.[47] Peristomal recurrence after laryngectomy has been considered above.

From Nasal Cavity and Paranasal Sinuses. Carcinomas of the nasal cavity and paranasal sinuses metastasize to the submaxillary, retropharyngeal, and jugular nodes, deep and superficial.[6,35] Local symptoms will suggest a disorder before nodes appear in the majority of patients. All facial, intranasal, and intra-oral bulges should be punctured.

From the Lips. Carcinoma of the lip may be confused with an inflammatory process and neglected. Metastases to homolateral and submandibular nodes and to submental nodes in midline lower lip lesions occur late. Only 60% of enlarged nodes at these sites are found to be metastatic, with the remainder inflammatory.[18]

The most common head and neck target presented for ABC is the palpable cervical lymph node. Of 1200 consecutive aspiration biopsies of cervical lymph nodes, Zajicek and colleagues reported benign cytology in 51.6%.[70] Five hundred and eighty one were reported to be malignant, of which 20.7% were lymphomas and 79.3% contained metastatic carcinoma.

Nonsquamous Carcinoma

The origin of nonsquamous metastases to cervical nodes cannot often be identified by the usual clinical criteria. They are common, however, making up 48% of a reported series of aspirated metastatic cervical nodes.[16] Well-differentiated thyroid (see Chap. 4) and renal cell (see Chap. 8) carcinomas, special types of breast carcinoma (see Chap. 6) and melanomas (see Chap. 13), some germinal tumors (see Chap. 12), and oat cell carcinomas (see Chap. 7) all offer the possibility of immediate identification by examination of smears from metastatic nodes.

Adenocarcinomas with a mucinous component will suggest origin from the gastrointestinal (GI) tract and breast, (Fig. 3-11) particularly signet ring types (Color Plate I), but most differentiated adenocarcinomas will not show cytologic features that enable identification of origin.

Excellent samples for study are usually obtained, however; and the diagnosis of carcinoma, albeit not the specific type, is not difficult. In contrast, undifferentiated carcinoma may be obtained (Fig. 3-12), and the differential diagnosis includes lymphoma (see Chap. 14), melanoma (see Chap. 13), and undifferentiated carcinoma of the lung (see Chap. 7). The specific type may be difficult or impossible to obtain by cytology or histology.[1] For further discussion, see Chapter 7.

Considerations of Differential Diagnosis

In palpable lesions under the mandible and at the angle of the jaw, including preauricular and postauricular masses, primary or metastatic salivary gland

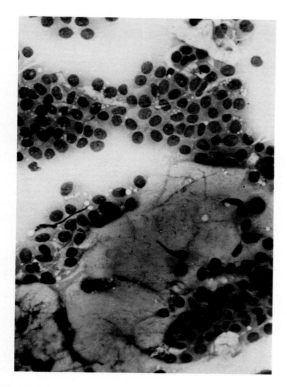

Fig. 3-11. Metastatic breast carcinoma, colloid type, cervical node. Opaque gray stroma is violet in R stain. It is surrounded by masses of uniform small cancer cells in an onolayer. (R stain, ×600)

tumors must be excluded (see Chap. 5). Mucoepidermoid tumors yielding primarily squamous material may suggest metastasis. More rarely, squamous carcinoma of the parotid or submaxillary gland may be difficult to distinguish from metastases, and a thorough search for a primary lesion is mandatory. Metastases to intraparotid lymph nodes occur from head and neck primary tumors (see Case 13-1, Chap. 13).

CYTOLOGIC VARIETIES OF SQUAMOUS CARCINOMA

Squamous carcinoma aspirated from a metastatic lymph node in which various portions of the node are sampled will yield a variegated cellular pattern.

Figures 3-13, 3-14, and 3-15 illustrate cytologic patterns in squamous carcinoma. All were obtained from a large submandibular mass in an 87-year-old woman with a prior squamous carcinoma of the lip treated surgically. Extremely well-differentiated squamous cells together with anucleated squames

appear benign and could be confused with an epidermoid or branchial cyst (see below). Less differentiated cells with condensed well-marked cytoplasmic borders are evident in Figure 3-14. Cytoplasm is blue with Romanowsky (R) stain.

Because of cytologic variability within each lesion, assessment of differentiation is made with caution.

Less differentiated but still clearly squamous carcinoma may be seen (Fig. 3-15). Malignant pearls (Fig. 3-16) as well as poorly differentiated nonsquamous cells may be present in the same smear. Anaplastic cells with giant round polymorphous nuclei as well as nucleoli, not easily recognized as squamous, may be accompanied by smaller keratinized squamous carcinoma cells.

Dispersed large carcinoma cells with a loose chromatin pattern (Fig. 3-17) are not readily recognized as squamous in origin, but were aspirated from cervical metastases occurring 1 year after laryngectomy for keratinized laryngeal carcinoma. This is an example of metastasis of the less differentiated portion of a primary tumor.

Smears modified by inflammation and necrosis are illustrated in Figures 3-5 and 3-18. Squamous carcinoma can usually still be identified in such smears.

With Papanicolaou (P) stain, nuclei have clear borders and well-defined chromatin clumping and hyperchromatism signaling malignancy (see Fig.

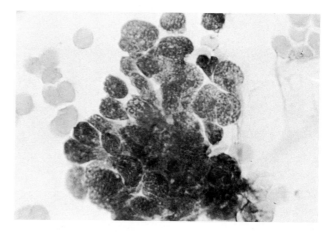

Fig. 3-12. Undifferentiated carcinoma, cervical lymph node. Metastasis from bronchogenic carcinoma. This is a cluster of immature cells with minimal identifiable cytoplasm. Nuclei are irregular in size and shape. Nuclear chromatin is reticulogranular, and nucleoli are not identifiable with this stain. (R stain, ×330)

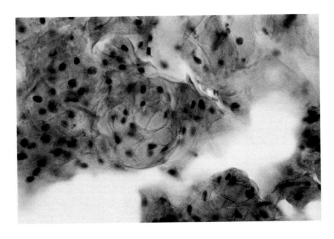

Fig. 3-13. Metastatic squamous carcinoma, submandibular mass. Highly keratinized cells together with anucleated squames. Well-differentiated with non-malignant characteristics. (P stain, ×400)

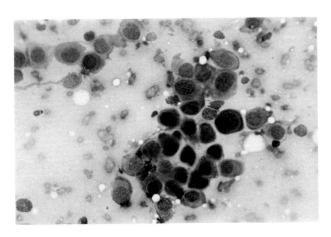

Fig. 3-14. Metastatic squamous carcinoma (same as Fig. 3-13). In this group of less differentiated squamous carcinoma cells, keratinization can still be detected in the royal blue cytoplasm with sharp cellular outlines. (R stain, ×600)

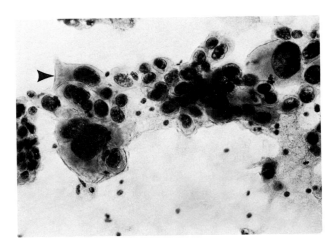

Fig. 3-15. Squamous carcinoma, cervical lymph node. Dense haphazard clusters of squamous carcinoma cells with marked variation in size. Giant cells are apparent. Note angular shape of keratinized tumor cell (*arrowhead*). (P stain, ×400)

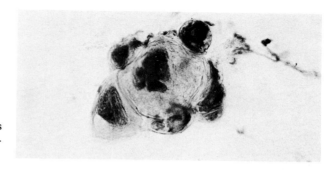

Fig. 3-16. Squamous carcinoma as seen in Fig. 3-15. This is a malignant pearl with abnormal nuclei and orangeophilic cytoplasm. (P stain, ×600)

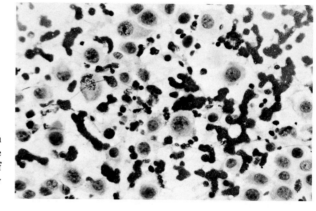

Fig. 3-17. Metastatic squamous carcinoma, cervical lymph node. There are dispersed large tumor cells with immature nuclei and prominent nucleoli *(arrows)* on a background of red blood cells and tissue fluid. Squamous origin not easily recognized. (R stain, ×600)

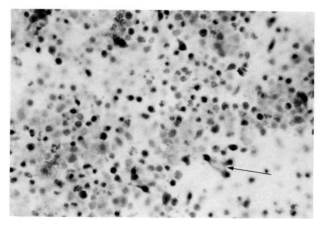

Fig. 3-18. Metastatic squamous carcinoma, cervical lymph node. Dispersed necrotic cells are in a thick background fluid. There are identifiable pyknotic squamous cells *(arrow)*. (P stain, ×400)

3-15). Nucleoli are usually not prominent but may be present (see Fig. 3-10). Cytoplasm is green and orange with the irregular shape characteristic of squamous cells.

In contrast, nuclear chromatin may be reticulogranular and the nuclear membrane less defined with R stain. Nuclear chromatin may be uniformly dense, suggesting early degeneration, or smudged, without clear chromatin and parachromatin spaces. Cytoplasm is pale to royal blue.

Nonkeratinized cells cannot clearly be distinguished as squamous by either stain. They are essentially nondifferentiated but differ from poorly differentiated metastases from some bronchogenic cancers or adenocarcinomas in which the cell yield is dispersed or the cells appear quite primitive (Fig. 3-19).

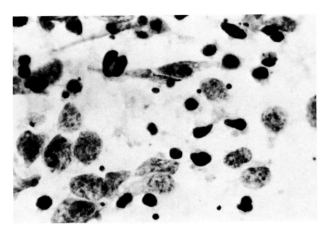

Fig. 3-19. Undifferentiated carcinoma, metastatic cervical lymph node. There are dispersed, elongated, primitive cells, intermixed with darkly staining lymphocytes, in oat cell carcinoma metastatic from the lung. Cytoplasm is not evident. (P stain, ×132)

Fig. 3-20. Nasopharyngeal carcinoma, cervical lymph node. There are numerous large, primitive cells reminiscent of lymphoblasts. There is, however, a tendency to cluster. The nuclear chromatin does not have the fine texture of primitive lymphoid cells. Surrounding the tumor cells are lymphocytes and polymorphonuclear leukocytes extracted from the host node. (R stain, ×132)

Nasopharyngeal Carcinoma

Tumors within the confined space of the nasopharynx are not targets for FNA. Primary tumors may be quite small, often not visualized, yet may produce prominent regional nodal and distal metastases. The epithelium of the nasopharynx is thin and varies from stratified squamous to pseudostratified as well as ciliated columnar. It is intimately backed by a lymphoid layer that has important pathogenetic significance. Epstein–Barr virus (EBV), which appears to be present in 75% of nasopharyngeal carcinomas, is conducted by the lymphocytes to the epithelial cells.[28,63] Nodal and other metastases will reflect the cell type of the original tumor. Keratinizing squamous carcinoma occurs in 25%,[62] does not contain EBV, and cannot be differentiated cytologically from metastases originating in the oral cavity, pharynx, or larynx. The metastasis is usually to regional nodes. Tumor cells in the metastatic site are often mixed with nodal lymphocytes[60,62] producing a distinctive cytologic pattern. It consists of dispersed large primitive cells intermixed with lymphocytes (Fig. 3-20). The undifferentiated cells may have an immature lymphoid appearance, but stand in contrast to the host lymphocytes of the metastatic node. Nuclei are large, chromatin is granular, and nucleoli may be inapparent in R but prominent in P.

Primitive cells dispersed and clustered are seen in Figures 3-21 and 3-22. Cells are fragile, elongated, and traumatized. Cytoplasm is absent. Compare with metastatic oat cell carcinoma (see Chap. 7). Clinical correlation is necessary.

In the absence of discernible primary, the differential diagnosis includes poorly differentiated non-Hodgkin's lymphoma (see Fig. 14-11A), poorly differentiated prostate carcinoma (see Fig. 11-29), poorly differentiated bronchogenic carcinoma (see Fig. 7-41), and germ cell tumors (see Fig. 12-10). Although nasopharyngeal cells are often dispersed, cell clumping can usually be detected and helps to distinguish this lesion from lymphoma and germ cell tumors.

Nasopharyngeal metastases may recur in remote sites including axillary and inguinal nodes. Anaplastic cells with marked variation in size and shape, including some giant forms, may be seen.

The World Health Organization classification further delineates a nonkeratinizing group, which accounts for 12% of nasopharyngeal carcinomas and histologically simulates poorly differentiated transitional cell carcinoma of the bladder[48,57] (Fig. 3-23).

The cytologist must be alert to the possibility of a primary tumor of the nasopharynx, particularly when typical keratinizing squamous carcinoma is not

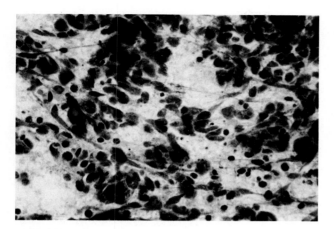

Fig. 3-21. Metastatic nasopharyngeal carcinoma, cervical lymph node. There are randomly dispersed, fragile, immature cells. Nuclei are elongated and distorted. Cytoplasm is not evident. (R stain, ×400)

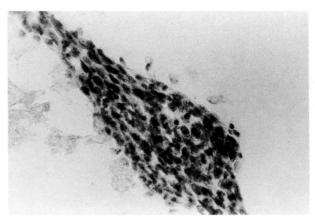

Fig. 3-23. Nasopharyngeal carcinoma, metastasis to cervical node. This is a somewhat poorly defined cluster of small, undifferentiated cells. Small nucleoli are present, but cytoplasm is poorly defined. (P stain, ×190)

evident on the smear. Nonkeratinizing squamous carcinoma originating from sites other than nasopharynx should be distinguishable from the nonkeratinizing tumors of the nasopharynx. Figure 3-24 is a section of metastatic lymph node with bands of nonkeratinized carcinoma and broad areas of necrosis. The needle traversing the gland will retrieve lymphocytes, tumor cells, and necrotic material. The histologic structure can be accurately surmised from the smear.

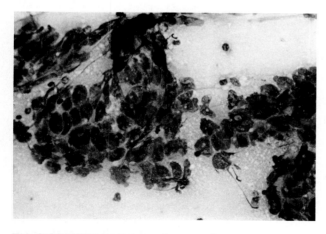

Fig. 3-22. Metastatic nasopharyngeal carcinoma, cervical lymph node. Primitive cells are clumped with some molding. Cells are larger than oat cells. (R stain, ×400)

The following list illustrates the potential findings of aspiration of metastatic cervical nodes:

Cytologic Types	Primary Tumors
Squamous carcinoma	Head and neck esophagus, lung
Undifferentiated carcinoma	Nasopharyngeal Prostate carcinoma
Small cell	Oat cell of lung
Adenocarcinoma	All glandular organs, including lung
Lesions specifically identifiable by cytology	Renal, thyroid salivary gland, breast, melanoma, seminoma

Supraclavicular and Suprasternal Targets

An enlarged transverse process may present as a hard mass at the root of the neck. Puncture with a fine needle is feasible, yielding bone marrow. This is a pitfall in clinical diagnosis.

The major palpable lesion in the region is the metastatic node. These nodes may be quite easy to find and aspirate or quite difficult because of the many muscle structures and the thickness of the covering skin, particularly after radiation. Raising the shoulder to relax the skin may help. Metastatic nodes are often behind the insertion of the sternocleidomastoid muscle at the clavicle, and it is necessary to

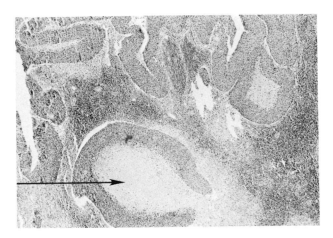

Fig. 3-24. Metastatic lymph node, tissue section. Non-keratinized squamous carcinoma has invaded a lymph node with a multifocal distribution. There are broad areas of necrosis *(arrow)*. Aspiration will primarily yield tumor cells with necrosis and, secondarily, a background of lymphocytes. (H & E stain, ×90)

palpate carefully, encircling the muscle. Always palpate directly behind the clavicle, where small, often hard, fixed nodes may be missed. Such findings often drastically change treatment plans, as illustrated in the following case reports:

■ *Case 3-3*
A woman aged 45 years was referred for consideration of postoperative therapy 1 week after removal of a fallopian tube carcinoma. A lymph node measuring less than 1 cm was palpated behind the left clavicle and yielded adenocarcinoma. This finding altered the treatment plan.

■ *Case 3-4*
A woman aged 80 years with severe low back pain was being prepared for laminectomy for "disk" disease. She was seen in consultation after 3 weeks of intensive radiographic and laboratory studies because of a mild anemia. A small firm-to-hard fixed node behind the left clavicle was palpated. Aspiration yielded adenocarcioma, resulting in cancellation of surgery. Further metastases became evident, but a primary tumor was not discerned.

The common primary tumors producing supraclavicular node metastases are breast, lung, stomach, colon, and testes. Figure 3-25 illustrates a supraclavicular node aspiration smear containing groups of tumor cells intermixed with lymphocytes. Of enlarged supraclavicular lymph nodes suitable for aspiration, 90% were malignant in one series,[52] 100% in another.[56]

The suprasternal notch is not a usual target; however, careful palpation may detect the tip of a lesion behind the sternum. It was the only palpable disease found in an elderly woman in whom aspiration yielded a cytologic smear diagnostic of Hodgkin's disease.

The clinical and cytologic spectrum of the enlarged lymph node is considered in Chapter 14.

THE AXILLA

Fastidious palpation of the axilla reaching well into the apex of the axillary pyramid is an important aspect of a complete physical examination. The variety of pathology of palpable nodes depends very much on referral patterns and the practice and interests of the examining physician. For example, pediatricians and infectious disease specialists will find mainly benign disease. Lymphoma is the most common cause of significant adenopathy seen by the he-

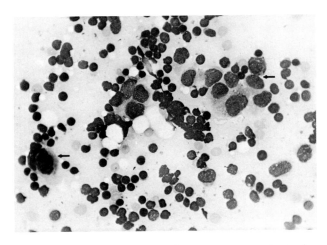

Fig. 3-25. Metastatic carcinoma to supraclavicular node. Groups of polymorphous tumor cells *(arrows)* are clearly distinguished from multiple lymphocytes. Lymphoglandular bodies are evident. (R stain, ×190)

matologist. In one general series only 24% of patients presenting with axillary lymphadenopathy had a malignant neoplasm.[34]

The major sources of metastases to axillary nodes are breast, lung, and melanoma although any primary may be present including thyroid, prostate, gastrointestinal tract, and head and neck tumors. In a significant number, no primary can be established. Undifferentiated cytology may arise from nasopharynx or prostate or may actually be an amelanotic melanoma (see Case 13-2). It is evident that FNA will provide valuable data in establishing a primary or directing the search for one.

In aspirating the axilla, careful palpation is critical. Nodes in the apex can be reached by relaxing the axillary skin. With the patient's arm lying close to the body wall with the one hand syringe, any palpable node can be needled. It is necessary to pinion a movable node against the rib cage with two fingers, allowing entrance for the needle between the fingers (Fig. 3-26). Often a 25-gauge needle is useful in piercing a small node with and without aspiration.

Axillary nodes must be distinguished from the axillary tail of the breast (see Chap. 6) and skin appendages.

CYSTIC LESIONS

Nonmalignant Cysts

Lateral Cysts

Most lateral cysts are of branchiogenic origin, appearing in the younger age group as rubbery elevations or swelling.[2] The mass is usually well circumscribed, and the size is reported to vary from the "size of a normal cervical lymph node to that of an orange."[7] The cyst may be present clinically along the entire length of the sternocleidomastoid muscle or at a variety of unexpected cervical sites.[7,10] The most common site is near the angle of the mandible, where confusion with salivary gland lesions may occur. Cystic Warthin's tumor may yield fluid that grossly resembles the content of a branchiogenic cyst. The cytology, however, is quite different (see Chap. 5). Recognition that most mobile, delimited, lateral cervical masses are likely to be branchiogenic cysts is important in anticipating cytologic findings. Branchiogenic cysts are mainly lined by squamous epithe-

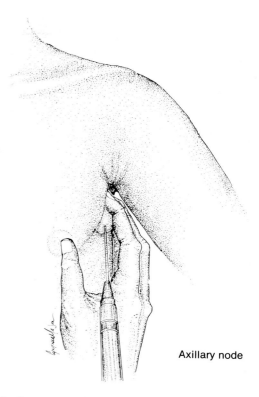

Axillary node

Fig. 3-26. Deep palpation of axilla with retraction of node and fixation against chest wall. Puncture may require a short 25-gauge needle if the node is small.

lium, but columnar type may also be present. A layer of lymphoid tissue lies beneath the lining epithelium. Inflammation is not uncommon (see below).

Puncture is ordinarily carried out with ease, yielding a turbid gray-yellow fluid. Its character is due to its cell content, which consists primarily of well-differentiated squamous cells, often degenerated; varying numbers of inflammatory cells (granulocytes, mononuclear cells, and histiocytes); and cholesterol crystals (Fig. 3-27). The background is murky on the slide. In a reported series of 71 lateral cysts, no cells were found in 3; columnar cells were not found; and benign squamous cells were found in 65. Thirty of the cysts contained cholesterol crystals.[17]

Cysts yielding water-clear acellular fluid are most likely of parathyroid origin.[24,67] Higher than normal parathyroid levels in the cyst fluid were reported.[67]

Aspiration of all cysts is followed by careful palpation and reaspiration of any residual mass. With

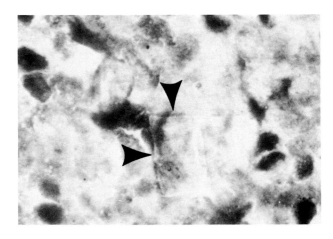

Fig. 3-27. Branchiogenic cyst. Cholesterol crystals, which may be nonspecific, are present in a variety of chronic cystic lesions (arrows). (R stain, ×190)

uncomplicated congenital cysts there is rarely residual tumefaction. The aspiration is therapeutic in a large proportion of cases.[67]

Midline Cysts

Midline cysts are largely of thyroglossal duct origin. The most common site is in or adjacent to the hyoid bone, but they may arise anywhere between the base of the tongue and the suprasternal notch. They are lined by both squamous and ciliated columnar epithelia, which, together with inflammatory cells and cholesterol crystals, compose the cytologic yield.

Heterotopic Thyroid Gland Tissue

Nodules with thyroid gland tissue, benign or with malignant alterations, may be found along the thyroglossal duct remnant, from the base of the tongue to the normal thyroid gland region. The tissue is easily recognized on smear as thyroidal in the clinical context described.

Malignant Cysts

Lateral Congenital Cysts

Most authorities discount the possibility in lateral congenital cysts.[64,65] Cases, however, have been reported.[5]

Midline Cysts

Midline cysts are well-defined entities. Smithers has collected 29 cases; in 24, the pathology was papillary.[59] The findings of papillary carcinoma on smear should be diagnostic (see Chap. 4).

Cavitary Carcinomas

Cystic breakdown of metastatic nodes occurs with squamous and papillary thyroid gland carcinomas.* The cyst fluid may not be distinctive but will usually contain easily identifiable malignant squamous cells (Fig. 3-28). Cystic papillary thyroid gland carcinoma metastases are a rare occurrence. Papillary thyroid gland cells can be expected. Metastatic thyroid gland cyst fluid had the brown color and staining qualities of thyroid gland cyst fluid (with R) in the one case we have seen.

In 6 of 98 samples from patients with proved metastatic cervical squamous carcinoma, only well-differentiated squamous cells were found on smear.[16] Well-differentiated squamous cells (see Fig. 3-12) indistinguishable from a benign proliferation may be extracted from a metastatic node. Cavitation of such a node will yield smears that may simulate those of a congenital cyst. In contrast, inflammation of a congenital cyst may result in sufficient atypia to raise a suspicion of carcinoma.[61]

Aspiration of a benign cervical cyst in which the clinical differential diagnosis was epidermoid carcinoma and branchial cleft cyst yielded the smears seen in Figures 3-29 and 3-30. Careful scrutiny is necessary to distinguish reactive histiocytic and giant cells from tumor cells.

These observations highlight the importance of clinical factors, including this patient's age and his-

---

* In 14 of 100 cases of metastatic cervical squamous carcinoma, fluid as well as cancer cells was found.[17]

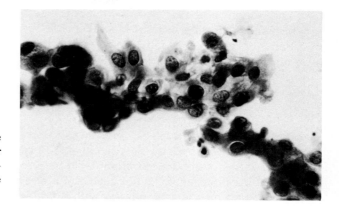

Fig. 3-28. Cavitary carcinoma, breakdown of metastatic squamous carcinoma in lymph node. This irregular cluster of malignant squamous cells was smeared from the sediment of cystic fluid, which was rich in cellular content. The chromatin is hyperchromatic. (P stain, ×190)

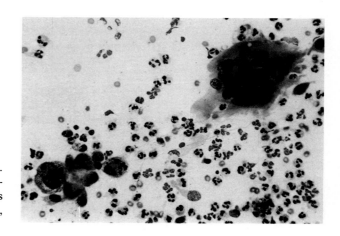

Fig. 3-29. Benign cervical cyst, subcutaneous mass. There is a background of inflammatory cells. One multinucleated giant cell and a cluster of atypical histiocytes simulating tumor cells are seen. Note phagocytes. (R stain, ×400)

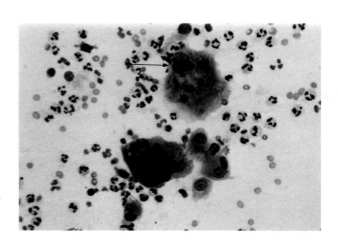

Fig. 3-30. Same case as in Fig. 3-29. Abnormal histiocytic cells with nucleoli simulate cancer cells *(arrow)*. (R stain, ×400)

tory, and the physical aspects of the mass. It is reasonable, when reporting such smears, to consider the possibility of malignancy and perform thorough head and neck examination before extirpation.

Carotid Body Tumors

The carotid body tumor is usually present as a slow-growing, ovoid, sometimes lobulated, rubbery to firm mass fixed at the fork of the carotid bifurcation. Size varies from 2 cm to 6 cm.[10,36] It lies beneath the sternocleidomastoid muscle and pulsates with it and is often attached to the carotid artery. It may, however, occur in a lower or posterior position. It is mobile horizontally but not vertically. The tumor is usually asymptomatic but may be tender to palpation, in contrast with metastatic nodes. If large enough, it may cause local symptoms.[4,20] The natural history is reviewed by Monro.[46]

Diagnosis

Extreme caution is recommended when a carotid body tumor is suspected.[15] Puncture must be carried out with minimal trauma to avoid hemorrhage. Pain may occur on palpation or suction. The complications of puncture include local hemorrhage, which may produce a compressive hematoma, or carotid artery emboli. Angiography to reveal the vascular pattern and widening of the carotid fork is recommended before attempting puncture.

Other tumors occur more commonly in the anterior cervical triangle and are best managed with aspiration and no angiography. Clinical judgment is essential. With negative angiography, aspiration can follow.

Cytology

The largest study group of carotid body tumors consist of 13 cases reported by Engzell, Franzen, and Zajicek.[15] All were stained with May–Grünwald–Giemsa, and primary description was based on this stain. Cells are dispersed or in loose clusters. The distinctive features are a marked variation in size of nuclei and the presence of oval or plump spindle cells. Naked nuclei may be evident, although abundant, poorly defined cytoplasm is normally present. Nuclear chromatin is finely granular and nucleoli and cytoplasm are poorly visible[15] (Fig. 3-31). Adenomatous structures may be present, and this leads to con-

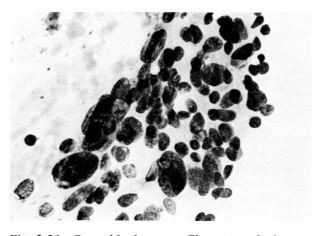

Fig. 3-31. Carotid body tumor. There is marked variation in cell size, with characteristic oval or plump spindle cells present. Nuclear chromatin is reticulogranular. Nucleoli and cytoplasm are poorly visible. Compare this to pheochromocytoma in Fig. 8-84. (R stain, ×190)

fusion with metastatic follicular thyroid carcinoma or adenocarcinoma (Fig. 3-32).

Hood and colleagues reported on three patients with paragangliomas, of which one had a carotid body tumor. In two of the cases, metastatic paraganglioma originating in the nasopharynx presented clinically as neck masses. The cytologic smears of the primary and metastatic cases were identical. The nuclei were pleomorphic with prominent nucleoli but

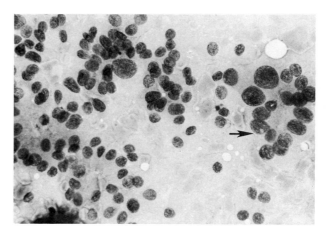

Fig. 3-32. Carotid body tumor. Cells show some variability in size but form adenomatous structures that can be confused with metastatic follicular thyroid carcinoma or adenocarcinoma (arrow). (R stain, ×190)

absent mitotic figures. Cytoplasm was abundant, eo-sinophilic, and reticular. The smears were stained with P.[29]

Differential Diagnosis

Metastatic adenocarcinoma, particularly of the thyroid gland, is a major consideration, since small adenomatous groups may be found. Squamous metastases pose no problem. Elongated cells (Fig. 3-31) can suggest neurofibromas, spindle cell melanoma, medullary carcinoma of the thyroid gland, and renal cell carcinoma.

Parathyroid Tumors

Experience in the identification of parathyroid tumors is limited. It is useful to quote the comments of Dr. Söderstrom published in the first edition of *Clinical Aspiration Cytology*.

> It should be possible to identify parathyroid glands with ABC. Presumably, all ambitious cytologists have looked at aspirates and imprints from known parathyroid glands and adenomas and stated that parathyroid cytology and thyroid cytology differ but that the difference is not easy to define. In my experience, the cytoplasm in parathyroid cells is either more narrow (chief-cells) or more diffuse and poorly stained (clear cells) than in thyroid follicular epithelium. Paravacuolar granules seem to be absent in parathyroid cells, which, in addition, do not cluster like thyroid epithelium cells, which form coherent follicular wall fragments. However, I have not been able to identify a parathyroid gland (not known beforehand) on cytologic criteria alone. Cytologic identification of parathyroid glands can be expected only of a cytologist with special experience.

Parathyroid glandular epithelium is seen in Figure 3-33.

FNA was carried out in seven cases of parathyroid adenoma and reported by Mincione and colleagues.[45] Echographic control was used. Several cytologic patterns were reported. In the chief-cell adenoma, ill-defined, occasionally microvacuolated, cytoplasm and slightly anisokaryotic nuclei with hyperchromatic chromatin and multiple nucleoli were seen. One oxyphilic adenoma consisted cytologically of numerous hyperchromatic naked nuclei of varying size with multiple small nucleoli. All of these were stained with P, and cytoplasm was not visible in the

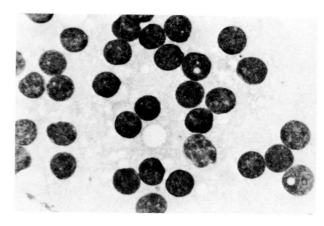

Fig. 3-33. Parathyroid glandular epithelium, smear from a surgical specimen. The cytoplasm is very poor in structural details. (MGG stain, ×1200)

oxyphilic cells. Garza and colleagues presented a case with a functional parathyroid carcinoma that appeared clinically as a hard, fixed, painless mass located in the right lobe of the thyroid.[23] The cells appeared in rounded, somewhat irregular clusters. The nuclear:cytoplasmic ratio was large. There was hyperchromatism, anisonucleosis, and ill-defined borders. There was no suggestion of microfollicles as reported in Lowhagen's[39] cases. In the case of parathyroid carcinoma reported by Guazzi and colleagues, it was stressed that marked cellular atypia may be present in both benign and malignant glands and a cytologic diagnosis of parathyroid *neoplasm* was rendered rather than carcinoma. In their case,

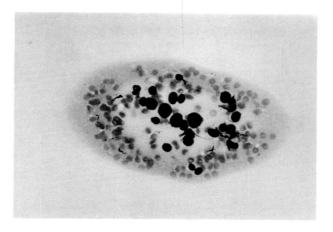

Fig. 3-34. Parathyroid adenoma. This is a cluster of well-defined cells with marked anisonucleosis. Cytoplasm is sparse. (R stain, ×400)

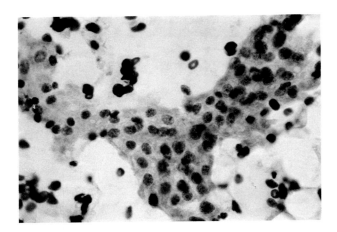

Fig. 3-35. Same case as in Fig. 3-34. Cells form a syncytium with a background of pink reticular cytoplasm. Small nucleoli are seen. (P stain, ×600)

there were clusters of somewhat bizarre carcinoma cells with clear cytoplasm.[27]

In a patient with adenoma, marked anisonucleosis with inapparent cytoplasm is seen with R (Fig. 3-34). From the same case, nuclei form a syncytium with pink reticular cytoplasm (Fig. 3-35).

Hazards of Neck Puncture

In puncture of targets in the neck, trauma to closely placed vital structures such as the carotid artery may occur, but it is not a significant risk. Aspiration of blood requires only local pressure after the needle is withdrawn. Carotid body tumors are an exception, as discussed above. Tumor spread after FNA has also been determined to be negligible. In a review of 656 of 1101 reported cases, no evidence of tumor spread was demonstrated.[16]

REFERENCES

1. Ackerman LV, Butcher HR: Surgical Pathology. St Louis, CV Mosby, 1964
2. Bailey H: The clinical aspects of branchial cysts. Br J Surg 10:565, 1923
3. Benergee TK, Gottschalk PG: Unusual manifestations of multiple cranial nerve palsies and mandibular metastases in a patient with squamous cell carcinoma of the lip. Cancer 53:346, 1984
4. Berdal P: Carotid body tumours. In Hamberger CA, Wersäll J (eds): Disorders of the Skull Base Region, pp. 313–319. Stockholm, Almqvist and Wiksell, 1968
5. Bernstein A, Scardino PT, Tomaszewski M–M et al: Carcinoma arising in a branchial cleft cyst. Cancer 37:2417, 1976
6. Berthrong M: Pathologic spectrum of neoplasms of the nasal cavity and paranasal sinuses. JAMA 219:339, 1972
7. Bhaskar SN, Bernier JL: Histogenesis of branchial cysts. A report of 468 cases. Am J Pathol 35:407, 1959
8. Comess MS, Beahrs OH, Dockerty MB: Cervical metastasis from occult carcinoma. Surg Gynecol Obstet 104:607, 1957
9. Conley JJ: How to examine the oro-naso-laryngopharynx for cancer. JAMA 215:456, 1971
10. Coulson WF: Surgical Pathology. Philadelphia, JB Lippincott, 1978
11. Czerniak B, Woyle S, Damagala W et al: Fine needle aspiration cytology of intraocular malignant melanoma. Acta Cytolpathol 27:157, 1983
12. Das DK, Das J, Natarajan R et al: Meibomian gland carcinoma initially identified by cytology. Diagn Cytolpathol 2:154, 1986
13. DeRosa G, Zeppa P, Traufa F et al: Acinic cell carcinoma arising in a lacrinal gland. Cancer 57:1988, 1986
14. Devine KD: Cancer in the neck without obvious source. Mayo Clin Proc 53:644, 1978
15. Engzell U, Franzen S, Zajicek J: Aspiration biopsy of the neck. II. Cytologic findings in 13 cases of carotid body tumor. Acta Cytol 15:25, 1971
16. Engzell U, Jakobsson PÅ, Sigurdson Å et al: Aspiration biopsy of metastatic carcinoma in lymph nodes in the neck. A review of 1101 cases. Acta Otolaryngol 72:138, 1971
17. Engzell U, Zajicek J: Aspiration biopsy of tumors of the neck. I. Aspiration biopsy and cytologic findings in 100 cases of congenital cysts. Acta Cytol, 14:51, 1970
18. Evans RW: Histological Appearances of Tumours. Edinburgh, E & S Livingston, 1968
19. Fagan MF, Rone R: Esthesioneuroblastoma: cytologic features with differential diagnostic considerations. Diagn Cytopathol 1:322, 1985
20. Farr HW: Carotid body tumors. Am J Surg 114:614, 1967
21. Feldman PS, Kaplan MJ, Johns ME et al: Fine-needle aspirations in squamous cell carcinoma of the head and neck. Arch Otolaryngol Head Neck Surg 109:735, 1983

22. Frable WJ, Frable MS: Thin-needle aspiration biopsy. The diagnosis of head and neck tumors revisited. Cancer 43:1541, 1979

23. Garza de la S, Garza de la FF, Bates FN: Functional parathyroid carcinoma. Diagn Cytolpathol 1:232, 1985

24. Ginsberg J, Jaung EM, Walfroh PG: Parathyroid cysts. JAMA 240:1506, 1978

25. Godwin J: Cytologic diagnosis of aspiration biopsies of solid or cystic tumors. Acta Cytol 8:206, 1964

26. Godtfredsen E, Lederman M: Studies on the cavernous sinus syndrome. Acta Neurol Scand 41:43, 1965

27. Guazzi A, Gabriele H, Guadagni G: Cytologic features of a functioning parathyroid carcinoma. A case report. Acta Cytol 26:709, 1982

28. Henderson BE, Loure EW, Jing JS et al: Epstein-Barr virus and nasopharyngeal carcinoma: Is there an etiologic relationship? JNCI 59:1393, 1977

29. Hood IC, Qizibash AH, Young JEM et al: Fine needle aspiration biopsy cytology of paragangliomas. Acta Cytol 27:651, 1983

30. Ibrahim R, Aufdemorte TB, Duncan DL: The diagnosis of radiolucent lesions of the jaw by fine needle aspiration biopsy. Acta Cytol 29:419, 1985

31. Jobst S, Ljung B–M, Gilken F: Cytologic diagnosis of olfactory neuroblastoma: Report of a case with multiple diagnostic parameters. Acta Cytol 27:299, 1983

32. Khang-Loon H: Primary meningioma of the nasal cavity and paranasal sinuses. Cancer 46:1442, 1980

33. Knowles DM, Jakobiec F: Orbital lymphoid neoplasms: A clinicopathologic study of 60 patients. Lab Invest 42:129, 1980

34. Krementz ET, Cerise EJ, Foster DS et al: Metastases of undetermined source. Curr Probl Cancer 4:5, 1979

35. Larsson L, Mårtensson G: Maxillary antral cancers. JAMA 219:342, 1972

36. LeCompte PM: Tumors of the carotid body and related structures. In Atlas of Tumor Pathology, Sect. IV, Fasc. 16. Washington, D.C., Armed Forces Institute of Pathology, 1951

37. Lee Y–TN, Terry R, Lukes R: Lymph node biopsy for diagnosis: A statistical study. J Surg Oncol 14:53, 1980

38. Ljung BME, Larsson SG, Hanafer W: Computed tomography-guided aspiration cytologic examination in head and neck lesions. Arch Otolaryngol 110:604, 1984

39. Lowhagen T: In Zajicek J: Aspiration Biopsy Cytology, Part I. Basel, S Karger, 1979

40. Lund HV: Tumors of the skin. In Atlas of Tumor Pathology, Sect. I, Fasc. 2. Washington, D.C., Armed Forces Institute of Pathology, 1957

41. Martin HE, Morfit HM: Cervical lymph node metastasis as the first symptom of cancer. Surg Gynecol Obstet 78:133, 1944

42. Maximow AA, Bloom W: A Textbook of Histology. Philadelphia, WB Saunders, 1944

43. Mendenhall WM, Million RR, Bova FJ: Analysis of time-dose factors in clinically positive neck nodes treated with irradiation alone in squamous cell carcinoma of the head and neck. Int J Radiat Oncol Biol Phys 10:639, 1984

44. Millon RR: Management of the neck node metastases. JAMA 220:402, 1972

45. Mincione GP, Borrelli D, Cicchi P et al: Fine needle aspiration cytology of parathyroid adenoma: A review of seven cases. Acta Cytol 30:65, 1986

46. Monro RS: The natural history of carotid body tumors and their diagnosis and treatment, with a report of 5 cases. Br J Surg 37:445, 1949

47. Ogura JH, Biller HF: Cancer of the head and neck: Hypopharynx and larynx: Surgical management. JAMA, 221:77, 1972

48. Pearson GR: Recent advances in research on the Epstein–Barr virus and associated diseases. In Fairbanks VF (ed): Current Hematology and Oncology. Chicago, Year Book Medical Publishers, 1986

49. Perez–Reyes N, Farhi DC: Squamous cell carcinoma of head and neck in patients with well-differentiated lymphocytic lymphoma. Cancer 59:540, 1987

50. Podoshin L, Gertner R, Fradis M: Accuracy of fine needle aspiration biopsy in neck masses. Laryngoscope 94:1370, 1984

51. Pollock PG, Koontz FP, Viner TF et al: Cervicofacial actinomycosis. Arch Otololaryngol 104:491, 1978

52. Ramzy I, Rone R, Scholtenover SJ et al: Lymph node aspiration biopsy. An analysis. Diagn Cytopathol 1:39, 1985

53. Rees WV, Robinson DS, Holmes EC et al: Altered lymphatic drainage following lymphadenectomy. Cancer 45:3045, 1980

54. Rubin P: Cancer of the head and neck. Hypopharynx and larynx. JAMA 221:68, 1972

55. Rubin R: Cancer of the head and neck—nasopharyngeal cancer. JAMA 220:390, 1972

56. Russ JE, Scanlon EF, Christ MA: Aspiration cytology of head and neck masses. Am J Surg 136:342, 1978

57. Shanmugaratnam K, Sobin L: Histologic typing of upper respiratory tract tumors. International histologic typing of tumors. No 19, Geneva, WHO, 1978

58. Slamovitz TL, Cahill KV, Sibony PA et al: Orbital fine-needle aspiration biopsy in patients with cavernous sinus syndrome. J Neurosurg 59:1037, 1983

59. Smithers DS: Tumors of the thyroid gland. Edinburgh, E & S Livingston, 1970

60. Svoboda DJ: Nasopharyngeal carcinoma: Pathologic classification and fine structure. JAMA 220:394, 1972

61. Warson F, Bommaert D, DeRoy G: Inflamed branchial cyst: A potential pitfall in aspiration cytology (letter to the editor). Acta Cytol 30:201, 1986

62. Weiland LH: The histologic spectrum of nasopharyngeal carcinoma. In de The G, Ito Y (eds): Nasopharyngeal Carcinoma: Etiology and Control. IARC Scientific Publications No 20 Lyon France, International Agency for Research on Cancer, 1978

63. Weiland LH, Neel III HB, Pearson GR: Nasopharyngeal carcinoma. In Fairbanks VF (ed): Current Hematology and Oncology, Chicago, Year Book Medical Publishers, 1986

64. Willis RA: Pathology of Tumours, 3rd ed, Scarborough, Ont., Butterworth & Co, 1960

65. Willis RA: The Spread of Tumours in the Human Body, 3rd ed, Scarborough, Ont., Butterworth & Co, 1973

66. Wilson JA, McIntyre MA, Tan J et al: The diagnostic value of fine needle aspiration cytology in the head-neck. J Coll Surg Edinb 39:375, 1955

67. Young JEM, Archibald SD, Shierk KJ: Needle aspiration biopsy in head and neck masses, Am J Surg 142:484, 1981

68. Zajdela A, de Maublanc MA, Schlienger P et al: Cytologic diagnosis of orbital and periorbital palpable tumors using fine-needle sampling without aspiration. Diagn Cytopathol 2:17, 1986

69. Zajicek J: Aspiration Biopsy Cytology, Part I. Basel, S Karger, 1974

70. Zajicek J, Engzell O, Franzen S: Aspiration biopsy of lymph nodes in diagnosis and research. In Ruttiman A (ed): Progress in Lymphology. Stuttgart, Georg Thieme Verlag, 1967

CHAPTER 4

Aspiration Biopsy Cytology of the Thyroid Gland

TORSTEN LÖWHAGEN

JOSEPH A. LINSK

Aspiration biopsy cytology (ABC) of the thyroid gland is a highly effective tool, yielding a morphologic diagnosis in a broad spectrum of palpable and nonpalpable thyroid diseases. The method was introduced by Söderström in 1952.[58] It was established as a valuable adjunct to clinical assessment by a series of papers beginning in 1962[16,19,45,46,64,68] and has now been the subject of increasing numbers of reports on new morphologic and imaging observations, comparative studies with other methods, and conservative dissent.[1,8,9,26,32,48,50,56,62]

All palpable thyroid disorders can be aspirated; however, with experience nonpalpable disorders can yield useful cytologic information as well. For example, thyroiditis and hyperthyroidism may not be discerned on palpation but may become targets for aspiration based on suggestive but inconclusive clinical isotope and laboratory studies.[29] Palpable abnormalities of the thyroid gland mainly occur as three different problems — diffuse swelling of one or both lobes, multinodular enlargements, and solitary nodular enlargements (Fig. 4-1A). Diffuse enlargement and multiple bilateral nodules usually indicate benign disease.

A small proportion of patients with solitary nodules will have a malignant tumor requiring treatment.[61] However, what appears on clinical examination to be a solitary nodule is often a predominant superficial part of a multinodular gland in which the nodularity is not detectable. Thyroiditis may also be clinically indistinguishable from other uninodular or multinodular goiters (Fig. 4-1B).[44]

In general, clinical data and scintiscan evaluation of thyroid nodules are not accurate enough for use in separating patients requiring medical therapy versus surgical therapy.[27,34] Before the advent of ABC, diagnostic surgery tantamount to major excision was frequently required. The role of ABC has been to assist in the selection of patients suitable for surgery and to morphologically confirm functional disorders such as hyperthyroidism and hypothyroidism resulting from thyroiditis.

TECHNIQUE

Palpation

Detection of gross enlargements requires no special technique. The patient lies with a pillow supporting the neck to help relax the anterior neck muscles. The physician stands preferably on the side opposite the lobe to be aspirated. The target lesion is then localized after searching palpation of the gland with one or two fingers. Changes in gland form and consistency, as in thyroiditis, can be appreciated. Finger-tip palpation may reveal small (less than 1 cm) surface masses that occasionally are papillary carcinoma.

Puncture

The gland is fixed between the investigating fingers and the trachea. The carotid and jugular bundle is deflected laterally (Fig. 4-2). The aspiration is carried

Diffuse enlargement

Multi-nodular

Solitary nodule

A

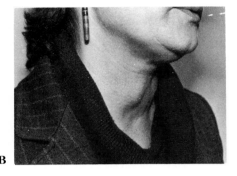

B

Fig. 4-1. **A.** Principal palpable alterations of the thyroid gland. Diffuse and multinodular enlargement favors hyperplasia (colloid goiter). A neoplasm usually presents as a solitary nodule. However a dominant nodule in a simple colloid goiter and sometimes a unilateral thyroiditis may appear as a solitary nodule clinically indistinguishable from a neoplasm. **B.** This patient's nodular thyroid was found to be Hashimoto's thyroiditis on fine needle aspiration. (From Linsk JA, Franzen S: Fine Needle Aspiration for the Clinician, p. 72. Philadelphia, JB Lippincott, 1986)

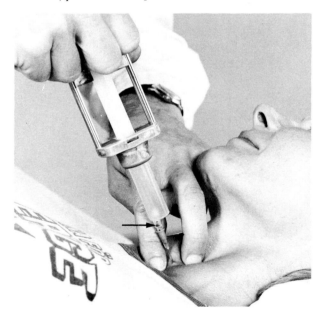

Fig. 4-2. Performance of aspiration. Note that nodule is pinched securely. The plunger is up, providing a full 10 cc of suction. Aspirate appears just above the needle (*arrow*), at which point the plunger is released and aspiration terminated.

out as described in Chapter 1. The suction is released when blood appears in the syringe tip. Some pressure on the puncture site may prevent hematoma. If hematoma appears, it will do so immediately or rarely after 1 or 2 minutes.

Fixation and Staining

Solid nodules of glandular tissue (in contrast with cysts) usually yield droplets of glistening, gelatinous, semisolid material. Blood or tissue fluid, however, may dilute the cellular yield.

Smears are made as described in Chapter 1. Where cystic fluid or excess blood is aspirated, it is expelled on one or two slides. Tilt the slide and allow the fluid to run off on to another slide, revealing particles that are then scraped off and smeared on a separate slide. The maneuver must be done quickly where blood is present to avoid coagulation. Cyst fluid is usually turbid brown and may vary in volume up to 10 ml. It should be promptly centrifuged and the sediment smeared. Further details about fixation and staining are described in Chapter 1.

PRINCIPAL LESIONS OF THE THYROID GLAND

Thyroiditis
Acute suppurative thyroiditis
Subacute thyroiditis (granulomatous and deQuervain)
Chronic lymphocytic thyroiditis (auto-immune thyroiditis, Hashimoto's disease, struma lymphomatosa)
Fibrosing thyroiditis (Riedel's struma)

Hyperplasia
Diffuse and nodular nontoxic goiter (colloid goiter)
Diffuse toxic goiter (Graves' disease)

Benign Tumors
Adenomas of follicular, trabecular, and solid type

Hürthle Cell Tumor

Malignant Tumors
Follicular carcinoma
Papillary carcinoma
Medullary carcinoma
Poorly differentiated "insular" carcinoma (columnar cell carcinoma)
Anaplastic carcinoma
Malignant lymphoma
Metastasis from other sites

The ABC method provides a spectrum of diagnoses that closely parallel the above listed disorders. In most cases it is quite possible to cytologically categorize a thyroid lesion into one of the three main disease entities: thyroiditis, hyperplasia, and neoplasia. In practice, this means a substantial narrowing down of possibilities, often sufficient for the further management of the patient.

NORMAL THYROID GLAND CYTOLOGY*

The normal thyroid gland yields a rather scant but usually typical cytology consisting of multicellular fragments of follicular walls, normally without distinct borders between individual cells (Figs. 4-3 and

* From Nils Söderström's contribution to the first edition.

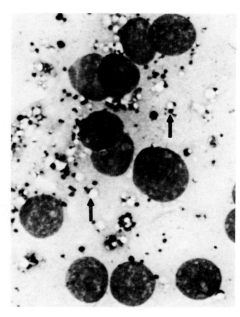

Fig. 4-3. Nontoxic nodular goiter. Detail of a follicular wall fragment with cells rich in lysosomal (paravacuolar) granules. (MGG stain, ×1000)

4-4). The nuclei are circular and sharply outlined. Distinct nucleoli are seldom seen. The slightly opaque cytoplasm stains grayish violet with May–Grünwald–Giemsa (MGG). It is poor in structural details but usually contains some small bluish black lysozymal granules bound to vacuoles (see Fig. 4-3). Only in some cases are these granules numerous enough to become conspicuous, but when looked for they are present in most cases. They may also be seen in other types of glandular epithelium. I use the label paravacuolar granules for such granules (contrast with granules in Subacute Thyroiditis below). An absolutely specific sign of thyroid gland tissue is when some follicular fragments have retained their bowl shape with some colloid left on the bottom (see Fig. 4-4).

In reality, thyroid cytology is much more characteristic than this description may suggest because the thyroid tissue the cytologist examines is more often from a multinodular goiter and presents the typical cytologic picture of this pathologic structure. Typical features of this pattern are microfollicles (Fig. 4-5) and cyst macrophages, which may look like any desquamated free macrophages but often show a chess board pattern of cytoplasmic granulation,

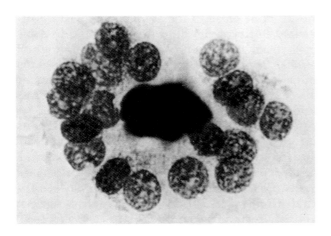

Fig. 4-4. Thyroid aspirated from case of multinodular goiter. The follicular wall fragment is an apparently bowl-shaped structure with some colloid left in the center. (MGG stain, ×1000)

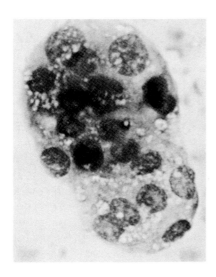

Fig. 4-5. Microfollicle from multinodular goiter. (MGG stain, ×800)

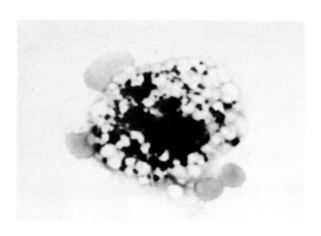

Fig. 4-6. Cyst macrophage with "chessboard" pattern in the cytoplasm. This is a fairly typical finding in nodular goiters. (MGG stain, ×1000)

which is fairly typical for the thyroid gland (Fig. 4-6). A multinodular thyroid parenchyma is usually present when the aspirate is rich in colloid, viscid, and yellow to the eye. In smears, it forms a hyaline background substance that usually stains dark blue with MGG.

THYROIDITIS

Acute Thyroiditis

Acute suppurative thyroiditis with and without abscess formation is rare. Clinically, the gland will have the signs and symptoms of acute inflammation. On puncture, the aspirate may be frank pus or may be dominated by granulocytes and fibrin. This finding must be distinguished from necrosis with its associated inflammatory cells, seen in highly malignant tumors (see Anaplastic Carcinoma below). Where the aspirate is liquified, repeat puncture from the periphery of the mass may reveal diagnostic malignant cells.

Subacute Granulomatous Thyroiditis

Subacute granulomatous thyroiditis (deQuervain's disease or giant cell thyroiditis) is rather unusual. It occurs most often in young individuals and is associated with a virus infection.[23,28,40] In its active phase, fever, a rapid sedimentation rate, and tenderness of the thyroid gland are useful findings. The gland may be asymmetric, simulating a tumor from which the above symptoms may help distinguish it. The needle may meet some elastic resistance as it enters the gland; and the puncture may be more painful than usual, a useful diagnostic point. Painless active subacute thyroiditis has been described[53]

   Key Cytologic Features of Subacute Thyroiditis
   1. All kinds of inflammatory cells including macrophages
   2. Numerous, large, multinucleated giant cells (Figs. 4-7, 4-8, and Plate I D)
   3. Degenerative follicular cells often with intravacuolar granules (Fig. 4-9)
   4. Epithelioid cells (see Fig. 4-7)
   5. Background precipitate of protein and debris

Quantitatively reduced glandular epithelium with degenerative changes will be present. Objective evidence of early degenerative change in the epithe-

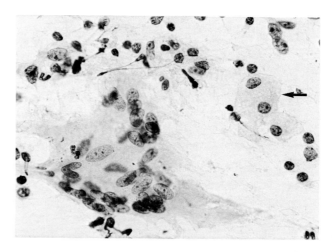

Fig. 4-7. Subacute thyroiditis (de Quervain's disease). A multinucleated giant cell with elongated epitheloid nuclei and irregular abundant cytoplasm are seen. Histiocytes *(arrow)*, lymphocytes, and other inflammatory cells make up the infiltrate. (P stain, ×600)

lium are cytoplasmic vacuoles with intravacuolar granules. (Fig. 4-9) Contrast with the paravacuolar granules described by Söderström and depicted in Figure 4-3. Multinucleated giant cells are also seen now and then in Hashimoto's thyroiditis and in papillary carcinoma. However, the combination of degenerative epithelium and multinucleated giant cells is characteristic of subacute thyroiditis.

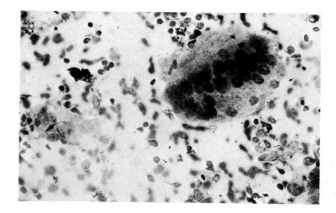

Fig. 4-8. Subacute thyroiditis (de Quervain's disease). A multinucleated giant cell and mononuclear inflammatory cells are seen. (MGG stain, ×600)

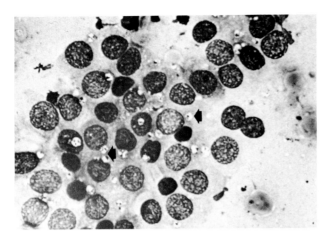

Fig. 4-9. Subacute thyroiditis. Thyroid epithelium with intravacuolar granules indicating regressive changes *(arrows)* (see Fig. 4-3). (MGG stain, ×1000)

Chronic Lymphocytic Thyroiditis

Chronic lymphocytic thyroiditis (Hashimoto's struma) is a common thyroid gland disorder, more so in women.[5] About half of childhood goiters prove to be lymphocytic thyroiditis.[24,42] Symmetric involvement is the rule but asymmetry may occur, with localized nodular enlargement (Fig. 4-1B). In clinical examination, the gland with Hashimoto's struma is usually more firm than the one with diffuse colloid

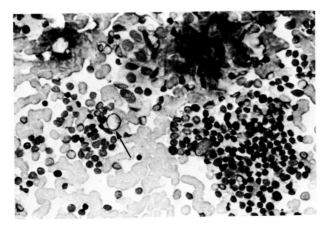

Fig. 4-10. Chronic lymphocytic thyroiditis (Hashimoto's disease). Numerous lymphocytes form a pattern similar to a lymph node aspirate. A transformed lymphocyte with enlarged nucleus is indicated *(arrow)*. A cluster of tightly bound epithelial cells is present above the lymphoid cells. (MGG stain, ×600)

goiter. The febrile syndrome of subacute thyroiditis is absent; however, in Hashimoto's struma there are often signs and symptoms of a hypofunctioning gland.

Key Cytologic Features of Chronic Lymphocytic Thyroiditis
1. Hypertrophic (Hürthloid) follicular cells
2. Lymphocytes and occasional plasma cells
3. Fragments of stroma with smudged lymphocytes — lymphoid tangles
4. Occasional multinucleated giant cells

Puncture usually yields droplets of bloody material with little evidence of colloid. The typical smear is cellular, and the number of small and transformed lymphocytes may be striking. (Fig. 4-10) Lymphocytes may be in clusters or dispersed. Scattered throughout the lymphoid background are clusters of large oxyphilic epithelial cells with well-defined cytoplasm and uniform nuclei (Hürthle cells) (Figs. 4-11A and B, 4-12A and B, and Color Plate I). The epithelial cells may form suggestive follicles (see Fig. 4-11A and B). In addition, depending on the intensity of the lymphoid infiltration of the thyroid gland tissue, normal or slightly hypertrophic follicular cells may be present, as may plasma cells. This may be an earlier stage in the evolution of the lesion. Occasional clusters of hyperplastic follicular cells with large, marginal colloid containing cytoplasmic vacuoles are found (Fig. 4-13).

Multinucleated giant cells often present in subacute thyroiditis and in papillary carcinoma can also be seen in Hashimoto's struma, although rather infrequently. Scattered agglutinated tangles of strungout lymphoid nuclear material and stroma are useful diagnostic clues (Fig. 4-14). Where the lymphoid cells are scant and Hürthle cells prominent, a differential diagnosis with Hürthle cell neoplasms must be considered (Fig. 4-15). This problem is complicated by the rather frequent and impressive atypicality seen in the oxyphilic cells (see Fig. 4-11A and B). In Hashimoto's disease when the number of lymphocytes is increased with minimal epithelial component, welldifferentiated lymphoma must be considered. Intensive lymphocytic infiltration may blur the distinction between lymphocytic thyroiditis and lymphoma. The majority of lymphomas, however, are poorly differentiated and should cytologically be readily recognized. Intensely lymphoid cytologic smears must be scrutinized carefully and interpreted in conjunc-

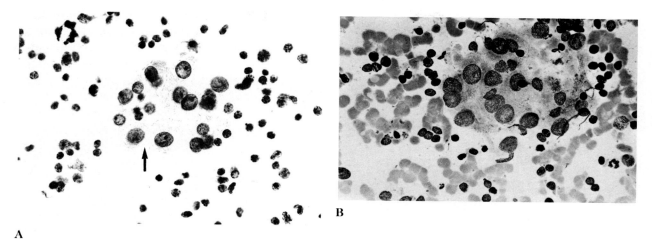

A

B

Fig. 4-11. A. Lymphocytic thyroiditis (Hashimoto's disease). In the background are many lymphocytes and scattered plasm cells. A cluster of enlarged oncocytic epithelial cells (Hürthle cells; *arrow*) is present with well-demarcated and abundant granular cytoplasm and prominent nucleoli. Such atypical epithelial cells have been confused with neoplastic cells by unaware observers. (P stain, ×600) **B.** Lymphocytic thyroiditis as seen in **A.** (MGG stain, ×600)

tion with the clinical picture. Immunologic staining technique on aspirated cells has proven to be of great value in distinguishing monoclonal malignant infiltrates from heteroclonal reactive infiltrates of auto-immune type (see Chapter 18).

The association of Hashimoto's struma and carcinoma has been reviewed by Doniach (in Smithers) and evidence of an etiologic relationship was deemed

inconclusive.[57] However, Crile treated 373 patients with Hashimoto's struma with thyroid substitutions and concluded that there was a relationship with carcinoma.[14]

Hashimoto's thyroiditis has been found in association with follicular neoplasms, papillary carcinoma, multinodular goiter, and lymphoma.[25,30,34] Most such associations have been noted in selected

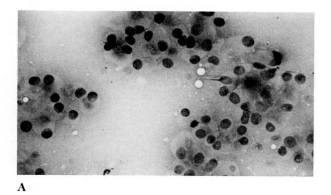

A

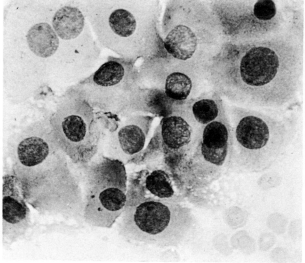

B

Fig. 4-12. A and **B.** Hürthle cells. In the absence of lymphoid cells, Hürthle cells obtained from lymphocytic thyroiditis cannot be distinguished from neoplastic Hürthle cells (see Fig. 4-15). (MGG stain: **A** ×400; **B** ×1000)

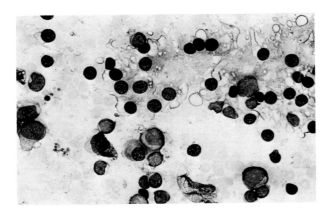

Fig. 4-13. Lymphocytic (autoimmune) thyroiditis. Shown are transformed lymphocytes and large follicular cells with delicate cytoplasmic vacuoles (flaming or foaming cells). This often seen feature, is considered a result of autoimmune factors. Identical follicular cells are seen in hyperthyroidism (Graves' disease). (MGG stain, ×600)

surgical pathology files. The incidence of neoplastic disease in association with Hashimoto's thyroiditis (HT) is extremely low[26] and does not make HT a surgical lesion in the uncomplicated case. Malignant thyroid lymphoma developed in 4 of 829 patients with HT.[30] In the series of Compagno & Oertel, lymphoma was almost always associated with lymphoid thyroiditis.[13] Clinical findings at diagnosis, follow-up, and response to therapy are critical deciding factors.

Fig. 4-14. Lymphoid tangle. Cohesive stromal microbiopsies are often obtained from lymphocytic thyroiditis. In smearing, the fragile lymphoid cells are smudged producing this type of structure. (MGG stain, ×160)

In a case reported by Guarda,[25] a patient with HT progressing to lymphoma failed to respond to treatment with thyroid suppression, and the gland continued to enlarge. Such clinical findings should dictate prompt re-aspiration and ultimately surgery if there is doubt. It is the clinical experience of one of us (J.L.) that patients with palpable HT are best managed by serial fine needle aspiration (FNA) at appropriate intervals. The presence of thyroid antibodies[28,34] does not eliminate the need for cytologic examination.[25,34] Not all HTs have antibodies. In the absence of both antibodies and hypothyroidism, repeat aspiration and careful scrutiny are in order.

Riedel's Struma

Riedel's struma is a sclerosing fibroblastic disorder of the thyroid gland, sometimes associated with a similar tissue reaction in the mediastinum, retroperitoneum, and orbit. It is a poor target for ABC because the yield is acellular or contains a few fibroblast-like cells. This disorder must be distinguished from infiltrating carcinoma, in which a fairly rich cell population will be obtained.

HYPERPLASIA

Diffuse and Nodular Nontoxic Goiter (Colloid Goiter)

In our experience, more than 85% of thyroid disorders causing enlargement (goiter) are classified cytologically as one or the other kind of nontoxic hyperplasia. This disorder is diffuse in its earliest phase. As a result of excessive accumulation of colloid, followed by degenerative changes, cyst formation, and fibrosis, the gland transforms to a nodular type. A palpable "solitary" nodule may be part of a multinodular gland in which other nodules are present but not palpable. Clinical determination that a gland is multinodular may be difficult. Further, the clinical determination that a solitary nodule is a neoplasm rather than a colloid nodule is also difficult (see above). Colloid nodules are, however, easily diagnosed by ABC, and the cytologic findings accurately reflect the histopathology (Fig. 4-16). Distended follicles yield large amounts of colloid that stains violet-purple with MGG and tan with Papanicolaou (P).

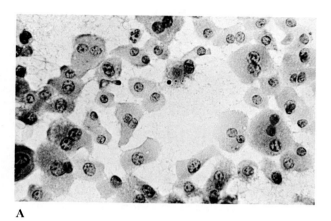

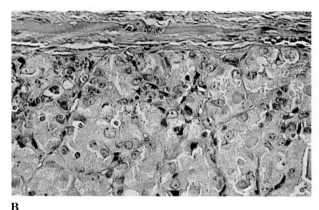

A

B

Fig. 4-15. A. Hürthle cells. This smear is from a Hürthle cells tumor. (P stain, ×400) B. Tissue section. Hürthle cells adenoma. (H & E stain, ×400)

The presence of colloid is highly significant, because carcinoma rarely produces colloid; and it is strong evidence for a benign colloid nodule.

Key Cytologic Features of Diffuse and Nodular Goiter

1. Abundant colloid
2. Small follicular cells in sheets or in preserved follicules of various size (Plate I B)
3. Cytoplasm, scanty and wispy
4. Numerous bare nuclei
5. Phagocytes[10]
6. Occasional Hürthle cells

Aspirate content consists of clusters and sheets of follicular cells or varying sizes sucked away from the distended follicular wall (Figs. 4-17 and 4-18). There are small regular cells with round to oval smooth nuclei and often well-defined bluish cytoplasm with MGG. Single cells may be scattered and must be distinguished from lymphocytes (Fig. 4-19). Dense groups of cells intermixed with a pink (acid mucopolysaccharide) stroma as well as free, irregular nuclei may be seen. Pink stroma is to be distinguished from purplish background of colloid in MGG (R) (Fig. 4-20). The normal thyroid gland unit is a spherical follicle of approximately 200 μmm, and these are occasionally extracted and seen in benign smears

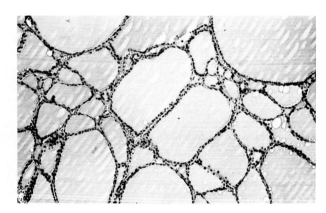

Fig. 4-16. Colloid goiter. Tissue section. Follicles are enlarged and distended with colloid. Epithelium is flattened. Puncture will rupture follicles, extracting colloid and fragment of colloid wall. (H & E stain, ×160)

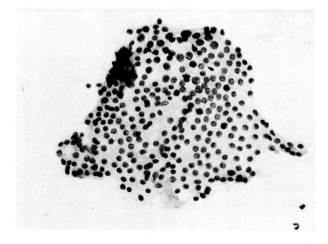

Fig. 4-17. Benign follicular epithelium. Sheet of cuboidal epithelium from wall of a distended follicle. Cells are in a regular pattern with nuclei equally separated by cytoplasm, forming a monolayer. (P stain, ×400)

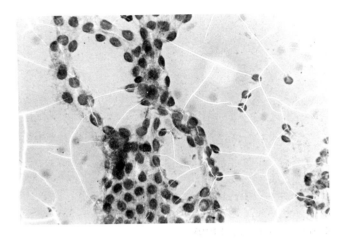

Fig. 4-18. Benign follicular epithelium. Aggregates of benign thyroid cells with round to oval nuclei and granular cytoplasm. Cracked colloid is present in the background, on which isolated cells cannot be distinguished readily from lymphocytes. (MGG stain, ×600)

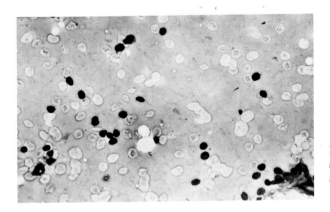

Fig. 4-19. Colloid goiter. Single cells and a few clusters are seen on a colloid background. At this power, lymphocytes and benign thyroid epithelium cannot be distinguished. (MGG stain, ×90)

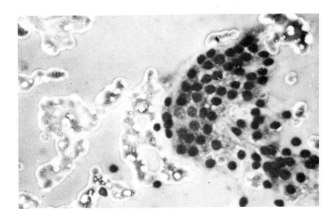

Fig. 4-20. Colloid goiter. Monolayer of monomorphic follicular cells represents a segment of distended follicle. In the background are colloid and red blood cells. (MGG stain, ×600)

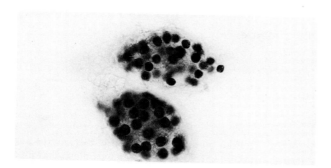

Fig. 4-21. Intact thyroid follicles.

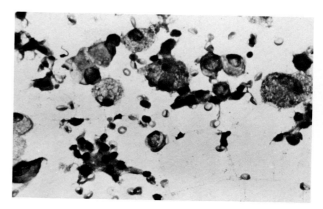

Fig. 4-23. Cystic colloid goiter. Fluid with foaming histiocytes and degenerated epithelial cells with pyknotic nuclei. (MGG stain, ×600)

(Fig. 4-21 and see Fig. 4-5). Overall cellularity of the scan is scanty to moderate compared to neoplastic aspirates, which are richly cellular.

Secondary changes in the hyperplastic parenchyma due to hemorrhage and cyst formation produce macrophages that are foamy in P (Fig. 4-22) and have black pigment in MGG (Figs. 4-23 and 4-24). Hemorrhage with cyst formation is more often responsible for cold nodules on scintigram than neoplasm. Fluid will be aspirated when there has been cystic degeneration. Cysts may present as rather sudden swellings in a pre-existing multinodular goiter or in an apparently normal gland. The fluid is usually brown and varies from a few drops to several milliliters (Fig. 4-25). On drying, it may form a typical cracked geometric pattern (see Fig. 4-18). On smear, direct or after centrifuging, there may be a uniform purplish stain (colloid) in Romanowsky (R) contain-

ing many or a few scattered pigment-laden macrophages (see Fig. 4-6). Small to large clusters of monomorphic atrophic cells are also seen in the smear together with red cells and lymphocytes (see Fig. 4-24). Differential diagnosis with cystic papillary carcinoma must be considered (see below).

There is a question of the origin of the hemosiderin-containing "histiocytes" in the cystic material. In some aspirates, the evolution of these cells appears to be traced from epithelial cells showing first vacuolization, then inclusion of pigment, and finally the appearance of a large "histiocyte" containing hemosiderin (Fig. 4-26). One study has demonstrated the development of the foam cell from monocytes.[10] An important point in examining preparations from manifestly nontoxic multinodular goiters is that some cell aspirates will be taken from nodules with a microfollicular structure, and these are difficult to separate cytologically from a follicular neoplasm. In this circumstance, accumulation of iodine in the nodule on scintigram usually excludes neoplasms; however, follow-up study is indicated.

Toxic Goiter (Hyperthyroidism, Graves' Disease, or Basedow's Disease)

Toxic goiters are often divided into primary hyperplasias or secondary uninodular or multinodular hyperplasias. Primary hyperthyroidism is characterized by diffuse enlargement of the gland. Sometimes nodular goiters can harbor areas of "autonomous" toxic nodules. The diagnosis of hyperthyroidism is usually

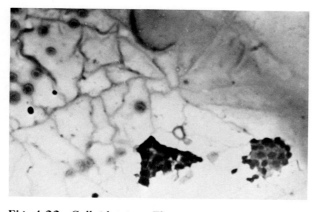

Fig. 4-22. Colloid goiter. There are several histiocytes with lacy, foaming cytoplasm. Dark isolated stripped nuclei (lymphocytes, epithelial cells, or histiocytes) are also seen. (P stain, ×60)

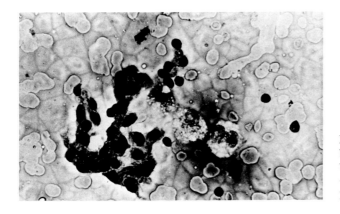

Fig. 4-24. Colloid goiter. A few foaming histiocytes and benign thyroid cells densely granular cytoplasm staining black with MGG and positive for iron stain. Thyroid follicular cells can incorporate hemosiderin. (MGG stain, ×600)

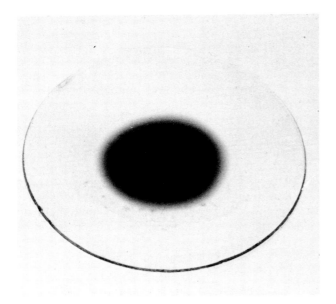

Fig. 4-25. Watch glass with cystic fluid. Thyroid cyst fluid is expelled into a watch glass, and particles are sequestered out for smearing.

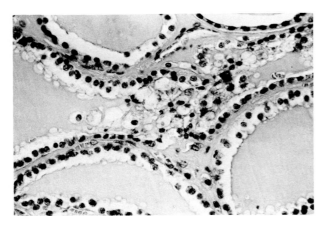

Fig. 4-26. Evolution of histiocytes. Tissue section. Ruptured follicle. Histiocytes "appear" to be forming in and emerging from the epithelium in the center of the section. The source of these histiocytes has, however, been attributed to monocytes.[10] (H & E stain, ×400)

a clinical one best made by biochemical tests, and cytologic or histologic confirmation is rarely required. An experienced cytologist, however, may suspect toxic goiter from a cytologic aspiration smear.

Key Cytologic Features of Toxic Goiter
1. Abundant blood; little or no colloid
2. Enlarged cells with finely granular cytoplasm
3. Peripheral cytoplasmic vacuoles—"flare cells" (Color Plate I A)

Aspiration biopsy smears of puncture material are usually rich in blood with little free colloid. The follicular cells have abundant finely granular cytoplasm with large peripheral vacuoles containing a colloid that stains pale red with MGG, justifying the name "flare cells." Morphologically, such cells cannot be distinguished from those described in Hashimoto's thyroiditis (see Fig. 4-13). Such changes must clearly be interpreted within a clinical context. Cell nuclei are round but can vary markedly in size. Naked and dispersed nuclei are common. Lymphocytes can be numerous, pointing to an auto-immune etiology and a tendency to progress to a hypofunctional type of lymphocytic thyroiditis (Hashimoto's disease). In patients with toxic goiter, smears containing a marked increase in lymphatic cells suggest a nonsurgical approach.

TUMORS

On physical examination alone, it is seldom possible to diagnose thyroid gland cancer in an isolated nodule.[36] Clinically suspicious solitary cold nodules on scintigram have traditionally been considered potential cancers requiring surgery. However, cysts, hemorrhages, and localized thyroiditis are also recording cold areas on scintigram. It is important to identify such lesions, as these patients rarely will benefit from surgery. In this dilemma, ABC has become the method of choice in accurate and cost-effective selection of candidates who might benefit from surgery in contrast with those who require suppressive or other forms of medical therapy.

The identification or exclusion of cancer is one of the most important pieces of information yielded by the ABC method. The variety of neoplastic growths is impressive, but each can generally be recognized by the cytologic patterns discussed below.

Follicular Adenoma and Follicular Carcinoma

An adenoma is a benign encapsulated tumor without invasion into capsular vessels. Histologically, adenomas can be solid, tribecular, or composed of follicles of variable size. Mixed types of adenomas with focally varied architecture including foci of Hürthle cells are frequently seen. In general, adenomas as well as nearly all thyroid carcinomas do not accumulate radionuclides, and they are deemed "cold" on scintigram.

Distinguishing follicular adenoma from follicular carcinoma or mixed follicular–papillary carcinoma, although valuable, is not crucial, since patients with these neoplasms will usually be candidates for surgery. Histologic examination for vascular and capsular invasion is required for all well-differentiated neoplasms. Neoplasms are sometimes treated with suppressive thyroid gland medication. However, failure to respond will result in surgery.

Follicular adenomas are most common in females. They may appear at any age but are most frequent in middle-aged women. The median age is about 10 to 15 years below that for a follicular carcinoma. Adenomas may be precursor lesions to follicular carcinomas.[65,66] They may be long present, then enlarge suddenly because of hemorrhage. On palpation, they may feel firm, elastic, or soft. They appear rounded and smooth.

On gross examination, all surgical specimens are encapsulated in contrast with papillary carcinomas (see below). On sectioning, recent or old hemorrhage may be found associated with infarctive or degenerative changes and a central fibrotic scar. Cystic degeneration may occur, leading to confusion with benign degenerated nodules.[66]

In smears, a pattern of repeated microfollicular structures with a monotonous appearance is seen (Fig. 4-27). Generally, there are monomorphic nuclei that resemble those of normal follicular cells (Color Plate II). With P, nucleoli are small and regular. In some follicular adenomas, nuclei may vary in size, sometimes, paradoxically, more than nuclei from follicular carcinomas (Fig. 4-28A and B). Other adenomas yield more solid cell clusters with fewer follicular structures. Follicular structures with foaming epithelial cells are seen (Fig. 4-29A and B). Hürthle cell adenomas consist of large oxyphylic (Hürthle)

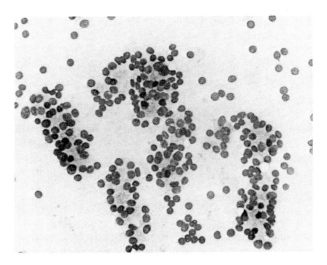

Fig. 4-27. Follicular neoplasm. Equal-sized acinar groups are found repeatedly forming a monotonous follicular pattern. Colloid is absent. Aspirates from follicular carcinomas and follicular adenomas give the same pattern and cannot be distinguished cytologically. (MGG stain, ×400)

cells with finely granulated cytoplasm, sharp cell borders, and a round nucleus that may be atypical. Cytologically, there are few or no preserved follicular structures. Cells are dispersed and in clusters. None or very little colloid is present in the smear.

Key Cytologic Features of Follicular Neoplasms
1. Smear rich in blood with scant or no colloid
2. Numerous equal-sized microfollicular groups in clusters (microbiopsies or dispersed throughout the smear)

Follicular carcinomas make up about one fourth of all thyroid gland carcinomas. They occur at any age, although incidence increases as age increases. They are several times more common in women than in men. A long history of goiter or mass is rather frequent, suggesting that follicular carcinomas may be clinically indolent and show clinical evidence of growth over a period of years. With this long history, FNA may be an excellent technique once thyroid gland enlargement is discovered. As with adenomas, follicular carcinomas may suddenly increase in size because of retrogressive changes accompanied by bleeding in the tumor.

Clinically, the tumor appears as a solitary mass, firm to hard and somewhat irregular. It is encapsulated, and, depending on capsular thickness, the fine needle may meet resistance on entering. Usually it does not.

The gross appearance of the surgical specimen is similar to adenoma. On microscopic section, there is a wide range from nearly solid tumor tissue with occasional follicular structures to abundant, well-formed, neoplastic follicles resembling normal thy-

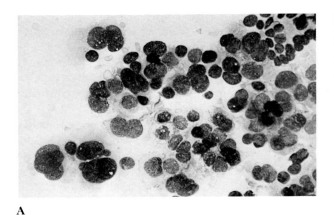

A

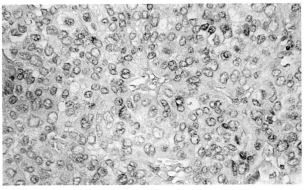

B

Fig. 4-28. A. Follicular adenoma with markedly atypical cells. A follicular arrangement can vaguely be discerned. There is strong variation in nuclear size and shape. (MGG stain, ×600) B. Histology demonstrated a densely cellular encapsulated follicular tumor (adenoma). The patient is alive and well 14 years later without evidence of disease.

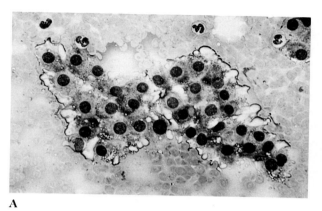

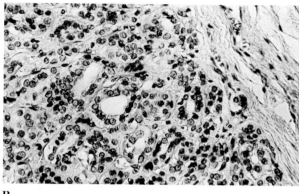

A

B

Fig. 4-29. A. Aspirate from adenoma with follicular formation of flare cells identical to those seen in toxic goiter. (MGG stain, ×600) **B.** Tissue section from the same tumor. (H & E stain, ×400)

roid gland follicles. Well-differentiated macrofollicular types contain colloid, but most tumors are composed of small acini devoid of colloid. Trabecular columns of densely packed cells may be present. The decisive pathologic changes are the presence of capsular and blood vessel invasion (Fig. 4-30). Sections taken from the interior of the specimen may give a picture very similar to adenoma.

The aspirate will contain blood and many epithelial cells singly, dispersed in small groups, and in follicles (Fig. 4-31). In general, follicles may be more easily fractured and dispersed than in adenomas. The finding of follicles is crucial to the diagnosis. Cytologic variation may range from minimal atypia to highly abnormal cells. Follicular carcinomas, like follicular adenomas, are characterized in smears by the presence of microfollicular structures that may be very similar to those seen in atypical aspirates from

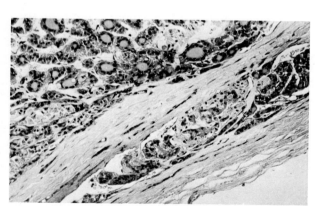

Fig. 4-30. Follicular neoplasm with angioinvasive growth. Tissue section. This phenomenon is considered the only reliable evidence in the diagnosis of follicular carcinoma. (H & E stain, ×160)

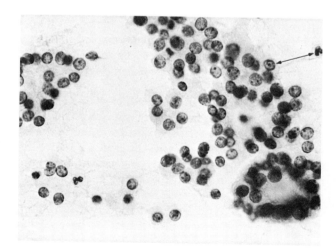

Fig. 4-31. Follicular carcinoma. Note that the size of the epithelial cell nucleus is equivalent to the polymorphonuclear leukocyte (*arrow*). This was a carcinoma that exhibited marked angioinvasion on section. Note rather small nucleoli. (P stain, ×600)

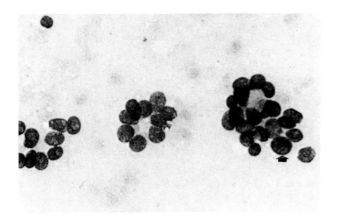

Fig. 4-32. Follicular carcinoma. Note small acinar groups of less than 10 cells and rather large cells with prominent nucleoli *(arrow)*. An angioinvasive tumor. (MGG stain, ×600)

hyperplastic cellular nodules (see Fig. 4-28A and B). Follicular structures seem to be less well defined in tumors with malignant potential. Malignant tumors may, however, retain well-preserved follicular structures, at times composed of enlarged cells with large nucleoli (Fig. 4-32).

Some authors examine such parameters as overlapping of nuclei, large nuclear size, and coarse chromatin texture and find them to be of help in differentiating carcinomas and adenomas.[35,37] Where such findings are diffuse and accompanied by large nucleoli and occasional follicles, they suggest an increased risk for malignant neoplasm (Fig. 4-33A, B, and C).

Cell size has been used in determining more precise diagnoses. In planimetric studies carried out on air-dried aspirates from follicular tumors, there were statistically significant differences between mean nuclear areas of benign and malignant neoplasms, but also a certain overlapping. It was concluded that high and low risk for malignancy in follicular neoplasms could be determined with a rather high degree of statistical probability.[9] A later study using videoplan image analysis failed to confirm these differences.[48] The question of determining high- and low-risk groups for follicular neoplasms by these methods must be left open.

Kini and colleagues determined the cytologic diagnoses of follicular carcinoma to be 75% accurate. The major differentiating criterion was the enlarged nucleus observed by light microscopy on P-stained smears.[35] Atkinson and associates had difficulty separating follicular lesions on the basis of cell size and nuclear size because of overlap.[4]

The cytologic differences in smears between benign and malignant follicular tumors are often so subtle that a diagnosis of follicular neoplasm must be rendered with the understanding that surgery will be performed for histopathologic diagnosis to differentiate carcinomas and adenomas.[37,47] A further compelling reason for proceeding with surgery in all follicular neoplasms is the possibility of anaplastic carcinoma arising at that site (see below).

HÜRTHLE CELL TUMORS

Oxyphilic Adenoma

Hürthle cell tumors are made up of cells of large, finely granular cytoplasm. The nuclei are equally large and round with prominent nucleoli. In most tumors, a vague follicular architecture can be discerned both in histologic sections and aspiration smears. In general, the biologic behavior of Hürthle cell tumors cannot be predicted on the basis of cytologic features. Only a minority of the tumors are malignant with dissemination and distant metastases. In aspirates from this tumor, oxyphilic (Hürthle) cells with variable degrees of nuclear atypia are obtained. Because nuclear atypia is an unreliable cytologic criterion of malignancy in these tumors, the cytologic diagnosis of oxyphilic cell or Hürthle cell tumor is given; and the lesion is removed for histopathologic diagnosis. Hürthle cells seen in aspirates from palpable lesions of the thyroid may puzzle the untrained observer. If oxyphilic (Hürthle) cells are seen in thyroid smears in the midst of a much larger population of normal follicular cells, their presence should not be given too much significance. Oxyphilic cells in varying numbers, when associated with lymphocytes, may be part of the morphologic evidence of chronic lymphocytic thyroiditis (see above). When a cellular smear exclusively contains Hürthle cells without a mixture of other follicular cells or inflammatory cells, it is highly probable that true neoplasm is present. Surgical intervention is then recommended.

Oxyphilic thyroid tumors may be impossible to distinguish from oxyphilic parathyroid tumors (Fig.

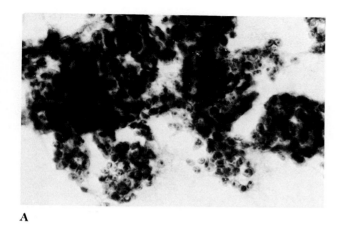

A

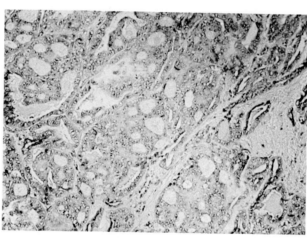

B

C

Fig. 4-33. **A.** Poorly differentiated (but not anaplastic) carcinoma. Dense clusters with a loosely acinar pattern and crowding of cells. Prominent nucleoli. The cell features contrast with those in common follicular tumors and papillary carcinoma. (P stain, ×400) **B.** Same aspirate in higher power. Note the markedly pronounced nucleoli. The tumor had a rapid malignant course. (P stain, ×600) **C.** Tissue section from same tumor. At first glance, the tumor appears well differentiated; however it forms a cribiform rather than a typical follicular pattern.

4-34A and B). Failure to store iodine further limits the differentiation. Immunoperoxidase staining technique will identify thyroid cells.

The specific cytologic diagnosis of follicular carcinoma (of Hürthle or non-Hürthle cell type) can usually be made from smears obtained at metastatic sites including lung, skeleton, and skin. The following case report illustrates the indolent course of local recurrence.

■ *Case 4-1*
A 59-year-old woman had a history of thyroidectomy and left radical neck dissection 20 years before presentation. She noted a nodule adjacent to the left lateral neck scar. On examination, there were two nodules in the skin measuring
less than 0.5 cm in their greatest dimension. FNA with a 25-gauge needle yielded scant cellular content but with cells forming follicles (Fig. 4-35). At surgery, several deposits of follicular carcinoma were found.

Papillary Neoplasms

Most pathologists agree that histopathologic separation of papillary adenomas from papillary carcinomas is difficult and of little clinical importance because papillary adenomas have a malignant potential and may form metastases. Papillary adenomas and papillary carcinomas are also grouped together cytologically.

At least 50% of all malignant tumors arising in the thyroid gland are papillary carcinomas. This

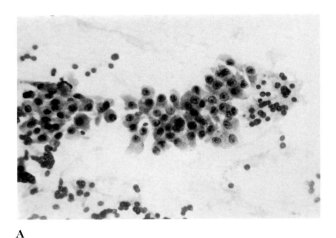

A

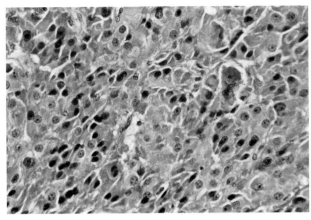

B

Fig. 4-34. A. Large loose cluster of Hürthle cells aspirated from a lateral neck lymph node. Metastatic Hürthle cell carcinoma. There is mild variation in nuclear size. Nucleoli are prominent. Cytoplasm is abundant with indistinct cell borders. (P stain, ×600) B. Tissue section. From original thyroid tumor. There are rare incomplete follicles. Distinction from oxyphillic parathyroid tumor requires immunoperoxidase staining. (H & E stain, ×800)

cancer affects patients of all ages, including children. There appears to be an etiologic relationship between radiation therapy directed to the head and neck of children and young adults and the later development of thyroid cancer.[57] As with all thyroid gland cancers, papillary carcinoma is more often diagnosed in women than in men. It is usually a solitary, firm or hard nodule, although the nodule may be soft as a result of cystic degeneration. On gross examination, the fibrous capsule is often imperfect, and the tumor

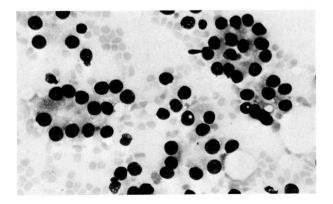

Fig. 4-35. Metastatic follicular tumor. The smear has no absolute criteria of malignancy except the presence of these few well-differentiated follicles in a dermal site. (MGG stain, ×600)

may infiltrate diffusely as a result, in contrast with follicular carcinoma. In a recent study, 10% of 859 papillary carcinomas were encapsulated.[49] Puncture is easily accomplished even though the mass may be hard and resistant. It may also be gritty, suggesting calcification (psammoma bodies or calcospherites; Fig. 4-36A and B). The rare presence of psammoma bodies in the cytologic smear, even without suspicious cytologic changes, suggests papillary carcinoma.

Papillary clusters suggesting carcinoma must be reviewed cautiously in pregnant women.[7] Papillary carcinoma may be simulated. On the other hand, papillary carcinoma may co-exist with other more clinically evident disorders including hyperparathyroid tumors or hyperplasias and benign hyperplastic thyroids. The diagnosis of papillary carcinoma should not be excluded when typical cytology is present.[6]

PAPILLARY CARCINOMA

Key Cytologic Features of Papillary Carcinoma

1. Monolayered sheets of enlarged cells with dense metaplastic cytoplasm (Figs. 4-36A, 4-37, and 4-38B)

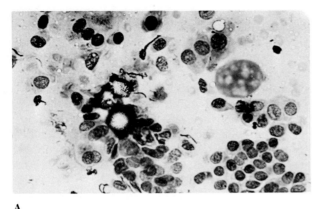

A

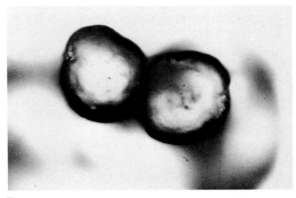

B

Fig. 4-36. A. Papillary carcinoma. Psammoma body with finely laminated center and black periphery. A lightly stained globule of dense colloid of "papillary" type is also seen *(arrow)*. (MGG stain, ×600) **B.** Psammoma bodies. (MGG stain, ×1000)

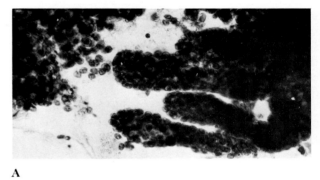

A

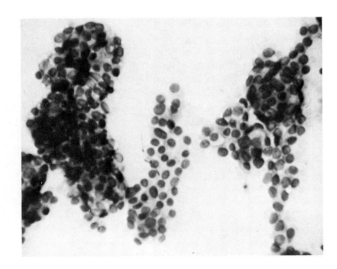

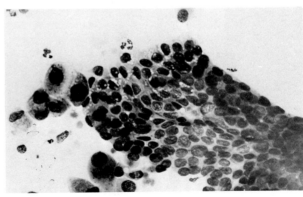

B

Fig. 4-37. Papillary carcinoma. Well-defined papillary cell cluster with crowding and overlapping of cells. (MGG stain, ×400)

Fig. 4-38. A. Papillary carcinoma. Three-dimensional papillae without fibrovascular cores. (MGG stain, ×400) **B.** Flat monolayer sheet with well defined smooth border *(upper right)*. Phagocytes and vacuolated tumor cells are features in aspirations from papillary carcinoma. (MGG stain, ×600)

2. Branching papillaroid clusters partly with well-defined smooth borders (Fig. 4-38 and Color Plate III)
3. Intranuclear cytoplasmic inclusions (Fig. 4-39)[59]
4. Globules of dense colloid and strands of viscous, ropy colloid (Fig. 4-40)
5. Psammoma bodies (Fig. 4-36 and Color Plate III)
6. Macrophages as evidence of cystic change (Fig. 4-38)
7. Lymphocytes and multinucleated histiocytic giant cells

Fine needle aspirates from palpable tumors are generally an accurate reflection of the histopathologic picture. The usual yield is a mixture of papillary cell clusters (Figs. 4-37 and 4-38), follicular cell clusters, free cells, and, occasionally, fluid from the cystic varieties. Individual cells usually have denser cytoplasm than follicular carcinoma cells. Cell borders are sharply demarcated. Nuclei are large, and, in more than two thirds of the smears, a varying number of cells display intranuclear cytoplasmic inclusions (pseudonucleoli) (Fig. 4-39).[59]

Pseudonuclei are important diagnostic but not pathognomonic features. They have now been described in follicular carcinoma,[20,41] follicular adenoma,[69] microfollicular adenoma,[20] Hürthle cell carcinoma,[51] and medullary carcinoma.[40]

Multinucleated giant cells, foamy histiocytes, and lymphocytes may be present in small numbers. Psammoma bodies, although infrequently seen, are, of course, of considerable help in the diagnosis when present (see above). Some smears will show colloid in ropy strands as well as globoid droplets characteristic for these tumors (Figs. 4-40 and 4-41). This contrasts with the normal colloid, which spreads out as a background (Color Plate II).

Papillary carcinomas must be considered in the differential diagnosis of thyroid cysts. Goellner and Johnson[21] reported on 66 papillary carcinomas in which 9 were cystic and identified cytologically. There were sheets of epithelial cells, enlarged pale nuclei, intranuclear inclusions, and psammoma bodies. Viscous colloid was not seen. Two additional lesions failed to show diagnostic cytology. They contained histiocytes and could therefore not be distinguished from degenerated colloid cysts. Metastatic lateral nodes may present without palpable thyroid gland disease and may be cystic. Examination of the fluid sediment should be decisive. The sediment of benign lateral cysts and of cavitating squamous carcinomas contains benign and malignant squamous epithelium and cellular debris (see Chap. 3). The fluid of cystic papillary carcinoma metastatic to lateral neck nodes may contain colloid and papillary cell groups.

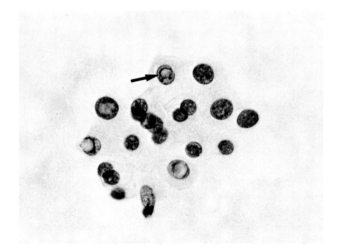

Fig. 4-39. Papillary carcinoma, intranuclear inclusions. This is a cluster of slightly polymorphous cells, several of which contain large intranuclear inclusions characteristic but not quite pathognomonic of papillary carcinoma. By electron microscopy these inclusions have been shown to be cytoplasmic. (P stain, ×600)

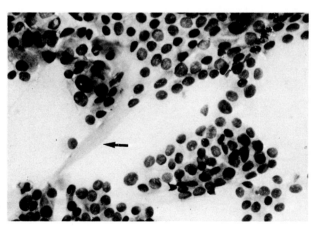

Fig. 4-40. Papillary carcinoma. Strip of ropy viscous colloid (arrow).

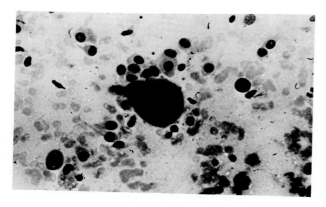

Fig. 4-41. Rounded dense globule of colloid. The elastic colloid of "papillary" type appears in this form when untouched by the smearing. (MGG stain, ×600)

Nuclear indentation or crease has recently been reported in papillary carcinoma and with further observation may become an important diagnostic criterion.[15]

MEDULLARY CARCINOMA

Medullary carcinoma constitutes about 5% to 10% of all thyroid carcinomas. Most of these tumors occur in patients over 40 years of age, but they are occasionally seen in younger adults and children. Occasionally, they occur in the context of multiple endocrine neoplasias.[33] Elevation of plasma calcitonin by radioimmunoassay in association with a nodular goiter seems to be pathognomonic of medullary carcinoma.[63] There may be history of diarrhea, palpitation, and flush, as in the "carcinoid syndrome."

The familial syndromes of medullary carcinoma have been well described.[57] Familial evidence can be useful in diagnosis, as illustrated in the following case report confirming the intimate relationship of clinical findings in cytology:

■ *Case 4-2*
A 47-year-old secretary, in a hospital pathology department, developed a goiter characterized by a primarily unilateral swelling of the thyroid gland (right side). She resisted counseling and any study other than a cursory examination. The mass was firm and angular. On further study, it
was learned that the patient's mother, age 80, had had a thyroidectomy several years previously. Review of the sections confirmed the diagnosis of medullary carcinoma. The patient was confronted with the findings and the potential familial character of the tumor and consented to aspiration biopsy. Cytology was characteristic of medullary carcinoma (see below). In preparation for surgery, chest x-ray films revealed bilateral pulmonary metastases.

The primary tumor may be quite small and difficult to palpate and puncture, although metastases, when they occur, may be considerably larger. Grossly, the tumor is hard, ill defined, and gray-white. On microscopic review of the tissue section, there are clusters of tumor cells growing in solid, trabecular, or glandular arrangements separated by a hyaline amyloid-containing stroma (Fig. 4-42). The cells may show variable morphology from spindle shaped to those seen in carcinoid tumors. Follicular structures may be formed in rare cases, but colloid is absent.

Key Cytologic Features of Medullary Carcinoma
1. Dispersed cell pattern (Fig. 4-43)
2. Nuclei, round-oval, occasional giant nuclei (Fig. 4-43)

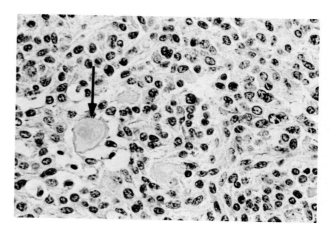

Fig. 4-42. Medullary carcinoma. Tissue section. Note amyloid plaque *(arrow)*. The tumor is solid with no follicular structures. Some spindle cells may be seen. (H & E stain, ×400)

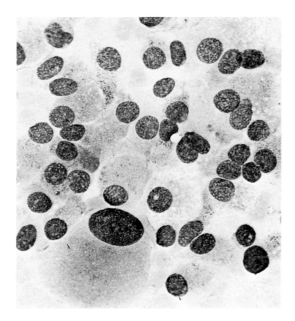

Fig. 4-43. Medullary carcinoma. The cells in this smear are quite characteristic of medullary carcinoma. There are no follicular structures. Cells are fairly uniform, with some polymorphism. The nuclei are eccentric, and the nucleoli are not visible. The cytoplasm contains fine red granulation, seen with MGG but not with P stain. (MGG stain, ×1000)

3. Cytoplasm often eccentric with fine red granularity (Figs. 4-43 and 4-44A)
4. Spindle cell forms and dome-shaped cells with eccentric nuclei (Fig. 4-45)
5. Amorphous pink–pale violet material (amyloid) (Fig. 4-44)

On examination of cytologic smears, the material may be bloody; and the cell picture often varies greatly among patients. The most consistent characteristic features are solid cell clusters and free oval to elongated cells. Scattered over the smear are quite large cells with markedly enlarged and hyperchromatic nuclei. The cytoplasm of the tumor cells is sometimes speckled with small red granules (Fig. 4-43).[60] Although classic, this is not always seen. There may be cells with scanty, eccentric cytoplasm (Fig. 4-44A and B) as well as spindle cells with wispy, sometimes granulated cytoplasm (Fig. 4-45A and B and Color Plate III).

In aspirates from medullary carcinoma, there is no colloid but there may be varying amounts of amorphous material that stains much the same as colloid with conventional stains. This cannot be recognized specifically as amyloid but is considered diagnostically helpful in the overall cytologic picture.[45] Problem cases may be clarified by silver stain (Sevier–Munger) and immunocytochemical staining for calcitonin, both positive in medullary carcinoma.[52]

Cells with intracytoplasmic vacuoles suggesting mucus production are exceptional features in some tumors (Fig. 4-46A and B).

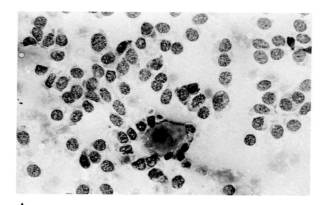

A

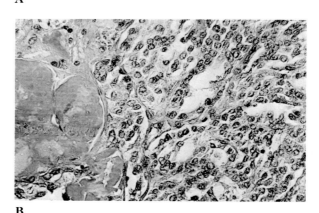

B

Fig. 4-44. A. Medullary carcinoma. Dispersed cells with round nuclei and ill-defined cytoplasm. There is also one plaque of dense amorphous material. (MGG stain, ×600) **B.** Tissue section of same case. Solid tumor composed of monotonous cells and large deposits of amyloid. (H & E stain, ×300)

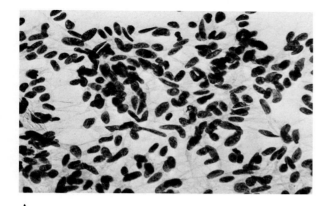

A

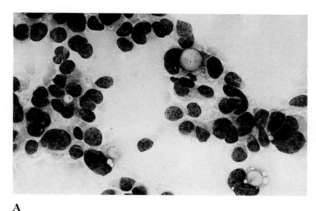

A

reports, however, have given evidence for a third category of thyroid carcinoma that falls somewhere between the aforementioned extremes, both regarding biologic behavior and morphologic appearance. This carcinoma was probably originally described by Langhan[39] in 1907 (as "Wuchernde struma") and reintroduced by Carcangiu and colleagues[12] in 1984 as "poorly differentiated (insular) carcinoma." Poorly differentiated carcinoma metastasizes to both

Fig. 4-45. A. Medullary carcinoma. Predominately spindle-shaped cells with oblong nuclei and delicate polar-extended cytoplasm. No organoid features. (MGG stain, ×600) **B.** Same aspirate in P stain. The fusiform cells produce a kind of streaming "fish-school" pattern. (P stain, ×600)

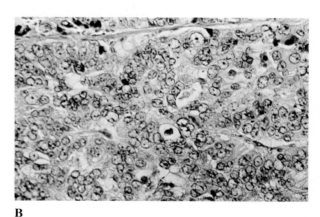

B

Fig. 4-46. A. Medullary carcinoma. Rather cohesive clusters of polymorphic large cells. An exceptional feature is the presence of intracytoplasmic vacuoles suggestive of mucus production. (MGG stain, ×600) **B.** Histology of the same case. Immunostain was strongly positive for calcitonin. The vacuoles did not stain. Precise nature is unknown. Medullary carcinomas produce a broad spectrum of cellular patterns. Although there is no typical single microscopic appearance of medullary carcinoma, there are several well-recognized patterns which this sequence of pictures illustrates.

POORLY DIFFERENTIATED (INSULAR) CARCINOMA

Current classifications of thyroid carcinoma usually contain two main groups — well differentiated (papillary, follicular) and undifferentiated (anaplastic) carcinoma. The former group is recognized as relatively indolent, amenable to surgery, and with a very good prognosis; while the latter group is highly aggressive and usually fatal within a short period of time. The view seems to have been that these two groups of malignancies are strictly separated. Recent

regional lymph nodes and to distant sites. Histologically, the tumor is characterized by large, well-defined solid nests of medium-sized cells that may form unevenly distributed small follicles and papillae. The tumor cells are only moderately pleomorphic, and the nuclei are hyperchromatic with inconspicuous nucleoli. A frequent and diagnostically important feature is the presence of necrotic foci in tumor nests often in a "peritheliomatous" arrangement. This is a feature rarely seen in well-differentiated carcinomas. Mitotic figures are usually present, although the

number is quite variable. Our experience with the cytomorphology of aspiration smears of these tumors is limited to a few cases where the histology and clinical course corresponds to what has been described. The aspirates have been cellular with tumor cells appearing in dense clusters with some nuclear overlapping without well-defined papillary or follicular structures. The nuclei were moderately pleomorphic, and in one case nucleoli were prominent. The amount of cytoplasm was variable and was readily identifiable. The aspirates contained necrotic material but no appreciable inflammatory component.

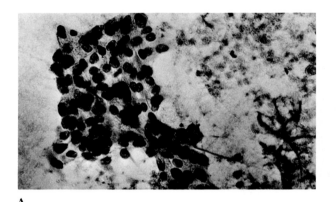

A

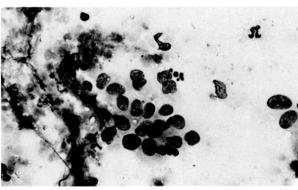

B

Fig. 4-47. A. Insular carcinoma. Cluster of hyperchromatic cells with rather prominent nucleoli. In this field of the aspirate there are no discernible organoid formations. In the background are numerous pale-staining necrotic cells and precipitate of stringy fibrinoid material. No inflammatory cells are seen. (P stain, ×600) B. An acinar formation of moderately polymorphic cells and necrotic debris, but only a few granulocytes. (MGG stain, ×600)

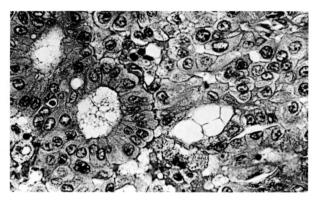

A

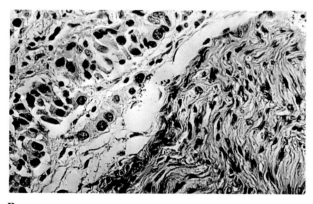

B

Fig. 4-48. A. Poorly differentiated ("insular") carcinoma. Tissue section. Solid tumor with some follicles of tall columnar cells. In other parts the tumor had the typical appearance of papillary thyroid carcinoma. (H & E stain, ×400) B. An unusual feature was marked perineural invasion. This 44-year-old man died of his disease within 2 years. (H & E stain, ×300)

Initially, all the cases were cytologically diagnosed as differentiated thyroid tumors without further classification. Except for the necrosis, there is no cytomorphologic similarity with anaplastic carcinoma. We hope that the awareness of this type of primary thyroid carcinoma will lead to more systematic reviews and accumulation of cytologic data. It seems quite feasible to recognize or at least suspect this type of thyroid carcinoma pre-operatively in aspiration smears (Figs. 4-47 and 4-48).

ANAPLASTIC CARCINOMA

About 10% of all thyroid carcinomas are anaplastic. They occur primarily in the elderly, usually appearing between the ages of 60 and 70. At diagnosis these tumors often have extensively infiltrated the thyroid gland and surrounding neck structures. They may, however, be present as a still-contained nodule; and failure to determine the nature of the tumor pre-operatively with thin needle puncture may result in an inappropriate operation. Clearly nodular and apparently diffuse goiters that may actually be composed of multiple ill-defined nodules may be of long standing. Transformation of such a goiter to anaplastic carcinoma appears to occur.[36] The antecedent goiter, however, is a neoplasm — often a well-differentiated carcinoma.[2] Cytologic aspiration can usually establish the presence of a colloid adenomatous goiter in contrast with a neoplasm and relieve concern about transformation to anaplastic carcinoma. Other opinions hold that anaplastic or undifferentiated carcinoma begins without antecedent and may be differentiated in part to contain the papillary or follicular elements that might have suggested a preexisting lesion.[18] Transformation of a preexisting nodule, when it does occur, produces sudden or rapid increase in nodule size. This clinical syndrome calls for immediate cytologic aspiration. All patients with longstanding thyroid gland nodules must be carefully evaluated. In a study of 308 solitary nodules by FNA, Jayaram found 5 anaplastic carcinomas.[31] Anaplastic carcinoma is the deadliest of all thyroid gland neoplasms, with few patients surviving 1 year after diagnosis. Rapid invasion of neck structures, including major vessels, occurs, producing a firm to hard fixed mass.

Key Cytologic Features of Anaplastic Carcinoma
1. Inflammatory cells, necrotic debris, and poorly preserved edematous cells and tissue fragments (Fig. 4-49)
2. Bizarre, large, pleomorphic, often multinucleated malignant cells with epithelial, lymphoid or mesenchymal appearance (Fig. 4-50)
3. Mitotic figures, numerous, often atypical (Fig. 4-51)

From the cytologic point of view, anaplastic carcinoma has a variety of morphologic patterns. Large cell carcinomas can be difficult to diagnose cytologically because of the large amounts of edema fluid, inflammatory cells, and necrosis obscuring the characteristic intact tumor cells (Fig. 4-49). Some tumors are sclerotic, making it difficult to obtain cell-rich aspirates. When satisfactory smears are obtained, the cell picture is usually highly variable (Fig. 4-52 and Color Plate IV). There are often clusters of oval to spindle-shaped cancer cells with pleomorphic or bizarre nuclei (Fig. 4-50). Mitotic figures and multinucleated giant cancer cells are common (Fig. 4-51). One variant of large cell anaplastic carcinoma is a malignant osteoclastomalike giant cell tumor that is equally lethal. Macronucleoli are commonly seen (see Figs 4-50 and 4-52). Necrosis consists of a proteinaceous background, nuclear strand, and intact but distorted neutrophils (Fig. 4-49). Spindle cell tumors

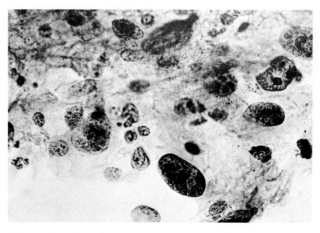

Fig. 4-49. Anaplastic carcinoma. Highly polymorphic malignant cells with prominent abnormal nucleoli. Necrotic debris and inflammatory cells. The pattern is similar to other giant cell tumors (lung, pancreas, sarcomas). (P stain, ×600)

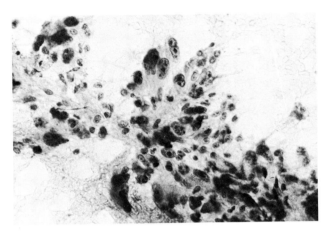

Fig. 4-50. Anaplastic carcinoma of sarcomatous type. Mesenchymal cluster of polymorphic malignant cells. (P stain, ×400)

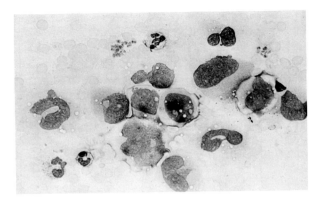

Fig. 4-52. Aspirate from a tumor in the thyroid area. Large high-grade malignant cells. Note arch-shaped nuclei. This pattern is not quite typical of a primary thyroid anaplastic carcinoma. Among other alternatives the differential diagnosis should also include anaplastic type of lymphoma (histiocytic). (MGG stain, ×1000)

yield nonspecific cytologic smears that must be differentiated from spindle cell sarcoma, melanoma, renal carcinoma, and medullary carcinoma (see above).

MALIGNANT LYMPHOMA OF THE THYROID GLAND

Malignant lymphoma of the thyroid is an extranodal lymphoma similar morphologically to lymphomas involving salivary gland (Chap. 5), lung (Chap. 7), gastrointestinal tract (Chap. 8), and testis (Chap. 12). The tumor presents as a nodular or diffuse goiter[3,11,13,57] and cannot be clinically diagnosed.

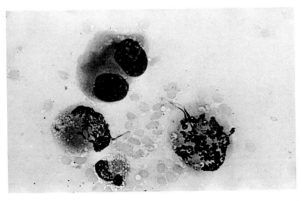

Fig. 4-51. Anaplastic carcinoma. Abnormal mitotic figures. (MGG stain, ×1000)

It has been debated whether small cell carcinoma truly exists. Woolner and colleagues[67] have suggested that most small cell carcinomas are primary lymphomas of the thyroid gland, a conclusion we strongly support. All small cell anaplastic tumors seen at Karolinska Hospital have primarily or on review been classified as lymphomas. Although it is difficult to distinguish small cell tumors from lymphomas histologically, the lymphoid differentiation is fairly evident cytologically. The cells are dispersed. Nuclei are poorly differentiated with fragile, sometimes vacuolated cytoplasm. Lymphoglandular bodies are present and give the smear a typical lymphoid background (Fig. 4-53). The cytologic picture does not differ from poorly differentiated lymphoma in other regions. It may also include mixed (centroblastic-centrocytic), monomorphic large cell, and immunoblastic types. Mature lymphocytic types are discussed in conjunction with Hashimoto's above.

The majority of thyroid lymphomas arise primarily in the thyroid and may extend to lymph nodes and other organs including the gastrointestinal tract. Extension of extrathyroidal lymphoma to involve the thyroid is distinctly unusual. Nevertheless, the presence of satellite or remote lymph nodes provides additional targets for aspiration and may clarify the clinical picture. Smears must be made carefully be-

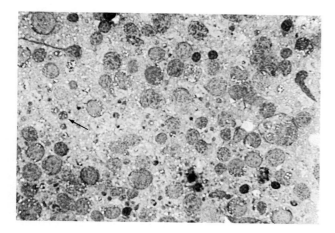

Fig. 4-53. Lymphoma, large cell type, poorly differentiated. There are dispersed immature large cells with fragile cytoplasm and numerous lymphoglandular bodies *(arrow)*. (MGG stain, ×600)

cause of the fragility of the lymphoid cells. We favor stains for identification of lymphoid cells, but both P and R stains should be used wherever possible.

UNEXPECTED CELL TYPES IN THYROID GLAND ASPIRATE

Histologically, squamous metaplasia occurs primarily in association with papillary carcinoma. Cytologic preparations rarely yield squamous cells. Epidermoid carcinoma is even more rare. Invasion of the thyroid gland by epidermoid laryngeal carcinoma can provide squamous material.

Palpable masses in the thyroid gland due to metastatic deposits were reportedly present in 30 of 2000 autopsied cancer cases.[55] In the same series, there were 19 primary cancers of the thyroid gland. Nonthyroid gland elements can be identified cytologically with some assurance from primary melanomas, oat cell carcinomas, and epidermoid carcinomas (see sections on these lesions).

Metastatic renal cell carcinoma to the thyroid was reported by Lasser.[38] We have reported two patients with clear cell metastases.[43] In each case nephrectomy was remote. The rare primary clear cell variant tumor of the thyroid must be kept in mind. The cytology of clear cell tumors is presented in Chapter 8.

Other undifferentiated cancers or adenocarcinomas, including those of the breast, may not give diagnostic etiologic findings but can usually be distinguished from well-differentiated follicular and papillary carcinomas.

Giant cell tumor metastatic to the thyroid gland may present a confusing cytologic pattern, as illustrated in the following case report:

■ *Case 4-3*
A 62-year-old man presented with rapidly enlarging firm masses in both lobes of the thyroid gland. The masses appeared on palpation to be intrinsic to the thyroid gland, and, in addition, there were discrete, hard lymph node enlargements adjacent to the thyroid gland bilaterally. On chest x-ray film there was a solitary 3 cm nodule in the right lower lobe. On cytologic smear there were masses of large anaplastic tumor cells with numerous macronucleoli (Fig. 4-54). The initial diagnosis of anaplastic thyroid gland carcinoma was subsequently determined to be bronchogenic carcinoma with metastases.

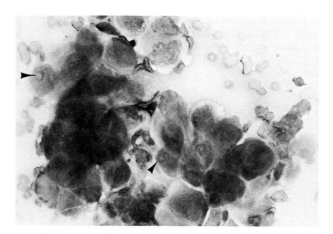

Fig. 4-54. Metastases to thyroid from large cell bronchogenic carcinoma. Aggregated large primitive cells with macronucleoli *(arrow)*. The findings are similar to some aspects of anaplastic carcinoma of thyroid. In the absence of accurate clinical history, the diagnosis of metastatic tumor is extremely difficult. (MGG stain, ×600)

FNA Versus Biopsy with Cutting Needle
(Table 4-1)

It is important to distinguish FNA biopsy, which produces smears, and core biopsy, which is a surgical procedure that uses Vim–Silverman and Trucut needles and produces tissue for histopathologic study.[54] The importance of this differentiation was highlighted in a report with commentary published by Esselstyn and Crile.[17]

In a study by Goldfarb and associates[22], the accuracy of the two techniques was the same. No attempt was made to differentiate benign and malignant follicular neoplasms and no lesion less than 1.2 cm was biopsied. The methods differ somewhat in rate of complications. FNA, with modest cytopathologic experience, should be free of hazards. In inadvertent puncture of a large neck vessel, the appearance of fresh blood in the syringe will indicate the need to stop the procedure and apply pressure. Postpuncture hematomas are prevented by local pressure (see the section on method). There have been no reports of tumor implantation after FNA, nor have any been seen at the cytology laboratory of the Karolinska Hospital.

Complications associated with core biopsy are also extremely infrequent. However, transient recurrent nerve palsy, hematoma, and tumor implantation have been noted.[17]

FNA cytology is of great value in the diagnosis and management of thyroid nodules. It has also become an important adjuvant in the evaluation of hyperthyroid and hypothyroid disorders. The diagnostic result is very much dependent on the amount and quality of aspirated material. The results should progressively improve with increased experience.

Table 4-1. Aspiration Biopsy of the Thyroid Gland: Short Guide for Microscopy; Main Cytologic Findings

COLLOID GOITER	FOLLICULAR TUMORS
Colloid — abundant Follicular cells — small Macrophages and cystic material	Equal-sized follicular clusters Abundant blood but scanty colloid
THYROTOXIC GOITER	**PAPILLARY TUMOR**
Colloid — scanty Blood — abundant Follicular cells — large Marginal vacuoles (fire flares)	Papilloid monolayered sheets Intranuclear inclusions Viscous colloid Macrophages and cystic
ACUTE SUPPURATIVE THYROIDITIS	**GIANT CELL ANAPLASTIC**
Granulocytes Proteinaceous material Necrosis Cave: anaplastic carcinoma	Necrotic fragments Granulocytes Bizarre tumor cells Mitosis — frequent
SUBACUTE GIANT CELL THYROIDITIS	**LYMPHOMA**
Follicular cells — small and retrogressive Histocytic giant cells — obligatory Lymphocytes and macrophages Epithelioid cells	Dispersed cell pattern Numerous mitotic figures Cytoplasmic fragmentation Lymphoid cells
CHRONIC LYMPHOID THYROIDITIS	**UNEXPECTED CELL TYPES**
Follicular cells — large and oncocytic Lymphocytes — numerous Histiocytic giant cells — rare	Bronchogenic Melanoma Breast Other

REFERENCES

1. Åkerman M, Tennvall J, Biörklund A et al: Sensitivity and specificity of fine needle aspiration cytology in the diagnosis of tumors of the thyroid gland. Acta Cytol 29:850, 1985
2. Aldinger KA, Samaan NA, Ibanez M et al: Anaplastic carcinoma of the thyroid. Cancer 41:2267, 1978
3. Aozasa K, Inoue A, Tajima K et al: Malignant lymphomas of the thyroid gland. Cancer 58:100, 1986
4. Atkinson B, Ernst CS, LiVols VA: Cytologic diagnosis of follicular tumors of the thyroid. Diag Cytopathol 2:1, 1986
5. Bastenie PA, Ermans AM: Thyroiditis and thyroid function. Clinical, morphological and physiopathological studies. In International Series of Monographs in Pure and Applied Biology, No. 36. Modern Trends in Physiological Sciences. Elmsford, NY, Pergamon Press, 1972
6. Beecham, JE: Coexistant disease as a complicating factor in the fine needle aspiration diagnosis of papillary carcinoma of the thyroid. Acta Cytol 30:435, 1986
7. Betsill W: Thyroid fine needle aspiration in pregnant women. Diag Cytopathol 1:53, 1985
8. Bondeson L, Bondeson A–G, Kindholm K et al:

Morphometric studies on nuclei in smears of fine needle aspirates from oxyphillic tumors. Acta Cytol 27:437, 1983

9. Boon ME, Löwhagen T, Willems J–S: Planimetric studies in fine-needle aspirates from follicular adenoma and follicular carcinoma of the thyroid. Acta Cytol 24:145, 1980

10. Budde R, Guenther M, Schaefer HE et al: Cytochemical investigations of phagocytes in thyroid gland cysts. Presented at the Seventh International Congress of Cytology, Munich, 1980

11. Burke JS, Butler JJ, Fuller LM: Malignant lymphomas of the thyroid: A clinical study of 35 patients including ultrastructural observations. Cancer 39:1587, 1977

12. Carcangiu ML, Zampi G, Rosai J: Poorly differentiated ("insular") thyroid carcinoma. Am J Surg Pathol 8:655, 1984

13. Compagno J, Oertel JE: Malignant lymphoma and other lymphoproliferative disorders of the thyroid gland: A clinicopathologic study of 245 cases. Am J Clin Pathol 74:1, 1980

14. Crile G Jr: Struma lymphomatosa and carcinoma of the thyroid. Surg Gynecol Obstet 147:350, 1978

15. Deligeorgi–Politi, H: Nuclear crease as a cytodiagnostic feature of papillary thyroid carcinoma in fine-needle aspiration biopsies. Diagn Cytopath 3:307, 1987

16. Einhorn J, Franzen S: Thin-needle biopsy in the diagnosis of thyroid disease. Acta Radiol 58:321, 1962

17. Esselstyn CB Jr, Crile G Jr: Needle aspiration and needle biopsy of the thyroid. World J Surg 2:321, 1978

18. Evans RW: Histological Appearances of Tumours. Edinburgh, E & S Livingston, 1968

19. Gershengorn NC, McClung MR, Chu EW et al: Fine-needle aspiration cytology in the preoperative diagnosis of thyroid nodules. Ann Intern Med 87:265, 1977

20. Glant MD, Berger EK, Davey DD: Intranuclear cytoplasmic inclusions in aspirates of follicular neoplasms of the thyroid: A report of two cases. Acta Cytol 28:576, 1987

21. Goellner JR, Johnson DA: Cytology of cystic papillary carcinoma of the thyroid. Acta Cytol 26:797, 1982

22. Goldfarb WB, Bigos TS, Eastman RC et al: Needle biopsy in the assessment and management of hypofunctioning thyroid nodules. Am J Surg 143:409, 1982

23. Greene JN: Subacute thyroiditis. Am J Med 51:97, 1971

24. Gribitz D, Talbot, NB, Crawford JD: Goiter due to lymphocytic thyroiditis (Hashimoto's struma): Its occurence in pre-adolescent girls. N Engl J Med 250:555, 1954

25. Guarda LA, Baskin HJ: Inflammatory and lymphoid lesions of the thyroid gland—Cytopathology by fine needle aspiration. Am J Clin Pathol 87:14, 1987

26. Hajdu SI, Melamed MR: Limitations of aspiration cytology in the diagnosis of primary neoplasms. Acta Cytol 28:337, 1984

27. Hamberger B, Gharib H, Melton AG et al: Fine-needle aspiration biopsy of thyroid nodule. Am J Med 73:381, 1982

28. Hay ID: Thyroiditis: A clinical update. Mayo Clin Proc 60:836, 1985

29. Henry JB: Clinical Diagnosis and Management by Laboratory Methods. Philadelphia, WB Saunders, 1979

30. Holm LE, Blomgren H, Löwhagen T: Cancer risks in patients with chronic lymphocytic thyroiditis. N Engl J Med 312:601, 1985

31. Jayaram G: Fine needle aspiration cytologic study of solitary thyroid nodule. Acta Cytol 29:967, 1985

32. Joensun H, Klemi P, Ecrola E et al: Influence of cellular DNA content on survival in differentiated thyroid cancer. Cancer 58:2462, 1986

33. Keider HR, Beaven MA, Doppman J et al: Sipple's syndrome: Medullary thyroid carcinoma, pheochromocytoma and parathyroid disease. Ann Intern Med 78:561, 1973

34. Kini SR, Miller JM, Hamburger JI: Problems in the cytologic diagnosis of the "cold" thyroid nodule in patients with lymphocytic thyroiditis. Acta Cytol 25:506, 1981

35. Kini SR, Miller JM, Hamburger JI et al: Cytopathology of follicular lesions of the thyroid gland. Diag Cytopathol 1:123, 1985

36. Lahey FH, Hare HF: Malignancy in adenomas of the thyroid. JAMA 145:689, 1951

37. Lang W, Atay Z, Georgii A: The cytological classification of follicular tumors in the thyroid gland. Virchows Arch Pathol Anat 378:199, 1978

38. Lasser A, Rothman JG, Calamia VJ: Renal cell carcinoma metastatic to the thyroid: Aspiration cytology and histologic findings. Acta Cytol 29:856, 1985

39. Laughan T: Über die epithelialen Formen der malignen Struma. Virchows Arch Pathol Anat 189:69, 1907

40. Levin SN: Current concepts of thyroiditis. Arch Int Med 143:1952, 1983

41. Lew W, Orell S, Henderson DW: Intranuclear vacuoles in non papillary carcinoma of the thyroid: A report of three cases. Acta Cytol 28:581, 1984
42. Ling SM, Kaplan SA, Weitzman JJ et al: Euthryoid goiters in children: Correlation of needle biopsy with other clinical and laboratory findings in chronic lymphocytic thyroiditis and simple goiter. Pediatrics 44:695, 1969
43. Linsk JA, Franzen S: Aspiration cytology of metastatic hypernephroma. Acta Cytol 28:250, 1984
44. Linsk JA, Franzen S: Fine Needle Aspiration for the Clinician. Philadelphia, JB Lippincott, 1984
45. Ljungberg O: Cytologic diagnosis of medullary carcinoma of the thyroid gland with special regard to the demonstration of amyloid in smears of fine needle aspirates. Acta Cytol 16:253, 1972
46. Löwhagen T, Sprenger E: Cytologic presentation of thyroid tumors in aspiration biopsy smears. Acta Cytol 18:192, 1974
47. Löwhagen T, Granberg P–O, Lundall G et al: Aspiration biopsy cytology (ABC) in nodules of the thyroid gland suspected to be malignant. Surg Clin North Am 59:3, 1979
48. Luck JB, Mumarr VR, Frable WJ: Fine needle aspiration biopsy of the thyroid: Differential diagnosis by videoplane image analysis. Acta Cytol 26:796, 1982
49. McConahey WM, Hay ID, Woolner LB et al: Papillary thyroid cancer treated at the Mayo Clinic 1946 through 1970:(Initial manifestations, pathologic findings, therapy and outcome). Mayo Clin Proc 61:978, 1986
50. Miller JM, Hamburger JI, Kini S: Diagnosis of thyroid nodules — use of fine needle aspiration and needle biopsy. JAMA 241:481, 1979
51. O'Morchoe PJ, Lee DC: Intranuclear cytoplasmic inclusion in carcinoma of the thyroid gland: Letters to editor. Acta Cytol 28:621, 1984
52. Rastad J, Wilander E, Lindgren P–G, et al: Cytologic diagnosis of medullary carcinoma of the thyroid by Sevier–Munger silver staining and calcitonin immunocytochemistry. Acta Cytol 31:45, 1987
53. Sanders LR, Moreno AJ, Pittman DL et al: Painless giant cell thyroiditis diagnosed by fine needle aspiration and associated with intense thyroid uptake of gallium. Am J Med 80.971, 1986
54. Schwartz HE, Weiburgs HE, Davis TF et al: The place of fine needle biopsy in the diagnosis of nodules of the thyroid. SGO 155:54, 1982
55. Shimaoka K, Sokal JE, Pickeren JW: Metastatic neoplasms in the thyroid gland. Cancer 15:557, 1962
56. Silverman JF, West RL, Finley JL et al: Fine-needle aspiration versus large needle or cutting biopsy in evaluation of thyroid nodules. Diag Cytopathol 2:25, 1986
57. Smithers D: Tumours of the Thyroid Gland. Edinburgh, E & S Livingston, 1970
58. Söderström N: Aspiration biopsy — puncture of goiters for aspiration biopsy. Acta Med Scand 144:237, 1952
59. Söderström N, Björklund A: Intranuclear cytoplasmic inclusions in some types of thyroid cancer. Acta Cytol 17:191, 1973
60. Söderström N, Telenius — Berg M, Åkerman M: Diagnosis of medullary carcinoma of the thyroid by fine needle aspiration biopsy. Acta Med Scand 197:71, 1975
61. Sokal JE: The problem of malignancy in nodular goiter — recapitulation and a challenge. JAMA 170:405, 1959
62. Stavric GD, Karanfilski BT, Kalamaras AK et al: Early diagnosis and detection of clinically nonsuspected thyroid neoplasm by the cytologic method: A critical review of 1536 aspiration biopsies. Cancer 45:340, 1980
63. Stepanas AV, Samaan NA, Hill CS et al: Medullary thyroid carcinoma. Cancer 43:825, 1979
64. Walfish PG, Hazani E, Strawbridge HTG et al: A prospective study of combined ultrasonography and needle aspiration biopsy in assessment of the hypofunctioning thyroid nodule. Surgery 82:474, 1977
65. Warren S, Meissner WA: Tumors of the Thyroid Gland. In Atlas of Tumor Pathology, Sec. IV, Fasc. 14. Washington, D.C., Armed Forces Institute of Pathology 1953
66. Willis RA: Pathology of Tumours. Scarborough, Ont, Butterworth & Co. 1960
67. Woolner LB, McConahey WM, Beahrs OH et al: Primary malignant lymphoma of the thyroid. Review of forty-six cases. Am J Surg 111:502, 1966
68. Zajicek J: Aspiration Biopsy Cytology. Basel, S Karger, 1974
69. Zirkin HJ: Follicular adenoma of the thyroid with intranuclear vacuoles in clear nuclei: A case report. Acta Cytol 28:587, 1984

CHAPTER 5

Aspiration Cytology of the Salivary Glands

JOSEPH A. LINSK

SIXTEN FRANZEN

This chapter will consider extra-oral and intra-oral palpable lesions of the parotid, submandibular, and minor salivary glands. These glands are composed of tubuloalveolar structures embedded in a mixed supporting stroma that yields a rich variety of lesions. The minor glands are distributed over the extra-oral and intra-oral mucosa.

Pre-operative assessment of salivary gland masses using only clinical criteria is difficult.[48] There is no question in our personal experience that pre-operative cytologic evaluation is of significant value in clinical management. Traditional diagnosis of parotid masses requires either superficial or total parotidectomy depending on the extent of the lesion. Incisional or excisional biopsy increases risk of recurrence, and most pathologists will hesitate to offer a frozen section diagnosis. Of 140 patients undergoing parotidectomy for diagnosis, 11% had malignant and 63% had benign lesions. Twenty-seven percent had nonneoplastic disorders.[21] In contrast of 910 salivary lesions diagnosed by fine needle aspiration (FNA), 278 were spared surgery; and most of these regressed.[11]

These studies as well as our personal experiences make it clear that inflammatory disorders may simulate tumors (see Subacute and Chronic Sialoadenitis below). Benign tumors including Warthin's and pleomorphic adenoma may be identified and surgery postponed to a more optimal time if necessary; or surgery may be eliminated in an elderly or medically compromised patient. Clinical judgment is critical.

Cytologic interpretation requires considerable experience, primarily because of the remarkable number of histopathologic types. Although each pathologic subtype may yield a characteristic diagnostic cytologic pattern on aspiration, morphologic variability of the cellular components with mixing and overlapping of tumors and nontumorous disorders, as well as tumor classes, creates cytologic problems.[13-16,50] Ductal, acinar, mucinous, oncocytic, stromal, and inflammatory cells participate in various combinations to produce a variety of cytologic patterns. Figure 5-1 provides a guideline and a correlation of anticipated lesions with particular cell types and groupings seen on smear.

The lesions listed below will be encountered in daily clinical practice and must be considered in differential diagnosis.

Inflammatory Disorders
Acute suppurative sialoadenitis
Subacute and chronic sialoadenitis
Lymphadenitis of intraglandular lymph nodes

Granulomatous Alterations
Sarcoidosis

Auto-immune Disorders
Benign lymphoepithelial lesions (Sjögren's disease, Mikulicz's disease, Godwin's disease)

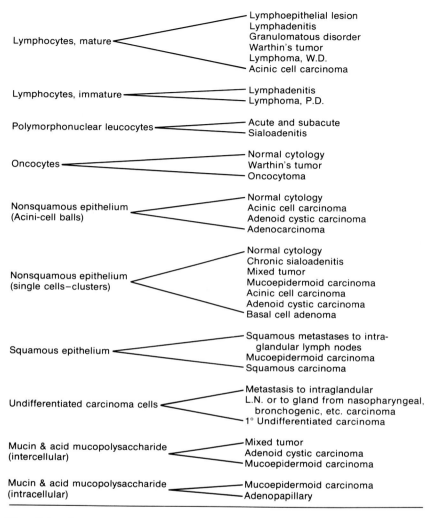

Fig. 5-1. Guidelines to differential cytologic diagnosis of salivary gland lesions.

Benign Neoplasms
Adenolymphoma (Warthin's tumor)
Oncocytoma
Pleomorphic adenoma

Malignant Neoplasms
Malignant mixed tumor (carcinoma in pleomorphic adenoma)
Adenoid cystic carcinoma
Acinic cell carcinoma
Mucoepidermoid carcinoma
Squamous carcinoma
Adenocarcinoma (adenopapillary carcinoma)

Carcinoma metastatic to salivary gland
Undifferentiated carcinoma
Lymphoma

Cysts
Simple (retention) cysts
Cystic mixed tumors
Cystic Warthin's tumor
Cystic mucoepidermoid carcinoma
Branchial cyst in salivary gland

Unusual Lesions
Lipoma
Extracranial meningioma

METHOD

Aspiration with fine needles and a one-hand syringe is carried out precisely as described in Chapter 1 for palpable lesions. Extra-oral masses are fixed, usually easily defined, and offer no special difficulties. Intra-oral lesions require a light source. Local anesthesia is usually not required but a topical spray may be used. The use of the fine needle allows repeated punctures, particularly helpful in sampling small lesions. Romanowsky (R) stain is most valuable because tinctoral qualities of stroma and cytoplasm are important determinants in diagnosis (see the sections on pleomorphic adenoma and adenoid cystic carcinoma). Wet fixation and stain with Papanicolaou (P) is more valuable in delineating squamous differentiation, which is not predictable before aspiration. Both stains should be used.

Complications are virtually nonexistent, and prompt pressure will prevent local hematoma.

CLINICAL CONSIDERATIONS

Bilateral parotid gland swelling (so-called Mikulicz's syndrome) due to fat infiltration, metabolic disorders, lymphoepithelial lesions, mumps, sarcoidosis, lymphoma, and bilateral tumors may be mistaken for normal facial contours.[1,23,33,37,47]

Patients with enlarged or prolapsed submandibular glands are often referred for evaluation of possible tumor. These glands are easily confused with metastatic or reactive lymph nodes (see Chap. 3).

Of 110 patients undergoing surgery for submandibular mass, 7% had malignant and 8% benign tumors. Eighty-five percent of the masses were inflammatory.[21] Diagnosis by FNA clearly reduces the amount of surgery.

Tumors may present as nodules under 1 cm. In addition, cysts, granulomas, intraglandular lymphadenopathy, and localized sialoadenitis may present

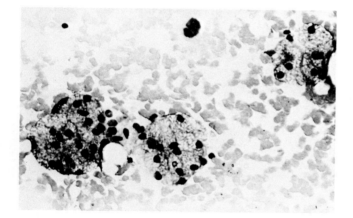

A

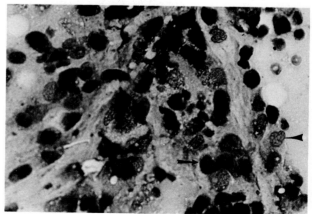

B

Fig. 5-2. A. Benign glandular epithelium, submandibular gland. There are glandular acini with frothy cytoplasm. (R stain, ×200) B. Subacute sialoadenitis, parotid. This is a cellular smear composed of moderately enlarged, partially degenerated epithelial (ductal) cells *(arrow)* intermixed with plasma cells and atypical mononuclear cells *(arrowhead).* This smear suggests a neoplasm. (R stain, ×300)

as palpable disease indistinguishable from neoplasm. Localized hypertrophy of glandular tissue, possibly due to intermittent ductal obstruction, may become a target for aspiration and yield benign epithelium, as illustrated in Figure 5-2.

Aspiration of a small nodular swelling with a 25-guage needle may easily yield diagnostic cytologic smears. Early discovery and prompt aspiration of extremely small lesions are valuable clinical manuevers.

INFLAMMATORY DISORDERS

Acute Sialoadenitis

In acute sialoadenitis the salivary gland is diffusely swollen and may be tender. The skin may suggest cellulitis. The cytologic picture reflects the pathologic changes. Acute suppurative sialoadenitis, ordinarily unilateral, initially yields a granulocytic exudate with necrotic debris identical to acute suppurative reactions in abscesses at other sites. Later in the course, lymphocytes, plasma cells, and phagocytic macrophages predominate. Fragments of acinar or ductal epithelium must be present to identify salivary gland origin. Acute and subacute lymphadenitis, including cat-scratch disease, may be wrongly diagnosed in the absence of salivary gland tissue. Differential diagnosis with tumors is generally not a problem.

Subacute and Chronic Sialoadenitis

Subacute and chronic sialoadenitis have greater importance from the standpoint of aspiration biopsy cytology (ABC) for two reasons. First, a firm mass occupying all or part of the parotid or submandibular gland may clinically simulate a tumor. There is often no history of prior inflammation. Second, there is a possibility of confusion with a tumor cytologically.

Smears consist of groups of atypical epithelium, sometimes in sheets with scattered inflammatory cells in the background. The epithelial cells represent hyperplastic and proliferated duct epithelium. Glandular elements have been destroyed (Fig. 5-2B). Smears that contain clusters of clearly benign epithelium — albeit somewhat atypical — that do not have the characteristics of any of the cytologic patterns of the tumor subtypes (discussed below) should be dealt with cautiously. The diagnosis of chronic

sialoadenitis should be considered.[50] Intraglandular lymph nodes may be the site of reactive or inflammatory lymphadenapathy, simulating a primary salivary gland lesion. Cytology is dealt with in Chapter 14.

Granulomatous Sialoadenitis

Sarcoidosis involving the salivary gland is an infrequent target for FNA, in our experience. It is usually part of a systemic or regional syndrome (uveoparotid fever) (Heerfordt's disease), involving bilateral salivary glands, and therefore may not require surgical biopsy for diagnosis. It may, however, present as a solitary mass in one parotid gland.[30]

Other specific and nonspecific acute and chronic granulomatous inflammations will yield florid smears with numerous semilunar epithelioid cells and multinucleated giant cells (Fig. 5-3).

BENIGN NEOPLASMS AND AUTO-IMMUNE DISORDERS

Cysts and Lymphoepithelial Alterations

Tense salivary gland cysts simulate tumors. Evacuation of cysts must be followed by careful palpation and re-aspiration of residual tumor. Cell-free fluid

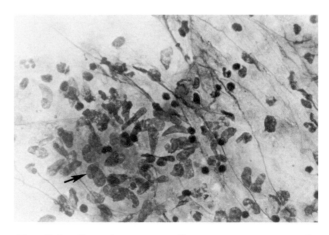

Fig. 5-3. Granulomatous inflammation, acute, angle of jaw. A cluster of semilunar epithelioid cells partly surrounds a Langhans' giant cell *(arrow)*. Tissue, cell debris, and acute inflammatory cells are in the background. (R stain, ×150)

cannot automatically be deemed benign in the presence of a palpable mass (see the section on mucoepidermoid tumors). Cysts occur as part of the structure of mucoepidermoid tumors and, occasionally, pleomorphic adenomas. Adenoid cystic tumors rarely present cystic targets for aspiration despite the designation *cystic*. Cysts yielding cell-free fluid with no residual tumor are termed *retention cysts*.

There is an intimate relationship between lymphoid and epithelial components of the salivary gland. Benign heterotopic salivary tissue occurs in lymph nodes detached from the gland, and a number of cases have been recorded and collected.[5] Such abnormal deposits occur in lymph glands within the salivary glands as well and may appear in both lymphoepithelial cysts and solid lesions.[5] Cytologic features include a background of small mature lymphocytes admixed with epithelial-cell aggregates composed of mucinous, foamy, and granular cells. Cells resembling metaplastic squamous cells are also present, and squamous cells may be potentially aspirated.[49] Characteristic branchial cleft (branchiogenic) cysts may occur within the parotid gland, and the fluid cytology is discussed in Chapter 3.[27] With lesions at the angle of the jaw, it may not be possible to define the location clinically as parotid or extraparotid in origin.

Benign lymphoepithelial lesion (Mikulicz's disease, Godwin's disease) is a solid, tumorlike alteration composed of diffuse or nodular lymphoid tissue within which are sequestered epithelioid clusters termed *epimyoepithelial islands*.[25,37] These islands do not have any glandular configuration, probably represent metaplastic change in ductal epithelium, and are composed of myoepithelial and altered ductal cells.[5,17]

On aspiration, both lymphocytic and epithelial cells may be obtained.[50] The epithelial islands may be quite sparse, and true identity of the lesion may require clinical correlation (Sjögren's syndrome) and, ultimately, tissue biopsy. The epithelial groups are composed of benign, bland, cuboidal cells with no distinguishing features. The lymphocytes may be sparse and scattered rather than densely carpeted as in lymphoma or lymphadenitis. They are a mixed population of small, intermediate, and transformed cells. Cytologic experience with these lesions is limited, and differential diagnosis with sialoadenitis, lymphoma with residual epithelial groups, and intraparotid lymphadenitis must be considered.

Papillary Cystadenoma Lymphomatosum

Papillary cystadenoma lymphomatosum (Warthin's tumor, adenolymphoma) is an awkward, but descriptive, designation of this benign tumor. It is a cystic papillary adenoma supported by a dense lymphoid stroma (Fig. 5-4A). It occurs predomi-

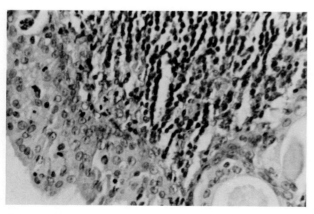

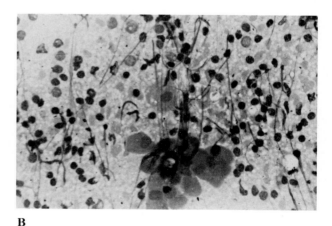

A   B

Fig. 5-4. A. Papillary cystadenoma lymphomatosum (Warthin's tumor). Tissue section. There is a multilayered band of oncocytic cells backed by a lymphoid stroma. (H & E stain, ×400). **B.** Aspirate from same tumor as in **A.** There are scattered lymphoid cells with strands of chromatin due to cell fragility. There is a sharply delineated cluster of oncocytic cells. (MGG stain, ×400).

nantly in the parotid gland but must be considered in the differential diagnosis of submandibular masses.[14] On palpation, it is distinctly softer than the common rubbery mixed tumor or firmer malignant tumor. It has been described as "doughy" but may be rubbery or firm, not unlike lymphoma tissue.[14] Distended cystic varieties may be firm. Occurring at the periphery of the gland, it may be confused with a lymph nodal or metastatic nodule, particularly since aspiration yields lymphoid and epithelial cells. The tumor may vary from less than 1 cm to several centimeters. These are tumors of middle age but must be considered at any age. They may be present for many years before identification and may occur bilaterally, synchronously, or sequentially.

Aspiration yields drops of tissue fluid, with or without blood, within which are tissue fragments. Cystic fluid is often the major finding, and the cytopathologist must be alert to the diagnostic clues in this material.[14]

Cytologic Findings

Solid tumors yield a mixture of lymphocytes and groups of tightly cohesive oncocytic cells in single and multilayered, sharply delineated clusters (Fig. 5-4 and Color Plate V C). These epithelial cells have abundant, opaque, occasional granular cytoplasm with well-defined cell borders. Cytoplasm is steel gray with R stain and pale tan with pink granulation in P stain. Nuclei are round to oval. Nucleoli are not evident, and the nuclear:cytoplasmic ratio is low. Lymphocytes are small, mature cells with a mixture of larger intermediate and early types from follicular centers (Fig. 5-5). They are scattered singly in the background rather than in dense sheets or clusters. Mast cells associated with monolayers of oncocytes were identified in 16 of 20 aspirates stained with R.[7] Pools of mucin with entrapped lymphocytes and amorphous debris are commonly seen.

Cystic fluid contains cell debris, some inflammatory cells, many lymphocytes, and cholesterol crystals in varying amounts. Isolated oncocytic cells as well as cell clusters may be found with ease or after careful search.

With R stain, isolated degenerated oncocytic cells may stain blue and simulate squamous cells. Such a smear may be mistaken for material from a cavitary squamous carcinoma (see Chap. 3). A few intact identifiable squamous cells may be seen. Mu-

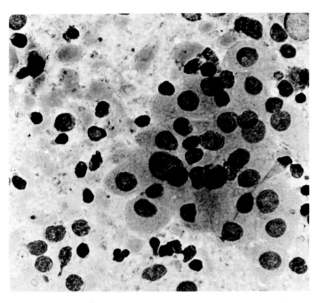

Fig. 5-5. Papillary cystadenoma lymphomatosum (Warthin's tumor), submaxillary gland. Monolayered cluster of oncocytes and lymphocytes are present in a background of precipitated fluid. (R stain, ×300)

coepidermoid carcinoma must also be considered in the differential diagnosis. A rare malignant tumor is seen in Figure 5-6.

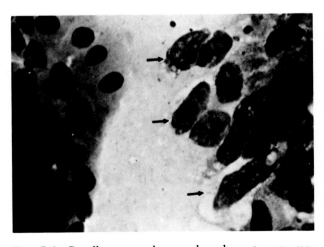

Fig. 5-6. Papillary cystadenoma lymphomatosum, malignant type. The first aspirate yielded fluid with oncocytes and lymphocytes. Aspiration of the residual mass yielded poorly differentiated carcinoma cells (*arrows*) with benign oncocytes. (R stain, ×400)

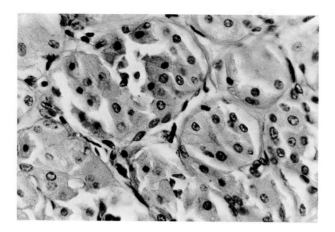

Fig. 5-7. Oncocytoma, parotid. Tissue section. Cells are in small lobules separated by delicate fibrous septae. Cytoplasm is granular, with small round nuclei suggesting normal hepatocytes. (H & E stain, ×150)

Oncocytoma

The oncocytoma is a benign delimited tumor, primarily of the parotid gland, composed of large oval and angular cells with abundant granular cytoplasm and small, regular, round to oval nuclei. The cells are arranged in small lobules and acini (Fig. 5-7). On aspiration we have observed two patterns. In one pattern, cells are arranged in a syncytium with nuclei scattered in a voluminous gray cytoplasmic background (Fig. 5-8A) In Fig. 5-8B cell borders are more evident. This picture may be indistinguishable from the epithelial component of a Warthin's tumor (Color Plate V C). Differential diagnosis must depend on the presence or absence of lymphocytes, cell debris, crystals, and fluid. A second pattern consists of dispersed oncocytic cells with markedly granular cytoplasm and small eccentric nuclei. Small nucleoli may be present (Fig. 5-9).

With P, cytoplasm is tan-gray with fine pink granulation. Intact alveolar structure are not present in cytologic smear.

Pleomorphic Adenoma

Pleomorphic adenoma (mixed tumor) is the most common benign tumor of the salivary gland, occuring in both the major and minor glands. It may appear at any site within the parotid gland, at the angle of the jaw, and in pre-auricular and postauricular locations. The presentation may vary from an inconspicuous nodular elevation measuring less than 1 cm to a massive, bosselated, deforming tumor present for years. On palpation, it is rubbery to firm, usually smooth, and movable. Intra-oral tumors may arise at any location in the mucosa, including the soft palate. This unexpected location was illustrated in Case 3-2.

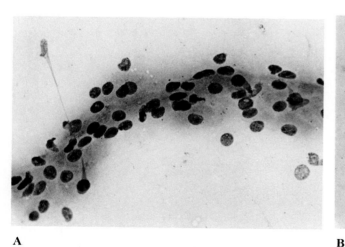
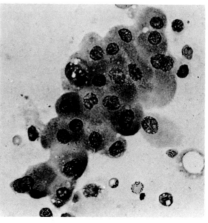

A   B

Fig. 5-8. A. Oncocytoma, parotid. Round-to-oval bland nuclei are scattered through a syncytium of fairly homogenous gray cytoplasm. **B.** A cellular cluster from a solid tumor. The cells have abundant, well-preserved, sharply delineated cytoplasm, finely granulated. Similarity to oncocytes in Warthin's tumor is evident. (R stain: A ×150; B ×400)

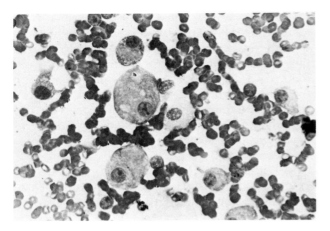

Fig. 5-9. Oncocytoma, parotid. This smear was extracted from a tumor in a 14-year-old girl. Cells are dispersed, with granular cytoplasm, eccentric, round nuclei, and small nucleoli. (R stain, ×150)

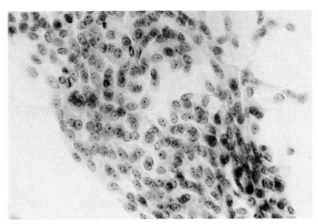

Fig. 5-10. Pleomorphic adenoma, parotid. Rather delicate epithelial cells with bland oval nuclei and small nucleoli dominate the smear. Cytoplasm is pale and poorly defined. Stroma is not evident. (P stain, ×150)

An understanding of the heterogeneous histology of this tumor is essential to an appreciation of the various cytopathologic patterns.[17,18] The epithelium is bland and varies in cellularity. The chondroid and myxoid stroma produce striking tinctorial reactions with R (Color Plate VI A). The tumor is of epithelial origin, histogenetically.[46] The stromal mucin has a high content of glycosaminoglycan that stains blue with alcian blue.[3,45] Studies have suggested that the origin of the dense chromatous secretion, prominently demonstrated in aspiration smears (see below), is the tubuloacinar rather than the connective tissue cells of the gland.[45]

Cytologic Findings

The aspirate consists of a thick, gelatinous mixture of cells and stroma. The smear is often opaque and covers a small area. With R, there is a dense pink to red stroma within which pale blue, round to oval nuclei are imbedded (Color Plate VI A). This visual pattern is almost instantly diagnostic.

Nuclei with narrow rims of pale cytoplasm are more clearly seen in the thinner stroma. Nuclei vary from bland and finely granular types with small pale blue nucleoli to dense nuclei that may suggest lymphocytes or plasma cells.[4]

Cellular mixed tumors may yield only epithelial cells, (Fig. 5-10), and the absence of typical stroma in R may create diagnostic difficulties. Such tumors,

however, usually provide a typical histopathologic picture with characteristic stroma that was not extracted by the aspiration.

With P, a pale indistinct background in contrast with dense pink stroma is present. Nuclei are mainly oval and finely granular with inconspicuous nucleoli. They are eccentrically placed in a poorly defined homogenous cytoplasm. Cells are both dissociated and arranged in clusters with overlapping as well as monolayer patterns (Figs. 5-10 and 5-11). Therefore, the instant identification of mixed tumor is not as easily

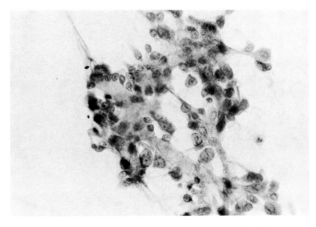

Fig. 5-11. Pleomorphic adenoma, parotid. These are atypical epithelial cells in a random cluster. Nucleoli are inconspicuous. Histologically, this was a benign mixed tumor. (P stain, ×150)

accomplished with P as with R. Figure 5-12 illustrates a monolayer with R.

Acinar and tubular patterns recognized histologically are not usually evident as such on smears, possibly because of distortion by the closely related stroma.

Cystic spaces may be detected by aspiration of cell detritus and fluid together with squamous material, which usually lines the cyst.[9] This may be a pitfall in diagnosis of pleomorphic adenoma.[40] (Also see Color Plate VII E and F.)

This tumor may proliferate areas of squamous, oxyphilic, adenoid cystic, mucoepidermoid, and acinic cell tumor, introducing pitfalls for the cytopathologist who is reviewing a small sample. In practical terms, this is rare; and the percentage of accurate cytologic diagnosis of pleomorphic adenoma is extremely high both in our experience and in a number of reports.[15,32,36,38,40,43,50]

MALIGNANT NEOPLASMS

Malignant Mixed Tumor

Clinical evidence suggests that malignant mixed tumor or, more specifically, carcinoma in pleomorphic adenoma arises by transformation of part of a pre-existing pleomorphic adenoma.[18,34] Patients are generally older, and they report recent rapid enlargement of a long-standing quiescent mass analogous to the transformation of a thyroid adenoma to anaplastic carcinoma (see Chap. 4). Seventy-five percent of malignant mixed tumors arise in the parotid gland, and 40% of patients exhibit metastases to regional lymph nodes at some time during the clinical course.[20]

The usual elements of pleomorphic adenoma must be evident in some portion of the aspirated material to orient the cytopathologist. The common type yields cells diagnostic of a fairly typical adenocarcinoma with polymorphous, irregularly arranged hyperchromatic cells (Fig. 5-13), perhaps in juxtaposition with the typical stroma of pleomorphic adenoma (Fig. 5-14). Squamous components as well as spindle cell types may be present. Atypical cytologic findings in pleomorphic adenoma may suggest malignant transformation but must be interpreted cautiously if benign structures are well represented. Histologic differentiation is also difficult, and malignant varieties should display destructive growth and invasion.[9] So-called benign mixed tumor (pleomorphic adenoma) may metastasize.[17,18] The occurrence of adenoid cystic carcinoma arising in a pleomorphic adenoma of the parotid has been diagnosed cytologically.[22]

Adenoid Cystic Carcinoma

Adenoid cystic carcinoma is a slow-growing, but insidiously recurring neoplasm with a distinctive histopathology (Color Plate VI F) that occurs in a variety of

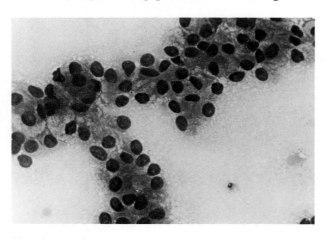

Fig. 5-12. Pleomorphic adenoma, parotid. Needle aspirate. Cells are in a monolayer, distributed in a papillary configuration. Nuclei are fairly dense and round to oval. Nuclear structure and nucleoli are not evident. Cytoplasm is finely vacuolated. (R stain, ×330)

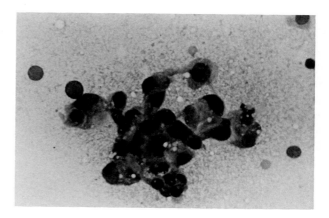

Fig. 5-13. Malignant mixed tumor. A cluster of pleomorphic hyperchromatic cells in random cluster. It has none of the features of other specific types of malignant salivary tumor. Compare with Fig. 5-17. (R stain, ×800)

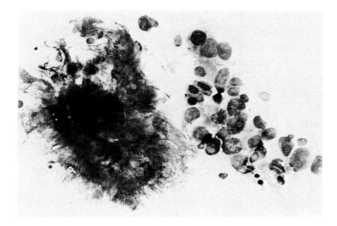

Fig. 5-14. Malignant mixed tumor (carcinoma in pleomorphic adenoma), parotid. This smear shows dense stroma typical of a benign pleomorphic adenoma, together with a cluster of polymorphous tumor cells. (R stain, ×600)

glandular tissues. Similar pathologic findings have been reported in salivary glands, trachea, lacrimal glands (see Case 3-3, Chap. 3), nasopharynx, palate, paranasal sinuses, bronchi, skin, breast, Bartholin's gland, and cervix.[12,16,17,18,19,36]

Pre-operative diagnosis and differentiation from other salivary gland neoplasms is important, because this tumor invades regional lymphatics early and planned radical surgery is necessary for cure.[35]

In the salivary gland, this tumor has been histologically divided into tubular, cribiform, and solid types.[41] The tubular and cribiform patterns are composed of cylindric and spherical layers of epithelial cells enveloping mucoid secretions. Although this tumor is called *cystic*, it is rarely so morphologically. The relationship of histology to clinical course has been considered by several observers. In a series of 34 adenoid cystic carcinomas from 877 salivary gland tumors, Foote and Frazell concluded that patients with the solid type had no worse a prognosis than those with the cylindric types.[18] On the other hand, the solid pattern has been associated with more aggressive behavior in several other series.[10,16,41] Therefore, the identification of histologic type by cytologic smears may be clinically useful. The delineation of cytologic patterns associated with the histologic grouping has been presented in detail by Eneroth and Zajicek.[16]

Cytologic Findings

Aspiration yields solid particles in a drop of fluid. Individual cell cytology is deceptively benign. Small, closely packed, fairly regular cells with ovoid nuclei and minimal hyperchromatism and small or absent nucleoli are present. In P, chromatin is pale and finely granular with a thin, nuclear membrane. Cytoplasm is pale, wispy, and poorly defined. Cells appear in small groups with a few single cells. In R, cells appear as small tight clusters and plugs. There may be large, dense groups with pink intercellular substance. The nuclear : cytoplasmic ratio is high. Nuclear chromatin is structureless, and nucleoli are not seen. In isolated cells, cytoplasm is skimpy, gray-blue, and generally eccentric. Characteristic of smears (Fig. 5-15 and Color Plate VI D) is the mucoid globule staining pink with R and surrounded by cells. With P, these globules are colorless. This pattern is extracted from the tubular and cylindric histologic patterns and may indicate less aggressive tumor growth (see above). The solid variety yields large irregular clusters composed of fairly uniform small cells without mucoid globules (Fig. 5-16).

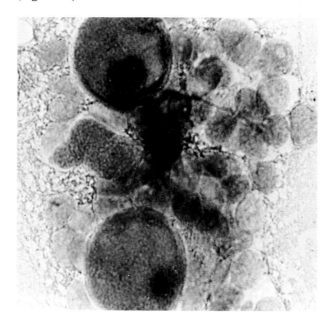

Fig. 5-15. Adenoid cystic carcinoma, parotid. The characteristic cytologic picture of this tumor. The epithelial cells have bland, uniform ovoid nuclei. Nucleoli are inconspicuous and cytoplasm is skimpy. Two large (pink staining) mucoid balls are evident. (R stain, ×1200)

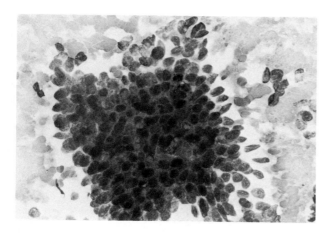

Fig. 5-16. Adenoid cystic carcinoma, metastatic facial mass, parotid primary. This is a dense, irregular cluster of small tumor cells. Nucleoli and cytoplasm are inconspicuous. The background extracted from vascular facial tissue consists of red blood cells. (P stain, ×150)

Acinic Cell Carcinoma

Acinic cell carcinoma is a low-grade tumor that occurs in the parotid and minor salivary glands.[8,13,18,42] Clinical features may suggest another benign tumor or a nonneoplastic hyperplasia or proliferation. Benign mixed tumor is the usual pre-operative diagnosis. The tumors are firm but not hard, single, and 2 cm to 4 cm.[18] Signs of malignancy including rapid growth, large size, infiltration, pain, and paralysis are rarely seen. Benign characteristics may be deceptive, since recurrences over many years may occur and ultimately lead to death.[18] The late results with tumor arising in minor salivary glands are more promising.[8]

Histologically, the tumor may closely resemble normal salivary gland tissue because of coarse cytoplasmic granulation that simulates normal secretory granules.[9,18] Less commonly, the cytoplasm is clear, and the nongranular cytoplasm is easily distinguished from normal serous acini but can be confused both histologically and cytologically with Warthin's tumor or oncocytoma, metastatic renal cell carcinoma, and clear cell variant of mucoepidermoid carcinoma.[50] It must therefore be considered in evaluations of cytologic material from metastatic sites, since the primary lesion occurring years previously may be forgotten.[44] Lymphoid stroma, organized into follicles, may be present in one third of tumors and may lead to confusion with metastatic lymph nodes cytologically (see below). These tumors may remain cytologically benign even in metastatic sites.

Cytologic Findings

Normal salivary cytology consists of glandular cells singly and in alveolar groups with foamy, granular cytoplasm and basal nuclei (see Fig. 5-2). Cell regularity is maintained in the alveoli. Fat may be intermixed. Acinic cell tumors, in contrast, yield more cells. Alveolar structures are not well maintained (Fig. 5-17). Cytoplasm may be abundant, granular, and occasionally foamy (clear cells), although stripped nuclei may also be present (Fig. 5-18 and 5-19). Nuclei are bland but there may be variation in size and shape. Polarity is lost in contrast with normal acini. Nucleoli are small or inconspicuous. In P, depending on differentiation, cells appear singly and in irregular clusters. Nuclei are variably enlarged and round to oval with regular sharp nuclear membranes.

The cytology of basal cell adenoma simulates the solid type of adenoid cystic carcinoma. The various cytologic patterns have been presented by Hood and associates.[28] Because of the aggressive biologic behavior of adenoid cystic tumors and the necessary extensive surgery, differentiation of the two tumors is extremely important.

Aspiration is particularly useful in establishing local recurrences, which may present as minimal induration at the resection margin. A 25-gauge needle inserted superficially into the skin may yield typical small, cuboidal, rather bland carcinoma cells. Late recurrences usually spread through regional lymphatic channels and leave tumor deposits in the skin of the face, scalp, and neck (see Chap. 3).

An almost identical cytologic pattern was identified in a benign trabecular adenoma by Layfield.[31] We have had an opportunity to review this case. Tinctorial differences in the mucus balls, staining blue in the trabecular adenoma and pink in the adenoid cystic carcinoma, may be a distinguishing feature. May–Grünwald–Giemsa (MGG) or other R stain is mandatory. (Also see Color Plate VII E and F.)

Benign eccrine cylindroma as reported by Bondeson and colleagues[6] may simulate the location and cytology of adenoid cystic carcinoma. This is another pitfall.

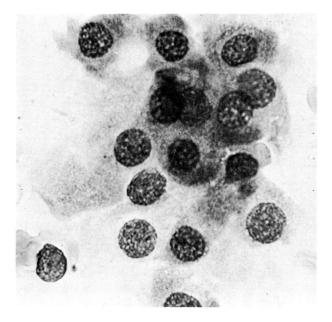

Fig. 5-17. Acinic cell tumor, parotid. A cluster of mono-morphic tumor cells with abundant, slightly ragged cyto-plasmic outlines. Note coarse granular peppering and com-pare to oncocytes in Figure 5-8A. (R stain, ×330)

Chromatin is moderately coarse and clumped, and there are multiple small nucleoli in some cells. Cyto-plasm is not well seen. Lymphocytes infiltrating the tumor may be abundant in some smears. Malignancy is readily diagnosed.

Decisive features in the diagnosis of well-dif-ferentiated acinic cell carcinoma are the presence of atypical acinous structures and the presence of acid alcian blue-negative and diastase resistant, periodic acid Schiff (PAS)-positive cytoplasmic granules.[39]

The less-differentiated variety is distinguished in part from solid adenocarcinoma by the presence of few larger, eosinophilic nucleoli in the latter. How-ever, lack of distinguishing features of acinic cell car-cinoma may prevent specific differentiation from ad-enocarcinoma in some cases.

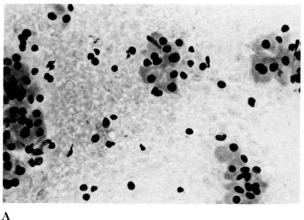
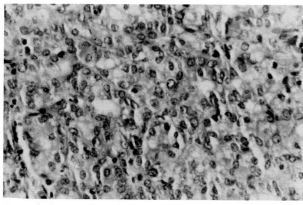

A   B

Fig. 5-18. A. Acinic cell tumor. There is abundant cytoplasm suggesting an oncocytic tumor. Cells are in small atypical acinous structures. (R stain, ×400) B. Tissue section. Cells are in poorly preserved acinous structures. (H & E stain, ×360)

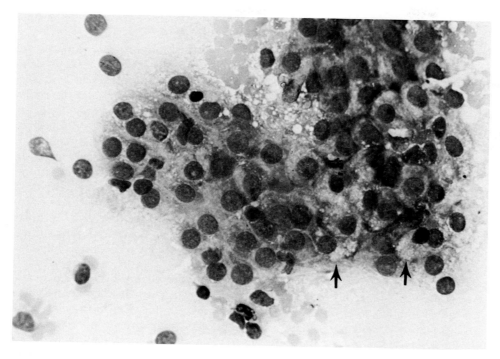

Fig. 5-19. Acinic cell tumor, parotid. Needle aspirate. This is a cellular smear in which some cells retain foamy cytoplasm and simulate benign salivary gland cells *(arrows)*. (See Fig. 5-2A). Cells have lost their polarity. Nuclei are enlarged and contain one or more small nucleoli. (R stain, ×132)

Mucoepidermoid Tumor

The mucoepidermoid tumor is of duct epithelial origin, and cell modification results in mucus-secreting, epidermoid, and intermediate or hybrid cells.[18] Review of multiple tumors or multiple sections of the same tumor will demonstrate heterogeneity of the cellular proliferation. Mucus-secreting cells may have clear cytoplasm, cytoplasmic vacuoles, or minimal evidence of mucin. Clear cell varieties may be confused with acinic cell or metastatic renal or bronchogenic carcinoma. Epidermoid cells may vary from small cuboidal to large typical squamous carcinoma cells.[9] Intermediate cells can be confused with small, nonkeratinizing squamous cells. They have clear cytoplasm and large, centrally located nuclei. Scattered epithelial cells in pools of mucin, characteristic of mucinous carcinoma of breast and gastrointestinal (GI) tract, are not seen. Free mucin occurs in pools, however, with cysts, which can be aspirated. Aware-

ness of the varieties of histologic structure helps clarify cytologic diagnosis.

Mucoepidermoid tumors are all malignant, including the low-grade variety.[18] This latter type may present as a small painless mass not diagnosable before cytology is reviewed. It never causes facial nerve paralysis.[18] When the mass is painful, larger, firm, and fixed, with taut covering skin, facial nerve paralysis is frequent; and it signals a higher grade tumor.[18] These clinical findings suggest a more aggressive tumor and are important data in evaluating cytologic smears.

Pre-operative cytologic determination of diagnosis, degree of differentiation, and extent of epidermoid component, coupled with clinical findings, will aid the clinician in deciding on preoperative radiation and extent of surgery.

Tumors of lesser grade are composed equally of well-differentiated squamous, mucus-secreting, and intermediate cells. Poorly differentiated epidermoid cells predominate in the more highly malignant variety.[9]

Distinction of this type from primary (Fig. 5-20) or metastatic epidermoid carcinoma may be difficult. Nine of 24 malignant tumors in a series of 160 salivary gland lesions diagnosed cytologically were metastatic squamous carcinoma.[43] Complete head and neck examination and clinical correlation are necessary to refine the diagnosis.

Cysts, lined with epidermoid and mucous cells, may be prominent. They contain mucin, cell detritus, and inflammatory cells. They may be confused with the cysts of adenolymphoma.[50] *Failure to demonstrate malignant elements in cyst fluid does not exclude diagnosis of mucoepidermoid carcinoma in the presence of a residual palpable mass after evacuation.*[36,40]

Cytologic Findings

Aspiration of cystic fluid must be followed by careful palpation and re-aspiration of any residual tumor. Cystic fluid must be scanned for clusters of oncocytic cells to establish a diagnosis of adenolymphoma,

since the surgical approaches to mucoepidermoid and adenolymphoma differ.[50]

Well-differentiated tumors or portions of tumors will yield mucus-secreting cells singly or in clusters. Nuclei are eccentric and small and lack atypia. Cytoplasm is clear or foamy and distended with mucin. Well-differentiated squamous cells are present in tight clusters with ill-defined cytoplasm or singly and spread out with central, finely granular, oval nuclei and well-defined opaque, gray, smooth, finely vacuolated cytoplasm (Fig. 5-21). Transition types from intermediate to squamous differentiation may be seen. A background of granular and vacuolated pink mucin may be present with R. In areas there are large clear vacuolizations. Dark pink stroma with imbedded nuclei similar to pleomorphic adenoma may be seen.

Moderately (intermediate) differentiated tumors do not form any clearly defined group cytologically. The cell types—mucus-producing, intermediate, and squamous—may appear in varying proportions with variable pleomorphism (Fig. 5-22).

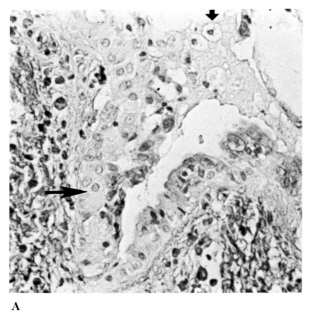
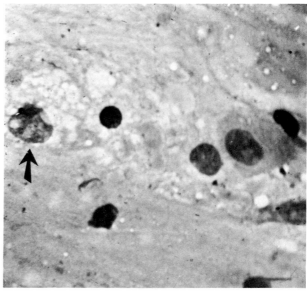

A   B

Fig. 5-20. A. Mucoepidermoid carcinoma, parotid. Tissue section. Epidermoid carcinoma cells *(long arrow)* and vacuolated mucus-producing cells *(short arrow)* are seen within an area that yields a mixed population on aspiration. (H & E stain, ×600) **B.** A high power view. Vacuolated mucus-producing cells *(arrow)* should not be misinterpreted as macrophages. Epidermoid cells are also present on a dense background of precipitated homogeneous mucus. (R stain, ×800)

Poorly differentiated cells in clusters characteristic of many undifferentiated carcinomas are not diagnostic of mucoepidermoid carcinoma.[43] This diagnosis can be suggested in the report, with histologic confirmation required (Fig. 5-23). Epidermoid carcinoma is readily identified in the high-grade tumors, and mucin or mucous cells must be present to distinguish primary or metastatic squamous carcinoma from mucoepidermoid carcinoma. This neoplasm is a lethal cancer that may yield focally benign cytology. Cell types and relationships should raise suspicion so that the possibility is included in the cytologic report.[50,51] On the other hand, highly malignant smears must raise the possibility of metastatic carcinoma, and the report should reflect this so that a careful examination of the head and neck region for a primary lesion is carried out. (See Color Plate VII A and B.)

Focal adenomatoid hyperplasia occurring primarily in minor salivary glands is included in the differential diagnosis of mucoepidermoid carcinoma.[2] The cytologic yield includes a mucoid background, acini, bare nuclei, and sheets of squamous cells.

Adenocarcinoma

As with squamous carcinoma, solid and trabecular adenocarcinomas do not yield cytologic smears with features that distinguish primary salivary gland carci-

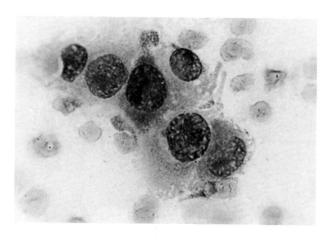

Fig. 5-22. Mucoepidermoid carcinoma, poorly differentiated. This is a cluster of polymorphous tumor cells with granular chromatin and enlarged nucleoli. Cytoplasm is poorly defined and frothy, which may suggest the diagnosis, although poorly differentiated adenocarcinoma cannot be excluded. (R stain, ×800)

noma from a metastatic lesion (Fig. 5-24). There is, however, some overlap, as discussed in the section on acinic cell carcinoma; and such smears require careful scrutiny. In the absence of compelling local signs, symptoms, and history, the patient should be studied thoroughly to exclude a primary site before radical surgery. Such smears may offer clues on primary sites such as breast and lung.

An exception to this difficulty is the fairly common pattern of adenopapillary carcinoma described by Zajicek.[50] Smears contain cells with pink, mucoid, intracytoplasmic globules when stained with R (Color Plate VII C) and pale typical signet ring cells when stained with P (Fig. 5-25 and Color Plate VII D), in contrast with extracellular globules of adenoid cystic carcinoma. These smears do not contain the blue-purple background and sparse cluster of cells of the mucinous carcinomas in the breast and GI tract (see Chaps. 6 and 8).

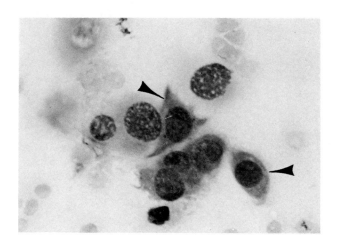

Fig. 5-21. Mucoepidermoid carcinoma, moderately differentiated, parotid. This cluster is a mixture of squamous cells *(arrows)* with dense nuclei and angular cytoplasm and mucus-producing cells with granular chromatin and frothy cytoplasm. (R stain, ×300)

Undifferentiated Carcinoma

The histogenetic origin of undifferentiated carcinoma may be identified by the presence of more differentiated elements in the tumor. This might apply to high-grade mucoepidermoid carcinomas. However, the origin of a smear composed entirely of un-

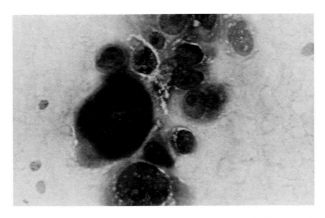

Fig. 5-23. Epidermoid carcinoma in parotid. Cluster of polymorphous poorly differentiated, but identifiable squamous cells cannot be specifically categorized as part of a mucoepidermoid, primary epidermoid, or metastatic squamous carcinoma. (R stain, ×1000)

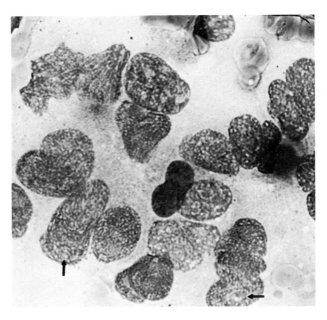

Fig. 5-24. Adenocarcinoma, poorly differentiated, parotid. This is a cluster of somewhat polymorphous tumor cells forming an acinus. Nuclear chromatin is granular and nucleoli (*arrows*) are large and blue in this stain. (R stain, ×600)

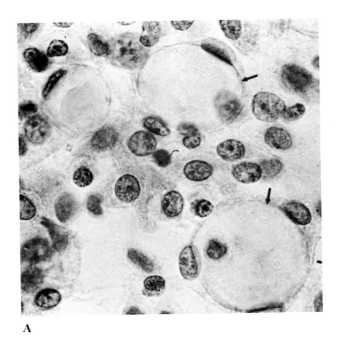

A

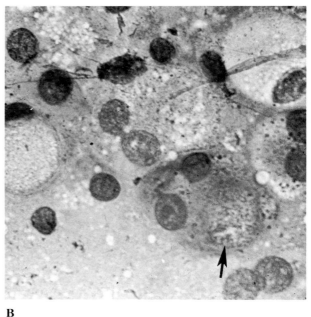

B

Fig. 5-25. A. Adenocarcinoma, mucus-producing adenopapillary type. Note large intracellular globule of mucin (*arrow*) and prominent nucleolus. (P stain, ×600) **B.** Adenocarcinoma, mucus-producing, adenopapillary type, signet ring cells (*arrows*). This is same material as **A.** (R stain, ×600)

differentiated carcinoma cells cannot be distinguished. All primary sites from which metastatic neoplasm could originate must be considered. In addition, we have seen one patient with breast carcinoma metastatic to the parotid gland, with the primary tumor not identified until necroscopy.

Malignant Lymphoma

ABC can identify lymphoma and permit limited biopsy to determine diffuse versus nodular structure. Treatment with radiation instead of radical surgery can follow.

The literature describing primary lymphoma of the salivary glands is sparse. Hyman and Wolff collected 31 cases before 1976 and then added an additional 33.[29]

More recently, 40 lymphomas in salivary gland tissue out of a total of 2340 salivary gland tumors (1.7%) were reported. Most arose within intraparotid lymph nodes. Four occurred in association with benign lymphoepithelial lesions. In none of these was the sicca syndrome present. FNA was not used in

diagnosis, and many were treated with superficial parotidectomy and enucleation.[23]

The cytologic range of diagnosis is reviewed in Chapter 14. Non-Hodgkin's lymphoma cells intermixed with salivary ductal tissue is seen in Figure 5-26. The differential diagnosis included lymphoepithelial lesions (see above), sialoadenitis, adenolymphoma, and benign hyperplastic intraparotid lymph nodes.

UNUSUAL LESIONS

A study of ectopic salivary tumors was presented in 1956.[26] The cytopathologist must be prepared to diagnose salivary lesions in unusual sites and unusual tumors in salivary gland sites. Gnepp and colleagues reported on small cell tumors of salivary gland.[24]

Several unusual lesions have been presented by Katz.[30] Tumors of the supporting tissues, as well as unexpected inflammatory lesions, may yield surprising cytologic smears.

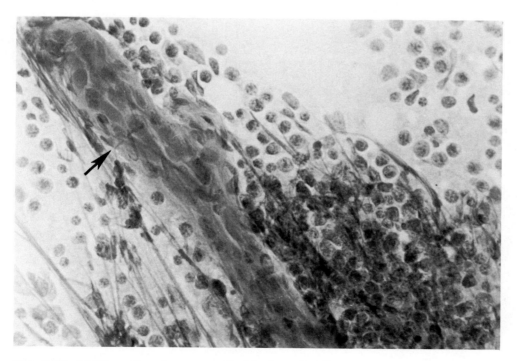

Fig. 5-26. Malignant lymphoma, non-Hodgkin's, well-differentiated lymphocytic. A column of ductal epithelium *(arrow)* is invested with a broad sheet of small-to-medium lymphocytes. This cytologic picture must include lymphoepithelial lesion in the differential diagnosis. (P stain, ×190)

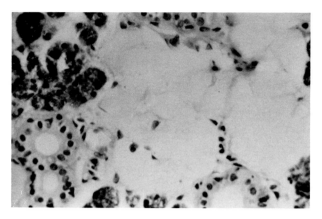

Fig. 5-27. Lipoma of parotid. Tissue section. Typical lobulated fat within the substance of the parotid. (H & E stain, ×600)

Lipoma

A lipoma may present as an intraparotid lobulated mass and can be identified by FNA. The cytology is not differentiated from that of soft tissue lipomas, which are described in Chapter 15 (Fig. 5-27). Such a lesion may be spared surgery.

Extracranial Meningioma

An extracranial meningioma may present as a firm subcutaneous mass behind the ear, simulating a tumor of the posterior lobe of the parotid gland. ABC is identical to that obtained by puncture of intracranial meningiomas. Cytology is illustrated in Figure 5-28.

CLINICAL APPLICATION

Inspection and palpation, followed by FNA of salivary gland targets, yield a rich diagnostic harvest. Knowledge of the pathologic diversity of these lesions is an important prerequisite to undertaking cytologic diagnosis. Cytologic separation of benign and malignant lesions and subdivision into pathologic type requires extensive experience.[50]

Pre-operative identification of suspected neoplasms offers distinct clinical advantages by answering the following questions, which have been summarized from the literature by Webb.[48]

Is the lesion a neoplasm?
Is surgery or radiotherapy indicated?

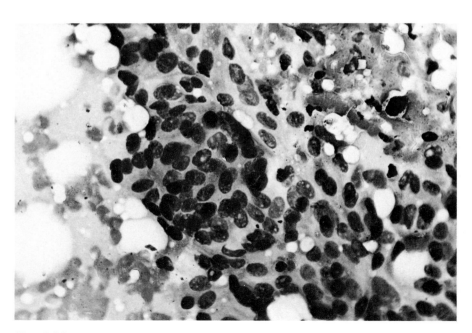

Fig. 5-28. Extracranial meningioma. This was a cellular lesion yielding clusters of oval cells that formed whorls identical to intracranial meningioma. (R stain, ×132)

How radical should surgery be?
Is pre-operative radiotherapy advisable?

No surgeon or oncologist can object to receiving this information as an aid in evaluating a mass. Finally, in the context of a known primary lesion or systemic disease (sarcoidosis, Sjögren's disease), ABC offers extremely useful information.

REFERENCES

1. Anderson WAD, Kissane JM: Pathology. St Louis, CV Mosby, 1977
2. Aufemorte TB, Ramzy I, Holt GR et al: Focal adenomatoid hyperplasia of salivary glands. Acta Cytol 29:23, 1985
3. Azzopardi JG, Smith OD: Salivary gland tumors and their mucins. J Pathol Bacteriol 77:313, 1959
4. Berge T, Söderström N: Fine-needle cytologic biopsy in diseases of the salivary glands. Acta Pathol Microbiol Scand 58:1, 1963
5. Bernier JL, Bhaskar SN: Lymphoepithelial lesions of salivary glands. Histogenesis and classification based on 186 cases. Cancer 11:1156, 1958
6. Bondeson L, Lindhom K, Thorstenson S: Benign dermal eccrine cylindroma—A pitfall in the cytologic diagnosis of adenoid cystic carcinoma. Acta Cytol 27:326, 1983
7. Bottles K, Löwhagen T, Miller TR: Mast cells in the aspiration cytology differential diagnosis of adenolymphoma. Acta Cytol 29:513, 1985
8. Chen S–Y, Brannon RB, Miller AS et al: Acinic cell adenocarcinoma of minor salivary glands. Cancer 42:678, 1978
9. Coulson WF: Surgical Pathology, Philadelphia, JB Lippincott, 1978
10. Eby LS, Johnson DC, Baker HW: Adenoid cystic carcinoma of the head and neck. Cancer 29:1160, 1972
11. Eneroth CM, Franzen S, Zajicek J: Aspiration biopsy of salivary gland tumors. A critical review of 910 biopsies. Acta Cytol 11:470, 1967
12. Eneroth CM, Hjertman L, Moberger G: Adenoid cystic carcinoma of the palate. Acta Otolaryngol 66:248, 1968
13. Eneroth CM, Jakobsson P–Å, Zajicek J: Aspiration biopsy of salivary gland tumors. V. Morphologic investigation on smears and histologic sections of acinic cell carcinoma. Acta Radiol (suppl) 310:85, 1971
14. Eneroth CM, Zajicek J: Aspiration biopsy of salivary gland tumors, II. Morphologic studies on smears and histological sections from oncocytic tumors (45 cases of papillary cystadenoma lymphomatosum and 4 cases of oncocytoma) Acta Cytol 9:355, 1965
15. Eneroth CM, Zajicek J: Aspiration biopsy of salivary gland tumors. III. Morphologic smears and histologic sections from 368 mixed tumors. Acta Cytol 19:440, 1966
16. Eneroth CM, Zajicek J: Aspiration biopsy of salivary gland tumors. IV. Morphologic studies on smears and histologic sections from 45 cases of adenoid cystic carcinoma. Acta Cytol 13:59, 1969
17. Evans RW: Histological Appearances of Tumours, 2nd ed. Edinburgh, E & S Livingston, 1968
18. Foote FW Jr, Frazell EL: Tumors of the major salivary glands. In Atlas of Tumor Pathology, Sec. N., Fasc. II. Washington, DC, Armed Forces Institute of Pathology, 1954
19. Frable WJ, Goplerud DR: Adenoid cystic carcinoma of Bartholin's gland: Diagnosis by aspiration biopsy. Acta Cytol 19:152, 1975
20. Frazell EL: Clinical aspects of tumors of the major salivary glands. Cancer 7:637, 1954
21. Gallia LJ, Johnson JT: Incidence of neoplastic versus inflammatory disease in major salivary gland masses diagnosed by surgery. Laryngoscope 91:912, 1981
22. Geisinger KR, Reynolds GD, Vance RP et al: Adenoid cystic carcinoma arising in a pleomorphic adenoma of the parotid gland. Acta Cytol 29:522, 1985
23. Gleeson MJ, Bennett MA, Cawson RA: Lymphomas of salivary glands. Cancer 58:699, 1986
24. Gnepp DR, Corio RL, Brannon RB: Small cell carcinoma of the major salivary glands. Cancer 58:705, 1986
25. Godwin JT: Benign lympheopithelial lesions of parotid gland (adenolymphoma. Chronic inflammation, lympho-epithelioma, lymphocytic tumor, Mikulicz disease)—report of 11 cases. Cancer 5:1089, 1952
26. Harrison K: A study of ectopic salivary tumours. Ann R Coll Surg Engl 18:99, 1956
27. Hoffman E: Branchial cysts within the parotid gland. Ann Surg 152:290, 1960
28. Hood IC, Qizilbash AH, Salama SSS et al: Basal-cell adenoma of parotid—Difficulty of differentiation from adenoid cystic carcinoma on aspiration biopsy. Acta Cytol 27:515, 1983
29. Hyman GA, Wolff M: Malignant lymphomas of the salivary glands. Am J Clin Pathol 65:421, 1976

30. Katz AD: Unusual lesions of the parotid gland. J Surg Oncol 7:219, 1975
31. Layfield LJ: Fine needle aspiration cytology of a trabecular adenoma of the parotid gland. Acta Cytol 29:999, 1985
32. Layfield LJ, Tan P, Glascow BJ: Fine-needle aspiration of salivary gland lesions. Arch Pathol Lab Med 111:346, 1987
33. Levin PA, Falko JM, Dixon K et al: Benign parotid enlargement in bulimia. Ann Intern Med 93:827, 1980
34. Livols VA, Perzin KH: Malignant mixed tumors arising in salivary glands. Cancer 39:2209, 1977
35. Marsh WL, Allen MS Jr: Adenoid cystic carcinoma. Cancer 43:1463, 1979
36. Mavec P, Eneroth CM, Franzen S et al: Aspiration biopsy of salivary tumors. I. Correlation of cytologic reports from 652 biopsies with clinical and histological findings. Acta Otolaryngol 58:471, 1964
37. Mikculicz J von: Concerning peculiar symmetrical disease of lachrymal and salivary glands. Med Classics 2:165, 1937
38. O'Dwyer P, Farrar WB, James AG et al: Needle aspiration biopsy of major salivary gland tumors. Cancer 57:554, 1986
39. Palma O, Torre AM, deCristofaro JA et al: Fine-needle aspiration cytology in two cases of well-differentiated acinic-cell carcinoma of the parotid gland. Acta Cytol 29:516, 1985
40. Persson PS, Zettergren L: Cytologic diagnosis of salivary gland tumors by aspiration biopsy. Acta Cytol 17:351, 1973
41. Perzin KH, Gullane P, Clarimont AC: Adenoid cystic carcinomas arising in salivary glands. A correlation of histologic features and clinical course. Cancer 42:2650, 1978
42. Perzin KH, LiVols VA: Acinic cell carcinomas arising in salivary glands. Cancer 47:1434, 1979
43. Qizilbash AH, Sianos J, Young JEM et al: Fine needle aspiration biopsy of major salivary glands. Acta Cytol 29:503, 1985
44. Sidhu GS, Forrester EM: Acinic cell carcinoma: Long term survival after pulmonary metastases. Cancer 40:756, 1977
45. Takenchi J, Sobue M, Yoshida M, et al: Glycosaminoglycansynthetic activity of pleomorphic adenomas, adenoid systis carcinoma and non-neoplastic tubloacinar cells of the salivary gland. Cancer 42:2020, 1978
46. Thackray AC, Lucas RB: Tumors of the major salivary glands. In Atlas of Tumor Pathology, Sec 16. Washington, DC, Armed Forces Institute of Pathology, 1974
47. Watt J: Benign parotid swellings: A review. Proc R Soc Med 70:583, 1977
48. Webb AJ: Cytologic diagnosis of salivary gland lesions in adult and pediatric surgical patients. Acta Cytol 17:51, 1973
49. Weidner N, Geisinger KR, Sterling RT et al: Benign lymphoepithelial cysts of the parotid gland. Am J Clin Pathol 85:395, 1986
50. Zajicek J: Aspiration Biopsy Cytology, Part I, Basel, S Karger, 1974
51. Zajicek J, Eneroth CM, Jakobsson P: Aspiration biopsy of salivary gland tumors. IV. Morphologic studies on smears and histologic sections from muco-epidermoid carcinoma. Acta Cytol 20:35, 1976

CHAPTER 6

Breast Aspiration

JOSEPH A. LINSK

SIXTEN FRANZEN

The breast is a surface organ, easily palpated and readily aspirated. Pathologic changes of its glandular and stromal components are extremely common. In 1951, Frantz and colleagues studied breast pathology in 225 autopsies of women with no clinical evidence of breast disease. Gross and microscopic cystic disease, including apocrine metaplasia and abnormal epithelial proliferation, were found in more than 50% of the cases.[27] In an autopsy study by Nielsen and colleagues, 1 invasive and 14 *in situ* carcinomas (25.4% of 77 cases) were found.[65] In addition to the enormous reservoir of benign pathology, cancer of the breast is the most frequent malignancy and the leading cause of cancer death in women.[5]

Breast palpation is a singular challenge to the clinician because of the large number of borderline lesions, and clinical breast evaluation must address the following crucial question: *"When is a palpable breast abnormality a tumor rather than an inflammatory or hyperplastic disorder?"* The answer to this question in traditional management of breast tumors has been surgical biopsy. In this tradition, 300 patients with a low clinical index of suspicion of malignancy were surgically biopsied. Twenty-five (8.5%) palpable lesions were pathologically malignant and 275 were benign.[77] Contrast this approach with the aspiration biopsy cytology (ABC) diagnostic method in which, with combined clinical, radiographic, and cytologic findings, diagnostic error is less than 1%.[49] Further, dividing palpable lesions into groups — clinically benign, questionable, and clinically malignant — allows a more rigorous evaluation of negative smears. Patients with clinically benign lesions that yield negative smears should be maintained under careful clinical follow-up without surgery.[43,47,84,99] The level of accuracy for benign palpable lesions is high.[78] This will spare the large burden of surgical procedures.

Excellence in obtaining and interpreting cytologic smears clearly improves with experience. The relationship of clinical factors to cytology is highlighted by the results in a study of 689 primary malignancies reviewed by Barrows and colleagues. The most significant factor that determined success of the procedure was the particular physician performing the aspiration.[6] Relating cytologic findings to multiple clinical aspects of the lesion and the patient further strengthens diagnostic confidence and establishes an important link with the clinician who is responsible for the surgical or nonsurgical decision. Griffith and co-workers have stressed the efficiency of aspiration as an office procedure.[35]

The cytologic aspiration technique evolved haltingly as an adjuvant diagnostic method over several decades.[31,58,73,100] The literature began to multiply in the early 1970s. Reports of clinical studies now appear on a regular basis, demonstrating the efficacy and consistency of the method in diagnosis of both benign and malignant disease.[1,6,25,26,30,34,42,49,50,53,67,82,84,88,95,98,99,103–105]

Aspiration has been appropriately linked to inspection and palpation.[16] Because of its simplicity, aspiration may be considered a probing extension of the physical examination. The clinical factors contributing to the final diagnosis have been elucidated in detail.[52]

Aspiration of breast lesions provides material that allows the clinician to diagnose, clarify, or exclude the vast majority of the following pathologic disorders.

Benign	Malignant
Gross cysts	Ductal carcinoma (NOS)*
Benign mammary dysplasia (BMD)	Colloid carcinoma
Fibroadenoma	Apocrine carcinoma
Papilloma	Medullary carcinoma
Lipoma	Papillary carcinoma
Granular cell myoblastoma	Lobular carcinoma
Fat necrosis	Cystosarcoma phyllodes
Abscess	Lymphoma
Unilateral gynecomastia	Carcinosarcoma
	Angiosarcoma
Vague nodular masses (NOS)*	Paget's disease (smears obtained by scraping)
	Nonpalpable disease

Identification of benign and malignant smears has been considered to be the only information derived from fine needle aspiration (FNA).[8,76] As the clinician gains experience, however, delineation of cell types and prediction of histologic pattern often become possible.

FNA findings provide morphologic data and, more importantly, valuable guidelines for clinical decisions.

Indications for Aspiration

To confirm an obviously clinically inoperable cancer (including inflammatory carcinoma) for possible radiation therapy

To distinguish diffuse suppurative disease from neoplastic disease

To identify positively clinically operable lesions as cancer, thus obviating frozen section, reducing pre-operative anxiety and anesthesia time, and aiding in convincing a reluctant patient of the need for surgery

To add negative cytologic findings to the clinical impression of benign disease in patients with vague nodularities and ill-defined thickenings[34]

To add negative cytologic findings to the clinical decision to excise questionable masses without frozen section

To establish clearly the diagnosis of fibroadenoma in at least 70% of young women, which allows postponement of surgery to a more opportune time (e.g., school holidays)[53]

To establish clearly the diagnosis of cyst with negative cytology of the fluid and of material aspirated from any residual mass

To establish clearly the diagnosis of granular cell myoblastoma[28,54,92]

To analyze estrogen receptor on needle aspirates from human mammary carcinoma[81,82]

To establish a positive diagnosis of cancer before initiating pre-operative irradiation or chemotherapy[97]

NEO-ADJUVANT CHEMOTHERAPY

Studies have shown that surgical biopsy or mastectomy results in generalized tumor cell proliferation, leading to increasing numbers of resistant cells.[83] For that reason, chemotherapy has been initiated preoperatively or perioperatively in several well-studied series.[66,72] Diagnosis in these cases should be rendered by FNA. This may become an important mandatory use for the method.

BREAST ANATOMY AND TOPOGRAPHY: PALPATION AND PUNCTURE

An understanding of the specific anatomic structure of the breast at different ages is useful in evaluating nodularity and thickenings that might become targets for ABC. Nodularity, not usually present in the young breast, results from enclosure of fat lobules by suspensory (Cooper's) ligaments.[14] Aspiration will yield fat. In the fat-depleted breast, the glandular structures are prominent and finely nodular but firm. It is important to be aware that nodularity, including tumors, may be present in any part of the breast, since

---

* Not otherwise specified.

tongues of flattened glandular tissue extend to the sternum, clavicle, and axilla.

In the upper outer quadrant, tissue may be thickened and nodular from aggregation of glandular and connective tissue.[36] However, it is also the most common site for carcinoma. In the inframammary ridge, often thickened by connective tissue, carcinoma must be excluded.

In office and clinic practice, the single most common source of cytologic smears is the "lumpy breast." Such breasts are considered clinically benign but have created patient anxiety, as well as physician concern a cancer may be present. The aspiration is carried out to confirm a benign impression.[34] A benign aspirate coupled with clinical impression and careful follow-up will eliminate many surgical biopsies. For the laboratory pathologist, the smears are often disappointing. The content ranges from fat to a few epithelial cells to sheets and clusters of regular epithelium accompanied by bipolar stripped nuclei. The latter pattern may be difficult to distinguish from fibroadenoma (see below). A well-defined, sharply circumscribed mobile mass in a young woman is characteristic of fibroadenoma and aids in the distinction.

Most breast targets for aspiration are discovered by palpation,[91] and its importance cannot be overemphasized. The technique of palpation is often a personal method developed by each clinician. Some considerations based on personal experience are listed below:

1. Ask the patient to point out the area of concern or the area noted by the primary physician, if any.
2. Be aware that small lesions palpable in the upright position may become imperceptible with supine position.
3. Soaping the breast to break surface tension may be revealing (Fig. 6-1).
4. Examine anterior and posterior folds as well as deeper structures of the axilla with patient's arm lying loosely near the body or resting on examiner's shoulder.

Experience has demonstrated that attention to the following simple rules and observations will yield the most optimum smears. *Failure to obtain representative material is the major pitfall in ABC of the breast.*

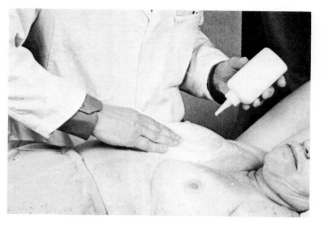

Fig. 6-1. Soaping of the breast followed by stroking palpation will highlight minimal or elusive lesions.

1. When the mass is deep within the breast, initially puncture the covering fat, approach the mass, and then aspirate in the usual manner.
2. When the skin is dimpled and attached to the outer surface of the mass, insert the needle directly through the skin into the tumor.
3. Firm, dense, gritty, and scirrhous tumors may be tough and resistant and require considerable force to penetrate. This may be painful, possibly indicated by a grimace only. Firm grasp of the mass, particularly if deep within the breast, and forceful movement and aspiration may be necessary for extraction.
4. Resistance to puncture may be imperceptible with benign lesions, particularly lipomas, but may vary according to the extent of the fibrous stroma.
5. Approach vague superficial nodules, inflammatory carcinoma, and suspected pagetoid skin by a superficial subepidermal aspiration with an angled 25-gauge needle. If the surface has been broken by Paget's disease or tumor erosion, scraping with a scalpel and smearing will give diagnostic material morphologically similar to aspirated cells. With small nodularities, multiple punctures will produce a higher yield of positive material.
6. Local anesthesia is useful but not necessary in lesions around the nipple and in gynecomastia.
7. Occasionally, clinically obvious breast carci-

noma fails to yield diagnostic cells. Aspiration of a palpable node may settle the issue.

8. Gross cysts tense with fluid may present as masses indistinguishable from solid tumors, although they usually feel more rubbery. After fluid evacuation, aspirate all residual masses with a fresh syringe and needle.

9. We have noted, with sufficient frequency to record it as a useful clinical observation, that the rapid appearance of blood at the puncture site suggests malignancy.

10. Finally, as with all aspirations, clinical data must accompany referred slides.

PRACTICAL ASPECTS OF ASSESSING BREAST ASPIRATION SMEARS

The perspective of the microscopist, cytotechnologist, or cytopathologist must be seasoned by factors other than pure morphology. If not, there will be considerable frustration, disappointment, and bias (overreading and underreading) in formulating a diagnosis.

The history and physical character of the mass are useful and often crucial data. Table 6-1 summarizes such data.

Table 6-1. Correlation of Clinical Data and Pathology

CLINICAL FINDINGS	EXPECTED PATHOLOGIC DIAGNOSIS
Mobile, firm, spherical mass in the young patient	Fibroadenoma
Tender, ill-defined mass posttrauma	Fat necrosis or hematoma
Large, fixed fungating mass	Cancer
Subareolar inflammatory mass[80]	Abscess
Ill-defined nodularity in young patient (only 1% to 2% of all breast cancer occurs in patients 30 years of age or younger)[68]	Benign tissue
History of metastatic melanoma and bronchogenic carcinoma	Suspected metastasis
Known non-Hodgkin's lymphoma	Suspected lymphoma

Submitted slides should be accompanied at the very least by the designation "clinically benign" or "clinically malignant." Additional information, as in Table 6-1 adds to the accuracy of reading.

The Unsatisfactory Smear

Smears with unrecognizable or traumatized cells or cellular and other debris may be deemed unsatisfactory. However, fibrotic lesions, acellular portions of fibroadenomatosis, fat, and acellular fluids may be quite satisfactory within a particular clinical context and do not necessarily require surgery.

The lumpy or nodular breast of young women in particular may yield minimal cells, which confirms the clinical impression and reinforces follow-up rather than surgery.[84,99] Unsatisfactory smears or those "suspicious for cancer" far more commonly originate from clinicians attempting the method for the first time or with limited experience. Surgeons with limited commitment to the method who have already decided to proceed with surgery will more often submit inadequate material.[6] The microscopist should maintain open communication with the clinician in an effort to improve the yield and quality of the smears. Particularly in beginning to report out smears, the temptation to provide a diagnosis to confirm the usefulness of the method even with less than optimal material must be resisted. Multiple smears stained with both methods coupled with clinical data provide the best foundation for optimum diagnosis.

Although surgical biopsy is mandatory in clinically malignant or suspicious but cytologically benign cases, as well as with a cytologic diagnosis listed as suspicious for cancer, reliance on surgery rather than careful attention to all of the above factors sharply reduces the usefulness of the method.

MINIMAL CANCER AND BORDERLINE LESIONS: RADIOGRAPHIC CONSIDERATIONS

The smaller the primary tumor and the lower the percentage of axillary metastases, the higher is the cure rate.[4,22,36] Minimal cancer refers to the physical size of the lesion, which is generally less than 1 cm.[3] Minimal cancer is definitive malignancy and should not be confused with "precancerous" or borderline lesions, which include intraductal proliferations and lobular carcinoma in situ. These lesions are part of the continuum from benign alterations, which are part of the spectrum of florid fibrocystic disease, to atypical intraductal hyperplasia, which ultimately metastasizes.[9]

Cytologically, minimal cancer may be highly malignant; and in 50% of patients metastases occurred at least 2 years before the primary tumor was clinically detected.[96] Problems of cytologic interpretation parallel those of pathologic interpretation, and clearly there are morphologic changes in both disciplines that must be termed borderline. However, the cytologic diagnosis of cancer in minimal lesions is regularly made (Color Plate X). Histologic studies are still the reference point for establishing the diagnostic accuracy of cytologic smears. Borderline or precancerous lesions diagnosed cytologically must be confirmed by pathologic examination, even though false-negative histologic results in the presence of positive cytologic findings have been reported.[29]

Mammography is discussed as a radiographic tool in Chapter 2. From a clinical standpoint, mammography is clearly an adjuvant diagnostic tool, and primary reliance on it may be hazardous.[30] False-negative and false-positive reports are frequent.[12,30,47] When used in conjunction with clinical examination and aspiration cytology, however, it is extremely valuable.[49,84] It is the only technique for detecting nonpalpable disease, which would include most minimal cancers. However, minimal cancer is not necessarily nonpalpable. In one study, 17% of minimal breast cancers were detected by physical examination alone.[63] It is evident that cancer can be detected by both radiography and physical examination. Diagnosis, however, requires biopsy. A stereotaxic method requiring considerable instrumentation has been successfully employed.[67]

Both mammographic and physical criteria for significant disease are highly variable. Surgical biopsy on minimal evidence relieves anxiety but results in a large burden of surgery. FNA helps in the determination of when the surgical approach is warranted.[21,67]

When radiographic evidence indicates that a lesion is highly suspicious, (e.g., the presence of punctate calcifications), it should be surgically biopsied regardless of the findings on FNA. When surgery is contraindicated for medical reasons, however, FNA of such lesions may still be indicated. On the other hand, when a lesion appears to be benign on mammography, a clinical decision may be made to follow patient clinically and radiographically rather than doing a surgical biopsy. FNA may confirm this nonsurgical approach. Between these extremes are a variety of radiographic and clinical findings that give ambiguous signals about surgery. Here FNA is a useful adjuvant tool.

Mammography after needle aspiration can result in false-positive readings because of hematomas. Radiographic changes remain for 2 weeks after the procedure.[79]

BENIGN DISORDERS

Aspiration of the clinically and radiographically normal breast is not done, and the diagnosis of so-called normal epithelium is not a practical problem. Physical irregularity of a "normal" breast may, however, yield only fat and a few benign epithelial cells—generally naked nuclei—from ABC. This type of smear cannot be distinguished from minimal BMD (see below).

Cysts

Cysts usually yield a clear yellow-green or tan-brown fluid. Turbidity is sometimes present, indicating that there is cellular content requiring examination. Clear fluid can be safely discarded if there is no residual mass.[46] This fluid arises from simple cysts lined with a single layer of epithelium. Turbid fluid is spun down, and the sediment is smeared. This arises from papillary cysts, in which overgrowth hyperplasia and infolding have taken place (Fig. 6-2). This pathologic change may result in the presence of phagocytes and inflammatory cells. Plugs and groups of benign apocrine, nonapocrine, and foam cells may be seen (Figs. 6-3 and 6-4) in the smear. In patients with gross cysts, the incidence of carcinoma is increased 3.5 times over a period of 13 years.[39] While carcinoma is rarely present in the wall of cysts, in the presence of frank blood or inflammation, cystic fluid should be examined carefully for carcinoma.[95] The presence of carcinoma cells in the sediment of non-bloody breast cyst fluid, although clearly uncommon, is not unknown. Figure 6-5 shows clusters of carcinoma cells intermixed with acute inflammatory cells harvested from breast cyst fluid. Sometimes a small carcinoma in a cyst wall is not detected, but this becomes a matter of clinical follow-up, since surgery is not routinely recommended without positive cytologic findings and residual palpable disease. The presence of large clus-

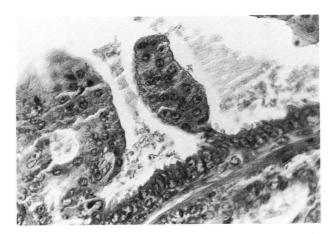

Fig. 6-2. Papillary cyst wall hyperplasia. Tissue section. Papillary masses may separate and become floating sediment in cyst fluid. (H & E stain, ×400)

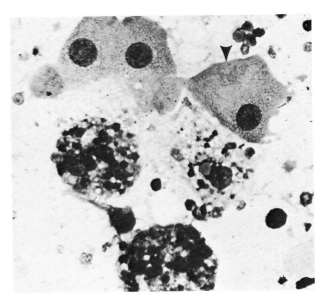

Fig. 6-4. Benign breast cyst, cellular content. There are several pigment-laden histiocytes along with granulated benign apocrine cells *(arrow)*. The background is turbid with debris and cell fragments. (R stain, ×600)

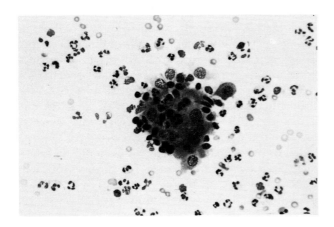

Fig. 6-3. Content of benign breast cyst with inflammatory reaction. Marked atypia of epithelial cells secondary to acute inflammation. Changes may simulate carcinoma (see Fig. 6-5). Histiocytic reaction may also simulate epithelial proliferation. (R stain, ×600)

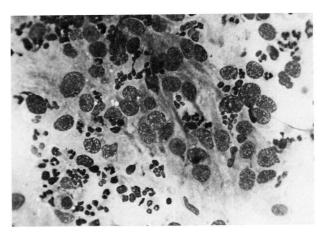

Fig. 6-5. Cystic breast carcinoma. Fairly dense, partially degenerated clusters of carcinoma cells intermixed with acute inflammatory cells. Note nucleoli in dispersed polymorphous cells (compare with 6-3). (R stain, ×300)

ters of *benign* ductal cells in cyst fluid smears has been found suspicious for carcinoma and may mandate surgery.[95] (See Color Plate VIII A.)

Benign Mammary Dysplasia

BMD consists of abnormal proliferation and cytologic variation in benign ductal and acinar epithelium and stroma. Various histopathologic varieties have been associated with increased hazard of malignancy, including a number of epithelial hyperplasias with and without papillary and cystic formations.[2,15,20,36,62,74,94]

Donnelly and colleagues have noted the relationship of benign breast lesions to subsequent cancer in women in Rochester, Minnesota.[17] In a series of cases followed over 30 years, Monson and coworkers demonstrated that chronic mastitis increased the risk of dying of cancer.[62] The positive association of benign breast lesions with carcinoma was discussed further by Black and colleagues.[10] A positive diagnosis of BMD is therefore important.

In evaluating the cytology of these benign smears, consider carefully the complexity of the histopathology of BMD. Gradual alteration from benign to hyperplastic to dysplastic and then to malignant complicates both histopathologic and cytopathologic diagnosis (Color Plate VIII B). Tissue sections as well as cytologic smears may show clearly benign, suspicious, and malignant areas. Carcinoma may be present in the wall of the benign cyst, in part of a hyperplastic duct, and adjacent to an area of sclerosing adenosis. Retrospective studies have demonstrated carcinoma in previously "benign" specimens.[70] Cytologic diagnosis of BMD and distinction from carcinoma must therefore rely on strict criteria. In histologically proven cases of BMD and FAN, 17.6% to 50% have been identified cytologically.[48,53]

Two cytologic problems must be addressed: the cytologic characteristics of BMD and distinguishing benign from malignant smears.

Cytologic Characteristics

The dominant cell types and structures of BMD include the following:

Apocrine Cells. Apocrine cells have abundant, well-preserved, gray-blue cytoplasm, usually densely stained with Romanowsky (R) stain and often show a fine granularity arranged in clouds and tails (Fig.

6-6). The nuclei are eccentrically placed and very regular in size and usually have a prominent central nucleolus. In Papanicolaou (P) stain the cytoplasm is eosinophilic. The cells are easily identified as benign in sheets; however, there may be marked atypicality (Fig. 6-7), which raises the question of carcinoma.

Cellular Smears. Areas of sclerosing adenosis, which may be confused both clinically and histopathologically with cancer, are the most common source of cellular smears (Fig. 6-8). They are often rich in clusters and have atypia that may suggest cancer (Fig. 6-9). Microacinar structures embedded in loose stroma may be extracted. Intact cells are rarely seen singly because of high cellular cohesion. Nuclei are small, oval, or round; in regular sheets; and usually clearly benign morphologically.

Sparse Smears. In a review of 210 cases of BMD, more than 80% of the cytologic smears contained few or no cells. The poor yield was attributed to fibrosis. Poorly delimited fibrotic nodules are the most common indication for surgery in BMD.[53] In contrast,

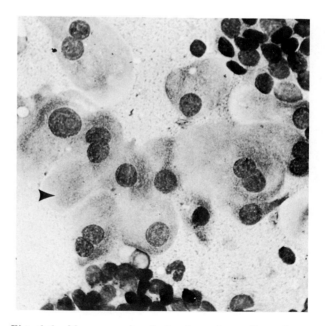

Fig. 6-6. Mammary dysplasia. Apocrine cells with eccentric nuclei, some double, have well-defined cytoplasm within which finely stippled granules are aggregated into swirled, pointed, and rounded shapes. Small uniform epithelial cells are present. (R stain, ×400)

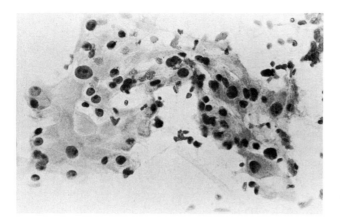

Fig. 6-7. Mammary dysplasia, atypical apocrine proliferation. Shown is an agglutinated cluster of apocrine cells with marked variation in nuclear size. The small nuclear:cytoplasmic ratio is present. Such a smear represents a pitfall in diagnosis because it can be mistaken for cancer. (P stain, ×190)

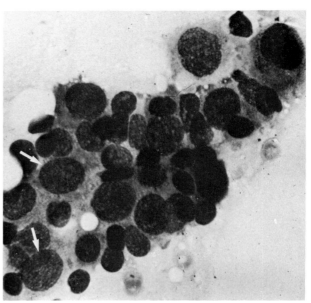

Fig. 6-9. Mammary dysplasia, sclerosing adenosis. Benign epithelium with variation in nuclear size. Nucleoli are present. The degree of atypia (arrows) is highly suspicious of malignant change in the absence of solitary, stripped, oval nuclei, which confers a benign diagnosis. (R stain, ×1000)

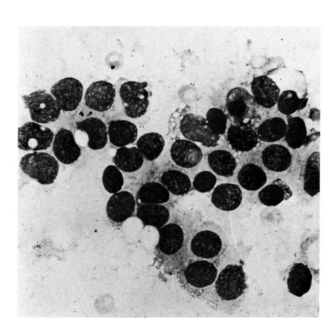

Fig. 6-8. Mammary dysplasia, sclerosing adenosis. Fairly uniform cells appear in a cluster. Cells are discretely separate here, with a vague suggestion of microacinar structures. (R stain, ×330)

Fig. 6-10. Benign mammary dysplasia, intraductal proliferation. Shown are masses of benign cells with interconnecting papillary cellular aggregates. (P stain, ×50)

when mammography was added to the selection process, 51% of 101 cases of BMD were diagnosed cytologically.[48]

Ductal-Type Clusters. Intraductal proliferation yields papillary and solid cell ductal-type clusters (Fig. 6-10). They are overstained with R. Very characteristic tubular and fingerlike projections may be seen (Figs. 6-11 and 6-12). Macrophages, regressive apocrine cells, and a "dirty" background may be present in the same smear.

Distinguishing Benign and Malignant Cells

Cytology of the major malignant lesions and their subtypes is considered below. The following, however, are useful guidelines:

1. Strictly single cells should be looked for initially (see Fig. 6-15).
2. Initial interpretation of clusters can sometimes be difficult (see Fig. 6-9).
3. *Small, naked bipolar nuclei are almost always present in benign smears* (see Fig. 6-14).
4. In malignancy, single cells almost always have round or irregular large- to moderate-sized nuclei and preserved dense cytoplasm. Nucleoli may be present in P smear and occasionally in R (see Fig. 6-30).

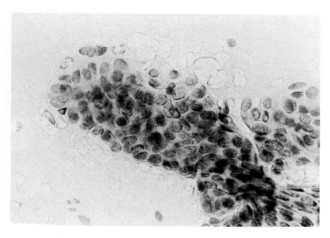

Fig. 6-11. Benign mammary dysplasia, intraductal proliferation. Single finger-like papillary projection of fairly uniform cells. This should be distinguished from duct papilloma (see Figs. 6-20 and 6-21). (P stain, ×150)

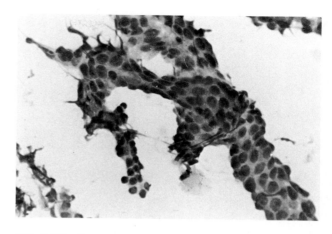

Fig. 6-12. Benign mammary dysplasia, intraductal proliferation. This is seen in interconnected three-dimensional structures. Benign cells are tightly cohesive. (R stain, ×150)

These rules of thumb allow discrimination of benign from malignant cells in more than 90% of cases.

The following comparative features of benign and malignant smears offer further diagnostic aids see Color Plates VIII through XV:

Benign	Malignant
Scanty cell yield	Cellular smears
Separation of fat and epithelial cells	Intermingled fat and epithelial cells (see Fig. 6-38)
Groupings of equal-sized acinar formations	Tubular adenopapillary formations
Regular sheets in an intact, cohered monolayer (Fig. 6-13)	Solid multilayered clusters with overlapping (see Fig. 6-28)
Apocrine cells and phagocytes (see Fig. 6-4)	Lymphocytes and plasma cells (see Fig. 6-47)
Clear background (see Fig. 6-13)	Possible necrosis or stippling (see Fig. 6-39)

Because of problems in sampling, all benign diagnoses must be accompanied by the statement that benign smears do not exclude cancer in the presence of clinically suspicious findings. Although a malignant diagnosis

Fig. 6-13. Mammary dysplasia. Uniform benign cells in an intact, cohered monolayer. Nucleoli are inconspicuous, and background is clear. (P stain, ×200)

can exist without palpable disease, the tumor is not usually a target for FNA (see the section on vague nodular masses). Diagnosis of nonpalpable disease requires imaging (see Chap. 2).

Fibroadenoma

The etiology, pathology, and histogenesis of the fibroadenoma have been described in numerous sources.[14,36,102] It is a benign lesion that by gradation may present with increasing cellularity of the stroma, ultimately verging into both benign and malignant cystosarcoma phyllodes (see below).[36,40] In its florid state, it consists of two cell populations: an epithelial and a stromal component. The stroma is often loose and extractable by aspiration. The yield of epithelium may be voluminous.

Fibroadenomas usually occur in women between pubescence and age 30, with maximal frequency in the 20s. In older women, fibroadenomas tend to lose their epithelial component, remaining as fibrohyalinized masses in which the outlines of the former epithelial pattern can be histologically discerned. Mucous degeneration of a fibroadenoma may occur and cytologically create confusion with a colloid carcinoma. It may also calcify; and, as a result,

there may be no cellular yield; such a lesion can be confused with a sclerotic (scirrhous) carcinoma in which the cell yield may also be scanty.[53]

Although the usual clinical presentation of a pure fibroadenoma is of one or more firm, movable, often spherical masses, it may be found on a smaller scale as part of the complex of BMD. It is important to be aware of this latter fact because BMD carries with it the potential for increased incidence of carcinoma, whereas pure fibroadenoma is not ordinarily a premalignant lesion. Carcinoma in or invading fibroadenoma, however, has been reported.[18,24,33,55]

In one series, aspiration of the mass yielded diagnostic material in 70% of histologically proved fibroadenomas.[53] This is an extremely useful positive diagnosis because it reassures patients, guides therapy, and allows planning of surgery. In older women, particularly those with medical problems, surgery can be eliminated or simple excision without frozen section can be encouraged.

Cytologic Findings

The cytology of fibroadenomas is as follows:

There are monolayers of benign cuboidal cells. The cellularity may be marked, much greater than usually seen in benign lesions (Fig. 6-14). Scattered oval naked nuclei are present, conferring (Figs. 6-14 and 6-15) a benign diagnosis. This is highly characteristic.[103]
Finger-like ductal structures are noted (Fig. 6-14).

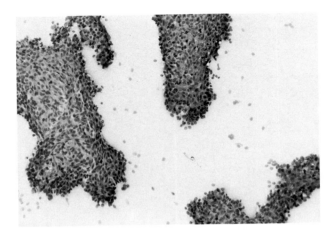

Fig. 6-14. Fibroadenoma. Shown are cellular masses of benign cuboidal cells with scattered, naked, oval nuclei. (R stain, ×50)

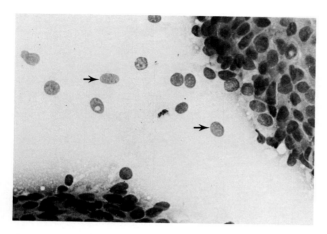

Fig. 6-15. Fibroadenoma. Shown is a cluster of benign epithelium with numerous, isolated, stripped, oval nuclei (*arrows*), which confirm a benign diagnosis. (P stain, ×150)

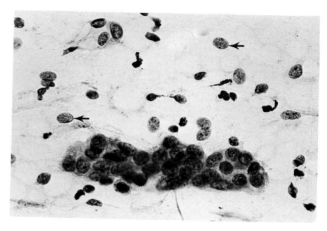

Fig. 6-17. Fibroadenoma. Shown is a cluster of mildly atypical cells with some overlapping. Naked nuclei (*arrows*) are prominent. (R stain, ×150)

Stroma, inconsistently present, in small or large masses (Fig. 6-16), stains eosinophilic with R, characteristic of mucopolysaccharides (Color Plate IX).

Occasional apocrine cells are present.

There are rare atypical epithelial cells in which cancer can be considered (Fig. 6-17).

Adenomas, which appear during pregnancy, are quite cellular with minimal stroma. Caution must be exercised. Figures 6-18 and 6-19 illustrate the cytologic difficulties.

The major distinguishing feature between fibroadenoma and BMD is the increased cellularity of the former, which also has more bipolar naked nuclei and more often has stromal fragments. However, a moderate number of cases of BMD show both clinical and cytologic features of fibroadenoma, including stromal fragments. Attention must always be directed to the clinical context.

In fibroadenoma, the presence of mild cellular atypia with marked cellularity may raise the possibility of carcinoma. The report must then reflect atypia, and surgery becomes mandatory.

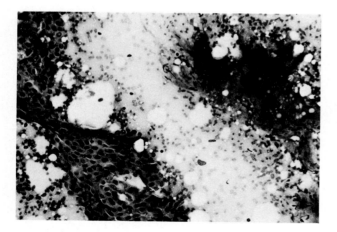

Fig. 6-16. Fibroadenoma. Stromal mass in juxtaposition to epithelial cluster. Elongated and comet-shaped nuclei are highly characteristic. Stripped bipolar nuclei are not very prominent in this area. (R stain, ×560)

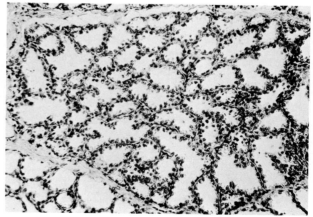

Fig. 6-18. Adenoma, pregnancy. Tissue section. There is minimal stroma, and individual cells can be easily extracted and dispersed. (H & E stain, ×180)

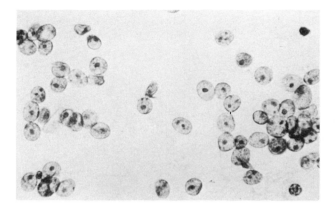

Fig. 6-19. Adenoma, pregnancy. There are dispersed naked nuclei with central nucleoli. Cellularity and immature appearance of the nuclei *may* suggest cancer. The central nucleolus is highly characteristic. (P stain, ×400)

Papilloma

The main clinical feature of duct papilloma is nipple discharge, which often occurs without palpable tumor. However, in Haagensen's series of 160 cases, tumor was palpated by patient or physician in 126.[36] The tumor may be quite small, varying from 0.3 cm to 2 cm.[36] It is almost always subareolar or near the areola. The method of aspiration has been discussed above. A small drop of clear or sanguinous fluid containing groups of cells may be obtained.

Cytologic Findings

Papillary tumors can often be recognized as a group but are difficult to distinguish as benign or malignant. The cytology is contrasted in the following list:

Duct Papillomas
Papillary clusters of monotonous, slightly enlarged, sometimes columnar ductal cells (Fig. 6-20)
Cellular smear with dirty appearance (Fig. 6-21)
Possible cytoplasmic degenerative changes with vacuolization
Foamy phagocytes
Naked oval nuclei or apocrine cells
Dense nuclei, indicating early degeneration (see Fig. 6-20)

Papillary Tumors
Papillary clusters and dense cellular masses of irregular, large, sometimes elongated cells

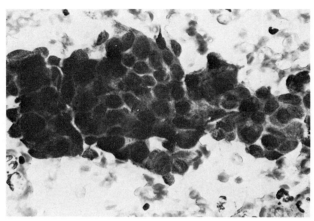

Fig. 6-20. Duct papilloma. This is a tightly agglutinated cluster of dark-staining cells that can be seen three dimensionally as a rounded structure. (R stain, ×560)

with hyperchromatic nuclei (Figs. 6-20 and 6-22) that have been stripped from papillary cores (Fig. 6-23; Color Plate IX D and E)
Variable atypia
Smear rich in blood
Inflammatory reaction (see Fig. 6-21)
Frequent phagocytes
Possible magenta bodies and granulation in the cytoplasm (see below)

Specific diagnosis of carcinoma must be made with caution in the absence of clear-cut characteris-

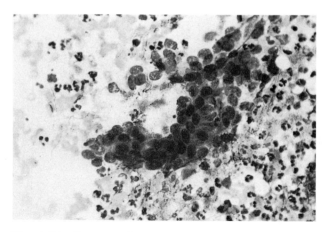

Fig. 6-21. Duct papilloma. This is a solid cluster of benign, fairly uniform cells forming a papillary structure. There is gross inflammation and turbid fluid in the background. (R stain, ×180)

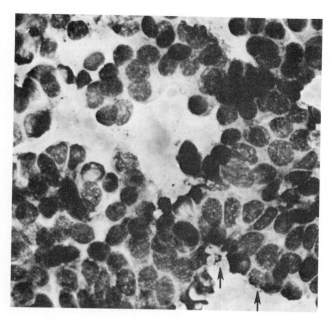

Fig. 6-22. Papillary carcinoma. There are cellular clusters of minimally irregular dense cells. Double columnar arrangement *(arrows)* results from stripping of connective tissue cores. (R stain, ×800)

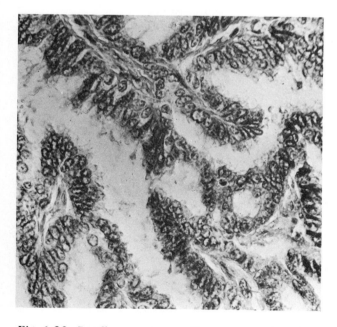

Fig. 6-23. Papillary carcinoma. Tissue section. Papillary fronds yield the cytologic pattern seen in Figure 6-22. Invasion is not evident in this portion. (H & E stain, ×600)

tics of malignancy. Suspicious lesions may be termed *papillary tumor*, and surgical excision may be required.

Breast Masses in Pregnancy and Lactation

Galactocele, adenoma, and carcinoma are all potential diagnoses that can be distinguished cytologically. Aspiration of milk is easily identified grossly. Lactating adenoma yields distinctive cytologic smears. Cells have dispersed naked nuclei, which are fairly uniform and round to oval, containing central prominent nucleoli (see Fig. 6-18). Cytoplasm is absent or filmy and vacuolated. Tissue section presents a delicate alveolar pattern with minimal stroma (see Fig. 6-17). This pattern might be confused cytologically with ductular-acinar carcinoma (see Fig. 6-37).

Ductal carcinoma yields the same cytologic pattern as in tumors of nonpregnant women. The main hazard is failure to realize that carcinoma occurs during pregnancy.[11]

Subareolar Abscess

Subareolar abscess may clinically simulate carcinoma.[80] Cytologic smears contain squames, multinucleated giant cells, acute inflammatory exudate, cholesterol crystals, metaplastic squamous cells, and atypical ductal epithelium.

The clinical picture of an inflammatory lesion often with characteristics of an abscess along with the full cytologic spectrum should sharply reduce the possibility of a false-positive diagnosis based on the atypical ductal cells. Preparation of many smears with both P and R stains is useful if not critical.

Lipomas

Lipoma, a benign tumor, usually occurs as an isolated mass in the breast. It was the pathologic diagnosis in 15 of 1713 mammary lesions reported in two sources.[29,103] The surgical pathology laboratory of the Presbyterian Hospital, New York City, reported 186 over 40 years.[36] Although the lipoma is an encapsulated delimited mass, it may be soft and may blend with the surrounding tissue.

On aspiration, the fatty smear does not differ from surrounding breast fat. It is therefore important to place the needle accurately and make several punctures. Mammography is helpful and may be decisive. Clinically, lipoma may be confused with cystosarcoma phyllodes.[36]

Granular Cell Myoblastoma

Granular cell myoblastoma has been renamed *granular cell tumor* because it is not of myoblastic origin.[36] It is a benign tumor occurring generally at a younger age, when carcinoma is not expected. However, it may be a well-defined mass causing dimpling of the skin and simulating neoplasm. In Haagensen's group of 14 patients, the tumor was in the upper inner quadrant in 11.[36]

ABC presents a quite diagnostic picture.[28,54] With R stain, nuclei are round to oval and cytoplasm is gray-violet and finely and uniformly granulated (Figs. 6-24A and B). With P stain, nuclei are more compact and regular and cytoplasm is pale, eosinophilic, and granular. Granules are PAS positive, diastase resistant.[92] Granular cytoplasm is the important morphologic clue. Tissue section (Fig. 6-24B, Plate VIII F) illustrates the source of the granular cells.

Fat Necrosis and Inflammation

Fat necrosis and inflammation may present as well-delineated masses. They are therefore prime cytologic targets, since a specific cytologic diagnosis will obviate surgery. In fat necrosis, disintegrating fatty tissue is seen as a stippled background. In addition, there are varying mixtures of inflammatory cells, histiocytes (mononuclear and multinucleated), lipophages (Figs. 6-25 and 6-26), and epithelioid cells.[103] Fat cells are dissociated in contrast with normal fat, which appears in clusters (see Color Plate VIII C and D).

Suppurative mastitis yields semisolid purulent material, mixed with inflammatory cells and occasionally bizarre histiocytes mimicking cancer cells. This may result in a false-positive reading.

Tuberculous Mastitis

In a report of nine cases of tuberculous mastitis by Gita,[32] all presented with a mass and three simulated carcinoma. Langhans' and foreign body giant cells, degenerating polymorphonuclear leukocytes, red blood cells, epithelioid cells singly and in clusters, foam cells, plasma cells, and caseous necrosis were present in varying degrees. Acid-fast bacilli were present in each case.

Unilateral Gynecomastia

Unilateral breast hyperplasia in the male may simulate a tumor. It occurs in pubescent and teenaged boys, and, although cancer is essentially nonexistent

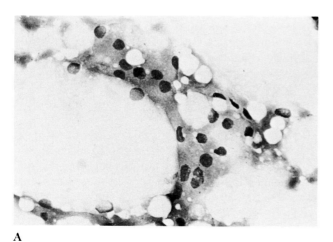

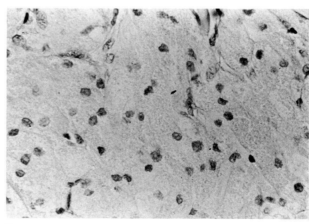

A

B

Fig. 6-24. Granular cell myoblastoma. **A.** Aspiration. Cells appear in a syncytium, invading fat. Nuclei are small and bland. Cytoplasm is granular and gray-violet, characteristic of the tumor. (R stain, ×400) **B.** Tissue section. Scattered nuclei and granular cytoplasm are identified as source of cells of the aspiration. (H & E stain, ×330)

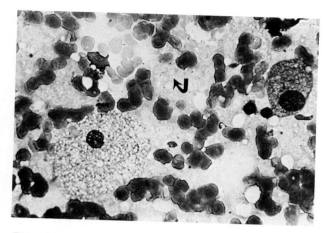

Fig. 6-25. Fat necrosis and inflammation. Stippled background *(curved arrow)* indicates disintegrating fat. Histiocyte and lipophage are evident. (R stain, ×560)

in males under the age of 20, aspiration may be requested or carried out because there is pain or tenderness or to relieve anxiety.[69] This disorder also occurs in adult life for many reasons and must be distinguished from cancer.[38,45,57,93]

The benign lesion presents as a firm, subareolar plaque that is movable below. Movement of the skin over the plaque may be difficult to discern. In male breast carcinoma, retraction of the nipple is present in the vast majority of patients. Nipple discharge occurs in many cases of carcinoma but has also been reported in gynecomastia.[14]

Histologically, there may be marked intraductal hyperplasia.[20] Papillary formation may be present, but duct papillomas have not been seen. There is marked stromal edema with formation of mesenchymal mucopolysaccharides. The stromal changes are similar to those seen in fibroadenoma of the female breast. Fibroadenomas only rarely have been reported in male breasts. Aspiration should be preceded by local anesthesia. The plaque may be tough and resistant, and vigorous penetration with a fine needle may be necessary.

Cytologic Findings

The cytologic picture consists of fairly tight groups of regular benign epithelial cells. The cytology is similar to that seen in BMD (see above) and is illustrated in Figure 6-27. Extraction may be difficult and painful and yield scant or no cells. Smears in carcinoma are identical to ductal carcinoma smears of the female breast.

Vague Nodular Masses

As described above, vague nodular masses may be physiologic fatty nodularity, glandular prominence in a fat-depleted breast, or small foci of BMD. The cytologic picture is largely that of exclusion.

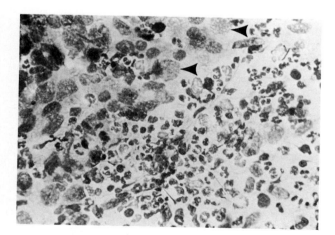

Fig. 6-26. Mastitis. Smear is composed of acute inflammatory cells and atypical histiocytes, which can simulate cancer cells *(arrows)*. (R stain, ×200)

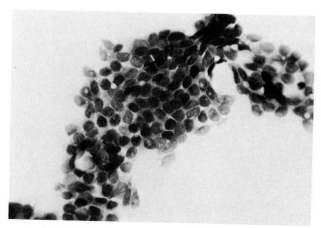

Fig. 6-27. Gynecomastia. This is a characteristic, tightly agglutinated group of benign epithelial cells. Nuclei are regular and round to oval with smooth chromatin and inconspicuous nucleoli. Cytoplasm is scanty. Cells are extracted in groups with difficulty. (R stain, ×132)

MALIGNANT LESIONS

In the cytologic diagnosis of breast malignancy, attention has been directed to two issues: the categorical determination that malignancy is present so that, in the appropriate clinical setting, surgery can proceed without frozen section or radiation can be prescribed; and the identification of histologic subtypes by cytologic examination.[98] Apart from contributions to the investigational aspects of histogenesis and the determination that certain types of cancer (colloid, medullary) carry a better prognosis, there is limited clinical gain in the preoperative cytologic identification of ductal carcinoma subtypes.[98] Case 6-1 illustrates an exception. Preoperative identification of lobular carcinoma, however, is currently useful in planning a diagnostic and surgical approach.

There has been no difference in the surgical approach to tumors identified as colloid and medullary compared to the large group classified as "not otherwise specified" (NOS).[23] The latter group made up more than 50% of 1000 cases reviewed histologically. It is possible, however, that with present and future trends reducing surgery to local excision and with chemotherapy programs demonstrating greater response by certain subtypes, this may become important.

One area of present importance may be the identification of specific cell type in bilateral synchronous and sequential breast cancer.[51] Therapeutic decisions may be guided by this identification, as illustrated in the following case report:

■ *Case 6-1*

A 59-year-old woman was seen in 1975 with a palpable left supraclavicular node. A left mastectomy had been carried out in 1967. On cytologic aspiration, the node was positive for carcinoma interpreted as ductal type (NOS) (Fig. 6-28). The patient was treated with radiation and hormones. In 1980, a mass was palpated in her right (remaining) breast. Aspiration yielded clear-cut apocrine carcinoma, cytologically quite distinct from the original aspirate from the left supraclavicular node (see Apocrine Carcinoma Fig. 6-42 below). This was considered a second primary tumor, and the woman was treated with surgery and radiation.

Ductal Carcinomas

Scrutinizing the smear for bipolar naked nuclei is essential to exclude a benign lesion regardless of cellularity and atypia (Fig. 6-9; see above). By contrast, the free-standing carcinoma cell will have intact cytoplasm. However, for the inexperienced cytopathologist, the following characteristics are most important.

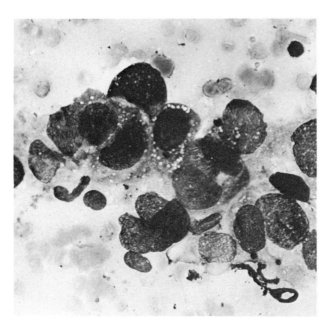

Fig. 6-28. Ductal carcinoma (NOS). Left supraclavicular node. This is a polymorphous smear. The nuclear chromatin is finely granular. There is marked variation in cellular size and shape, with overlapping and indentation of cells. (R stain, ×600)

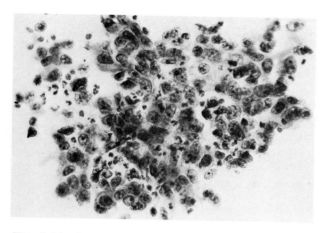

Fig. 6-29. Ductal carcinoma (NOS). Haphazard carcinoma pattern with overlapping. Variation in cell size and shape and prominent nucleoli are evident, but hyperchromasia is not dominant. There is an inflammatory reaction. (P stain, ×190)

Individual Cell Cytology

Enlarged irregular nuclei with hyperchromatism, nuclear variations, and nucleoli are seen in varying prevalence and intensity (Fig. 6-29; Color Plate VIII). Hyperchromatism in P stain may not be prominent and may fail to provide the guidelines that it does in exfoliative cytology. In R stain, nucleoli often do not play a significant diagnostic role. The chromatin pattern may be unremarkable, except that poorly differentiated cells will demonstrate fine granularity of nucleus. Nuclei in general are round to oval, but variation may be striking (Fig. 6-30). Naked nuclei are not seen; the exception is the acinar type as noted below.

Cytoplasm is usually well defined in single cells, although cell borders may be indistinct in larger cell aggregates (Figs. 6-29 and 6-31). It is almost always detectable, even when scanty or pale. *The presence of granulation in carcinoma cells strongly suggests breast carcinoma. It is uncommonly observed in other cell types.* Medullary carcinoma of the thyroid gland, carcinoid, acinic cell carcinoma of the salivary glands, and oncocytoma may harbor granulation in the cytoplasm with R stain. Clinically, they are not likely to be confused with breast cancer.

We have observed a reddish purple cytoplasmic inclusion termed a *magenta body,* which is highly characteristic of breast carcinoma when seen in a primary or metastatic site. This structure may vary from

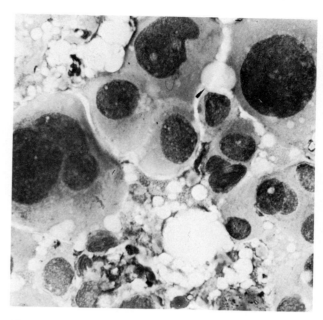

Fig. 6-30. Ductal carcinoma (NOS). Enlarged cells with striking variation in size. Nuclear chromatin is granular and cytoplasm is well defined. (R stain, ×800)

a paranuclear reddish area to a small but better-defined body to a large, very evident structure (Fig. 6-32; Color Plate X D). In a study of imprint cytology of the breast, structures characterized as "peculiar intracytoplasmic inclusions within vacuoles" were found in 0.43% of breast biopsies without carcinoma and in 20.2% of those with carcinoma.[41] They were found to be periodic acid–Schiff (PAS) positive.

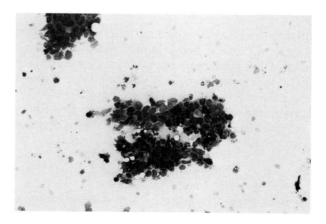

Fig. 6-31. Ductal carcinoma. There is a fairly tight aggregate of overlapping cells in which cytoplasmic borders are not readily evident. (R stain, ×200)

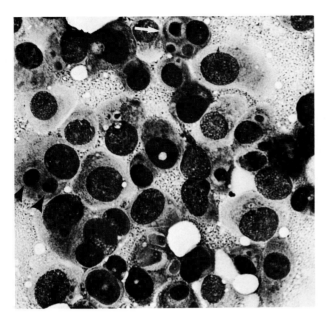

Fig. 6-32. Ductal carcinoma (NOS). Numerous cells contain large, well-defined inclusions staining pink to purple (magenta bodies; *arrows*). Such inclusions are pathognomonic of breast cancer cells and can be identified in metastatic sites. (R stain, ×800)

They were also described as intracytoplasmic lumina in breast carcinoma in a study using electron microscopy (EM).[7] The lumina had a clear, empty appearance or contained eosinophilic proteinaceous material that was PAS positive. They may be the ultrastructural bases for the magenta bodies.

Similar pink staining intracytoplasmic inclusions are found in hepatomas (see Chap. 8). This is a pitfall in the diagnosis of hepatic metastatic breast carcinoma. Squamous differentiation is illustrated in Color Plate XIII D.

Increased Cellularity

Increased cellularity is evident at once on scanning the smear (see Fig. 6-34). In fact, when the smear is made, individual tissue fragments seen as white particles, separate from the fat, will be an initial clue to the cellularity. A well-placed needle will provide up to thousands of tumor cells on one smear. *Attempts at categoric diagnosis of primary breast malignancy on sparse cellular aspirators should be discouraged.*

Cellular Pattern

Because ductal carcinomas are adenocarcinomas, they share the characteristic of nonpolarized, aggregated, overlapped cells. In general, in aspiration material, loss of the orderly, sheetlike, mosaic pattern seen so readily in fibroadenoma favors malignancy. Extraction of an intact polymorphous group is seen in Figure 6-33.

Breast cancer is distinctive in that a significant number of carcinomas (94 to 598) may appear as a monolayer of rounded cells with enlarged nuclei and well-defined cytoplasm (Fig. 6-34; Color Plates XI, XII, XIII, and XIV).[98] The cells do not overlap and suggest the orderly arrangement of benign disease. Such dissociated cells are extracted from ducts crowded with cells without stromal support. This particular pattern is useful in that aspiration of remote metastatic deposits can be identified as of breast origin. This is generally not true of the aggregated overlapping cell groups, which cannot easily be distinguished from other adenocarcinoma.

Cells of the monolayer pattern are often easily aspirated. In contrast, linear cords, acinar structures, and cell plugs composed of enlarged atypical nuclei, albeit somewhat sparse (Fig. 6-35), have generally been extracted forcibly from a fibrous stroma (Fig. 6-36) and are recognizable as a carcinoma pattern.

Separate categories labeled ductular–acinar and acinar have been described.[98] They are distinguished in part by the presence of naked nuclei from which fragile cytoplasm has been stripped on extracting or smearing (Fig. 6-37). These nuclei are large and immature and not unlike the stripped nuclei of

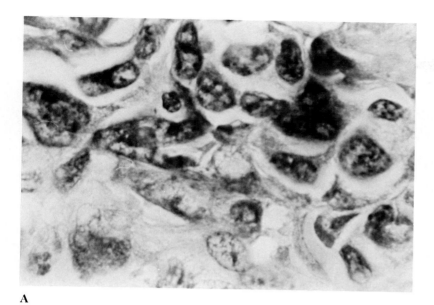

A

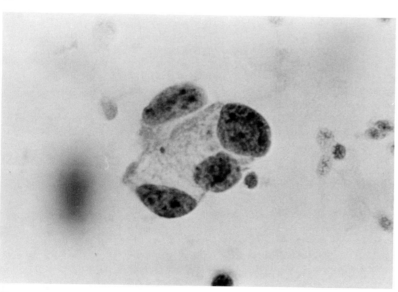

B

Fig. 6-33. Ductal carcinoma. **A.** Tissue section. Malignant cells in organoid groups. (H & E stain, original magnification ×800) **B.** Needle aspirate. Abnormal acinar cluster extracted from **A.** (P stain, ×1000)

poorly differentiated tumors of the prostate (see Chap. 11). The presence of naked nuclei in this subtype, which is a refinement of the original subclassification, has not caused a problem in the overall diagnosis of carcinoma.[103] The final diagnosis is based on the fully realized multifactorial picture. A poorly differentiated tumor with intact cytoplasm in cytologic smear is illustrated in Color Plate XIV A.

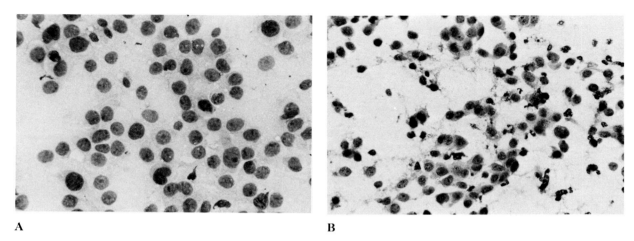

A   B

Fig. 6-34. **A.** Ductal carcinoma. Monolayer pattern. Large cell yield. Cells are dispersed. Variation in nuclei is easily perceived. Nucleoli are present but inconspicuous. Cytoplasm is pale, but cell borders are well defined in some of cells. (R stain, ×600) **B.** Same case stained with P. Cells appear smaller. Nucleoli are prominent. Cytoplasm is irregular. (P stain, ×600)

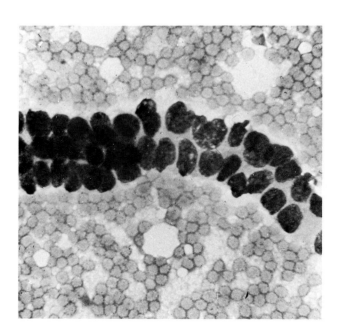

Fig. 6-35. Ductal carcinoma, scirrhous pattern. Carcinoma cells are seen in linear formation (Indian file pattern). Cytoplasm is scant or absent. (R stain, ×600)

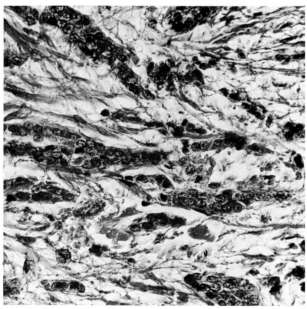

Fig. 6-36. Ductal carcinoma, scirrhous type. Tissue section. Carcinoma cells were trapped in dense fibrous stroma and extracted to produce the pattern in Figure 6-35. (H & E stain, ×400)

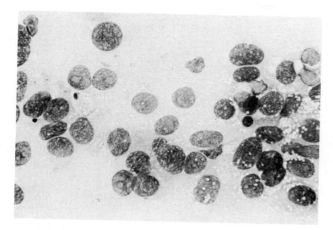

Fig. 6-37. Ductal carcinoma (NOS). This is the so-called ductular–acinar pattern, characterized by cells with fairly immature nuclei and stripped of cytoplasm. These cells are not likely to be confused with the stripped nuclei of BMD and FAN. (R stain, ×600)

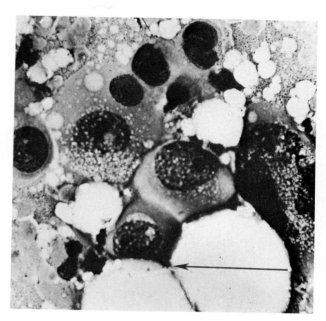

Fig. 6-38. Ductal carcinoma, invasive pattern. Groups of carcinoma cells are intermixed and appear to invade fat lobules *(arrow)*. Vacuolated cytoplasm is not uncommon and suggests a degenerative process. (R stain, ×1000)

Fat and Necrosis in Carcinoma

Additional evidence of malignancy is the intermixture of plugs of abnormal cells with readily identified fat (Fig. 6-38), suggesting invasive carcinoma. This is not seen in benign spreads. This pattern may be a surprise finding on microscopic examination, since the presence of obvious fat on gross examination of the unstained smear suggests a benign process. The aspirate may also consist of thickened semisolid material before it is smeared out. There may be much necrotic debris consistent with comedocarcinoma or intraductal carcinoma.[2] Scattered groupings of identifiable, often degenerated cancer cells are seen on smear (Fig. 6-39).

Colloid Carcinoma

Colloid carcinoma is distinguished by a background of mucinous material staining violet or purple with R (Fig. 6-40) and pink to clear with P (Fig. 6-41). This is important less because of the identification of a distinctive colloid carcinoma with its possibly altered prognosis than as a fairly pertinent indication that cancer is indeed present and that the cells should be reviewed diligently. Tumor cells may be somewhat sparse, in groups, or singly with eccentric nuclei and

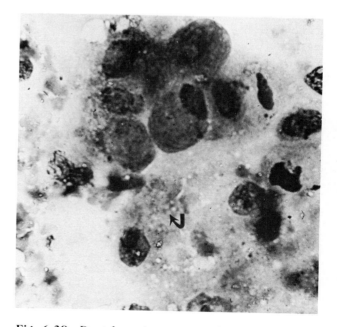

Fig. 6-39. Ductal carcinoma, comedo type. Groups of somewhat degenerated and faded carcinoma cells are present in a background of granular necrotic material *(curved arrow)* and cell fragments. (R stain, ×800)

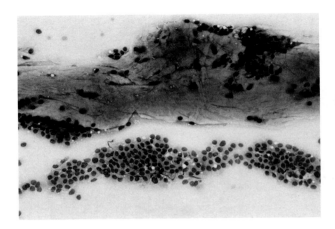

Fig. 6-40. Colloid carcinoma. Dense colloid (staining purple) has trapped small tumor cells. The band of tumor cells below is quite bland with minimal alterations suggesting malignancy. (R stain, ×300)

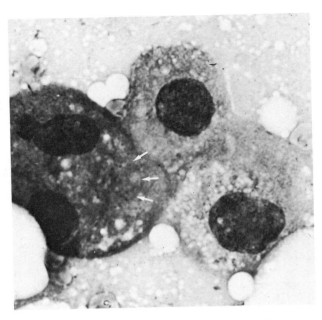

Fig. 6-42. Ductal carcinoma, apocrine type. High-powered view highlights granularity. Purplish inclusion is outlined *(arrows)*. (R stain, ×1000)

occasional vacuoles. Cells may appear columnar with oval to round nuclei or quite polymorphic and obviously malignant (Color Plate XI C, D, and E).

Apocrine Carcinoma

In apocrine carcinoma the cells lie singly and in small groups and sheets. Nuclei are round to irregular and eccentric. Cytoplasm is abundant, finely granular, and gray-pink or violet with R stain. Purplish inclu-

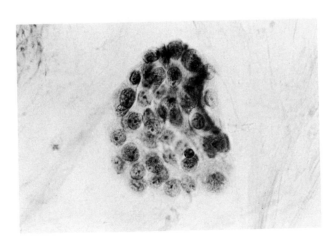

Fig. 6-41. Ductal carcinoma, colloid type. This is an isolated cluster of tumor cells in a mucus background staining pale orange. (P stain, ×200)

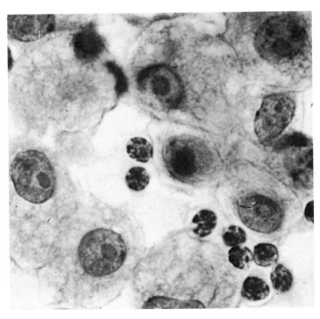

Fig. 6-43. Ductal carcinoma, apocrine type. Nucleoli are striking. Nuclear:cytoplasmic ratio is small. The abundant cytoplasm contains fine eosinophilic granulation. (P stain, ×1000)

sions may be seen (Fig. 6-42). With P stain, nucleoli may be prominent and the cytoplasm has a fine eosinophilic granulation (Fig. 6-43). Atypical, benign apocrine epithelium may be confusing (see Fig. 6-7).

Cytoplasmic tails and snouts are features shared by apocrine metaplastic and neoplastic cells. Nuclear and nucleolar abnormalities are distinguishing features (Fig. 6-42). Histologic study may be necessary. In a study by Mossler and colleagues,[64] apocrine carcinoma cells were found deficient in estrogen receptors. This would suggest that a diagnosis of metastatic apocrine carcinoma would call for chemotherapy rather than hormonal manipulation.

Signet ring cell carcinoma has been shown to be of lobular origin.[61] Some of the cells resemble apocrine carcinoma cells (Fig. 6-44), but the distinct rounded inclusion with central "iris" is easily identified (Fig. 6-45 and Color Plate XV C). The same cells are seen on tissue section (Fig. 6-46).

In reviewing smears with groups of small cells, look carefully for signet ring cells, which may be sparse but will confirm identity of lobular carcinoma (see Fig. 6-51).

Medullary Carcinoma

A cellular smear composed of medium to large, rounded, dissociated tumor cells with granular cytoplasm intermingled with lymphocytes is diagnostic of medullary carcinoma (Fig. 6-47; Color Plate XII A and B). Groups may be seen. The nuclear : cytoplasmic ratio is large, and cytoplasm is poorly defined. Smears are obviously malignant. Metastasis to the lung in a patient with prior mastectomy may be erroneously interpreted as poorly differentiated bronchogenic carcinoma.

Paget's Disease

Tumor cells have invaded the squamous epithelium of the nipple in Paget's disease (Fig. 6-48). Aspiration of a palpable mass or thickened subareolar tissue may yield typical large ductal tumor cells with dense compact nuclei. Abrasion of the nipple also yields tumor cells with enlarged irregular nuclei and granular chromatin. Cytoplasm is ill defined and finely vacuolated (Fig. 6-49). Squames or squamous cells may be present.

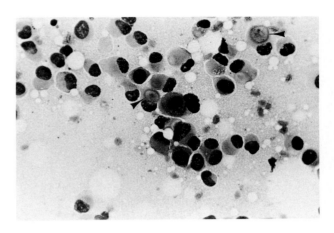

Fig. 6-44. Signet ring carcinoma. The smear resembles the apocrine type, but scattered cells contain inclusions with a characteristic concentric pattern *(arrows)*. (R stain, ×150)

Inflammatory Carcinoma

Figure 6-50 demonstrates invasion of the subepithelial tissue with tumor cells singly and in groups. In inflammatory carcinoma there is permeation of the lymphatics and dilatation of the superficial vessels to produce hyperemia. It is entirely a nonsurgical lesion. Oblique puncture as described above produces typical large ductal cancer cells (see Fig. 6-30) and should be carried out on any breast in which cellulitis is clinically suspected to avoid misdiagnosis and delay.

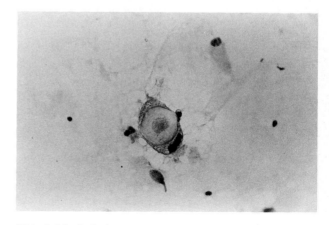

Fig. 6-45. Lobular carcinoma, signet ring type. Concentric pattern of the cellular inclusion distending the cell is clearly demonstrated. (P stain, ×330)

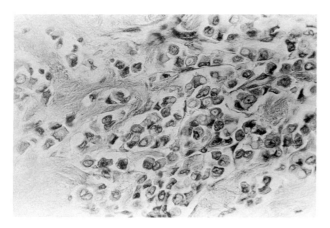

Fig. 6-46. Lobular carcinoma, signet ring type. Tissue section. The inclusions are clearly seen in multiple cells. (H & E stain, ×132)

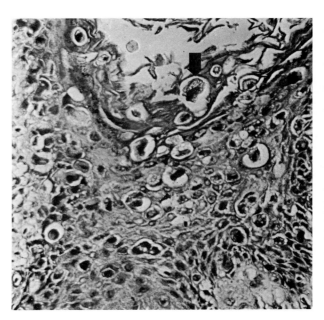

Fig. 6-48. Ductal carcinoma, Paget's disease. Tissue section. Tumor cells have invaded the squamous epithelium of the nipple (*arrow*). Cells can be scraped from the surface, or subsurface thickening can be aspirated. (H & E stain, ×600)

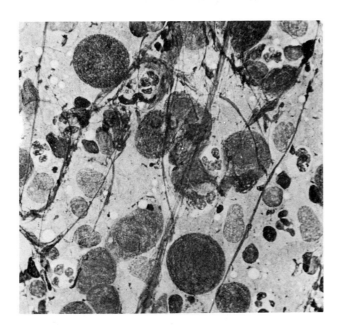

Fig. 6-47. Ductal carcinoma, medullary type. Cells are large, rounded, and dissociated, with granular immature nuclei and scant cytoplasm. The background is dirty, with a granular base containing numerous inflammatory cells. Cellular fragility is demonstrated by the nuclear strands, not unlike the picture of poorly differentiated non-Hodgkins lymphoma (see Chap. 14). (R stain, ×1000)

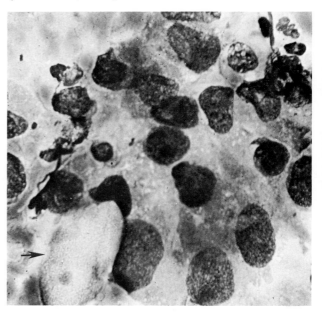

Fig. 6-49. Ductal carcinoma, Paget's disease. Lesion scrape. Cell borders are poorly defined, and the cells are somewhat degenerated. An anucleated squame is indicated (*arrow*). (R stain, ×1000)

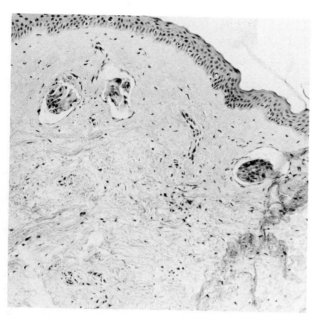

Fig. 6-50. Ductal carcinoma, inflammatory type. Tissue section. Invasion of dermal lymphatics by tumor. This produces a tense, inflammatory cutaneous reaction. FNA with a fine gauge needle will draw tumor cells out in sufficient numbers to prepare a diagnostic smear. (H & E stain, ×90)

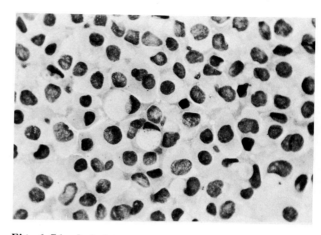

Fig. 6-51. Lobular carcinoma. Shown is a monolayer pattern similar to that in ductal carcinoma. Key features are rounded inclusions, molding nuclei (signet ring pattern) and indented nuclei. This pattern is less common than the cluster pattern seen in Figures 6-52 and 6-54. (R stain, ×250)

Lobular Carcinoma and Lobular Neoplasia

Preoperative identification of lobular carcinoma is clinically valuable beyond the simple diagnosis of cancer because of its well-known multicentricity and bilaterality.[36,37,51]

The clusters in lobular carcinoma are presumably aspirated from packed intralobular ducts (Fig. 6-51). In contrast, the monolayer of duct carcinoma is extracted from enlarged, cell-filled ducts.

Lobular neoplasia is a precursor lesion of lobular carcinoma and cannot currently be identified cytologically. It appears in nonpalpable deposits, often combined with some aspect of mammary dysplasia. Cells that show atypia and might be called precancerous appear in smears that carry the final diagnosis of mammary dysplasia. They may be of lobular origin, and the indication for surgical biopsy will depend on clinical factors as well as cytologic findings.

The cytologic picture in lobular carcinoma is as follows:

Monolayer of small cells with intact cytoplasm (see Fig. 6-51) similar to the monolayer of duct carcinoma

Tightly packed clusters of small cells (Figs. 6-52 and 6-53)

Minimal atypia and hyperchromatism

Absence of naked bipolar nuclei

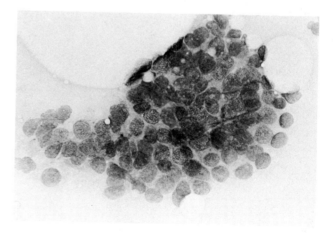

Fig. 6-52. Lobular carcinoma. Tightly packed cluster of small cells with inconspicuous cytoplasm. Nuclei are rounded and fairly regular. Nuclear chromatin is granular and bland. Nucleoli are not observed. (R stain, ×250)

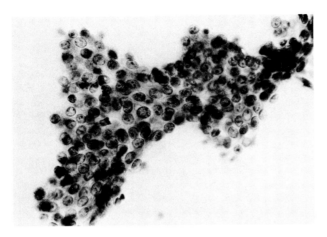

Fig. 6-53. Lobular carcinoma. The cell pattern is similar with P stain. Cell atypicality is not striking, although nucleoli are more prominent than with R stain. Cytoplasm is not prominent. (P stain, ×200)

Presence of indented and lobulated nuclei (Fig. 6-54)

Presence of angular nuclei with flattened surface (see Fig. 6-50)

Presence of signet ring cells (see Figs. 6-44, 6-45, and Plate XI D).

Occult Breast Carcinoma with Axillary Metastases

The problem of metastases with unknown primary tumor is considered in Chapter 3. The specific problem of axillary metastases in women is considered here.

Axillary lymph node metastases with unknown primary tumor in women strongly suggests occult breast carcinoma.[71] Even locally noninvasive occult breast carcinomas may give rise to axillary metastases.[75] DNA studies of such tumors may reveal aneuploidy (see Chap. 18) and dictate more vigorous therapy.

In a study by Patel and colleagues of 29 patients with occult breast carcinoma and enlarged axillary nodes, poorly differentiated adenocarcinoma was diagnosed in 21, medium differentiated in 3, medullary carcinoma in 2, lobular carcinoma in 2, and squamous carcinoma in 1.[71]

Adhering to cytologic criteria, it should be possible to identify breast origin of the metastases on aspiration in colloid (see Fig. 6-39), apocrine (see Figs. 6-42 and 6-43), medullary (see Fig. 6-46), lobular (see Fig. 6-43), and carcinoma with magenta bodies (see Fig. 6-32) in many cases.

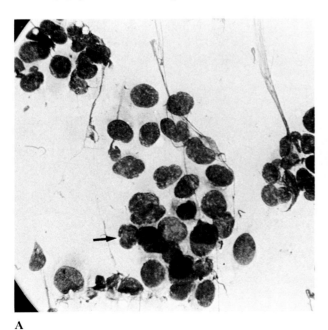

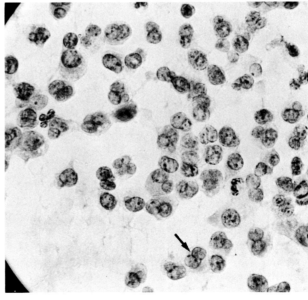

A
B

Fig. 6-54. A. A cluster of lobular carcinoma cells illustrating indented nuclei (*arrow*). (R strain, ×390) **B.** Lobular carcinoma. A monolayer of fairly small cells, occasionally with indented nuclei (*arrow*). (P stain, ×300)

Positive mammography will usually identify the primary tumor, which, however, is not excluded by negative radiographic studies. Cytologic smears of axillary nodes will also identify lymphoma, melanoma, thyroid tumor, and some primary tumors of lung, ovary, and gastrointestinal tract. Clinical data and correlation are extremely important.

Axillary Tail Versus Metastatic Nodes

Most palpable masses in the axilla are due to enlarged neoplastic or inflammatory nodes. Masses in the tail of the breast may be difficult to distinguish from low-level nodes.

Aspiration yielding lymphoid smears rules out carcinoma. If tumor is aspirated, differential diagnosis between metastatic nodes and axillary tail must be established.

The monolayer type of breast carcinoma and amelanotic melanoma may pose a problem (Figs. 6-34A and B and 13-12). Fat, which would not ordinarily be present in aspiration from a metastatic node, suggests a primary breast tumor. The technique for axillary aspiration is presented in Chapter 14.

Male Breast Carcinoma

Clinical aspects of male breast carcinoma are discussed in the section on gynecomastia. Cytologic material is extracted with ease, and the cytology is identical to that seen in ductal carcinoma (NOS). This must be contrasted with the benign epithelial cells that appear in rather tight groups in gynecomastia, as illustrated in Figure 6-27.

Postradiation Breast Alterations

Radiation induces alterations in the breast that simulate tumors. Hyperemia and edema occur early in treatment. Clinical correlation distinguishes this change from inflammatory carcinoma. After treatment, fibrosis may occur, particularly in large breasts; late clinical fibrosis remains in 20% of patients.[101] Changes may present as a "mass" that is indistinguishable from recurrent cancer or a new primary benign or malignant lesion. Aspiration during the course of radiation will produce varying patterns, including necrosis (Fig. 6-55). Fat necrosis that occurs postradiation simulates carcinoma clinically and on mammography.[90] Recurrent cancer is significantly

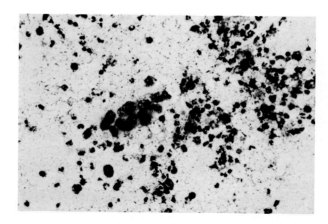

Fig. 6-55. Ductal carcinoma. Radiation necrosis. Cancer cells are ill-defined and degenerated. There is surrounding debris and inflammatory cells. (R stain, ×400)

more common in the irradiated breast in which incomplete excision has been carried out.

Pain or alteration in shape or texture should prompt FNA. Although the yield is often scanty, it may be surprisingly cellular. Extraction of carcinoma cells may require a fairly vigorous aspiration maneuver; the cells obtained are often stripped and distorted, but identifiable single cells or clusters are usually present (Fig. 6-56).

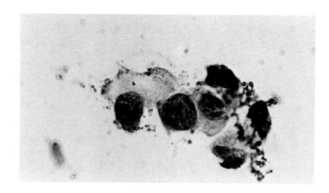

Fig. 6-56. Carcinoma. Postradiation smear. Single cells and small irregular clusters are extracted with rather vigorous aspiration. Cells tend to be angular and distorted. It is not always clear whether such cells can proliferate. (P stain, ×400)

Lymphoma

Primary malignant lymphoma of the breast is rare. Only fourteen instances of the disease were found in a review of 11,277 primary malignant breast tumors.[56] In addition to being indistinguishable clinically from breast carcinoma, all had a diffuse histologic pattern and were cytologically composed of less differentiated cells. Eighteen localized non-Hodgkin's lymphomas of the breast were observed at Memorial Hospital in New York from 1970 to 1984.[85]

Well or moderately differentiated lymphoid smears may be obtained by aspiration of an intramammary multicentric or metastatic mass in a patient with known non-Hodgkin's lymphoma. Intramammary lymph nodes have been identified on mammography[19] and may be the source of benign lymphoid smears or the site of spread of non-Hodgkin's lymphoma. Not infrequently, upper-outer quadrant nodules yield lymphoid smears, and the possibility of a low-lying axillary node must be considered and discussed with the clinician. Cytologic patterns may be identical to any of the types described in Chapter 14. Clinical data may be decisive.

Hodgkin's disease involving the breast is equally rare. Six cases were found in a series of 2365 cases diagnosed and treated at M.D. Anderson Hospital.[60]

Cystosarcoma Phyllodes

The name *cystosarcoma phyllodes* is retained because of its long history of use and familiar morphologic features. The histology presents a spectrum from benign, the most common, to infrequent malignant sarcomas.[40] Aspiration cytology clearly reflects tumor histopathology.

The characteristic changes identifying the lesion are in the stroma, and the dominant cytologic finding on smear is the presence of highly cellular stromal fragments (Fig. 6-57). In contrast, the stroma in fibroadenoma is largely an acellular substance staining pink with R (Color Plate IX A).

Individual tumor cells are dissociated, with irregularly rounded to oval nuclei and wispy, fragile, elongated cytoplasm. Tumor cells are seen in a dense aggregate entrapped in an acid mucopolysaccharide substance in Figure 6-58. The malignant variety

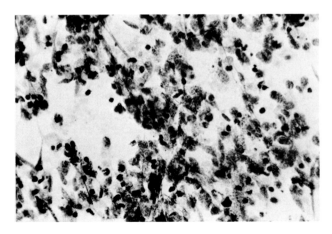

Fig. 6-57. Cystosarcoma phyllodes. Cellular smear composed of dissociated and aggregated fragile stromal cells. The cells are ill-defined and elongated with wispy cytoplasm. Nuclei have a fine chromatin pattern with P, but epithelial cells are largely obscured. (P stain, ×250)

shows enlarged hyperchromatic nuclei with prominent nucleoli and foamy, poorly delimited cytoplasm when stained by P. Mitoses and numerous tissue fragments may be present.[89]

Epithelial cells, stripped by the aspiration from the surface of the stromal projections of the tumor, form fairly tight aggregates with varying atypia. They are, however, benign.

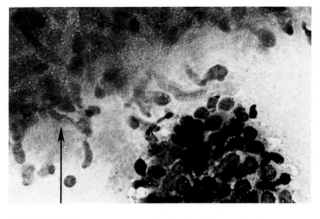

Fig. 6-58. Cystosarcoma phyllodes. The spindled tumor cells are enmeshed in pink-staining acid mucopolysaccharide substance *(arrow)*. A tight cluster of epithelial cells is in juxtaposition. (R stain, original magnification ×600)

Sarcomas and Carcinosarcoma

The cytology of sarcomas is presented in Chapter 15.

Metastatic Tumors in the Breast

Consideration of metastatic deposits in the breast must be given in patients with known prior tumors. The incidence has been reported to vary from 1.2% to 2.3%. The tumor mass may be clinically indistinguishable from primary cancer.

The most common metastatic tumor in the breast is breast carcinoma in the opposite breast. It can sometimes be distinguished cytologically from a second primary tumor. Lymphoma, melanoma, and tumors of the kidney can also be distinguished cytologically from aspiration of their metastasis in breast. Ten cases of metastatic tumor were reported by Charache.[13]

METHODS OF REPORTING

The designation *ductal carcinoma* is used for most malignant breast masses, although the report should categorically identify a tumor subtype such as colloid or apocrine carcinoma or fibroadenoma when possible.

A strictly negative aspiration will be essentially acellular or contain small amounts of fat, tissue fluid, and inflammatory cells. It is usually obtained from a superficial vague nodularity rather than a well-defined mass, although densely fibrotic masses (including sclerotic carcinomas and fibroadenomas) may yield similar scant cellular content.

Negative reports do not exclude cancer and, in the presence of suspicious palpatory or radiographic changes, cannot forestall surgical biopsy. This must be made explicit in the report until the referring physician is familiar with the use of ABC. From a practical standpoint, a negative aspiration in the presence of suspicious clinical findings is often clarified by repeat puncture. It is also useful to have the patient wait after the initial aspiration; and, for this reason, rapid staining enables quick review and report.

The aspiration procedure may generate much anxiety; and, by prearrangement with the referring physician, a direct negative report or even a positive report may be given to a waiting patient. Repeat aspirations and direct reporting are particularly of value when the patient has traveled a considerable distance.

FALSE-NEGATIVE AND FALSE-POSITIVE ASPIRATIONS

In the context of the modern combined and cooperative approach to diagnosis and treatment of breast cancer, the frequency of false-negative reports in any particular institution is of limited significance in assessing the use of ABC. The term *false negative* generally refers to failure to identify cancer cells when the targeted lesion is malignant. This is usually due to failure to reach the cancer with a needle. Misinterpretation of cancer cells as benign is another cause of false-negative results for cancer. Fibrosis, necrosis, and dilution contribute to a false-negative experience.[50] It occurs with decreasing frequency as cytologic experience broadens.[29] False-negative results also occur in histopathologic examination.[8,70] False-negative results may also refer to failure to identify a specific benign lesion, such as fibroadenoma or BMD.

The percentage of false-negative results varied from 7.4% to 26.7% in several large series.[29] Reduction of this percentage must come from steady increase in experience rather than any conscious effort, since the latter might lead to false-positive readings. It is important to emphasize that nothing is lost by the false-negative diagnosis because clinical findings always take precedence and surgery will proceed as indicated.[25]

False-positive results appear to occur from overreading atypical smears (Color Plate VIII), rendering a diagnosis on limited material, and simply insufficient experience at the microscope.[29,49,88] Any false-positive result must be considered unacceptable, and reduction to the zero level will require intensive selection by each cytopathologist commensurate with experience. Experience will dictate which specimens are unquestionably malignant and which continue to raise doubts, however minimal. Mastectomy after a positive cytologic report that later turns out to have been a false-positive report has obvious disadvantages. Even though a frozen section may be demonstrated to be benign and no trauma results, such a reading can lead to loss of confidence in the method. Cases of positive cytology and negative pathology have, however, subsequently developed into locally recurrent or metastatic carcinoma, a problem to which surgical pathology must address itself.[29] In the best and largest of five series, false-positive results were recorded in 0.1% of attempts. The range was 0.1% to 2.5% in the five series.[29]

It is well established that false-positive results may occur in the frozen section technique.[44,87] It takes place in a climate of urgency with extremely limited tissue samples and restricted staining methods. The preoperative cytologic diagnosis is made under optimum conditions with easy opportunity for repeat corroborative aspiration and has largely replaced frozen section in a number of centers. When initially employed, it is largely adjunctive until there is sufficient experience and confidence. Frozen section on breast carcinoma is now a rare event in Sweden.

HAZARDS AND SIDE-EFFECTS OF BREAST ASPIRATION

The single or multiple insertion of a thin needle into breast tissue causes an extremely low incidence of morbidity. Hematomas may result and are minimized by the application of pressure to the aspiration site. Although the breast itself contains no vital structures, penetration of the chest wall can occur with a poorly directed needle, producing pneumothorax.[1]

On the question of tumor dissemination, no effect on long-term survival was noted in a pointed study by Robbins and colleagues.[73]

REFERENCES

1. Abele JS, Miller TR, Goodson WH et al: Fine-needle aspiration of palpable breast masses. Arch Surg 118:859, 1983
2. Ackerman LV: Surgical Pathology, 3rd ed. St Louis, CV Mosby, 1964
3. Ackerman LV, Katzenstein AL: The concept of minimal breast cancer and the pathologist's role in the diagnosis of "early carcinoma." Cancer 39:2755, 1977
4. Aldersson MR, Halin I, Stauton MD: The relative significance of prognostic factors in breast carcinoma. Br J Cancer 25:646, 1971
5. American Cancer Society: Cancer Facts and Figures, 1980. New York, American Cancer Society, 1980
6. Barrows GH, Anderson TJ, Lamb JL et al: Fine needle aspiration of breast cancer: Relationship of clinical factors to cytology results in 689 primary malignancies. Cancer 58:1498, 1986
7. Battifora H: Intracytoplasmic lumina in breast carcinoma. Arch Pathol 99:614, 1975
8. Bauermeister DE: The role and limitation of frozen section and needle aspiration biopsy in breast cancer diagnosis. Cancer 46:947, 1980
9. Beahrs OH, Smart CR: Diagnosis of minimal breast cancers in the breast. Cancer 43:848, 1979
10. Black MM, Barclay THC, Cutler SJ et al: Association of atypical characteristics of benign breast lesions with subsequent risk of breast cancer. Cancer 29:338, 1972
11. Bottles K, Taylor RN: Diagnosis of breast masses in pregnant and lactating women by aspiration cytology. Obstet Gynecol 66:765, 1985
12. Burns PE: False-negative mammograms delay diagnosis of breast cancer. N Engl Med J 299:201, 1978
13. Charache H: Metastatic tumors in breast. Ann Surg 115:42, 1942
14. Cheatle GL, Cutler JL, Martinez D et al: Fat necrosis of the breast simulating recurrent carcinoma after primary radiotherapy in the management of early stage breast carcinoma. Cancer 52:442, 1983
15. Davis HH, Simons M, Davis JB: Cystic disease of the breast: Relationship to carcinoma. Cancer 17:957, 1964
16. Donegan WL, Lewis JD: Clinical diagnosis and staging of breast cancer. Semin Oncol 5:373, 1978
17. Donnelly PK, Baker KW, Carney JA et al: Benign breast lesions and subsequent cancer in Rochester, Minnesota. Mayo Clin Proc 50:650, 1975
18. Doremus WP: Carcinoma of the breast arising in a fibroadenoma. NY State J Med 65:918, 1965
19. Egan RL, McSweeney MB: Intramammary lymph nodes. Cancer 51:1838, 1983
20. Evans RW: Histological Appearances of Tumours, 2nd ed. Edinburgh, E & S Livingston, 1968
21. Feig SA: Localization of clinically occult breast lesions. Radiol Clin North Am 21:155, 1983
22. Fisher B: Cancer of the breast: Size of neoplasm and prognosis. Cancer 24:1071, 1969
23. Fisher ER, Gregario RM, Fisher B: The pathology of invasive breast cancer. Cancer 1:1, 1975
24. Fondo, EY, Rosen PP, Fracchia AA et al: The

problem of carcinoma developing in a fibroadenoma. Cancer 43:553, 1979

25. Frable WJ: Thin-needle aspiration biopsy—a personal experience with 469 cases. Am J Clin Pathol 65:168, 1976

26. Frable WJ: Needle aspiration of the breast. Cancer 53:671, 1984

27. Frantz VK, Pickeren JW, Melcher GW et al: Incidence of chronic cystic disease in so-called "normal breasts." Cancer 4:762, 1951

28. Franzen S, Stenkvist B: Diagnosis of granular cell myoblastoma by fine needle aspiration biopsy. Acta Pathol Microbiol Scand 72:391, 1968

29. Franzen S, Zajicek J: Aspiration biopsy in diagnosis of palpable lesions of the breast. Critical review of 3479 consecutive biopsies. Acta Radiol Ther 7:241, 1968

30. Frazier TG, Rowland CW, Murphy JT et al: The value of aspiration cytology in the evaluation of dysplastic breasts. Cancer 45:2878, 1980

31. Gibson A, Smith G: Aspiration biopsy of breast tumours. Br J Surg 155:236, 1957

32. Gita J: Cytopathology of tuberculous mastitis—a report of nine cases with fine needle aspiration cytology. Acta Cytol 29:974, 1985

33. Goldman RL, Friedman MB: Carcinoma of the breast arising in fibroadenomas, with emphasis on lobular carcinoma—a clinicopathologic study. Cancer 23:544, 1968

34. Grant CS, Goellner JR, Welch JS et al: Fine needle aspiration of the breast. Mayo Clin Pathol 61:377, 1986

35. Griffith CN, Kern WH, Mikkelsen WP: Needle aspiration cytologic examination in the management of suspicious lesions. Surg Gynecol Obstet 162:142, 1986

36. Haagensen CD: Disease of the Breast, 2nd ed. Philadelphia, WB Saunders, 1971

37. Haagensen CD, Lane N, Lattes R et al: Lobular neoplasia (so-called lobular carcinoma in situ) of the breast. Cancer 42:737, 1978

38. Hardy HD: Gynecomastia associated with lung cancer. JAMA 173:1462, 1960

39. Harrington E, Lesnick G: The association between gross cysts of the breast and breast cancer. Breast 7:1981

40. Hart WR, Bauer RC, Oberman HA: Cystosarcoma phyllodes. Am J Clin Pathol 70:211, 1978

41. Helpap B, Tschubel K: The significance of the imprint cytology in breast diagnosis. Acta Cytol 22:133, 1978

42. Henning K, Johansson J, Rimsten A et al: X-ray and fine needle biopsy in diagnosis of nonpalpable breast lesions. Acta Cytol 19:1, 1976

43. Hindle WH, Navin J: Breast aspiration cytology: A neglected gynecologic procedure. Am J Obstet Gynecol 136:482, 1983

44. Holaday WJ, Assor MD: Ten thousand consecutive frozen sections. Am J Clin Pathol 61:769, 1974

45. Hunt VC, Budd JW: Gynecomastia associated with interstitial cell tumor of the testicle. J Urol 42:1242, 1939

46. Kern WH, Dermer GB: The cytopathology of hyperplastic and neoplastic mammary duct epithelium—cytologic and ultrastructural studies. Acta Cytol 1:120, 1972

47. Kline T, Josh LP, Neal HS: Fine-needle aspiration of the breast: Diagnoses and pitfalls. Cancer 44:1458, 1979

48. Kreuzer G: Aspiration biopsy in proliferating benign mammary dysplasia. Acta Cytol 22:128, 178

49. Kreuzer G, Boquoi E: Aspiration biopsy cytology, mammography and clinical exploration: A modern setup in diagnosis of tumors of the breast. Acta Cytol 20:319, 1976

50. Kreuzer G, Zajicek J: Cytologic diagnosis of mammary tumors from aspiration biopsy smears: III. Studies on 200 carcinomas with false negative or doubtful cytologic reports. Acta Cytol 16:249, 1972

51. Leis HP, Cammarata A, LaRaja R et al: Bilateral breast cancer. Breast 7:13, 1981

52. Linsk JA, Franzen S: Fine Needle Aspiration for the Clinician. Philadelphia, JB Lippincott, 1986

53. Linsk J, Kreuzer G, Zajicek J: Cytologic diagnosis of mammary tumours from aspiration biopsy smears: II. Studies on 210 fibroadenomas and 210 cases of benign mammary dysplasia. Acta Cytol 16:130, 1972

54. Löwhagen T, Rubio CA: The cytology of the granular cell myoblastoma of the breast: Report of a case. Acta Cytol 21:314, 1977

55. McDivitt RW, Stewart FW, Farrow JH: Breast carcinoma arising in solid fibroadenomas. Surg Gynecol Obstet 125:572, 1969

56. Mambo NC, Burke JS, Butler JJ: Primary malignant lymphomas of the breast. Cancer 39:2033, 1977

57. Mann NM: Gynecomastia during therapy with spironolactone. JAMA 184:778, 1963

58. Martin HE, Ellis EB: Biopsy by needle puncture and aspiration. Ann Surg 92:169, 1930

59. Massie FM, McClellan JT: An analysis of 1500 breast biopsies. Ann Surg 26:509, 1960

60. Meis JM, Butler JJ, Osborne BM: Hodgkin's disease involving the breast and chest wall. Cancer 57:1859, 1986

61. Merino MJ, LiVols VA: Signet ring carcinoma of the female breast: A clinicopathologic analysis of 24 cases. Cancer 48:1830, 1981

62. Monson RR, Yen S, MacMahon B et al: Chronic mastitis and carcinoma of the breast. Lancet 2:224, 1976

63. Moskowitz M: Minimal breast cancer redux. Radiol Clin North Am 21:93, 1983

64. Mossler JA, Barton TK, Brinkhous AD et al: Apocrine differentiation in human mammary carcinoma. Cancer 46:2463, 1980

65. Nielsen M, Jensen J, Andersen J: Precancerous and cancerous breast lesions during lifetime and at autopsy. Cancer 54:612, 1984

66. Nissen–Meyer R, Kjellgren K, Mansson B: Adjuvant chemotherapy in breast cancer: Recent results. Cancer Res 80:142, 1982

67. Nordenström B, Zajicek J: Stereotoxic needle biopsy and preoperative indication of nonpalpable mammary lesions. Acta Cytol 21:350, 1977

68. Noyes RD, Spanos WJ, Montague WD: Breast cancer in women aged 30 and under. Cancer 49:1302, 1982

69. Nydick M, Bustos J, Dale JH et al: Gynecolastia in adolescent boys. JAMA 178:449, 1961

70. Patchefsky AS, Potok J, Hoch WS et al: Increased detection of occult breast carcinoma after more thorough examination of breast biopsies. Am J Clin Pathol 60:799, 1973

71. Patel J, Nemoto T, Rosner D et al: Axillary lymph node mestastasis form an occult breast carcinoma. Cancer 47:2923, 1981

72. Regaz J, Baird R, Rebbeck P et al: Neoadjuvant (preoperative) chemotherapy for breast cancer. Cancer 56:719, 1985

73. Robbins GF, Brothers JH, Eberhart WF: Is aspiration biopsy of breast cancer dangerous to the patient? Cancer 7:774, 1954

74. Rogers LW, Page DL: Epithelial proliferative disease of the breast—a marker of increased cancer risk in certain age groups. Breast 5:2, 1979

75. Rosen PP: Axillary lymph node metastases in patients with occult noninvasive breast carcinoma. Cancer 46:1298, 1980

76. Rosen P, Hadhu SI, Robbins G et al: Diagnosis of carcinoma of the breast by aspiration biopsy. Surg Gynecol Obstet 134:837, 1972

77. Rupnick EJ, Williams EL, Johnson WH: Breast biopsy—an outpatient procedure using local anesthesia. Milit Med 9:743, 1978

78. Shabot MM, Goldberg IM, Schick P et al: Aspiration cytology is superior to tru-cut needle! Ann Surg 196:122, 1982

79. Sickles EA, Klein DL, Goodson WH III et al: Mammograph after needle aspiration of palpable breast masses. Am J Surg 145:395, 1983

80. Silverman JF, Lannon DR, Unverferth M et al: Fine needle aspiration cytology of subareolar abscess of the breast. Acta Cytol 30:413, 1986

81. Silverswärd C: Estrogen Receptors in Human Breast Cancer, Ph.D. dissertation, Institute for Tumor Pathology, Karolinska Institute. Stockholm, 1979

82. Silverswärd C, Humla S: Estrogen receptor analysis on needle aspirates from human mammary cancer. Acta Cytol 24:54, 1980

83. Simpson–Herren L, Sanford AH, Holmquest JP: Effects of surgery on the cell kinetics of residual tumor. Cancer Treat Rep 60:1749, 1976

84. Smallwood J, Herbert A, Guyer P et al: Accuracy of aspiration cytology in the diagnosis of breast disease. Br J Surg 72:841, 1985

85. Smith RS, Brustein S, Straus DJ: Localized non-Hodgkin's lymphoma of the breast. Cancer 59:351, 1987

86. Smithers DW, Rigby–Jones P, Galton DAG et al: Cancer of the breast. Br J Radiol(suppl) 4:1952

87. Sparkman RS: Reliability of frozen sections in the diagnosis of breast lesions. Ann Surg 155:924, 1962

88. Stavric GD, Tevcev DT, Kaftandjiev DR et al: Aspiration biopsy cytologic method in diagnosis of breast lesions. Acta Cytol 17:188, 1973

89. Stawicki ME, Hsiu JG: Malignant cystosarcoma phyllodes—a case report with cytologic presentation. Acta Cytol 23:61, 1979

90. Stefanik DF, Brereton HD, Lee TC et al: Fat necrosis following breast irradiation for carcinoma: Clinical presentation and diagnosis. Breast 8:4, 1982

91. Strax P: Results of mass screening for breast cancer in 50,000 examinations. Cancer 37:30, 1976

92. Strobel GL, Shah NT, Lucas JG et al: Granular cell tumor of the breast—a cytologic immuno-histochemical and ultrastructural study of 2 cases. Acta Cytol 29:598, 1985

93. Summerskill WH, Adrian MA: Gynecomastia associated as a sign of hepatoma. Am J Dig Dis 7:250, 1962

94. Swerdlow M, Humphrey LJ: Fibrocystic disease of the breast and carcinoma. Arch Surg 87:457, 1963

95. Takeda T, Suzuki M, Sato Y et al: Aspiration cytology of breast cysts. Acta Cytol 26:37, 1982

96. Tubiana M, Chauvel P, Renaud A et al: Growth

rate and natural history of human breast cancers. Bull Cancer 62:341, 1975

97. Wallgren A, Arner O, Bergström J et al: Pre-operative radiotherapy in operable breast cancer. Breast 42:1120, 1978

98. Wallgren A, Zajicek J: Cytologic presentation of mammary carcinoma on aspiration biopsy smears. Acta Cytol 20:469, 1976

99. Wanebo H, Feldman PS, Wilhelm MC et al: Fine needle aspiration cytology in lieu of open biopsy in management of primary breast cancer. Ann Surg 199:569, 1984

100. Webb AJ: Through a glass darkly (the development of needle aspiration biopsy). Br Med J 89:59, 1974

101. Welch JS: The postirradiated breast. Mayo Clin Proc 61:392, 1986

102. Willis R: Pathology of Tumours, 3rd ed. Scarborough, Ont., Butterworth & Co, 1960

103. Zajicek J: Aspiration Biopsy Cytology, vol I. Basel, S Karger, 1974

104. Zajicek J, Caspersson T, Jakobsson P et al: Cytologic diagnosis of mammary tumors from aspiration biopsy smears: I. Comparison of cytologic and histologic findings in 2111 lesions and diagnostic use of cytophotometry. Acta Cytol 14:370, 1970

105. Zajicek J, Franzen S, Jakobsson P et al: Aspiration biopsy of mammary tumours in diagnosis and research—a critical review of 2200 cases. Acta Cytol 11:169, 1967

CHAPTER 7

Aspiration Biopsy of the Thorax (Excluding Breast)

DAVID B. KAMINSKY

RAFFAELE PERRONE DONNORSO

The frequent incidence of neoplastic or infectious conditions originating in or secondarily involving the thorax complexed with an escalating financial burden for health care has focused attention on aspiration biopsy as the logical diagnostic modality for achieving expedient, minimally interventive, cost-conservative, tissue-equivalent diagnostic information that can often be initiated on an outpatient basis and rationally direct patient management. Osseous and soft tissue structural abnormalities of the chest wall and lesions of lungs, pleura, and mediastinum constitute appropriate and critical targets for analysis by aspiration biopsy. An external chest wall abnormality can be expediently sampled by puncture under direct visualization; interior lesions generally require radiographic image enhancement for guidance of the needle to its target (Fig. 7-1).

The primary indication for aspiration biopsy is to differentiate benign from malignant disease reliably, with minimal invasion, so that therapy can be expediently initiated. This is particularly relevant to the medically restricted patient who is considered inoperable, but for whom tissue-equivalent information is needed to direct further care. Implicit is the concept of triage in which the cellular composition of the lesion dictates assignment to radiation or chemotherapy, thoracotomy, surveillance, or continued diagnostic intervention.

Pre-operative knowledge of cell type and extent of disease may assist the surgeon with his procedural approach and with the organization of time and schedules in his private office, clinic, and the operating suite. A judgment of nonresectability may be rescinded when an aspirate confirmation of malignancy justifies preliminary irradiation for reduction of tumor burden, preparing the lesion for resection. The aspiration biopsy may contradict premature conclusions of inoperability.[8] Conversely, nonresectability can be substantiated as an early conclusion by the cytological demonstration of bilateral involvement by a primary pulmonary tumor such as bronchioloalveolar carcinoma or by confirmation of metastasis from an extrapulmonary source. The analysis of a metastatic repository by aspiration biopsy is facilitated by immunoperoxidase stains for tumor markers and by comparison with needle aspirates prepared from the primary surgical resection specimens. An adequate cellular sample obtained from a radiographically benign lesion may influence the clinician to follow the patient conservatively. This surveillance program relies on the conscientious participation of the patient as well as the physician's dedication to continuity of care and should never be invoked when a benign aspirate is reported in the context of a high clinical index of suspicion for cancer.

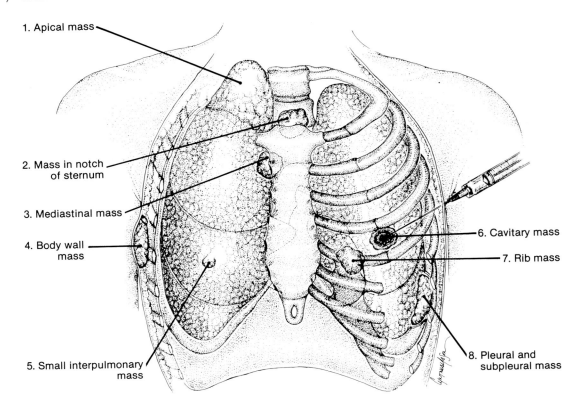

1. Apical mass

2. Mass in notch
   of sternum

3. Mediastinal mass

4. Body wall
   mass

5. Small interpulmonary
   mass

6. Cavitary mass

7. Rib mass

8. Pleural and
   subpleural mass

Fig. 7-1. Lesions of the thorax.

1. Apical mass (Pancoast's tumor) may be reached by puncture through the supraclavicular space. This may be accomplished with the patient in a sitting position.
2. Suprasternal mass (substernal thyroid, lymphoid mass) is easily and safely approached.
3. Mediastinal mass. Parasternal puncture above sternal angle on the right and in upper
   three interspaces on left is easily accomplished.
4. Body wall mass presents as a bulge.
5. Intrapulmonary mass requires fluoroscopic guidance.
6. Cavitary mass may yield necrosis, inflammatory exudate, or tumor. An attempt
   should be made to aspirate the shell.
7. Rib mass is easily entered.
8. Pleural and subpleural masses will usually require fluoroscopy.

■ *Case 7-1*

A 70-year-old man was initially evaluated by
fluoroscopically directed aspiration biopsy of a
solitary, smoothly contoured opacification in the
right upper lobe (Fig. 7-2), which produced
reactive bronchial epithelial cells in association
with macrophages and multinucleated histiocytes (Fig. 7-3). A diagnosis of probable granuloma was recorded, and the patient was assigned
to surveillance. He was lost to follow-up for 2

years, and there were no interim chest films.
When he reappeared for consultation, a chest
film revealed that the lesion had enlarged to
massive proportions (Fig. 7-4). The aspiration
biopsy was repeated with a two-needle tandem
approach for improved sampling (Fig. 7-5).
Multinucleated giant cells and reactive, atypical
bronchials persisted (Fig. 7-6); but in addition
there were aggregates of adenocarcinoma cells
(Fig. 7-7A and B), implicating origin in a scar
cancer or a sampling error sustained during the

original biopsy, which failed to disclose the presence of malignancy at the inception of its radiographic presentation. Retrospectively, it is clear that continued diagnostic intervention to establish the etiology of the original lesion should have superseded surveillance in an unreliable physician-patient relationship.

When the clinical diagnosis is infection, aspiration biopsy offers a particular advantage by facilitating the morphological identification and cultural isolation of specific etiologic agents. *Coccidioides immitis, Candida albicans,* and *Blastomyces dermatitidis* are readily identified by Papanicolaou (P) stain, enhanced by PAS stain, and confirmed by culture. Purulent exudate from an abscess can be obtained for innoculation to initiate aerobic and anaerobic cultures. The evaluation of the immunocompromised transplant or acquired immune deficiency syndrome (AIDS) patient for opportunistic pulmonary infection is expedited[5] by the technique in this critically ill population for whom immediate information must be obtained with minimal traumatic impact. *Pneumocystis carinii* and viral inclusions implicating herpes or cytomegalovirus[4,20] can be identified in material

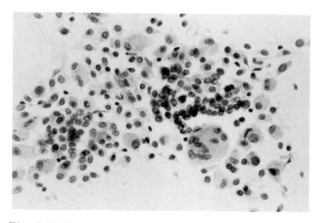

Fig. 7-3. Reactive bronchial epithelial cells associated with macrophages and multinucleated histiocytes. (P stain, ×500)

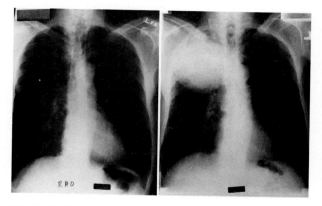

Fig. 7-4. Comparative chest films, original right upper lobe nodule compared with the lesion 2 years later.

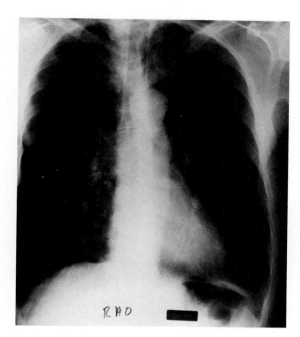

Fig. 7-2. Chest film demonstrating right upper lobe solitary nodule in a 70-year-old male.

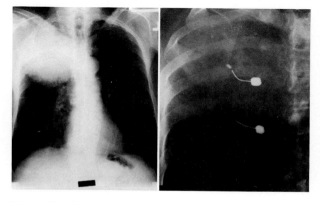

Fig. 7-5. Comparative chest films demonstrating the right upper lobe mass contrasted with tandem needles in the target.

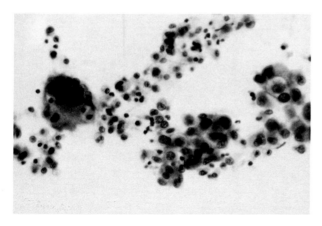

Fig. 7-6. Multinucleated giant cells in association with reactive, atypical bronchial epithelial cells. (P stain, ×500)

harvested by aspiration biopsy. When infection is encountered serendipitously during an aspiration biopsy procedure for a suspected tumor, additional material for special stains or cultures can be obtained from a second pass; or the saline wash of the needle may serve as an accessory reservoir for material to innoculate acid-fast and fungal cultures.

The following contraindications to performing aspiration biopsies of the thorax refer primarily to lung and are summarized by Johnston and Frable[20]:

Hemorrhagic diathesis
Anticoagulant therapy

Pulmonary hypertension
Arteriovenous malformations
Advanced emphysema
Hydatid cyst

When rational case selection accounts for these contraindications, serious complications can be averted and optimal diagnostic information contributed.

The accuracy of transthoracic aspiration biopsy varies with the expertise of the radiologist in puncturing the lesion and acquiring a reliable sample and with the competence and experience of the pathologist for cytomorphological interpretation. Both have improved with time and are the response to clinical demands. As early as 1966, Dahlgren and Nordenstrom,[9] who are credited with provoking recrudescent interest in the technique, achieved an accuracy of 90%, a statistic improved to 93% in the 1972 report that Dahlgren contributed with another Swedish colleague.[8] Ninety percent accuracy is reiterated by Koss in his recent textbook[26] based on material from 200 patients evaluated at Montefiore Medical Center and 1100 consecutive cases studied by his co-authors in Poland. Tao and colleagues[45] related their experience over time with transthoracic fine needle aspiration (FNA) biopsy at the Toronto General Hospital for the years 1967 through 1985. The diagnostic rate improved from 82.8% in the period from 1967 through 1978 to 93.5% in 1980. This improvement is related to enhanced expertise in needling combined with more skillful cytomorphologic interpretation. The

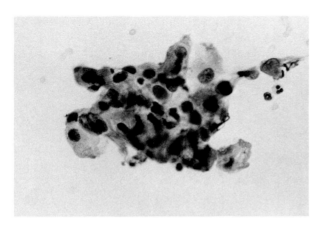

A

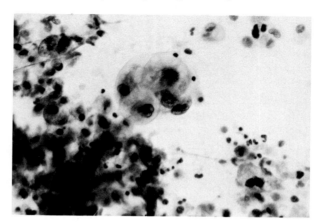

B

Fig. 7-7. **A.** Adenocarcinoma cells isolated from the lesion in Fig. 7-6. **B.** Mucus-producing adenocarcinoma isolated from the same lesion. (P stain, ×625)

Toronto series reflected that the chance with which an experienced radiologist could obtain a representative sample with a single puncture was 93%.

TECHNICAL CONSIDERATIONS

The standard procedure (or its variants for aspiration biopsy of pleura, lung, and mediastinum) is essentially a modification of the technique initially structured at the Karolinska[9] that depends on image-intensified direction of a 22-gauge flexible spinal needle into the radiographic target. Aspiration is accomplished by replacing the stylet with a 20-ml syringe to facilitate transfer of cells from the lesion to the interior of the needle. Biplane fluoroscopy and computed tomography provide for assessment of depth, angle of penetration, and confirmation of needle position.

The patient is selected as a candidate for the procedure based on his clinical history, the size and the location of the radiographic lesion, and exclusion of the contraindications discussed. The radiologist may elect to use computed or standard tomographic images to determine the size, depth, and character of the pulmonary lesion prior to initiating fluoroscopic guidance. He explains the objectives, technique, and complications to the patient and then obtains and documents informed consent. The procedure is scheduled with the cytopathologist (or with his cytotechnologist designate) who initiates a dialog with the patient before the procedure begins, generally explaining his role in the processing and diagnostic interpretation of the specimen.

The patient is positioned horizontally on the fluoroscopy table (or the mobile platform for the CT scanner). Chest roentgenograms are reviewed momentarily by radiologist and pathologist, and the puncture site is selected. The skin is prepared with Betadine cleansing and anesthetized with 1% Xylocaine, which is then directed into deeper tissues. The radiologist carefully selects the trajectory approach, which may be particularly strategic if the lesion is less than 2 cm in diameter. As he punctures the skin with the needle (stylet in place), his continuous explanation invites the cooperation and understanding of the patient, whose intimidation and requirements for sedation are reduced. The needle is guided to a predetermined depth by contemporaneous display on the television monitor. The needle is readjusted until there is unison of motion of needle tip and the lesion during respiratory excursions. When the needle tip satisfactorily resides in its destination, the stylet is removed and a disposable 20-ml syringe with a clear hub (but without Luer lock) is attached. The plunger is withdrawn to create a vacuum, and the needle is oscillated gently. The evacuation process is canceled the moment material enters the hub. The vacuum is allowed to re-equilibrate before the needle is removed from the tissue in order to avoid displacement of the material from the needle to the internal barrel of the syringe. The needle and syringe are given to the pathologist (or cytotechnologist), who removes the needle to reintroduce air into the syringe. This provides an expulsive force to displace the aspirated material from the core of the needle onto glass slides. Droplets are released onto slides at the end opposite the frosted zone for label application. Additional slides are placed in contact with the droplets until surface tension displaces the liquid in a circular monolayer, concentrating the screening area to one locale. The slides are separated in a perpendicular direction to avoid smearing the cells and are immediately immersed in 95% ethanol for fixation. A minimum of 3 minutes is allowed for fixation, and then the slides are stained by a rapid variant of the P technique while the patient is maintained in position in the radiology suite. Additional slides may be air-dried for May–Grünwald–Giemsa (MGG) stains. The quality of the analytical sample is rapidly assessed, and a diagnosis is rendered immediately, if possible. If the material is inadequate or an additional sample is required for special stains or cultures, a second pass is conveniently accomplished. The needle is rinsed in saline to provide a reservoir from which cell membrane or cytospin preparations can be made or cultures innoculated. Immediate and 2-hour postbiopsy chest films are taken to search for complications. Outpatients are carefully evaluated before release or are admitted for optional observation or treatment of complications.

An enthusiastic, experienced, motivated, and dexterous radiologist is pivotal to the success of this procedure. He should intuitively cultivate, establish, and maintain a close professional rapport with the cytopathologist to ensure unconditional cooperation. Within the context of this relationship, there must be an exchange of clinical and radiographic expectations for tissue-equivalent information. If the lesion is radiographically malignant but the cytological appearance is insubordinate, the discrepancy should stimulate dialog that triggers a decision tree of further investigative effort. The radiologist should choose a simple 22-gauge flexible spinal needle, no devices. He can elect to place two to three needles simultaneously to ensure adequate sampling in one maneuver, particularly if perilesional hemorrhage or pneumothorax develops. The interventional radiologist should have the expertise to insert a chest tube for immediate treatment of symptomatic pneumothorax. When pulmonary and hepatic lesions are contemporaneous, he should analyze the hepatic lesion first to reduce mortality and morbidity. When the pulmonary lesion is situated near critical structures or is located ambiguously in the periphery, he should select computed tomography as the preferred imaging modality. It is his prerogative to schedule and accomplish cases on an outpatient basis whenever possible, and certainly on the same day as admission, because the aspiration biopsy diagnosis can direct the remainder of the workup conservatively.

If the chest wall contains a visible or palpable lesion, image intensification may be eliminated and direct biopsy accomplished. An attenuated rib may be easy to puncture with a 22-gauge needle, but accidental penetration or deflection through the cortical perimeters and into lung can easily occur and must be avoided. Dense bone may require substitution of a 20- or 18-gauge, short, rigid needle, whereas superficial soft tissue or cutaneous nodules are best handled with a 25-gauge fine, truncated needle.

The recrudescent interest in aspiration biopsy of the lung, particularly for the analysis of the peripheral lesion, developed in a milieu of comfort and confidence in sputum and bronchial brush exfoliative cytology. Reluctant, traditional attitudes required educational revision; and literature directed to the community hospital[21] emphasized the need to select aspiration biopsy first, rather than the familiar sputum screen and bronchoscopic approach. An implicit reticence to use aspiration biopsy as a primary diagnostic maneuver survived this reorientation for years, and clinicians reluctantly acquiesced to the procedure only when bronchoscopic brushing cytology and sputum analysis failed. When the fertility of these alternative procedures was ultimately objectively scrutinized, needle aspiration prevailed, providing incentive to investigate with the needle first. Landman and associates[28] compared 100 bronchial brushings and 80 percutaneous needle aspiration biopsies of lung and concluded that although both procedures are valuable and complementary, the diagnostic accuracy of needle biopsy surpasses bronchial brush cytology when the lesions are peripherally situated, less than 2 cm in diameter, arise in the upper lobes, or do not originate from bronchial epithelium. Dahlgren and Lind[8] compared the needle aspirate sample with concomitant sputum in 125 patients who had at least one satisfactory sputum specimen. Of 101 patients with malignant tumors, 93% were diagnosed by needle aspiration and 64% by sputum cytology. Central origin favors diagnosis by sputum cytology because of access to the bronchial lumina, whereas a peripheral location often precludes exfoliation into the bronchial system and eventual expectoration. The proximity to the periphery[26,39] confers a high degree of accessibility to the needle probe and promotes aspiration biopsy diagnosis of minute and asymptomatic lesions.

A recent study of 1758 needle cytology procedures from the Tokyo Medical College[23] reaffirms aspiration biopsy of the peripheral nodule as the initial diagnostic procedure of choice. Percutaneous aspiration biopsy identified 90.95% of peripheral cancers in contrast to sputum cytology, which detected 50.57%, and bronchial brush cytology, which diagnosed 76.05%. Furthermore, the diagnostic value of aspiration cytology is not dependent on the histological type of the tumor, contrary to sputum analysis in which cellular dissociation and exfoliation are functions of the tissue type. All pulmonary carcinomas may be cellular donors to needle acquisition, but small cell undifferentiated carcinomas and adenocarcinomas hesitate to uncouple their cells to dispersion in sputum,[8] whereas squamous and large cell undifferentiated carcinomas are enthusiastic exfoliators.

COST CONTAINMENT

The implementation of the Tax Equity and Fiscal Responsibility Act (TEFRA) on October 1, 1983, converted government reimbursement to hospitals for Medicare patients from a cost-per-test basis to payment per case at discharge. The new fiscal consciousness mandated reduced lengths of stays, pre-admission testing and re-assignment to outpatient services that continue to be paid on a cost basis. Hospitals are

Table 7-1. Cost Comparison: Conventional Biopsy vs. Aspiration Biopsy in Lung Cancer*

EXPLORATORY THORACOTOMY			FLUOROSCOPIC LUNG BIOPSY	
Action	Unit Cost	Total Cost	Action	Cost
Office visit	60.00 ⎫		Office visit	60.00
Outpatient chest x-ray	45.60 ⎭	105.60	Outpatient chest x-ray	45.60
Hospital admission				
Laboratory work			Inpatient aspiration biopsy	
CBC	21.10 ⎫		Fluoroscopically directed biopsy	228.00
Urinalysis	14.00		Radiologist's aspirate charge	95.00
Biochemical panel	65.00		Pathologist's interpretation charge	127.00
Protime	16.25 ⎬ 152.85		Hospitalization (1 day)	245.00
Partial thromboplastin time	21.00			
Bleeding time	15.50 ⎭			
Electrocardiogram (over 40 years)	41.00	41.00		
Presurgical hospitalization (1 day)	245.00	245.00		
Thoracotomy				
Surgeon's fee	2,400.00 ⎫			
Anesthesiologist	540.00			
Anesthesia mask	23.00			
Operating room (60 minutes @ $90/0.25 hours)	360.00 ⎬ 3,703.00			
Frozen section consultation	100.00			
Tissue processing	180.00			
Recovery room	100.00 ⎭			
Postoperative hospitalization (6 days)	1,470.00	1,470.00		
Total charge		$5,717.45	Total charge	$800.60
			(Reduction in cost = 86%)	

* For a patient presenting with a persistent cough.
(From Kaminsky DB: Acta Cytol 28(3):333–336, 1984)

now closely scrutinized by federal and third-party payors for appropriateness of admissions and expedient management with judicious, efficient testing. Aspiration biopsy, in the context of the Medicare fiscal policy,[22] offers a diagnostic modality that responds to the mandates of this legislation by providing a vehicle for efficient, precise, rapid, cost-effective patient evaluation that is adaptable to outpatient care, eligible for reimbursement, and billable by the pathologist. Its applications are as universal as the shortages of bed space, operating room facilities, and therapy units and the ramifications of the House Committee on Financial Affairs (HCFA) modulations of the cost of medical care. Its assimilation has conserved the time and energies of physicians, nurses, and histotechnologists, thereby reducing workload and staffing patterns, conserving hospital costs without threatening the quality of care, and liberating schedules and resources to permit expansion into new technologies of image analysis and flow cytometry. Reduction in patient inconvenience, anxiety, and invasive testing translates to improved quality of care.

Table 7-1 excerpted from a special issue of *Acta Cytologica* dedicated to aspiration biopsy[22] contrasts the diagnostic related groups (DRGs) cost of a diagnostic thoracotomy for evaluation of a peripheral pulmonary lesion (that cannot be adequately assessed by fiberoptic bronchoscopy or sputum cytology) with a fluoroscopically directed aspiration biopsy. If the comparison ignores the empirical concept that the thoracotomy may prove to be therapeutic as well as diagnostic and addresses only the diagnostic phase of invasive surgery, the fiscal aspects may be more meaningful. This assumption is justified by the discovery of a lesion whose histoarchitecture defines a nonsurgical treatment (for example, oat cell carcinoma) or represents a metastasis, an inoperable cancer, or an infectious process. The percutaneous transthoracic aspiration biopsy of lung or mediastinum can provide this information and can be initiated as an outpatient. Often the patient is admitted for observation following the procedure, accounting for the additional cost of a day's hospitalization in the accountancy, but the current trends are for outpatient surveillance in a "holding area" with eventual discharge without an admission. The cost of a fluoroscopically directed aspiration biopsy of lung with one day's subsequent admission is $800.60 compared with $5,717.45 for a conventional operative biopsy, representing a conservation in costs of at least 86%. If the result of the thoracotomy were one of the nonsurgical entities described above, the cost of acquiring this information by thoracotomy is unconscionable, given an expedient and safe alternative.

COMPLICATIONS OF THE FLUOROSCOPICALLY DIRECTED FNA BIOPSY OF LUNG AND MEDIASTINUM

The rejuvenated interest in aspiration biopsy as a diagnostic modality in the United States is attributable to technological advances in biopsy instruments and parallel refinement of radiographic image intensification for the guidance of the needle to its target. The flexible 22-gauge Chiba spinal needle with stylet or its modifications became the appropriate tool to use with biplane fluoroscopy as did the 9800 General Electric whole body CT-Scanner to obtain a cellular sample for the cytological identification or tumor or infection. This technological collaboration reduced traditional time delays, expense, serious complications, and reliance on conventional histological processing; and it gave flexibility to the diagnostician, minimized costs and inconvenience to the patient, and a financial advantage to hospital administrators and utilization review committees by reducing lengths of stay or reassigning evaluation outside the hospital.

But procedural progress is not without criticism, skepticism, and technical confusion, which seem to be inseparable derivatives of complications attributable to the method. Pneumothorax, hemorrhage, infection, air embolism, subcutaneous emphysema, translocation of tumor cells along the needle track, and death are the skeptics' vocabulary for advising caution. These caveats are often expressed out of context as misplaced modifiers that only delay progress. America is years behind Sweden in the evolution and acceptance of this technique because of attitudinal conservativism and confusion in the definition of terms. "Percutaneous aspiration biopsy of the lung" has been the generic identity for three diverse techniques, only one of which utilizes a fine

needle; but complications published under this comprehensive designation may, by imprecision in terminology, discredit FNA biopsy when a cutting needle was the offending instrument.[4,32] A critical look at complications underscores the necessity to clarify which procedure is scrutinized by the accruing statistics.

Pneumothorax

Pneumothorax is considered the most common "complication" of aspiration biopsy of the lung and mediastinum and has been reported in as many as 49% of cases, but averages quoted are generally in the vicinity of 20% to 30%. The significant statistic, however, is not the incidence, but rather the absolute number or percentage of symptomatic patients requiring therapy, which generally involves the insertion of a chest tube, a therapeutic event that many interventional radiologists are competent to perform without the assistance of a thoracic surgeon. Experience has dictated that pneumothorax is to be expected when a procedure violates the pleuroparenchymal relationship as the lung is penetrated; *symptomatic* pneumothorax is the *complication*.

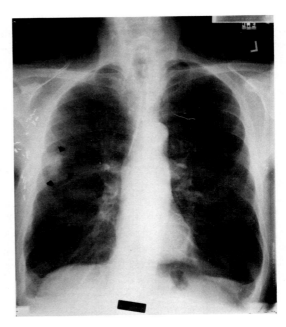

Fig. 7-8. Chest film of a 73-year-old woman with a radiographic density in the right upper lobe taken prior to aspiration biopsy.

Pneumothorax is suspected when the patient becomes dyspneic, but is generally detected as silent pulmonary deflation by chest roentgenography in immediate or delayed postbiopsy films. The incidence is partly a reflection of intensity of the search, which is determined by the radiologist's compulsivity and attention to detail.

■ *Case 7-2*

Figure 7-8 depicts the posterior-anterior (PA) chest film of a 73-year-old woman with a radiographic density in the right upper lobe. Metallic axillary clips annotate prior lymphadenectomy for treatment of metastatic melanoma. An aspiration biopsy was performed percutaneously under fluoroscopic guidance to obtain cells intended to clarify whether the lesion represented metastatic melanoma or a new pulmonary primary. The needle was inserted to the periphery of the mass, representing the site of cellular harvest (Fig. 7-9). The survey film taken 15 minutes postbiopsy demonstrated a 20% asymptomatic pneumothorax (Fig. 7-10); atelectasis displaces and intensifies the parenchymal markings including the tumor that was elucidated as a second primary papillary adenocarci-

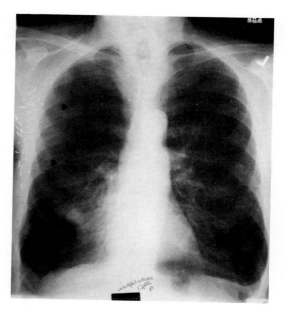

Fig. 7-10. Postbiopsy chest film demonstrating pneumothorax *(arrows)*.

noma of lung (Fig. 7-11A and B). The pneumothorax resorbed spontaneously and did not require a chest tube. Subsequent lobectomy histologically confirmed papillary adenocarcinoma of lung (Fig. 7-12).

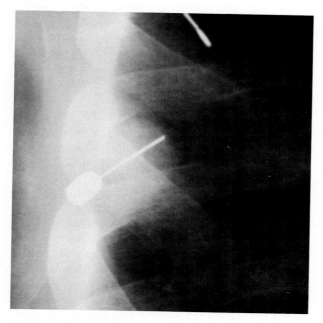

Fig. 7-9. The needle is situated at the periphery of its target.

Familiarity sometimes substitutes reassurance for contempt; progress in the technical and clinical comprehension of percutaneous aspiration biopsy of the thorax has resulted in a more relaxed or forgiving attitude toward its most frequent complication. The literature supports this notion.

Pneumothorax was the only significant complication in Fontana's[17] series, affecting about half the patients and requiring intrapleural suction in about a third of these. Dahlgren and Lind[8] referred to "pneumothorax of some degree," particularly in reference to patients with chronic obstructive pulmonary disease (COPD), but did not emphasize this as a serious complication in their experience of more than 3000 aspirates. Their nonchalance may have been intentional to protect the technique from the skeptics. Sinner[40] published the overwhelming Swedish effort of 4000 transthoracic needle biopsies performed on 2450 patients at Stockholm's Karolinska sjukhuset,

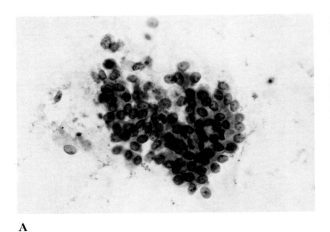
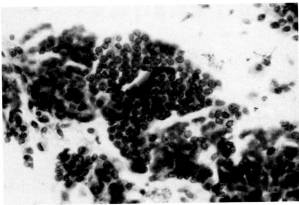

A   B

Fig. 7-11. A and B. Papillary adenocarcinoma of lung. (P stain: A ×625, B ×500)

Department of Thoracic Radiology. He focused his report on complications of 163 needle biopsies performed on 106 patients, which defined primary or secondary malignancy. In 16 cases, a small pneumothorax was observed with a 1 to 2 cm rim of apical air. These remained asymptomatic and were spontaneously resorbed within several days. In nine cases the pneumothorax was considered moderate at 4 cm. Three patients were dyspneic and required suction of air. One case with a large symptomatic pneumothorax required a chest tube after suction failed to reinflate the lung.

Lalli and colleagues[27] reported the Cleveland Clinic's experience with aspiration biopsy of 1223 patients with chest lesions. Pneumothorax was the primary complication, but was limited to 24.2% of patients. Only 4.4% of all patients required treatment, which generally involved simple aspiration utilizing a syringe, needle, and three-way valve.

A diversity of complex factors influences the occurrence and perhaps severity of pneumothorax. Berquist and associates[3] evaluated complications of transthoracic needle biopsy in relation to location and type of lesion for 430 patients at the Mayo Clinic for the period 1968 through 1977. The needle probe was an 18-gauge instrument that was intended to provide material for both histologic and cytologic study with the implication that cytology was performed if the procedure failed to supply tissue. This is technically not "fine needle" aspiration biopsy, so the results must be interpreted in this context. Pneumothorax occurred in 49% of patients, and 43% of these (21% of the entire group) required chest tube therapy. Curiously, a "large number of these patients" were asymptomatic, but chest tubes were placed simply because the volume of the pneumothorax exceeded 25%. Tension pneumothorax was recorded with an incidence of 1.4%. The incidence of pneumothorax was less than 10% in patients under 50 who were in good health. Cavitating lesions predisposed to a higher incidence of pneumothorax (75%) than solid (49%) or infiltrative (39%) lesions. Biopsy of mediastinal lesions provoked a higher rate of pneumothorax (66%) than pulmonary lesions (49% for peripheral, 51% for central). Pleural-based lesions were comparatively innocuous (11%). It is fascinating that in 19 patients who had a previous thoracotomy, the incidence of pneumothorax was reduced to 17% when

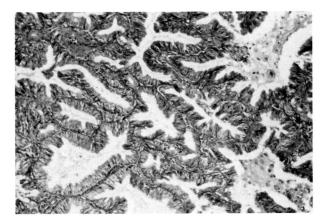

Fig. 7-12. Histologic section. Confirmation of papillary adenocarcinoma of lung. (×160)

the biopsy was performed on the same side as the surgery because of the protective effect of the adhesions or pleural scarring.

The choice of imaging modality[42] and technique[11] may influence pneumothorax. Gobien and associates[19] introduced CT-assisted fluoroscopically guided aspiration biopsy of central hilar and mediastinal masses. The combined image-intensification permits exquisite definition of the lesion and its relationship to critical mediastinal and hilar vessels, allowing for increased accuracy and fewer complications. Pneumothorax was the only complication encountered in 4 of 17 patients. Only one required a chest tube.

Incidence of pneumothorax is related to the coexistence of chronic obstructive pulmonary disease, which predisposes the parenchyma to collapse following traumatization of blebs and bullae and is directly proportional to the number of "passes" required to obtain diagnostic material. Zornoza[54] contributed the experience of the interventional radiologist as another factor, citing his own declining rate of pneumothorax from an overall incidence of 14% to 6% in his last 50 cases. Cormier[7] proposed that breathing 100% oxygen during the procedure decreases the number and size of pneumothoraces by expediting the rapid absorption of small leaks immediately following lung puncture.

Sinner's classical 1976 paper[40] specifically addresses the complications of the fluoroscopically directed transthoracic FNA biopsy of lung from the experienced perspective of 5300 aspirates on 2726 patients. Although the average incidence of pneumothorax was 27.2%, the occurrence of symptomatic pneumothorax requiring treatment was at the acceptable rate of 7.7%.

The recurrent theme in clinical experience with pneumothorax following percutaneous lung aspiration is that occurrence is expected based on violation of the pleuroparenchymal relationship and related to diverse factors, but that the sequelae are either inconsequential or manageable with a chest tube and do not invalidate the disproportionate advantages of the technique.

Hemorrhage

Hemorrhage is not a serious threat to the patient who sustains a percutaneous FNA biopsy of lung or mediastinum, particularly if there is appropriate case se-

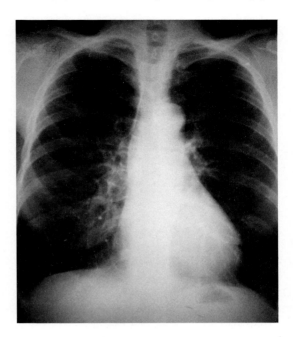

Fig. 7-13. Chest film of a 57-year-old female with chronic obstructive pulmonary disease who presents with a stellate opacification in the right upper lobe.

lection to exclude individuals with hemorrhagic diathesis, pulmonary hypertension, anticoagulant therapy, and arteriovenous malformations. Hemorrhage is generally a radiographic phenomenon without significant clinical expression or consequence. There may be transient expectoration of blood-tinged sputum or mucus; cough followed by self-limited hemoptysis; and rarely, as isolated events usually associated with penetration by cutting needles, endobronchial hemorrhage, hemothorax, or fatal exsanguination.[1,32] Two clinical cases demonstrate the common presentation of hemorrhage radiographically.

■ *Case 7-3*
A 57-year-old woman with chronic obstructive pulmonary disease developed a stellate, irregular, solitary opacification in the right upper lobe (Fig. 7-13). The dense center and serrated periphery were considered evidence for an infiltrative cancer (Fig. 7-14), and aspiration biopsy under fluoroscopic control (Fig. 7-15) was elected as the diagnostic procedure of choice because of the COPD. Perilesional asymptomatic hemorrhage

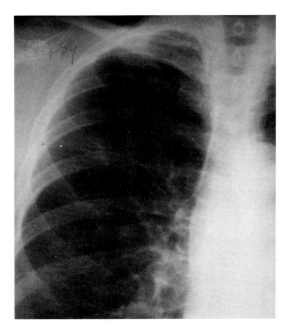

Fig. 7-14. Chest film, focus on stellate irregular solitary opacification in the right upper lobe, prebiopsy.

developed coincidental to the first puncture, resulting in enlargement and enhancement of the target with accentuation of the irregular periphery (Fig. 7-16). This configuration is attributable to silent intralesional and perilesional hemorrhage that does not manifest as hemoptysis, is self-limited, and eventually is resorbed. However, its occurrence precludes additional punctures because of obscuration of the target and delays diagnosis if the harvest did not incorporate a satisfactory sample. In this case, only macrophages (Fig. 7-17) were obtained; and the procedure had to be abandoned without a firm diagnosis. Carcinoma was subsequently confirmed histologically.

■ *Case 7-4*
A 70-year-old woman with recent history of cough presented with a left upper lobe mass situated near the hilum (Fig. 7-18). Fluoroscopy guided the 22-gauge needle through treacherous vascular terrain to its target at the periphery of the lesion (Fig. 7-19). The aspirated cells were arranged in papillary clusters and exhibited

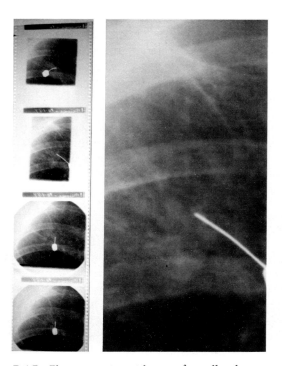

Fig. 7-15. Fluoroscopic guidance of needle placement.

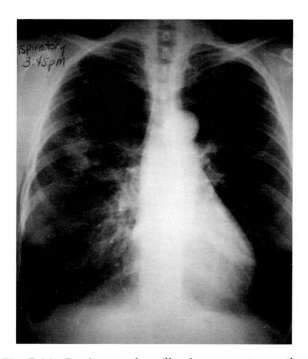

Fig. 7-16. Postbiopsy chest film demonstrating perilesional hemorrhage.

arbitrarily arranged cells with variability in nuclear size, contour, chromatinic texture, number, and position of nucleoli. One cell was captured in mitotic division (Fig. 7-20). A diagnosis of adenocarcinoma was fortunately made with the first puncture because the immediate postbiopsy film (Fig. 7-21) demonstrated enhancement of the aspiration site due to perilesional hemorrhage and the patient coughed a small quantity of blood-tinged sputum.

The perspective on hemorrhage is unclear in the literature because consequential bleeding that came to be associated with FNA biopsy of the thorax was, in reality, the result of tissue trauma inflicted by cutting needles[32]; this impression must be rectified because a 22-gauge flexible spinal needle skillfully placed in a pulmonary target under image intensification is not an instrument of trauma and rarely produces clinically consequential bleeding. Sinner's[40] data specifically utilizing fine needles reflected a mean incidence of local pulmonary bleeding of 11% and of hemoptysis in 5% of central and in 2% of peripheral lesions. He regarded the local bleeding as inconsequential because no patient required treatment; and the bleeding was completely resorbed after a few days, up to 1 to 2 weeks. He reiterated Zelch and Lalli's[30] contention that incidental puncture of significant intrathoracic blood vessels, the right atrium, arteriovenous aneurysms, and adjacent vital organs is innocuous. There was no mortality from hemorrhage in this Scandinavian series of 5300 biopsies performed on 2726 patients.

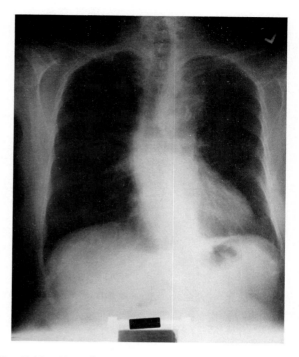

Fig. 7-18. Chest film of a 70-year-old female with recent history of cough and left upper lobe mass near the hilum.

Berquist's[3] data on 430 patients from the Mayo Clinic raised some interesting points about the hemorrhagic complications of aspiration biopsy; however these must be considered in the context of his 18-

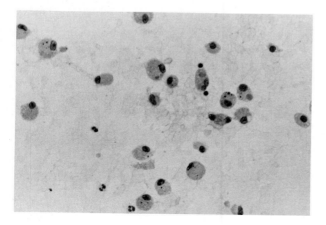

Fig. 7-17. Pulmonary macrophages. (P stain, ×500)

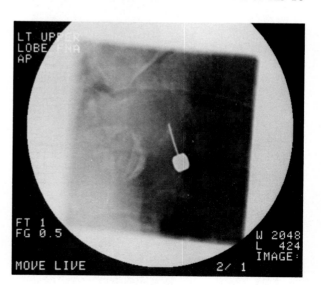

Fig. 7-19. Fluoroscopic guidance of the needle into its target.

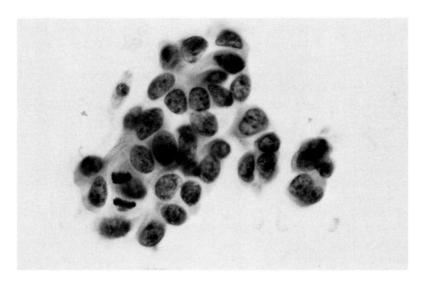

Fig. 7-20. Adenocarcinoma of lung illustrating mitotic division. (P stain, ×800)

gauge needle intended primarily to obtain tissue for histologic study, but failing an adequate sample to provide cells for cytological analysis. The aspiration trajectory in the vicinity of prior thoracotomy raised the incidence of hemoptysis from 11% generally to 42%. Hemoptysis was more frequent in patients with infiltrative (24%) and cavitary (17%) lesions than with solid nodules (10%) and in central rather than

peripheral positions. Two deaths were attributed to hemorrhage (hemoptysis). The first resulted in aspiration of blood into the entire pulmonary tree, and the second terminated a succession of intermittent bouts of hemoptysis with severe bleeding on the fifth day following the procedure. Both lesions were friable, partially necrotic, and cavitary, which explains their fragility and reaction to the trauma of needle penetration. Air-fluid levels in a cavitary lesion may portend communication with a bronchus and provide an avenue for blood to enter the pulmonary tree.

Hemorrhage must be particularly avoided in the immunocompromised patient whose opportunistic pulmonary infections have already embarrassed respiratory function. Matthay and Moritz[30] at Yale commented in their extensive paper on invasive procedures for diagnosing pulmonary infection that none of the smaller series describing aspiration biopsy of the lung in the compromised host documents an incidence that exceeds Sinner's statistics from Stockholm. Castellino and Blank[5] discovered a 3% hemoptysis rate, defined as transient but heavily blood–tinged sputum, in 108 pulmonary biopsies of immunocompromised patients utilizing an 18-gauge thin-walled needle.

When aspiration biopsy of lung, mediastinum, or pleura is performed with the proper instrument skillfully guided by image-intensification for carefully selected patients, hemorrhage is not a significant complication. A conscientious postbiopsy chest film will often demonstrate enhancement of the tar-

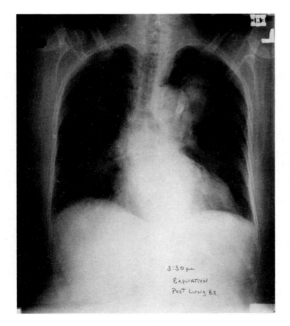

Fig. 7-21. Postbiopsy chest film demonstrating perilesional hemorrhage.

get due to perilesional hemorrhage, which is frequently the only, and silent, manifestation of bleeding from the puncture. Expectoration of blood-tinged sputum or hemoptysis is generally transient and self-limited. Death from endobronchial hemorrhage is exceedingly rare following a fine needle biopsy and has been documented to occur when a cutting needle has been injudiciously selected as the biopsy tool.

Infection

The general impression conveyed by the literature is that infection induced or spread by transthoracic FNA biopsy does not occur or is clinically insignificant. Sinner[40] does not mention infection accruing from the procedure as a complication among his 5300 aspirates. Lalli[27] specified that there were no instances of infection induced by aspiration biopsy in his series of 1296 procedures at the Cleveland Clinic. Linder and colleagues[29] do not include infection as a complication of mediastinal aspiration biopsy for 32 procedures performed on 29 patients during a 6-year period at Duke University Medical Center. Even more encouraging is Castellino and Blank's[5] use of an 18-gauge needle to perform pulmonary aspiration biopsies in 108 presumed infectious episodes in 82 immunocompromised patients without the introduction or spread of infection; they reported only pneumothorax and limited hemoptysis as complications. At the Eisenhower Medical Center, which services patients from a desert environment, there has been successful demonstration of coccidioidomycosis by pulmonary aspiration biopsy without transfer of the organisms to the pleural cavity, body wall, or skin. The technique, therefore, is presumed safe as regards infection.

Air Embolism

Using roentgenograms of the skull, Westcott[50] exquisitely demonstrated air in the basilar artery of a 62-year-old man with chronic lymphocytic leukemia and consolidative pneumonia who coughed immediately following insertion of an 18-gauge needle under fluoroscopic control to obtain microbes for culture. When the coughing subsided, the stylet was withdrawn, the syringe attached, and the needle removed while aspiration was continuous. The patient coughed a small amount of blood and became unresponsive, incontinent, and bradycardic. He then entered irreversible asystole and death supervened in 30 minutes. The postmortem examination demonstrated a puncture wound that entered a bronchus and peripheral pulmonary vessel in the consolidated lung. Endobronchial blood was discovered in association with a frothy mixture of air and blood in the cardiac chambers and arteries.

Westcott elaborated that air in the right heart and pulmonary arteries generally causes no difficulty but that as little as 2 to 3 ml injected into the pulmonary veins may be fatal. He described two situations in which air embolism can occur at the time of pulmonary aspiration — (1) atmosphere to lung: If the needle tip lodges in a pulmonary vein, no stylet or stopcock intercedes, and the atmospheric pressure exceeds pulmonary pressure; (2) bronchus or lung to pulmonary vein: If the needle punctures a cavity, air cyst, or bronchus, and then the pulmonary vein, in a rigid, consolidated lung that prevents venous collapse, air embolism can occur if the pressure in the cavity, air cyst, or bronchus exceeds venous pressure. Coughing incidental to the unfortunate puncture of the vein may raise intrathoracic pressure sufficiently to push air into the vein. The situation may be avoided if the needle is kept closed with a stylet or the operator's finger and the patient refrains from straining or coughing.

Sinner[40] referenced Westcott's report of air embolism as the only strictly documented occurrence, despite circumstantial clinical evidence lacking pathologic proof in other illusory instances, and estimated the relative risk per patient at 0.07%.

Subcutaneous Emphysema

The dissection of air into the subcutaneous tissues (Fig. 7-22) following needle aspiration biopsy of pulmonary lesions is a rather benign complication that attracts little attention or clinical concern. It is generally a limited process that spontaneously resorbs. Detection may be by palpation of crepitance or roentgenographic display.

Sinner[40] reported seven cases (2726) of moderate to marked subcutaneous emphysema accompanying a large pneumothorax, but no supplementary treatment was required.

Klein and Gamsu[25] described pneumomediastinum as an unusual complication of the procedure

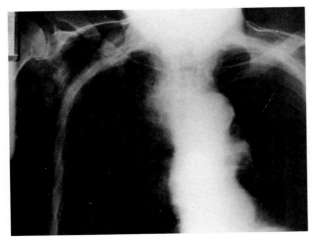

Fig. 7-22. Chest film of a 56-year-old male demonstrating subcutaneous emphysema as a complication of aspiration biopsy of lung.

and related its development to possible origin from subcutaneous emphysema initiated in the chest wall with subsequent extension and dissection of gas downward to the mediastinum along fascial planes. An alternate explanation was the traumatic rupture of alveoli with an interstitial route of dissection.

Translocation of Tumor Cells

The risk of mechanical tumor spread by FNA biopsy of pulmonary lesions was analyzed in the classical paper by Sinner and Zajicek.[41] The technique demonstrated malignancy in 1264 patients, but in only one did malignancy develop at the biopsy site. Based on this incidence and a correlative study of reported comparable survival rates for patients with pulmonary cancers diagnosed by aspiration biopsy in contrast to other technique, these scientists concluded that FNA biopsy may be reasonably and safely used when clinically indicated. Their personal search of the literature revealed no definite reports of local tumor extension caused by the aspiration procedure. Furthermore, a close look at their references implicates large bore cutting needles as the instruments responsible for needle track translocation of tumor cells and for the unjustifiable penumbra over the reputation of FNA biopsy. Reports that accentuated implantation metastasis conveniently failed to mention the size of the needle. Dutra and Geraci[14] disparag-

ingly reported a single occurrence of subcutaneous implantation metastasis and recommended that needle biopsy of the lung be restricted to patients in whom there is evidence for inoperability. They used a Vim–Silverman needle. Wolinsky and Lischner[51] reported the second documented occurrence of needle track implantation of tumor after percutaneous biopsy of an adenocarcinoma of the lung and provided a dramatic photograph of a large subcutaneous neoplasm on the emaciated chest of the 70-year-old victim. A Franklin–Silverman needle was the culprit. A more reliable report[31] that truly focuses on the rarity of the complication with *bona fide* thin needles demonstrated the occurrence of tumor in the chest wall 13 months after a transthoracic aspiration biopsy of a composite bronchogenic tumor with partial squamous and partial spindled (carcinoid, intermediate oat-like) patterns. It was justified by the position of the implant metastasis at the site of the biopsy and by concordance of cell architecture in aspirated samples from the lung lesion and in the skin biopsy. The modern perspective does not encourage abandonment of the technique, but rather emphasizes translocation of tumor as an exceedingly rare event.[13]

An interesting, but somewhat confusing model was developed by Struve–Christensen[43] for the experimental histologic study of the needle track produced by FNA of pulmonary specimens resected at surgery for malignant disease. After demonstrating that 89% of the needle tracks contained malignant cells, he elaborately defended the technique by proposing that the insufflated lung provided less friction to the needle than the atelectatic, lifeless lung; that the living lung destroyed tumor cells translocated to the tracks; and that tumors exposed to needles for diagnosis were so soon eradicated that time for a neoplastic foothold was preempted.

Engzell and associates[15] designed an experimental and clinical investigation to assess the occurrence and nature of needle track translocation with implantation metastases following aspiration biopsy (in general) using 18-gauge and 22-gauge needles, respectively. They concluded that the gauge of the needle may be the critical factor. The smaller needle produces a smaller puncture wound and less trauma; and it transports less stroma, an element that may be essential to carryover and implantation.

Two other studies looked at survival as a means

for analyzing the threat of needle aspiration regarding tumor cell translocation. Berg and Robbins[2] classical work at the Memorial Hospital in New York City analyzed survival times for 1412 patients subjected to radical surgery; 370 patients with breast carcinoma diagnosed by FNA biopsy were compared to an equal control group. No statistical differences in morbidity or mortality were determined for a 10-year period, indicating that the procedure had no deleterious affect on survival in mammary carcinoma. Von Schreeb[37] reported a series of 150 cases of renal cell carcinoma treated by nephrectomy. In 77 cases, preoperative needle aspiration with secondary injection of contrast media was performed, with the remaining cases serving as controls. An analysis of the 5-year survival rate disclosed no significant differences between the two groups.

It is imperative that the contemporary clinical scientist comprehends the context in which FNA biopsy is accused of tumor cell translocation and implantation metastasis and that the rarity of this event be maintained in perspective to support, rather than condemn, a technique that offers advantages of a more realistic magnitude.

Fatalities

Reported deaths secondary to percutaneous transthoracic FNA biopsies are exceedingly rare, although the generic references to the technique tend to provoke a defamatory regard because of the implicit confusion with cutting instruments. The historical scrutiny by Norenberg and colleagues[32] accomplished a linear exposé of reports, partially clarifying the confusion and permitting exoneration of fine needles. This paper emphasized that reported fatal complications have occurred almost exclusively with cutting needles (generally Vim–Silverman or Franklin–Silverman modifications) rather than with fine needles or the high speed air drill trephines; furthermore, the common pattern of death was massive endobronchial hemorrhage. Pearce and Patt[33] reported in the same year a case of fatal pulmonary hemorrhage after percutaneous lung biopsy. The needle was 18-gauge, certainly not the contemptible cutting needle, but also not the refined 22-gauge flexible spinal Chiba needle that is generally safer. Death was due to massive endobronchial hemorrhage, but the underlying pathology was pulmonary

thromboemboli and infarcts that certainly predisposed to a hemorrhagic exit.

Contrasted with grim commentaries on the large bore needles was the cited report from Dahlgren and Lind[8] of more than 3000 FNA biopsies without fatality. The recommendation was for the use of fine needles and the condemnation of cutting needles. Sinner's[40] most careful scrutiny for complications in 2726 patients disclosed no mortalities.

Perspectives on Complications

Ten years ago in America, cytopathologists with a vision about aspiration biopsy and an intuitive compulsion to convince a skeptical medical community about its advantages found themselves defending and promoting the technique by exploring the true occurrence of complications in the hands of pioneer individuals. Their progress was often thwarted by publication of outrageous clinical anecdotes and inflexible opinions about "needle biopsy" that were unrealistic because they erupted from the misfortunes of the cutting needle, which structurally, philosophically, and theoretically bore little resemblance to the 22-gauge fine needle instrument in use throughout the world today. We now realize that hemorrhage is generally a radiographic phenomenon or expresses itself clinically by the transient appearance of blood-tinged sputum or mild hemoptysis. We understand that pneumothorax is an expected concomitant of a procedure that violates the pleura, but is generally asymptomatic or responsive to evacuation or a chest tube. Translocation of infectious agents is not reported and implantation metastases from tumor cells in transit are rare, with most anecdotal reports confusing fine needles with the responsible cutting needles. Air embolism can be avoided by a meticulous technique that prevents atmospheric air from entering pulmonary veins or by controlling cough and excluding the candidacy of the patient who cannot cooperate with breathholding. Subcutaneous emphysema is a curious event but requires no therapeutic intervention. Fatalities can occur but are considerably less likely with fine needles than with cutting instruments and are generally related to endobronchial hemorrhage. Careful selection of patients with conscientious consideration of disqualifying contraindications should reduce the small numbers of deaths that may happen. As medical sci-

Table 7-2. Cytologic Classification of Carcinoma

KOSS[26]	JOHNSTON AND FRABLE[20]
Epidermoid	Epidermoid
Keratinizing (WD)	Keratinizing
Nonkeratinizing (PD)	Nonkeratinizing
Large cell	Large cell anaplastic
Small cell	Small cell anaplastic
Oat cell	Adenocarcinoma
Adenocarcinoma	Bronchogenic acinar and
Bronchoalveolar	papillary
Bronchogenic adenocar-	Bronchoalveolar
cinoma	Giant cell carcinoma
Mucoepidermoid	Mucoepidermoid carcinoma
Giant cell	

DAHLGREN[52a]	LINSK AND FRANZEN
Squamous cell	Squamous
Small cell anaplastic	Adenocarcinoma
Adenocarcinoma	Bronchoalveolar
Bronchoalveolar	Undifferentiated
Large cell anaplastic	Large cell type
Bronchial adenomas	Intermediate lymphomatoid
Cylindromas	Small cell type (oat cell)
Carcinoid tumors	Giant cell
	Bronchial gland
	Carcinoid
	Mucoepidermoid
	Adenoid cystic

entists, we must be optimistic about a technique that offers overwhelming advantages of rapid, precise, cost-effective, tissue-equivalent diagnostic information with minimal invasion and patient discomfort.

CLINICAL SPECTRUM AND ASPIRATION CYTOLOGY OF THORACIC DISEASE

Clinical examples of infectious and neoplastic disease will be considered according to anatomic region, in descending order of targeted frequency within the thorax, lung, mediastinum, pleura, chest wall, and ribs. For carcinomas of the lung, Table 7-2 summarizes four pertinent and pragmatic classifications to assist in categorization of tumor according to cytomorphological features in the context of prescribing therapy.

The purpose of cellular analysis of a tumor by aspiration biopsy is to classify the lesion accurately so that the clinician or surgeon can reliably decide the course of therapy and establish prognosis without resorting to traditional, invasive biopsy with histopathologic efforts. If the pulmonary tumor is potentially resectable, cellular classification will verify that it belongs to surgical categories such as adenocarcinoma or squamous carcinoma or will reassign the lesion to a therapeutic trial that circumvents surgery (*e.g.*, oat cell carcinoma, which responds effectively to chemotherapy). Inoperable lung tumors may well provide the most frequent targets for aspiration biopsy. Patients in this category include those with the following disorders:

Lesions identified as inoperable by size or rib invasion on x-ray
Compromise of laryngeal nerves
Mediastinal and hilar adenopathy
Known primary neoplasm elsewhere with radiographic pattern of pulmonary metastases (Some solitary and multiple metastatic nodules may retain operability, however.)
Interstitial lung pattern suspicious for lymphangitic spread
Medical diseases that preclude surgery

The information obtained impacts directly on patient care and clinical judgment for the conduct of management. It is therefore imperative that every effort be expended to obtain an appropriate analytical sample, to interpret and classify it accurately and reliably, and to communicate the results expediently to other members of the health care team. This spirit of unconditional cooperation and responsibility cannot be emphasized enough. The cell yield is within the jurisdiction of the radiologist to provide and the cytopathologist to assimilate. Cell yield depends on the distribution of the tumor cells or other processes within the lung, the technical process of acquisition, the cytopreparatory expertise during processing, and natural influences on the lesion such as necrosis, inflammation, and hemorrhage. If puncture is made in an area in which tumor has infiltrated but not replaced normal lung tissue, the cell yield may be a composite of neoplastic cells, pigmented (carbon or hemosiderin) pulmonary macrophages, reactive pneumonocytes, and benign epithelium. If puncture is confined to a solid neoplasm, a luxuriant harvest of viable tumor cells in arrangements predicting histoarchitecture is expected. If infarction, necrosis, inflammation or hemorrhage affect the tumor, the cellular detritus of tumor diathesis contaminates the smear. State of preservation may be a function of tumor viability rather than fixation imposed at acquisition. The cytopathologist must search inquisitively among the elements of debris to discover tumor cells or micro-organisms and provide ocular dissection of fibrin meshworks in bloody smears to excavate the cancer. Mucus strands incorporating epithelial cells must be considered with caution for the separation of benign cells from mucinous carcinoma. The composition of the cell yield is different from typical exfoliative preparations for two reasons: (1) exfoliated cells are frequently modified adversely by their environment resulting in reduction of morphologic crispness; and (2) the architectural structure of some neoplasms precludes, inhibits, or promotes exfoliation, modulating what cells may be detected by exfoliative methods. All pulmonary tumors are however potential cellular donors to aspiration biopsy.

The most fundamental decision a cytopathologist must make when confronted with an aspiration biopsy from a thoracic lesion is whether the process is benign or malignant. Among the myriad of diverse primary epithelial malignancies, the only clinically significant distinction is oat cell versus non–oat cell carcinoma because of the favorable chemotherapeutic implications for small cell undifferentiated carcinoma and the reprieve from surgery. Identification of subtypes under the adenocarcinoma and undifferentiated classifications is a useful cytologic exercise that will help further refine clinicians' perceptions of the histologic and histogenetic complexity of these tumors, but whether they are distinguished by prognostic difference remains to be determined. Preliminary review of the Ludwig study of 800 bronchogenic tumors in Europe fails to show any correlation of subtypes with survival.* Separation of squamous carcinomas and adenocarcinomas into well-differentiated and poorly differentiated tumors should provide prognostic criteria in keeping with general principles of tumor biology. Clinical experience with bronchogenic carcinoma suggests, however, that staging may be more important than differentiation in predicting tumor conduct. Keep in mind also that one cytologic type, (e.g., squamous carcinoma) may have well-differentiated and poorly differentiated

* JS Willems, personal communication.

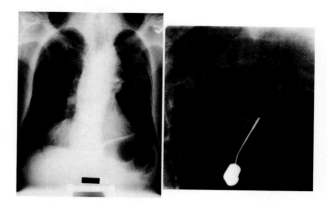

Fig. 7-23. Chest film of a 71-year-old male demonstrating lesion in right upper lobe contrasted with the needle in its target.

areas represented in a single smear. Although the broad range of bronchogenic carcinoma presently responds poorly to treatment, attention to cytologic diagnosis should be continued in anticipation of further therapeutic advances.

MORPHOLOGY

Through a series of brief encounters with patients whose lesions were analyzed by needle aspiration cytology, we intend to communicate the spectrum of disease that can be defined by aspiration biopsy and illustrate the appearance in this medium of common and unusual, neoplastic and inflammatory aberrations that affect the thoracic structures.

Squamous Cell Carcinoma

A 71-year-old man with a history of pulmonary adenocarcinoma presented with a radiographic opacification in the right upper lobe, which was examined by aspiration biopsy under fluoroscopic control (Fig. 7-23). The cellular harvest stained by the P technique is strikingly polymorphous and diverse. Dissociated cells vary from polygonal to stellate, with elongate forms exhibiting tapering, eosinophilic, homogeneous cytoplasm. Angular, oval, sometimes lobulated nuclei vary considerably in size but are universally, densely hyperchromatic and homogenized (Fig. 7-24). Foci of tenuous cellular cohesion in abortive mosaics appear in juxtaposition to epithelial

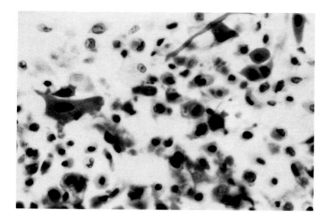

Fig. 7-24. Keratinizing squamous cell carcinoma. (P stain, ×625)

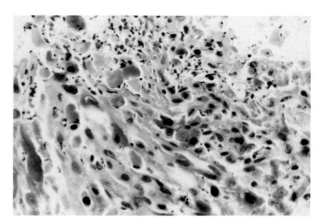

Fig. 7-26. Histologic section. Keratinizing squamous cell carcinoma. (×625)

pearls composed of cells embraced in concentric swirls (Fig. 7-25). Intercellular bridges are not evident, correlative with dyshesion. The background contains cytoplasmic fragments, inflammatory cells, and erythrocytes. Fiberoptic bronchoscopy subsequently obtained partially necrotic tissue whose residually viable component presents stratified spindle and polyhedral cells with hyperchromatic and angular, fusiform nuclei surrounded by abundant, eosinophilic, homogeneous cytoplasm (Fig. 7-26). The cellular identity between aspirated material and the traditional histopathology is self-explanatory and constitutes the classical cytomorphology of keratinizing squamous cell carcinoma. When these lesions progressively enlarge, they characteristically cavitate centrally and the probe, which reaches the interior of

the lesion, may isolate only necrotic tumor cells and the elements of a diathesis. As squamous cell carcinoma dedifferentiates, the cells may become smaller, the nuclei more fusiform and less hyperchromatic, and the tendency to dissociation more common, so that singly dispersed cells are seen among tenuous aggregates. Chromatinic detail may become more obvious, so that texture and nucleoli are visible.

Sometimes the only clue to squamoid differentiation is the pattern of organization within loosely cohesive groups that portrays a sense of stratification and mosaic (Fig. 7-27). The differential diagnosis provoked by cellular detail and arrangement is the intermediate variant of oat cell carcinoma. However, persistent search through the smears reveals suggestion of keratinization and pearl formation, with spin-

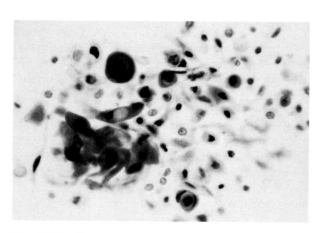

Fig. 7-25. Keratinizing squamous cell carcinoma. (P stain, ×625)

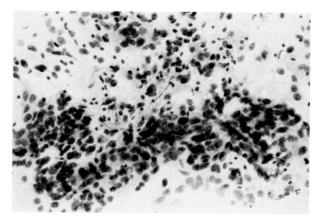

Fig. 7-27. Poorly differentiated squamous cell carcinoma. (P stain, ×400)

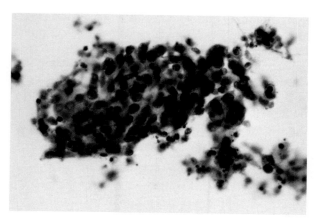

Fig. 7-28. Poorly differentiated squamous cell carcinoma. (P stain, ×500)

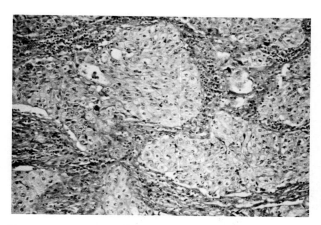

Fig. 7-30. Histologic section. Poorly differentiated squamous cell carcinoma. (×200)

dled and tadpole microforms. Poorly differentiated squamous cell carcinoma loses keratin-referable details. Spindles, tadpoles, and pearls disappear and are replaced by smaller cells with reduced, often minimal cytoplasm surrounding nuclei, which are variably hyperchromatic or vesicular, and almost carried within a syncytium (Fig. 7-28). Another panoramic view of the aspirate smear from the same case (a biopsy of a 59-year-old woman with cough, hemoptysis, and a tomographically confirmed left upper lobe radiodensity) shows poorly differentiated aggregates in a background of inflammatory cells and pulmonary macrophages (Fig. 7-29). Occasional, larger cells maintain their polygonal contours, as if just separated from their mosaic affiliates, and provide densely hyperchromatic, smooth nuclei so fa-

miliarly characteristic of squamous origin. These cells confer identity to the darkly clustered elements dispersed among inflammatory cells and fibrin. The lobectomy specimen incorporated a 3-cm mass with serrated margins and a centrally excavated zone. Histoarchitecturally, polyhedral cells relate in mosaic, insular aggregates separated into lobules by circumscribing, delicate connective tissue septa (Fig. 7-30). Individual nuclei appear vesicular with prominent eccentric nucleoli and are surrounded by abundant, clear cytoplasm. An occasional squamous tumor will donate cells that conform to arrangements that imitate adenocarcinoma so convincingly that a diagnostic error results.

A 58-year-old man was referred for a fluoroscopically directed aspiration biopsy of a mass in the

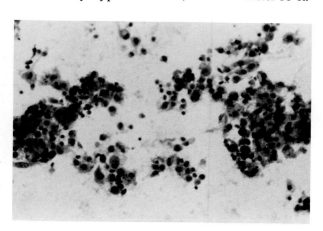

Fig. 7-29. Poorly differentiated squamous cell carcinoma. (P stain, ×400)

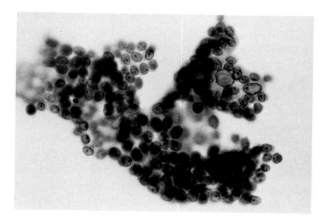

Fig. 7-31. Squamous cell carcinoma masquerading as papillary adenocarcinoma. (P stain, ×625)

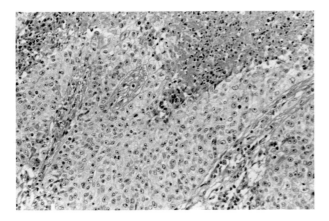

Fig. 7-32. Histologic section. Tissue correlate of Fig. 7-31, confirming squamous cell carcinoma. (×315)

right upper lobe considered to represent either carcinoma or fungoma. A cellular sample is organized in three-dimensional papillary clusters with the peripheral cells aligned serenely and cohesively in regular regimentation (Fig. 7-31). Minimal delicate cytoplasm surrounds large, oval, vesicular nuclei with prominent eccentric nucleoli. Occasional nuclei appear disproportionately large and considerably more "ground glass." These are considered features of papillary adenocarcinoma of lung, and the case was so designated pre-operatively. A right upper lobectomy resected at the apex incorporated a variegated, red-brown centrally excavated mass measuring 4.5 × 3.7 × 3.7 cm, containing a central accumulation of caseous, necrotic material. Microscopically, a solid proliferation of polyhedral cells imbricated in mosaics is altered by central necrosis (Fig. 7-32) imparting a serrated character to the interface edges of viable tumor with necrotic debris. Delicate cytoplasm surrounds and highlights vesicular nuclei with prominent nucleoli identical to the cytomorphology of cells expressed in the aspirate as papillary groups. Despite cellular identity, the architectural arrangement is persuasively squamoid and no glandular features could be discovered. No "pearls" of keratin or wisdom assist with this distinction in isolated cases.

The possible confusion of squamous with adenocarcinoma has particular significance when the lesion is peripherally situated, if the interpretation of cell type is influenced by the location of the lesion. Koss[26] reminds his readers that about 25% of peripheral lung cancers presenting as small roentgenographic abnormalities are squamous carcinomas and that squamous tumors constitute the second largest group of peripheral bronchogenic cancers. Peripheral squamous cancers are generally poorly differentiated, contributing to the differential confusion. The resemblance of our last case to adenocarcinoma may be a function of diminished differentiation in a peripheral (apical) carcinoma.

Adenocarcinoma

The interventional radiologist and cytopathologist actively involved in aspiration biopsy of the lung will encounter pulmonary adenocarcinoma as the most frequent peripheral primary lung tumor that may be variably associated with pleuroparenchymal scars or represent derivatives of bronchial glands, type II pneumonocytes, or Clara cells. The architectural spectrum includes papillary, mucinous, bronchioloalveolar, and arbitrary arrangements and embraces variable degrees of differentiation and overlap with squamoid expressions. Histochemical and immunoperoxidase stains may be useful in implicating mucus-production, glandular versus mesothelial origin, and metastatic versus primary adenocarcinoma; but the basic P stain will satisfactorily display nuclear features characteristic of adenocarcinoma (Plate XVI) and reveal patterns of cellular interrelationships that reflect histological structure and associated findings. The facility for identification of an adenocarcinoma is directly proportional to the degree of differentiation; as the lesions digress from specific criteria with loss of differentiation, there is a greater tendency to overlap with squamoid patterns, thereby in-

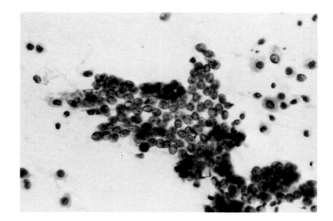

Fig. 7-33. Papillary adenocarcinoma of lung. (P stain, ×625)

troducing confusion that may be further extrapolated to truly complex tumors with composite patterns. Specific case studies illustrate patterns and perceptual difficulties.

A 76-year-old man presented with cough and radiographic evidence for multiple nodules in the left lung. A lingular biopsy was performed under fluoroscopic guidance and harvested a luxuriant cellular population that relates in papillary conformations (Fig. 7-33). Cellular boundaries are indistinct, so that the cellular arrangement is both honeycombed and syncytial. Cells are aligned in radial palisades at the periphery, imparting a smoothly scalloped and gently undulating margin to the papillations. Nuclei vary only subtly in size and are consistently round or oval with delicately granular nucleoplasm and slightly eccentric, punctate but intense nucleoli. Nuclear margins are crisply delineated, and there is minimal overlapping without molding. Cellular cohesion prevails, and there is little dissociation. This pattern is typical of well-differentiated *papillary adenocarcinoma* of lung and correlates identically with the histologic counterpart (Fig. 7-34). Papillary structures are created by the propagation of uniform, palisaded columnar cells along fibrovascular stalks. Uniform nuclei with vesicular nucleoplasm and prominent nucleoli are basally situated. Papillary projections extend into alveolar spaces where fibrin precipitates and macrophages are sporadically distributed in the background. Compare the cytological presentation of this tumor with a cellular aspirate from a 67-year-old male smoker evaluated for chest and back pain accompanied by weight loss (Fig. 7-35). The individual

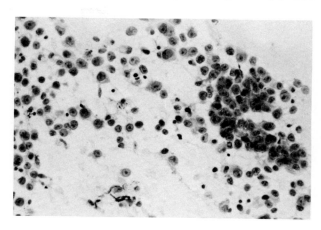

Fig. 7-35. Poorly differentiated adenocarcinoma. (P stain, ×400)

cells are reminiscent of constituent elements of the papillations in Figure 7-32 with respect to vesicular nucleoplasm and prominent nucleoli, but there are important differences. The individual cells are arranged rather haphazardly in sheets, cords, and abortive acini rather than in structured papillary aggregates. Nucleoli are large, dense, and coarse rather than sharp and punctate. The nuclear chromatin is more obtuse and the margins heavily accentuated rather than crisp, thin, and sharp. There is a tendency to molding and greater anisokaryosis accentuated by the immediate juxtaposition of cells in crowded congeries. Cytoplasm is more conspicuous and abundant but cellular edges are ambiguous. If the cells in papillations are considered well-differentiated adenocarcinoma, these elements contrast as poorly differentiated adenocarcinoma with dyshesion and deviation from uniformity as glaring attributes.

A 73-year-old woman presented with a radiographic density in the right upper lobe. An aspiration biopsy incorporated cellular aggregates distributed in a secretory background of fibrillar, pink-gray mucus (Fig. 7-36). The polyhedral and columnar cells vary considerably in size and present angular, erratic, densely hyperchromatic nuclei surrounded by eosinophic, vacuolated cytoplasm. Mucus vacuoles displace nuclei peripherally and distort cell contours. Cellular cohesion provides for the unification of small cell clusters entrapped within the interstices of the mucoid matrix. A right upper lobectomy incorporated a subpleural 2.2-cm mass histologically represented by neoplastic columnar and polygonal ebul-

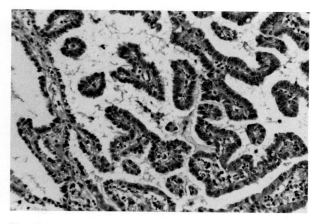

Fig. 7-34. Histologic section. Papillary adenocarcinoma of lung. (×315)

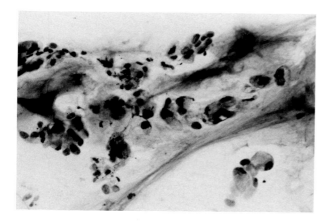

Fig. 7-36. Mucinous adenocarcinoma of lung. (P stain, ×500)

lient cells arranged in well-developed acini and sheet-like aggregates with central necrosis and strong cytoplasmic mucin-positivity (Fig. 7-37). This confirms the typical appearance of *mucinous carcinoma* whose cytological features include cytoplasmic and extracellular mucus accumulation, nuclear hyperchromasia, and abnormalities in peripheral contours.

Tao and associates[44] discussed not only the possibility of cytologic diagnosis of *bronchioloalveolar carcinoma* by transthoracic FNA biopsy, but offered criteria to identify and subclassify the lesion into nonsecretory, secretory, and poorly differentiated variants. Nonsecretory and secretory tumors are constructed of polygonal cells with round, uniform nu-

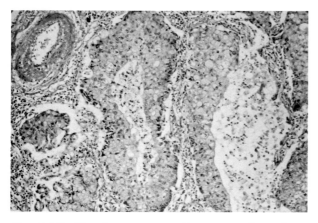

Fig. 7-37. Histologic section. Tissue correlate of mucinous adenocarcinoma of lung. (×160)

clei containing delicate chromatin granules, ranging from 25 to 35 μm in diameter. Nonsecretory cells relate in monolayered sheets with occasional tendency to papillary formation, whereas secretory cells are arranged in three-dimensional clusters with overlapping nuclei. Nonsecretory cytoplasm is delicately vacuolated while secretory cells have abundant foamy, transparent, or singly vacuolated cytoplasm. Poorly differentiated carcinoma contains cells with anisokaryosis, anisonucleosis, abnormally disturbed nuclear cytoplasmic ratios, coarsely granular chromatin, and prominent nucleoli. Cytoplasm is generally neither foamy nor vacuolated. Distinction from bronchogenic or metastatic adenocarcinoma is possible because the alternative differentials generally provide multilayered cells that exceed 35 μm in diameter and present multivacuolated cytoplasm with mixtures of secretory or nonsecretory elements in the same cell clusters.

Figure 7-38 illustrates a cell cluster obtained by fluoroscopically directed aspiration biopsy from a lesion at the apex of the right lung in a 67-year-old man. The cells are essentially arranged in a monolayered sheet in a configuration suggesting the tip of a papillary structure. Communal exterior borders are smooth, and the cellular integration creates a serpiginous peripheral contour. Individual cells are polygonal with uniform, round nuclei containing a delicate chromatin pattern, visible but not particularly emphatic nucleoli, and sharply delineated nuclear margins. There is slight overlapping at the center. A diagnosis of bronchioloalveolar carcinoma was made, consistent with Tao's nonsecretory variant. A right upper lobectomy incorporated an indurated, partially cavitary, 4.5-cm mass at the apex. Histologic sections demonstrate neoplastic polygonal and columnar cells whose propagation relies on thickened alveolar septal walls for supportive scaffolding (Fig. 7-39). Attenuated septa serve as fibrovascular cores for a papillary growth pattern in delicately arborescent fronds, replete with cells containing intranuclear vacuoles (Fig. 7-40). Compare this case with the cytomorphology of poorly differentiated bronchioloalveolar carcinoma (Fig. 7-41). The cells attempt to cluster but have abdicated cohesive communal affiliation. Uniformity is replaced by considerable variability in cellular and nuclear size. The nuclear: cytoplasmic ratios are high, and the nuclear configurations aberrant with coarse textural characteristics; condensation of chromatin along folds in the mem-

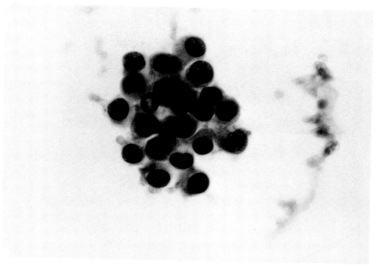

Fig. 7-38. Well-differentiated bronchoalveolar carcinoma. (P stain, ×800)

branes and along their edges; interchromatin clearing; and prominent, dark, eccentric nucleoli. Occasional cells along the periphery of the cluster are reminiscent of alveolar pneumonocytes. The histological counterpart (Fig. 7-42) reflects the characteristic proliferation along alveolar septal supports, but the cells are anaplastic rather than uniform and correspond to cells whose nuclear derangements are so clearly defined by the P stain. These characteristic features of bronchioloalveolar carcinoma will result in a correct tissue-equivalent diagnosis when there is direct correlation with the chest film and clinical presentation.

Poorly differentiated adenocarcinoma is composed of medium to large cells forming loosely cohe-

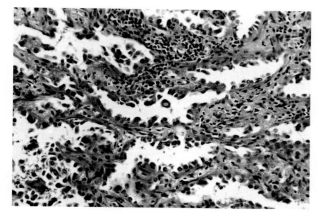

Fig. 7-40. Histologic section. Bronchoalveolar carcinoma illustrating intranuclear vacuolization. (×315)

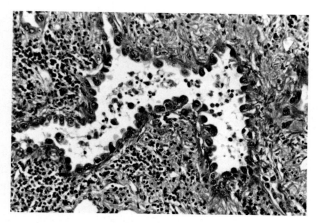

Fig. 7-39. Histologic section. Tissue correlate of well-differentiated bronchoalveolar carcinoma. (×500)

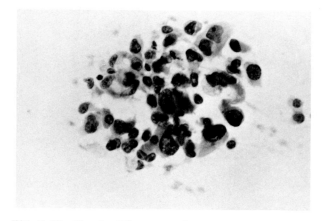

Fig. 7-41. Poorly differentiated bronchoalveolar carcinoma. (P stain, ×625)

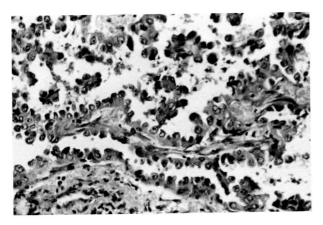

Fig. 7-42. Histologic section. Tissue correlate of poorly differentiated bronchoalveolar carcinoma. (×400)

sive, unstructured clusters without definitive acinar structures (Fig. 7-43). Individual cells provide classical parameters of malignancy including variability in nuclear size and shape, chromatin clumping, nucleolar prominence, overlapping, or molding (Fig. 7-44). Cytoplasm is pale, delicately vacuolated, occasionally abundant, and often without distinct borders. The arrangement may be so ambiguous that clues to glandular derivation may simply be by histochemical evidence for mucus production by an occasional cell or focal abortive acinar formations with some semblance of a central lumen. When adenocarcinoma dedifferentiates further, the cells become distinctly anaplastic and mimic large cell undifferentiated carcinoma. A 52-year-old man with history of grade II

transitional cell carcinoma of the urinary bladder was referred for aspiration biopsy of a right upper lobe lesion clinically suspect for a metastatic deposit. The needle extracted large, polyhedral cells with markedly irregular nuclei of variable size and contour with reticulated chromatin, hyperchromasia, angularity, and spiculation (Fig. 7-45A). Abundant cytoplasm appears delicately vacuolated and cyanophilic. Contrast these features with MGG counterparts of this poorly differentiated adenocarcinoma (Fig. 7-45B and C). Metastatic transitional cell carcinoma was effectively excluded, and the features of poorly differentiated primary pulmonary adenocarcinoma were illustrated. A right upper lobectomy was predicated on this tissue-equivalent information. Microscopically, the tumor appears as a well-circumscribed, but unencapsulated proliferation of anaplastic epithelial cells arranged in mosaic sheets, interdigitated with inflammatory cells. Individual polyhedral tumor cells present slightly granular or foamy cytoplasm containing mucus-positive vacuoles surrounding nuclei that correspond to those of the aspirated cells. An occasional abortive attempt at glandular formation is noted (not illustrated) but for the most part, cells are distributed in arbitrary sheets with the inflammatory infiltrate eclipsing a cohesive effort (Fig. 7-46). It is sometimes impossible to assign a specific category to a population of cells, and the subsequent availability of tissue does not easily resolve this enigma.

A 71-year-old man presented with an incidental finding of a left upper lobe coin lesion on a routine

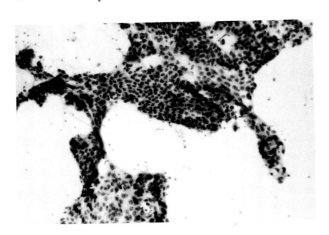

Fig. 7-43. Poorly differentiated adenocarcinoma. (P stain, ×160)

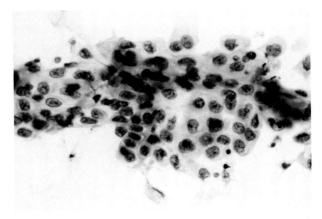

Fig. 7-44. Poorly differentiated adenocarcinoma. (P stain, ×500)

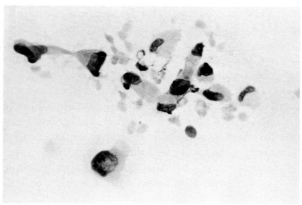

A

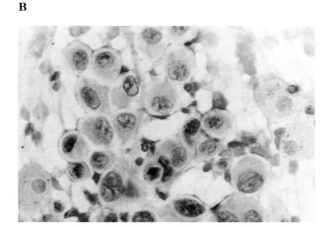

B

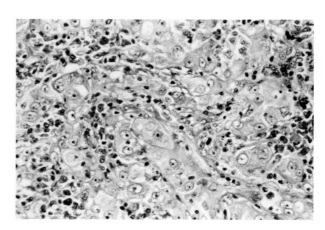

C

Fig. 7-45. A. Dyshesive, poorly differentiated adenocarcinoma. (P stain, ×800) **B.** Poorly differentiated adenocarcinoma. (Wright–Giemsa Stain, approximately ×625) **C.** Poorly differentiated adenocarcinoma. (Wright–Giemsa stain, approximately ×800)

screening chest film. The aspirated cells are uniform in size, with only subtle and minimal variability in nuclei, suggesting a small cell pattern to be distinguished from oat cell carcinoma (Fig. 7-47). Chromatin is delicate; nucleoli are visible, multiple, and occur in unpredictable numbers. No particular pattern is established to assign architectural organization to a glandular or squamoid derivative. This case was simply reported as poorly differentiated carcinoma of the non–oat cell variety. Histological sections prepared from the lobectomy (Fig. 7-48) demonstrate a localized but poorly delimited proliferation of polyhedral and elongate cells with vesicular nuclei and pale, slightly granular cytoplasm. The uniform nuclei exhibit clear vacuolization and prominent nucleoli without significant mitotic reproduction. The lobulated arrangement of the cells is further accentuated by central acantholytic necrosis and by delicate fi-

Fig. 7-46. Histologic section. Tissue correlate of poorly differentiated adenocarcinoma. (×500)

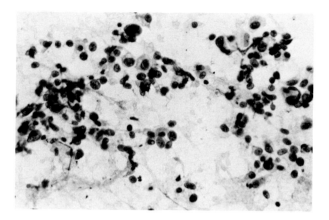

Fig. 7-47. Poorly differentiated carcinoma, "non-oat cell type." (P stain, ×500)

brous septal circumscription. Rare cells stain positively for mucin production (not illustrated) and a metastatic deposit in one lymph node suggested a papillary differentiation. Therefore, the histoarchitectural pattern is considered to be consistent with poorly differentiated carcinoma whose features are intermediate between squamous and glandular differentiation, confirming the non–oat cell status of the aspiration biopsy interpretation. This conclusion required considerable effort and was not reached by inspection of the aspiration biopsy exclusively.

Associated findings of interest in aspiration biopsies of pulmonary adenocarcinomas deserve brief comment. Adenocarcinomas, particularly in the presence of extracellular mucus, may exfoliate their

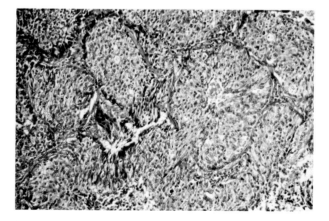

Fig. 7-48. Histologic section. Tissue correlate of poorly differentiated carcinoma, "non-oat cell type."

cells in conjunction with the appearance of multinucleated histiocytic giant cells. The aspirated material from a right lower lobe lesion in a 68-year-old woman with chronic cough demonstrates in panoramic views the co-existence of papillary adenocarcinoma with multinucleated histiocytes of variable size, number, and arrangement of constituent nuclei (Fig. 7-49A and B). The gosamer background is a fibrillar meshwork of mucus strands entrapping inflammatory cells and the dominant bimorphic population of concern. This case, so representative of our early experience, stimulated a futile search for etiological organisms such as *Coccidioides immitis* that could be provocative of a granulomatous reaction. Histology provided a reasonable explanation. The multinucleated histiocytes are identified within dilated, generally stellate acini lined by the proliferating adenocarcinoma, but containing inspissated mucus (Fig. 7-50). Organized granulomata were never discovered, and organisms could not be demonstrated by stains or cultures. The multinucleated histiocytes are evidently in response to extravasated mucus that seems to be the only associated stimulus. What is interesting, and perhaps contradictory, however, is Kini's[24] description of multinucleated histiocytes in association with papillary carcinoma of the thyroid in which mucus production is not a feature. Conjecturally, it could be the papillary nature of the neoplasm rather than an extracellular secretory product that is the offending mechanism.

Roggli and colleagues[36] published the first report documenting detection of asbestos bodies by FNA biopsy of lung. One of the two cases was from a 72-year-old man who developed a mass in the right upper lobe. He had a history of heavy cigarette smoking but no occupational exposure to asbestos. Aspiration biopsy procured malignant cells consistent with poorly differentiated adenocarcinoma in direct association with asbestos bodies, some of which intersected papillary aggregates of tumor cells (Fig. 7-51, Plate XVII). The three-dimensional cluster is formed by columnar cells with round to oval, hyperchromatic nuclei of variable size, surrounded by opaque cytoplasm with indistinct boundaries and is contrasted with an adjacent pneumonocyte. The asbestos body is a golden segmented fibrous filament with encrustations of iron and protein. The segmented quality is emphasized in another field where it co-exists with reactive cells (Fig. 7-52). The post-

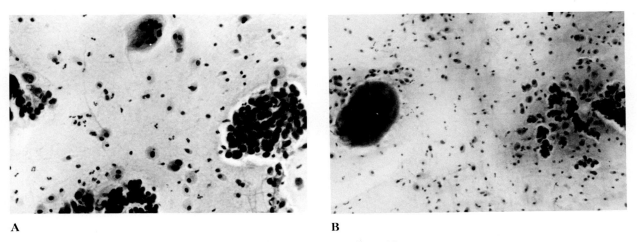

A   B

Fig. 7-49. A and B. Adenocarcinoma associated with multinucleated histiocytes.
(P stain, A ×315, B ×200)

mortem examination confirmed a 1.5-cm mass that microscopically is a poorly differentiated adenocarcinoma, with areas resembling giant cell carcinoma, apparently arising in a peripheral scar, intimately intermingled with asbestos bodies (Fig. 7-53). The light microscopic asbestos count was 4200 ABs/g. By scanning electron microscopy there were 58,000 asbestos fibers (coated and uncoated) per gram of lung tissue, with a mean asbestos length of 29.6 μm and core diameter of 0.50 μm. The level of asbestos burden is consistent with occupational exposure, although this history could not be elicited. The occurrence of asbestos bodies in fine needle aspirates of

lung may reflect considerable occupational exposure and predict that the patient is at risk for the development of asbestos-related disease. This becomes particularly valuable if the procedure is directed at lung parenchyma where asbestos body counts tend to be higher. Counts are generally lower within neoplasms, possibly because of their expansile nature. The more commonly recognized association of asbestos bodies with mesotheliomas should not subjugate recall of the occurrence of adenocarcinoma with these injurious fibers.

The distinction between pleural-based adenocarcinoma and mesothelioma may be an inordinately

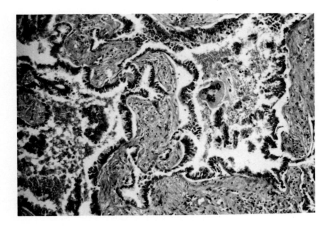

Fig. 7-50. Histologic section. Adenocarcinoma in association with multinucleated histiocytes. (×200)

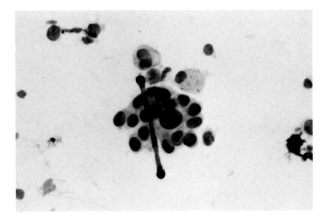

Fig. 7-51. Ferruginous body associated with adenocarcinoma. (P stain, ×625)

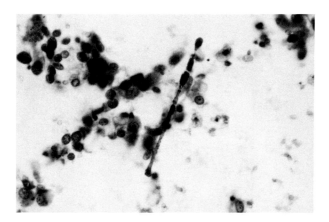

Fig. 7-52. Ferruginous body associated with reactive pneumonocytes and bronchial epithelial cells. (P stain, ×625)

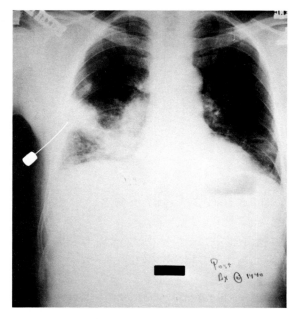

Fig. 7-54. Chest film of a 66-year-old male with pleuroparenchymal opacification involving the right lung (needle in target).

difficult diagnostic decision on aspirated cytologic material, but adjunctive utilization of immunoperoxidase stains for keratin and epithelial membrane antigen may enhance the resources for separating these lesions rationally. A 66-year-old man presented with radiographic pleuroparenchyma opacification involving the right lung (Fig. 7-54). Aspiration biopsy was accomplished with facility under fluoroscopic guidance because of proximity of the pleural-based lesion to the chest wall. The harvest presented papillary aggregates of uniform small polygonal cells with bland, monotonously similar nuclei containing prominent, eccentric nucleoli. Cohesion, a unified peripheral margin, and the arborescent papilliform architecture in a "clean" background provided additional features for a low-grade papillary adenocarcinoma (Fig. 7-55). The three-dimensional organization, cytoplasmic vacuolization and prominent nucleoli of another cluster are illustrated in Figure 7-56. This pattern could reflect mesothelioma, partic-

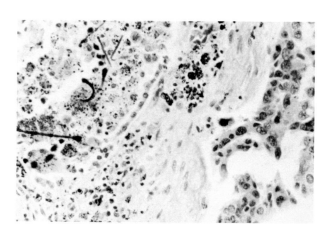

Fig. 7-53. Histologic section. Tissue correlate of ferruginous body deposition juxtaposed to invasive adenocarcinoma. (×500)

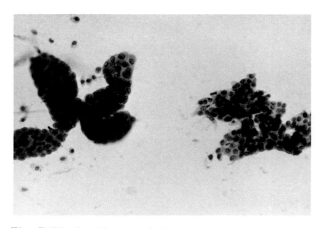

Fig. 7-55. Papillary epithelial aggregate subsequently proven by immunoperoxidase to represent papillary adenocarcinoma. (P stain, ×625)

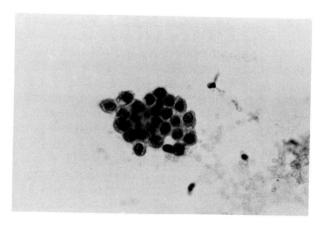

Fig. 7-56. Cellular aggregate subsequently proven by immunoperoxidase methods to represent adenocarcinoma. (P stain, ×625)

ularly in the context of pleural attachment and possible origin. Mucicarmine stains are negative. Therefore, immunoperoxidase reactions for epithelial membrane antigen (Fig. 7-57A) and keratin (Fig. 7-57B) were obtained. The illustrations reflect intense granular cytoplasmic staining for both reactions, supporting a diagnosis of adenocarcinoma rather than mesothelioma. The statistical frequency of adenocarcinoma and the utility of special stains applicable to cytological preparations will help to justify the diagnosis of pulmonary adenocarcinoma by transthoracic FNA biopsy.

Small Cell Tumors of Lung

The small cell tumors of the lung (Plate XVIII) that are diagnostically pertinent to the cytopathologist examining aspiration biopsies include undifferentiated carcinoma (oat cell and intermediate variants), bronchial carcinoid tumors (round and spindle cell variants), and lymphoma. Although metastatic small cell tumors of childhood could be included in this designation, the opportunity to study these neoplasms in our laboratories has not been forthcoming.

Small cell undifferentiated carcinoma of the lung can originate centrally or peripherally; and many tumors are considered derivatives of bronchial Kulchitsky cells, which confer the potential for amine precursor uptake and decarboxylation (apud) clinically manifest by endocrine disturbances resulting from the secretion of hormonally active substances. Bronchial carcinoids belong to this metabolic group and derivation, but their cytomorphology is distinct from the undifferentiated tumors. Two subgroups of small cell undifferentiated carcinoma are distinguished, primarily on the basis of cell size and characteristics rather than biological potential, and are designated "oat cell" carcinoma for tumors composed of smaller cells and "intermediate" for neoplasms with larger cells. Both variants are lethal but can respond effectively to chemotherapeutic regimens.

A 68-year-old woman presented with a well-circumscribed area of opacification in the right lower lobe (Fig. 7-58) incidentally discovered on routine

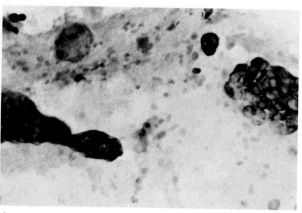

A

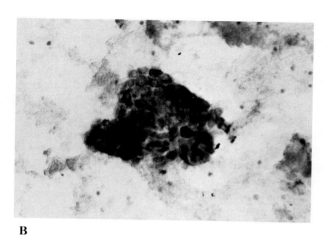

B

Fig. 7-57. A. Positive reaction for epithelial membrane antigen. B. Positive reaction for keratin. (Immunoperoxidase stain, ×400)

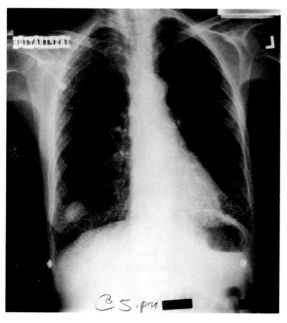

Fig. 7-58. Chest film of a 68-year-old female with a radiopaque lesion in the right lower lobe.

chest x-ray. An aspiration biopsy of this peripheral mass was accomplished under fluoroscopic control (Fig. 7-59) without incident and produced an extraordinary cellular harvest. The smears are characteristic of classical oat cell carcinoma (Fig. 7-60). There is a dense population of cells universally dispersed singly, randomly, and without architectural design, although occasional couplets or minute aggregates may signify an abortive cohesive effort. Where cells are in contact, molding is visible. This is a critical feature in separating this small cell lesion from lymphoma in which the cells are singly dissociated and never form aggregates. The fragility of the cells results in the characteristic "crush" artifact that is repeated in histologic sections and in the pyknotic obliteration of nuclear detail in some cells. Cytoplasm is so minimal that it is an ambiguous component of cellular architecture or removed mechanically by the process of smearing. Nuclei are uniform in their small size, resembling transformed lymphocytes in their dimensions, but contours may vary from round to oval or fusiform, with the elliptical or almondine configuration predominating. For those cells spared the degenerative hyperchromasia, nuclear texture is generally delicate with evenly distributed chromatin. Nucleoli are generally described to be indistinct or absent, but

our developing experience with superbly preserved material immersed in ethanol within seconds of acquisition, contradicts this concept. Nucleoli can be portrayed as sharp, round, central, or slightly eccentric single structures. Compare these cellular features with a smear from a patient with the "intermediate" variant of small cell undifferentiated carcinoma (Fig. 7-61) photographed at the identical magnification. The cells are distinctly larger; and nuclei are more variable in size, contour, and configuration. Dispersion prevails, but clustering is more obvious (single arrow). Nuclear molding is well illustrated by the tightly coupled large nuclei with coarsely textured chromatin and irregular margins (double arrows). Intermediate cells are also fragile, and may display pyknosis, "crush" artifact, and cytoplasmic disintegration (Fig. 7-62). The histologic correlate of oat cell carcinoma is a solid tumor composed of sheets and lobules of small, dark, often fusiform cells with pyknotic and crushed nuclei, ambiguous cytoplasm, no specific architectural preference, and variable necrosis (Fig. 7-63).

A second case illustrates with clarity of cellular detail the characteristics of the intermediate variant of small cell undifferentiated carcinoma and empha-

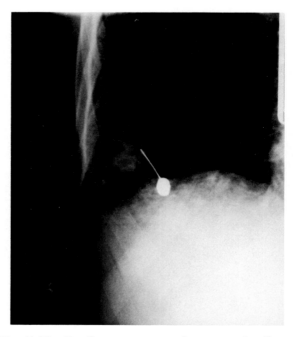

Fig. 7-59. Needle penetration of target under fluoroscopic guidance.

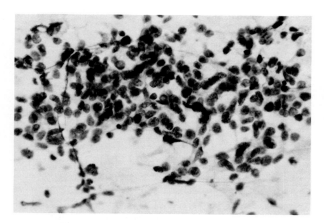

Fig. 7-60. Classical "oat cell" carcinoma. (P stain, ×625)

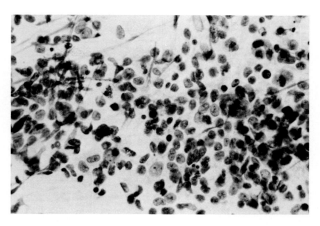

Fig. 7-62. Crush artifact and cellular fragility in undifferentiated carcinoma, "intermediate cell variant." (P stain, ×500)

sizes radiographically that a peripheral lesion may coexist with hilar involvement and incite a diagnostic challenge within the differential scheme of mediastinal lesions. A 60-year-old man presented with cough and chest pain. A PA chest film demonstrated a radio-opaque lesion in the right lung posteriorly (Fig. 7-64). Computed tomography confirmed its location, size, and peripheral contour, but also demonstrated nodular masses in the right hilum consistent with lymphadenopathy (Fig. 7-65). The pleuroparenchymal mass was investigated first under fluoroscopic guidance (Fig. 7-66) and produced a highly cellular, dispersed population of fragile cells with

minimal or no cytoplasm and round to fusiform nuclei with delicate chromatinic texture, occasional clustering, and molding (Fig. 7-67) characteristic of the intermediate variant of small cell carcinoma. The tissue correlate of this pattern (selected from another case, Fig. 7-68A and B) emphasizes the spindled, palisaded, and focally necrotic features of the tumor. An aspirate from the hilum demonstrated undifferentiated carcinoma, but the cells are smaller and qualify more appropriately as oat cell carcinoma (Fig. 7-69).

The principal diagnostic differential of oat cell carcinoma is malignant lymphoma, because there are

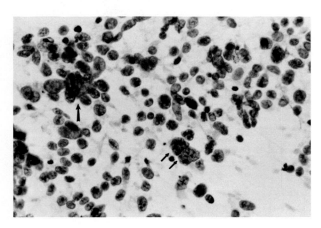

Fig. 7-61. Undifferentiated carcinoma "intermediate cell variant." (P stain, ×625)

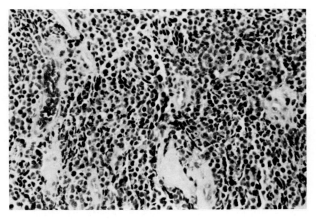

Fig. 7-63. Histologic section. Tissue correlate of "oat cell" carcinoma. (×400)

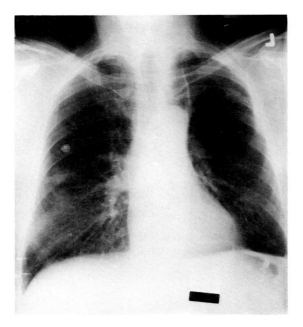

Fig. 7-64. Chest film of a 60-year-old male with a radiopaque lesion in the right lung posteriorly.

frequently overlapping features that include characteristic dispersion of cellular samples, small uniform cells with minimal cytoplasm, and slightly irregular nuclei containing delicate chromatin. The availability

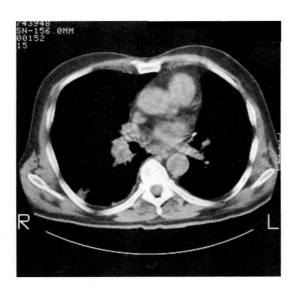

Fig. 7-65. Computed tomography emphasizing right posterior lesion and hilar lymphadenopathy.

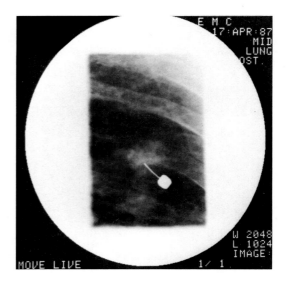

Fig. 7-66. Needle in target under fluoroscopic guidance.

of immunoperoxidase stains for common leukocyte and epithelial membrane antigens, and for kappa-lambda light chains and immunoglobulins, in conjunction with flow cytometry has simplified the distinction in difficult cases. A 65-year-old woman presented with weight loss, fever, anorexia, cough, and x-ray evidence of an opacification in the left upper lobe. It was elected to evaluate the infiltrative process by fluoroscopically directed FNA biopsy utilizing standard technique. The cellular harvest is a luxuriant population of dissociated cells with mini-

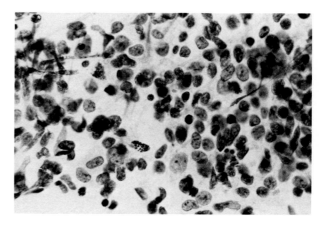

Fig. 7-67. Peripheral nodule demonstrating undifferentiated carcinoma, "intermediate cell variant." (P stain, ×625)

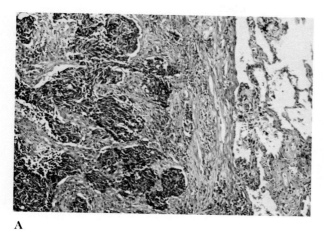

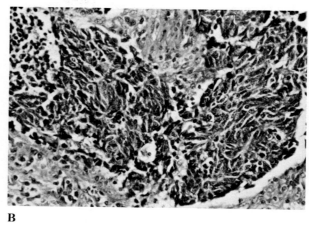

A   B

Fig. 7-68. A and **B**. Histologic section. Peripheral nodule representing tissue correlate of undifferentiated carcinoma, "intermediate cell variant." (**A** ×125, **B** ×400)

mal cytoplasm, round or slightly convoluted nuclei exhibiting indentations, coarse chromatin granules, punctate chromatin stippling, and smudge artifact (Fig. 7-70). Spindle forms, cohesion, crush artifact, and prominent nucleoli are not readily identified. Conventional immunoperoxidase methodology was utilized to establish monoclonality and B-cell derivation with the application of stains of kappa-lambda light chains. Therapy for lymphoma was initiated, but the patient died with fever and sepsis. The postmortem examination confirmed the presence of solid and nodular infiltrates in the lungs with evidence for radionecrosis. Microscopically the tumor is composed of a diffuse proliferation of small cells with hyperchromatic or pyknotic nuclei and minimal cytoplasm. There is no aggregation or molding, and necrosis interrupts the homogeneity of the process (Fig. 7-71). This case illustrates the correct diagnosis of poorly differentiated B-cell lymphocytic lymphoma and that it is possible to separate lymphoma from small cell undifferentiated carcinoma by aspiration biopsy cytology (ABC). Certainly, there are some primitive or high-grade lymphomas whose architectural features are so distinct that confusion with oat cell carcinoma is not of concern. They can be characterized morphologically and by flow cytometric methods with accuracy and confidence.

Bronchial carcinoid tumors (Plate XVI) are diagnosed initially by aspiration biopsy infrequently,

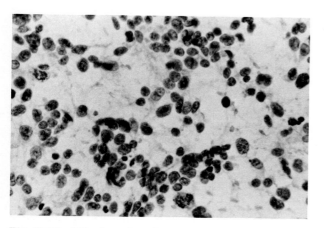

Fig. 7-69. Hilar lymph node representing classical "oat cell" carcinoma. (P stain, ×625)

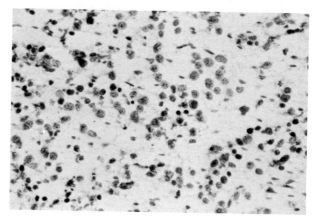

Fig. 7-70. Lymphoma of lung. (P stain, ×400)

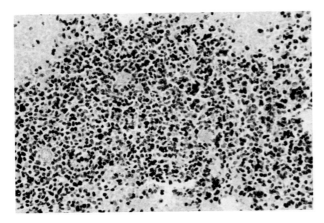

Fig. 7-71. Histologic section. Autopsy lung confirming aspiration biopsy diagnosis of lymphoma. (×400)

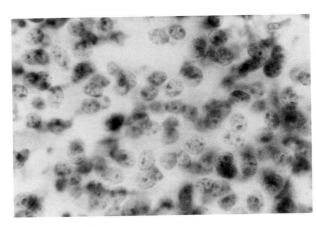

Fig. 7-73. Bronchial carcinoid. (P stain, approximately ×625)

perhaps because of their generally central origin and accessibility to conventional fiberoptic broncho-scopic biopsy. Frable[18] reported only two cases in his 1983 monograph from the Medical College of Virginia representing classical small cell carcinoid in festooning ribbons contrasted with a spindle cell variant. Our experience is similar, if not identical, but serves to illustrate pertinent diagnostic information. A 78-year-old woman was evaluated for a central bronchial lesion by conventional radiographically assisted aspiration biopsy at the Istituto Regina Elena (National Cancer Institute, Rome). A monotonous succession of round, uniform cells with minimal cy-

toplasm, resembling lymphocytes, but congregated in ribbons and amorphous clusters, is representative of the aspirated material (Fig. 7-72). Nuclei are identical in size, contour, crisply delineated nuclear margins, delicate chromatin, and prominent nucleoli accentuated by paranucleolar clearing (Fig. 7-73). Linear or trabecular enfilades, nuclear synchrony, and central origin are features of classical bronchial carcinoid. These cells stain positively by silver impregnation methods and by electron microscopy offer neurosecretory granules that permit assignment to an amine precursor and uptake decarboxylation classification.

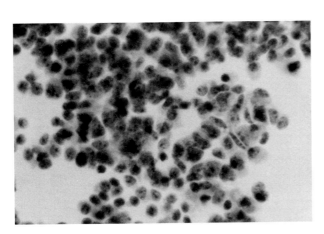

Fig. 7-72. Uniform small round cell population in trabecular cords consistent with bronchial carcinoid. (P stain, approximately ×500)

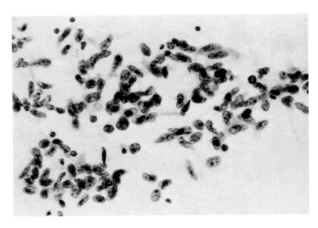

Fig. 7-74. Spindle cells in abortive rosettes consistent with spindle cell variant of bronchial carcinoid. (P stain, ×625)

A 70-year-old woman presented with roentgenographic evidence for a subpleural mass in the left lower lobe. An aspiration biopsy harvested a population of uniform, short spindle cells arranged in abortive rosettes (Fig. 7-74). The fusiform nuclei present crisp, well-delimited margins and coarse chromatin clumping without prominent nucleoli or visible mitoses. Cytoplasm is pale, indistinct or ambiguous, and occasionally fibrillar. The background is devoid of diathesis and fibrosis, and there is no evidence for amyloid. A spindle cell tumor, consistent with spindle cell carcinoid was offered as the pre-operative diagnosis that eventuated in a left lower lobectomy. The atelectatic parenchyma incorporated a well-circumscribed mass with a lobulated, granular but homogeneous cut surface, a serpiginous margin, and direct penetration and stellate retraction of the pleura, measuring $6 \times 5 \times 3.5$ cm (Fig. 7-75). Histologic sections verify a spindle cell proliferation in an organoid configuration created by the fascicular packaging of short, uniform, fusiform cells within lobules and rosettes created and emphasized by intersecting delicate fibrous septa (Fig. 7-76 and 7-77). Although this is not illustrated, the lesion proliferates in collision with the walls of peripheral bronchioles and is intimate with vascular spaces; argentaffin reactions are negative, but argyrophilic affinity is demonstrable. There are visible but infrequent mitoses. No metastatic tumor was detected in hilar lymph nodes.

Peripheral spindle cell carcinoid tumors can be successfully diagnosed by percutaneous aspiration

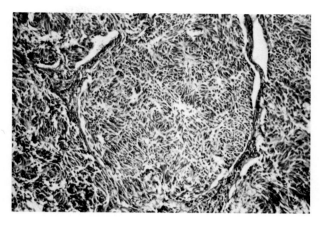

Fig. 7-76. Histologic section. Tissue correlate of spindle cell bronchial carcinoid emphasizing fascicular arrangement. ($\times 200$)

biopsy because of their accessibility to the needle probe and their unique morphology. Ranchod and Levine's publication[35] of 35 cases summarizes salient clinical and morphological features of the neoplasm, which has a slight female predilection, a definite but inexplicable preference for the right middle lobe, and an almost universal peripheral or subpleural origin, most likely from Kulchitsky cell proliferations in distal bronchioles. This lineage explains argyrophilic affinity, neurosecretory granules, and ectopic endocrine hormone production such as corticotrophin[52] secretion. Eighty percent of tumors are less than 2 cm in diameter, and size is not a reliable predictor of malignant potential. Although the constituent cells

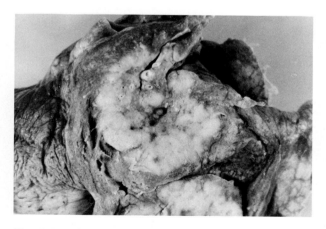

Fig. 7-75. Gross pulmonary lobectomy specimen incorporating spindle cell bronchial carcinoid.

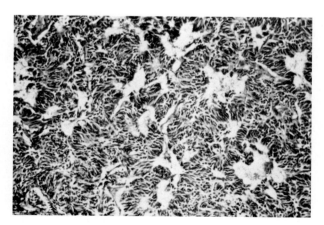

Fig. 7-77. Histologic section. Tissue correlate, spindle cell bronchial carcinoid emphasizing rosette arrangement. ($\times 200$)

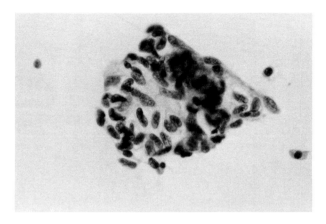

Fig. 7-78. Spindle cell neoplasm consistent with leiomyosarcoma. (P stain, ×625)

are spindle elements, two subsets may be encountered: short spindle cells in knotted *zellballen* separated by thin-walled vascular spaces produce the classical organoid pattern; long spindle cells resembling mesenchymal elements may form interlacing fascicles that must be distinguished from neural tumors and leiomyomas. Nuclei are generally coarsely stippled with inconspicuous nucleoli and infrequent mitotic activity. Large bronchioles have been described to be in close association or entrapped within the tumors; and some of these contain mucosal foci of Kulchitsky cell proliferation, sometimes associated with satellite tumorlets, implicating the derivation of the carcinoid from this progenitor cell. Although mitotic rate may reflect malignant potential, the histology does not; and a small percentage of metastases to regional nodes can be expected. Tumorlets should not be confused with metastatic satellites. The differential diagnosis for short spindle cell lesions includes oat cell carcinoma, tumorlets, and chemodectoma; for long spindle cell lesions, leiomyoma, schwannoma, fibrous mesothelioma, and metastatic neoplasms must be considered (spindled squamous and renal carcinomas, melanoma, and sarcomas).

Sarcoma

Sarcomas of the lung are either mesenchymal derivatives of pulmonary tissues or metastatic repositories and may reflect in their morphological identity a comprehensive spectrum of stromal anatomy from simplistic to complex. The following case illustrates the isolation and proper identification of leiomyosarcoma in the lung by percutaneous transthoracic aspiration biopsy. A 69-year-old man presented with a recurrent nodular radiographic opacification involving the anterior chest wall adjacent to a rib. Four years previously, a resected pulmonary lesion was interpreted as a leiomyoma because of its bland spindle cell architecture with innocuous nuclei and insignificant numbers of mitoses. Spindle cells isolated from the recurrent nodule by a 22-gauge needle under fluoroscopic control are aggregated in informal fascicles (Fig. 7-78). Nuclei are fusiform but suggestively lobated with punctately stippled or coarse chromatinic texture, parachromatin clearing, and slight pleomorphism. Nuclear lobulations and pleomorphism are more apparent in an adjacent field (Fig. 7-79) where tumor cells coexist with inflammatory elements and macrophages and offer a more epithelioid quality. The antecedent history was influential in the decision to categorize these cells as microcomposites of recurrent leiomyosarcoma, but confirmation could have been supported by immunoperoxidase stains for muscle cell markers. Histology corroborated the cell biopsy by presenting intertwined fascicles of spindle cells in a solid nodule replacing pulmonary parenchyma (Fig. 7-80). Although it is not illustrated, the comparison of the tissue from the recurrence with the previous primary lesion revealed increased atypia and mitotic activity in the recurrent nodule. Because of the myriad neoplasms that can present with spindle cells, it may be impossible to subclassify an aspi-

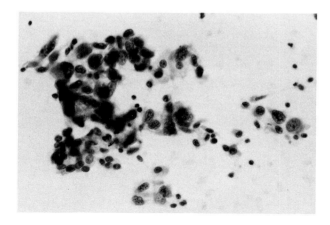

Fig. 7-79. Spindle cell neoplasm reflecting nuclear lobulations and pleomorphism, compatible with leiomyosarcoma. (P stain, ×500)

rated tumor as a specific type of sarcoma, and a nonspecific diagnosis that simply implicates a spindle cell sarcoma must suffice. Certainly specific entities such as cartilage, osteoid, actual bone, vascular structures, or fat may assist the direction of formulation for the final designation of the aspirated sarcoma.

Metastatic Cancers in Lung

Metastatic envoys to the lung from virtually any primary malignancy may implant and flourish, eventually reaching clinically pertinent proportions, radiographic expression, and identity. The confirmation of metastatic disease and its differentiation from primary malignancy and infectious processes simulating cancer can be accomplished through image-guided aspiration biopsy with facility and confidence. Although this distinction can be reasonably assured on morphological grounds alone, the adjunctive application of immunoperoxidase reactions may identify the parent tissue of origin through functional parameters, thereby reducing or clarifying the candidates within a specified differential. Selected examples of aspirated metastases illustrate these concepts.

A 74-year-old man with history of transurethral prostatic resection and hormonal manipulation for adenocarcinoma of the prostate presented 4 years postoperatively with radiographic densities in the right lower lobe. Serum measurements of acid phosphatase were elevated as circulating tumor markers. An aspiration biopsy from one of the nodules clini-

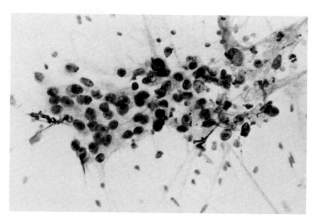

Fig. 7-81. Metastatic prostatic adenocarcinoma. (P stain, ×625)

cally suspect for a metastatic deposit produced an aggregate of cells arranged in a honeycombed network in a background of inflammation (Fig. 7-81). Nuclei are oval and slightly variable in size; tend to overlap but not ident or mold; and provide prominent, round, smooth, slightly eccentric nucleoli suspended in delicate chromatinic networks. The delicately vacuolated, pale cytoplasm provides the essence of columnar silhouettes, but margins are ambiguous and a syncytium is suggested. The composite cells are rather large and displayed essentially *en face*. The cluster certainly resembles the appearance of prostatic adenocarcinoma in direct aspirates of prostate, but could qualify as adenocarcinoma from another source, including lung. Immunoperoxidase stains for prostatic specific antigen superimposed on the P stain were positive, providing reassuring information in support of the interpretation that the harvested cells reflected metastatic prostatic carcinoma in lung.

A 60-year-old Mexican American woman was referred for aspiration biopsy under fluoroscopic guidance for the evaluation of intrapulmonary nodules considered to represent metastatic disease from an undetermined primary source. The sample presents papillary aggregates of loosely cohesive columnar cells exhibiting nuclei of variable size and contour with delicate chromatin and occasional intranuclear vacuoles (arrow, Fig. 7-82). Cytoplasm is dense with well-defined margins imparting a polygonal architecture to the cells rather than a columnar outline. Since primary pulmonary adenocarcinomas may be expressed in papillary configurations, the

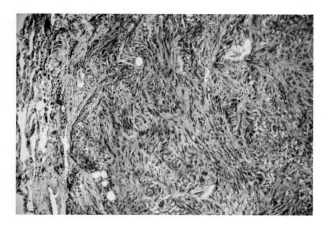

Fig. 7-80. Histologic section. Tissue correlate of leiomyosarcoma. (×125)

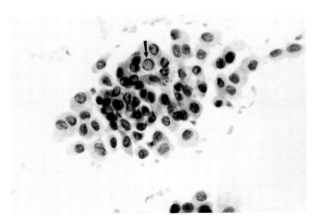

Fig. 7-82. Metastatic papillary adenocarcinoma of thyroid demonstrating intranuclear vacuolization. (P stain, ×500)

diagnostic task included the elimination of a pulmonary primary and the implication of a specific external source. In this situation, one can rely on the papillary architecture and the presence of intranuclear inclusions to direct the investigation to the thyroid as a possible origin for the tumor. Immunoperoxidase stains for thyroglobulin were positive, concluding the diagnostic effort.

A 62-year-old woman developed a pulmonary radiographic density several years following mastectomy for an infiltrating duct cell carcinoma of the breast. The routine chest film characterized the lesion as a large, spherical, posterior, pleurally based opacification accessible to the needle probe. The cells are arranged in cohesive clusters with communal affinity

that does not impose molding or nuclear deformity (Fig. 7-83). Nuclear : cytoplasmic ratios are markedly skewed, with minimal cytoplasm and nuclear dominance. Oval nuclei present delicate chromatin and single or multiple nucleoli outlined by crisp, thin envelopes. The arrangement of cells in packets and the nuclear features are certainly compatible with adenocarcinoma of the breast, correlative with patient history. A lobectomy was performed to resect this solitary metastasis and acquire hormonal receptor assays. The histological appearance is typical of duct cell carcinoma (Fig. 7-84) and presents polygonal elements in nests and sheets, separated by delicate fibrous septa that intersect and circumscribe arbitrary clusters. The pattern is indistinguishable from sections taken at the time of the initial breast biopsy.

Metastatic malignant melanoma is a particular challenge in aspiration biopsy of the lung because of its peculiar protean propensity for cell mimicry and its resemblance to cytomorphological formats of other tumors. When the characteristic melanin pigmentation is present and readily discernible from hemosiderin and other pigmentary granules, the decision is simplified; but amelanotic polygonal and spindle cells may prevail and simulate a variety of metastatic tumors. In these situations, immunoperoxidase stains for S100 protein may provide useful information, assuming there is clever insight into the possibility of a melanotic lesion disguised as a sarcoma or anaplastic carcinoma. An elevated, spherical, pigmented 1.5-cm nodule was discovered on the skin of the back of a 69 year-old man concurrent with radiographic evidence for a "coin" lesion in the left

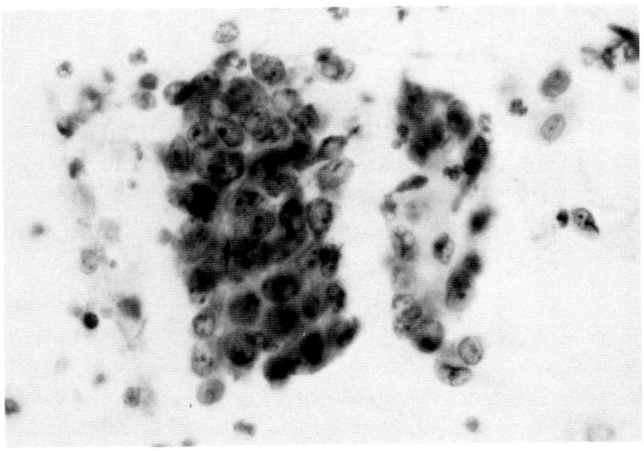

Fig. 7-83. Metastatic infiltrating duct cell carcinoma of breast. (P stain, ×800)

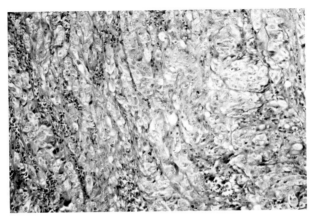

Fig. 7-84. Histologic section. Tissue correlate of metastatic infiltrating duct cell carcinoma of breast. (×200)

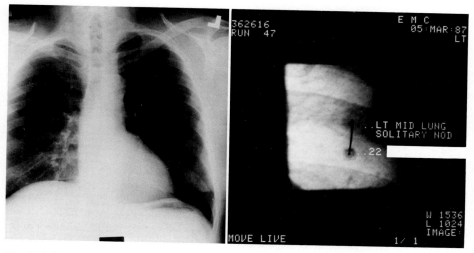

Fig. 7-85. Chest film of a 69-year-old male with a "coin lesion" in the left mid-lung field, contrasted with the aspiration biopsy documenting the position of the needle in its target.

mid-lung field (Fig. 7-85). Computed tomography (Fig. 7-86) enhanced the size, contour, and location of the nodule, which was considered to represent either a metastasis from cutaneous melanoma or a coincidental granuloma. A clinical decision to investigate the pulmonary lesion prior to surgical therapy for the melanoma primary resulted in a fluoroscopically directed aspiration biopsy utilizing a 22-gauge Westcott needle that pierced its target centrally (Fig. 7-85). The cellular harvest incorporated a bimorphic population. Background cells are primarily alveolar pneumonocytes (Fig. 7-87), but scattered single dominant polygonal cells are included. Note the dense, granular texture of the abundant cytoplasm reflecting the accumulation of melanin pigment and the peripherally displaced, hyperchromatic nucleus with translucent microvacuoles. In another field (Fig. 7-88) there are smaller cells with enlarged lobulated and indented nuclei, visible to slightly accentuated nucleoli and disturbed nuclear:cytoplasmic ratios. Focal

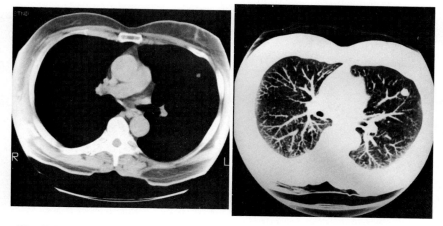

Fig. 7-86. CAT scan correlate of Fig. 7-85 demonstrating the solitary pulmonary metastasis in the left lung.

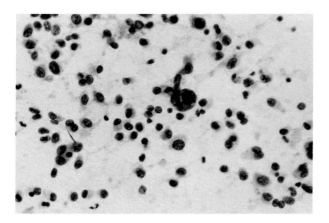

Fig. 7-87. Metastatic malignant melanoma demonstrating cytoplasmic pigment. Background cells are alveolar pneumonocytes. (P stain, ×500)

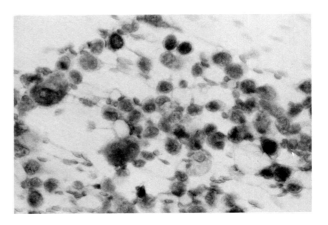

Fig. 7-89. Metastatic malignant melanoma with positive reaction for S-100 protein. (Immunoperoxidase stain, ×625)

abortive rosette formation is seen in association with occasional couplets. The dominant cell contains a nucleus replaced by a translucent vacuole with chromatin condensation at the periphery. For demonstration purposes, an immunoperoxidase stain for S100 protein is included to illustrate the red-brown cytoplasmic granularity that may occasionally override the nuclei (Fig. 7-89), realizing that the value of the stain is in the identification of the melanoma that expresses itself ambiguously. A 70-year-old man was discovered by chest roentgenography to harbor an infiltrative and destructive lesion in the left upper lobe. A flexible spinal needle was inserted under fluoroscopic control into the center of the lesion for extraction of the analytic population. Polyhedral and polygonal cells (Fig. 7-90) are arranged in loose aggregates. Adhesion is abdicated to informal juxtaposition without intercellular anchorage. Nuclei are irregular in size and contour but are universally hyperchromatic with areas of chromatinic transparency, nucleolar emphasis and irregularity, chromatin clumping, and density. Cytoplasm is variable in amount and conspicuously granular, but without unequivocal pigment. The clinician was advised of the diagnosis of metastatic malignant melanoma; he subsequently located the primary lesion on the skin of the left lower abdominal quadrant. The wide resection provided his-

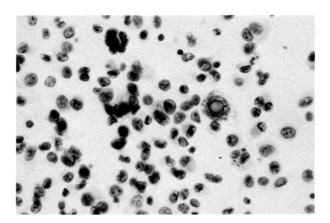

Fig. 7-88. Metastatic malignant melanoma demonstrating cellular polymorphism and intranuclear vacuole. (P stain, ×625)

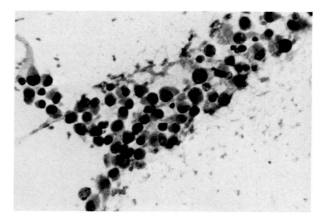

Fig. 7-90. Metastatic pleomorphic amelanotic malignant melanoma. (P stain, ×500)

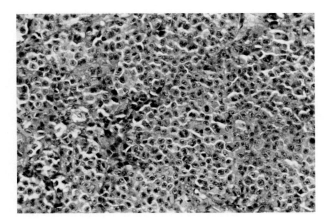

Fig. 7-91. Tissue section. Histologic correlate of metastatic malignant melanoma confirming FNA illustrated in Fig. 7-90. (×315)

tologic confirmation (Fig. 7-91). Polygonal melanoma cells are arranged in neval-like nests and sheets and conform identically to the cytomorphology of the aspirated cells. The aggressive proliferation resulted in invasion of the epidermis and reticular dermis (Clark's Level IV), achieving a depth of infil-

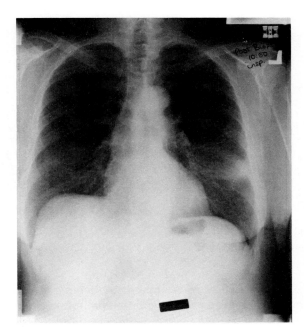

Fig. 7-93. Postbiopsy chest film of a 74-year-old woman demonstrating perilesional hemorrhage.

tration several micra in excess of Breslow's 0.76 mm critical value, portending metastatic predilection. Inevitably, death occurred with cerebral metastases.

Benign Conditions of Interest

The pulmonary hamartoma is perhaps the most common benign neoplasm of the lung and may account for a radiographic opacity that is inert or indolently

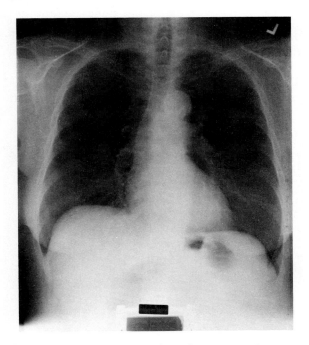

Fig. 7-92. Prebiopsy chest film of a 74-year-old woman with a sharply delineated lesion at the periphery of the left lower lobe.

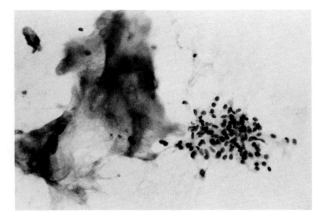

Fig. 7-94. Hyaline cartilage and reactive bronchial epithelial cells characteristic of bronchial hamartoma. (P stain, ×400)

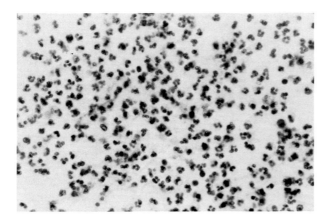

Fig. 7-95. Fibrinopurulent exudate consistent with origin from pulmonary abscess. (P stain, ×625)

expansive. The size is unpredictable but can reach rather large proportions, while maintaining a smooth peripheral contour. The radiographic differential includes granuloma and circumscribed cancer, but generally the appearance is so characteristic that hamartoma is predicted by the radiologist[53] before the slides are stained and coverslipped. A 74-year-old woman presented a well-circumscribed, spherical, uniformly opaque lesion with a sharply delineated margin at the periphery of the left lower lobe (Fig. 7-92). Aspiration biopsy resulted in perilesional hemorrhage that altered the contour and peripheral boundaries (Fig. 7-93). The needle captured the essential elements for a diagnosis of *hamartoma* (Fig. 7-94). The dominant structure is stellate, fibrillar, opaque hyaline cartilage

juxtaposed to a cluster of benign bronchial epithelial cells. Occasionally ciliated cells may be incorporated and macrophages isolated in the background.

Nonspecific *pulmonary abscess* is characterized by an essentially pure fibrinopurulent exudate in a background of detritus (Fig. 7-95). Specific etiologic organisms may not be visualized, and cultural identification is requisite. Abscess produces or results in areas of architectural degradation, so that anatomic lung components may not be reflected in the debris. The lesion can be well circumscribed and partitioned by a reparative response.

There may be unusual provocateurs of such a reaction. A 70-year-old woman presented with persistent cough and central mediastinal discomfort. A screening chest film demonstrated a mass in the left upper lobe that was clinically and radiographically considered to represent lung cancer. Verification was intended by aspiration biopsy, and the needle indeed followed a trajectory to the center of the mass. Instead of the anticipated carcinoma cells, the harvest was inflammatory (Fig. 7-96). The smears present a background of precipitated fibrin, which creates interstices for the entrapment of neutrophils. (The triangular structure at the center was overlooked as artifact.) A diagnosis of abscess was offered, and a second pass was performed for acquisition of material for cultures. Because of the discrepancy with the clinical expectation of tumor, a left upper lobectomy was performed predicated on the assumption that if tumor were present, it would be adequately treated; and if the lesion were exclusively abscess, the inflam-

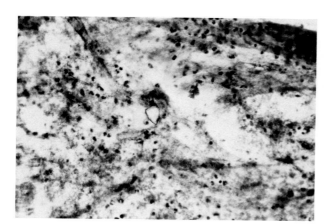

Fig. 7-96. Fibrinopurulent exudate and a triangular foreign object at the center. (P stain, ×400)

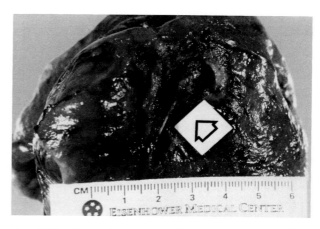

Fig. 7-97. Gross left upper lobectomy specimen illustrating an abscess cavity containing a toothpick encrusted with inflammatory debris.

matory tissues would be debrided. A 3-cm mass with central cavitation was discovered subpleurally (Fig. 7-97), and serendipitously at the center is a toothpick (arrow) encrusted with inflammatory debris. Microscopic sections through the inflammatory cavity and toothpick (Fig. 7-98) disclose striated layers of cellulose interdigitated with infiltrating aggregates of neutrophils penetrating from the exterior where they constitute the inflammatory crust. The triangular structure depicted in Figure 7-95 is considered retrospectively to represent a dislodged cellulose fragment from the wooden object. Apparently the patient had swallowed the toothpick, which perforated the esophagus and transmigrated into the pulmonary parenchyma where it lodged as an inflammatory stimulus that incited abscess formation.

Aspiration biopsy of lung can successfully isolate specific etiologic agents of infection for morphological and cultural identification, resulting in selection of appropriate therapy, reprieve of a cancer diagnosis, and often circumvention of surgery. A dynamic, active, 44-year-old man employed in the film industry presented with a "coin lesion" in the left upper lobe several weeks following completion of a motion picture filmed in the desert of southern California during a particularly windy season. His clinician referred him immediately for aspiration biopsy to exclude an early malignancy. Although the patient had smoked for several years, he had discontinued this 10 years prior to the appearance of the lesion. The aspirate produced a smear with a background of cellular detritus and inflammatory debris incorporat-

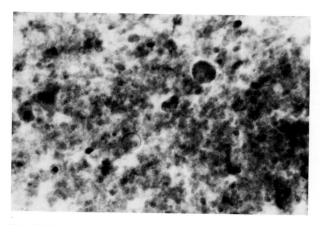

Fig. 7-99. Typical spherules of *Coccidioides immitis* in a background of inflammatory detritus. (P stain, ×625)

ing spherules of variable diameter exhibiting refractile walls and internal sporulation characteristic of *Coccidioides immitis* (Fig. 7-99, Plate XIX). This mycotic infection is an anticipated finding in the context of the history and radiographic appearance. The patient had been exposed to sandstorms for a 10-week period; the fungus grows in superficial soil as mycelial strands containing rectangular arthrospores, which are the infectious forms dispersed by wind and inhaled into the pulmonary milieu where they convert to spherules that reproduce by endosporulation. The typical host response is the evolution of a granuloma (Fig. 7-100), which may be well circumscribed, solitary, and eventually partitioned, ultimately becoming quiescent. The cellular expression of this may

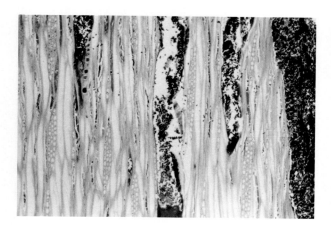

Fig. 7-98. Histologic section. Toothpick demonstrating cellulose structure interdigitated with infiltrating fibrinopurulent exudate. (×400)

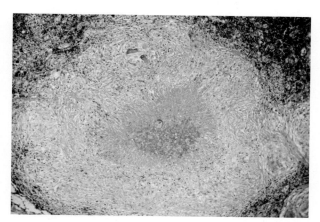

Fig. 7-100. Histologic section. Caseating granuloma produced by *Coccidioides immitis* demonstrating a central typical spherule. (×125)

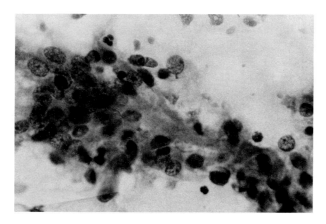

Fig. 7-101. Epithelioid histiocytes arranged in fascicles and pseudorosette clusters, consistent with origin from granuloma. (P stain, approximately ×625)

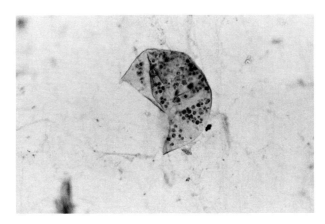

Fig. 7-103. Sputum cytology. Expectorated spherule of *Coccidioides immitis* demonstrating endosporulation. (P stain, ×625)

be epithelioid histiocytes arranged in fascicles or pseudorosette-like clusters (Fig. 7-101) or, simply, the proliferation of multinucleated histiocytes (Fig. 7-102). Sixty percent of humans infected by this fungus remain asymptomatic,[47] but present a chest film that could be confused with malignancy. Cough could result in the expectoration of spherules with endosporulation in the sputum (Fig. 7-103). Symptoms vary from a mild upper respiratory infection to a prolonged severe illness with dissemination to bones, joints, internal organs, or the meninges. This is particularly prevalent in AIDS victims who develop serious lymphadenitis without a significant granulomatous response. Postmortem studies demonstrate extensive inflammatory and necrotic replacement of pulmonary parenchyma by proliferating spherules in the context of detritus, correlating directly with the appearance of clinical smears from the infected lungs (Fig. 7-104). In the clinical example cited, the lesion was localized, and the patient was spared surgical intervention so that he could return to work on antibiotic therapy with minimal interruption in routine and with considerable relief of anxiety.

Infections with *Cryptococcus neoformans* and *Aspergillus fumigatus* (Plate XIX) are likely to produce a necroinflammatory process so that yeast forms, hyphae, and specific structures such as fruiting heads must be assiduously searched for in the

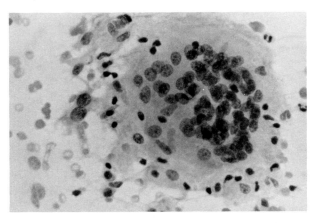

Fig. 7-102. Typical multinucleated giant cell from a granulomatous reaction. (P stain, approximately ×625)

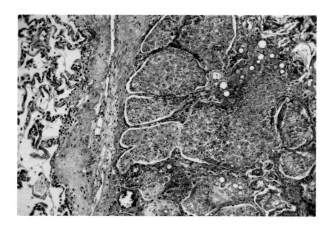

Fig. 7-104. Histologic section. Pulmonary parenchyma demonstrating proliferation of *Coccidioides immitis* spherules associated with a destructive inflammatory reaction. (×125)

exudate. The acquisition of typical forms of these fungal species by aspiration biopsy introduces some assurance that they are pathologic isolates and not contaminants from the upper tracts. *Cryptococcus neoformans* is best characterized by its polysaccharide capsule demonstrable with mucicarmine stains and India ink preparations, urease production, and antigenic markers by latex agglutination.[10] Typical cells vary from 2.5 × 7 to 3.5 × 8 μm and from round or oval to indented. The small organisms entrapped in the exudate of a clinical pulmonary aspirate (Fig. 7-105) are oval and concave. They produce typical cream-colored glistening colonies within 48 hours on media. Note the fully developed fruiting body of *Aspergillus* obtained with a 22-gauge needle from a pulmonary nodule under fluoroscopy (Fig. 7-106). The plane of focus deemphasizes the background inflammatory detritus in an effort to illustrate fungal structure.

Lung Aspirates in AIDS

Pneumocystis carinii pneumonitis (PCP) is the commonest life-threatening opportunistic infection in AIDS patients, occurring at least once in 65% to 80% of patients. Each episode has a 30% to 50% mortality rate associated with it.[16] The clinical features include nonspecific symptoms such as fever, chills, cough, and dyspnea, which occur with the same frequency as in non-AIDS immunocompromised patients; however, the duration of time between the onset of symptoms and the documentation of PCP is considerably longer, with a median of 28 days for AIDS

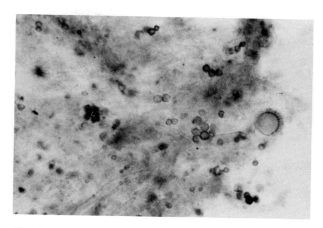

Fig. 7-106. Typical fruiting body of *Aspergillus fumigatus*. (P stain, ×200)

patients versus 5 days for non-AIDS individuals. Onset of symptoms is generally subtle with slow progression, rather than dramatic with a fulminant course. AIDS patients are usually not as hypoxemic as non-AIDS patients and may have normal arterial blood gases. The chest roentgenogram demonstrates the classical pattern of bilateral alveolar and interstitial infiltrates; but occasionally there may be only mildly increased diffuse interstitial markings, so that reliance on gallium scans becomes requisite for selection of patients for a definitive diagnostic procedure. Since the organism has not yet been isolated utilizing culture techniques and serological tests for antigen or antibody are unreliable, the diagnosis depends on the morphological identification of the organism in pulmonary secretions or tissue. Aspiration biopsy assumes a critical role.

Wallace and associates[46] reported their experience with percutaneous transthoracic aspiration biopsy of lung to investigate 16 occurrences of pneumonia in 14 patients with a confirmed or suspected diagnosis of AIDS. Standard techniques for acquisition were utilized but adjunctive processing included Gomori's methenamine silver staining for the demonstration of *Pneumocystis carinii*, immunofluorescent preparations for the identification of *Legionella*, and innoculation of the needle wash fluid for bacterial, *Legionella*, viral, mycobacterial, and fungal cultures. Aspiration biopsy successfully determined the presence of PCP in 10 of 11 (91%) confirmed cases and serendipitously identified other opportunistic organisms, including cytomegalovirus, *Mycobacterium*

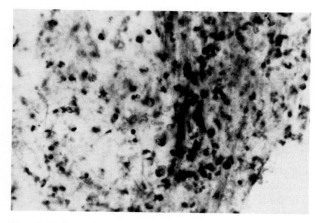

Fig. 7-105. *Cryptococcus neoformans* in an inflammatory background. (P stain, ×625)

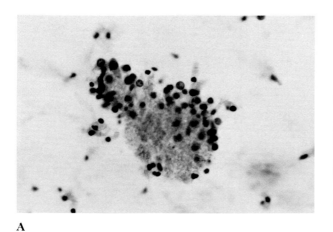
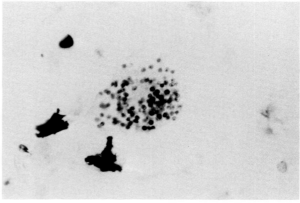

A   B

Fig. 7-107. **A.** Typical foamy intra-alveolar exudate characteristic of *Pneumocystis carinii* pneumonitis. (P stain, ×625) **B.** Typical ovoid spherical and cup-shaped structures of *Pneumocystis carinii* cysts measuring 4 to 6 micra in diameter. (Gomori's methenamine silver nitrate stain, ×625)

avium intracellulare, and pyogenic bacteria. A considerably earlier report by Chaudhary and colleagues[6] published prior to the AIDS pandemic endorsed aspiration biopsy of lung as a reliable procedure for the identification of *Pneumocystis carinii.* The diagnosis was established by the morphological identification of the organism in 121 of 202 patients with suspected PCP. Six patients whose aspirates did not contain the organism were demonstrated to be positive for the parasite at autopsy. Recommendations for improvement of diagnostic accuracy or yield included the consistent utilization of fluoroscopy for guidance (avoiding "blind biopsy"), and the combination of Gomori methenamine silver nitrate and toluidine blue O for enhancement of the organisms. Pintozzi and associates[34] reiterate the necessity for critical staining to establish the diagnosis cytomorphologically and insist that "the definitive diagnosis of pneumocystosis rests upon the use of methenamine silver nitrate or toluidine blue O stain, specific for the cysts, and the Giemsa, Wright, Wright–Giemsa and hematoxylineosin stains, specific for the sporozite. Identification of the trophozoite stage is difficult and too time-consuming since quick diagnosis is necessary because of the emergent nature of the laboratory diagnosis of PCP." They also indicate that in P-stained smears sporozoites stain purple; trophozoites are difficult to discern; and cyst walls do not stain, although outlines are sometimes visible. A character-

istic speckled pattern is imparted to the panorama of the smear by the stained sporozoites.

Watts and Chandler[48] exquisitely described the morphology of *Pneumocystis carinii* cyst forms by electron microscopic investigation of Gomori's methenamine silver nitrate (GMS) stained sections, demonstrating that the disputed darkly stained foci correspond to a focal thickening of the cell wall and are unrelated to sporozoites or other intracystic organelles. Cyst forms appear as spherical, ovoid, or cup-shaped structures measuring 4 to 6 micra in diameter (Fig. 7-107A). Cyst walls are sharply outlined with GMS and contain single or paired discretely staining foci; these are not visible in the foamy intra-alveolar exudate obtained by aspiration without staining enhancement (Fig. 7-107B). The corroborative histologic diagnosis of PCP generally requires the presence of classical features, which include a foamy or honeycombed intra-alveolar exudate associated with proliferating alveolar macrophages and an interstitial infiltrate of lymphocytes, plasma cells, and histiocytes (Fig. 7-108A). The intra-alveolar exudate may be focal or diffuse and partially organized. GMS stains demonstrate the classical cysts that are folded and crescentic (Fig. 7-108B); Giemsa stains enhance the demonstration of sporozoites within the cysts. The reliance on classical histologic features without confirmatory GMS stains may be dangerous. Weber and colleagues[49] determined in a study of 36 cases of

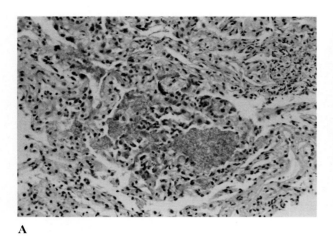
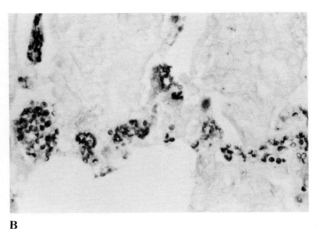

A   B

Fig. 7-108. **A.** Histologic section. Tissue correlate of intra-alveolar exudate character-ized by H & E stain. (×315) **B.** Histologic section. Tissue correlate of *Pneumocystis carinii* pneumonitis demonstrating the typical organisms by Gomori's methenamine silver stain. (×625)

PCP that in 47% of the proven cases, there were no classical histologic features that would permit a diagnosis of PCP without special stains. They described "atypical features" that included the absence of the typical exudate, the occurrence of epithelioid granulomas often incorporating cysts, multinucleated giant cells, interstitial fibrosis, dense septal inflammation and alveolar macrophage aggregation, focal calcification, and focal hyaline membrane deposition.

In considering the total number and mortality of American cases of AIDS reported to the Centers for Disease Control as of October 29, 1984, Kaposi's sarcoma without PCP accounted for 24% of cases, both Kaposi's sarcoma and PCP for 6%, PCP without Kaposi's sarcoma for 54%, and opportunistic infections without Kaposi's sarcoma and PCP for 17%.[12] Kaposi's sarcoma occurs disproportionately in the homosexual AIDS population; approximately 93% of all male cases of epidemic Kaposi's sarcoma in the United States have been diagnosed in homosexuals and bisexuals. Approximately 45% of all homosexual males with AIDS present with or eventually develop epidemic Kaposi's sarcoma during the course of their illness. By comparison, only 4% of parenteral drug users and 12% of Haitians develop Kaposi's sarcoma. It has been recorded in 29 women as of December 1984, and in four pediatric patients as of September 1984. Yet, only rarely does Kaposi's sarcoma cause

death directly; demise is generally the victory of the opportunistic infection.[12]

Kaposi's sarcoma is considered an angioproliferative disorder of endothelial cell derivation with a multicentric origin rather than metastatic dissemination from a primary site of development. It can therefore arise in the skin of the thorax or in the lungs, synchronously or metachronously, and can be diagnosed from either site by aspiration biopsy. A 27-year-old male homosexual presented with multiple elevated, purple, macular, and nodular cutaneous excrescences over the skin of the thorax and upper extremities (Fig. 7-109A). Development of respiratory distress prompted a chest film that demonstrated sporadic but consolidated densities primarily in the lower lobes (Fig. 7-108B). Although PCP was a clinical consideration, pleuropulmonary Kaposi's sarcoma was the principal diagnosis. An aspiration biopsy of lung performed under fluoroscopic guidance produced single cells and clusters of spindle elements with dense fusiform nuclei considered to represent the malignant endothelial cell constituents of Kaposi's sarcoma (Fig. 7-110; also see Fig. 13-41). An open lung biopsy was subsequently performed to search for concurrent opportunistic organisms and demonstrated replacement of pulmonary parenchyma by a fascicular proliferation of spindle cells creating vascular slit-like spaces containing linear

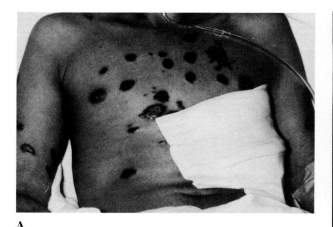

A

Fig. 7-109. A. Clinical photograph of a 27-year-old male homosexual with Kaposi's sarcoma. B. Chest film of a 27-year-old male homosexual with Kaposi's sarcoma of skin and lungs.

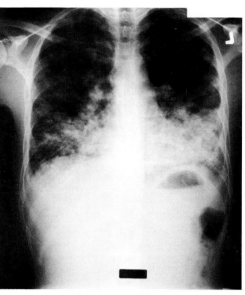

B

enfilades of erythrocytes and incontinent hemosiderin (Fig. 7-111). Kaposi's sarcoma was confirmed, but no opportunistic organisms were demonstrated. Origin in lung is generally associated with the vascular system, so that angiocentric proliferation, often incorporating residually recognizable blood vessels is common. Pleuropulmonary Kaposi's sarcoma is considered an ominous prognostic sign, and the patient died within 2 weeks of biopsy diagnosis.

Aspiration Biopsy of Mediastinal Lesions

Mediastinal FNA biopsy is more direct, less complicated, and certainly more economical than thoracotomy or mediastinoscopy. Patients tolerate the procedure because it reduces pain, anxiety, financial burden, inconvenience, and recuperation without compromising information about malignancy or infection. The procedure is endorsed as a safe, effica-

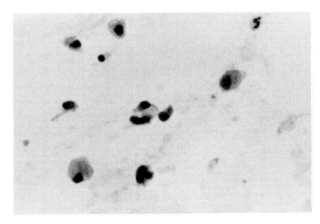

Fig. 7-110. Isolated single spindle cells characteristic of Kaposi's sarcoma. (P stain, ×800)

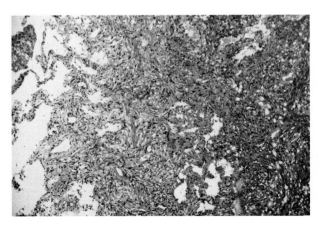

Fig. 7-111. Histologic section. Tissue correlate of pulmonary Kaposi's sarcoma. (×125)

cious, and cost-effective technique by Linder and associates[29] from Duke University Medical Center in their review of 32 mediastinal aspirates from 29 patients. They illustrated the efficiency of the technique by obtaining suitable material for cellular diagnosis in 26 of 29 patients (89.7%). Metastatic carcinoma, generally from an occult pulmonary primary, was identified in 64% of the patients. Difficult interpretations involved distinguishing among thymoma, lymphoma, and seminoma and between benign and malignant thymomas, a distinction that often depends on surgical identification of local invasion. No serious complications resulted from tissue penetration; two small pneumothoraces and two mild episodes of hemoptysis were recorded. Selection of candidates based on exclusion for contraindications resulted in a reduced complication rate.

In our experience, lesions of the mediastinum that may be investigated by aspiration biopsy as a preliminary diagnostic modality include metastatic carcinoma, lymphoma, soft tissue sarcoma, tumors of germinal epithelium, and thymomas. Successful acquisition of tissue-equivalent information may establish inoperability, surgical candidacy, radiosensitivity, a benign tumor, or infection.

A 72-year-old woman presented with respiratory distress, substernal pain, and weight loss. Physi-

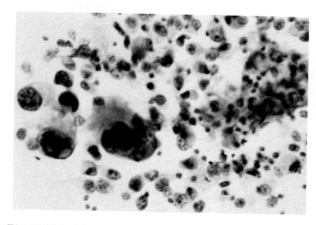

Fig. 7-113. Pleomorphic giant cell carcinoma involving mediastinum. (P stain, ×200)

cal examination disclosed a palpable, indurated, nodular mass in the region of the thyroid gland. Computed tomography disclosed a bosselated mass in the anterior mediastinum (Fig. 7-112) without definitive contiguity to the thyroid tumor. Aspiration biopsies were obtained under direct visualization from the thyroid and by CT-scan guidance from the mediastinum. Papillary carcinoma and anaplastic cells were detected in the thyroid. The mediastinal lesion was characterized by pleomorphic, gigantic cells with bizarre nuclear configurations, multilobation, coarsely textured chromatin, and parachromatin clearing dispersed in a diathesis (Fig. 7-113). Thyroidectomy and excision of the mediastinal mass

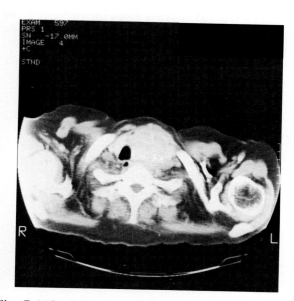

Fig. 7-112. CAT scan, mediastinum, of a 72-year-old female with respiratory distress and thyroid and substernal masses.

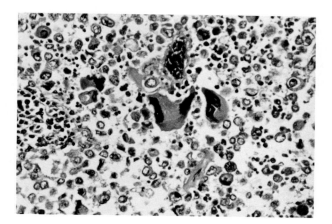

Fig. 7-114. Histologic section. Tissue correlate of mediastinal metastasis from thyroid giant cell carcinoma. (×200)

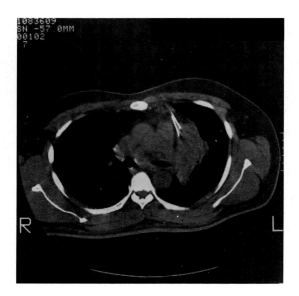

Fig. 7-115. CAT scan of a 45-year-old male demonstrating a 7-cm mediastinal mass with tandem position of needles preparatory to aspiration biopsy.

were performed concurrently; and a mixed carcinoma was verified as a primary thyroid malignancy with papillary, follicular, and giant cell components. The tissue from the mediastinal metastasis consisted of bizarre cells with deviant meganuclei (Fig. 7-114) associated with necrosis and inflammation as the

basis for the diathesis recorded in the smears. Although metastasis from occult lung carcinoma may be the most frequent source of metastatic cancer in the mediastinum, melanoma, direct extension from thyroid, and unpredictable other sources may result in malignant mediastinal repositories.

Although non-Hodgkin's lymphoma may present considerable diagnostic challenge for classification, separation from thymoma, and distinction from other small round cell cancers of the mediastinum, flow cytometry and surface markers may assist in the refinement of the final diagnosis. Hodgkin's lymphoma may be readily identified when Dorothy Reed–Sternberg cells prevail in the aspirate, but the cytopathologist must be reticent to make this diagnosis when a seductive polymorphous pattern occurs without these classical cells. A 45-year-old man consulted his physician because of respiratory distress accompanying flu-like symptoms. A chest film disclosed the unsuspected presence of a 7-cm mediastinal mass. Following an unsuccessful fiberoptic-bronchoscopic effort to obtain diagnostic tissue, examination by CT-scan directed aspiration biopsy was conducted with a dual-needle tandem approach (Fig. 7-115). Ethanol-fixed, P-stained smears demonstrated lymphocytes, plasma cells, occasional eosinophils, and infrequent binucleate cells considered to be diagnostic Dorothy Reed–Sternberg cells (Fig. 7-116). The ovoid, slightly reniform, binucleate pat-

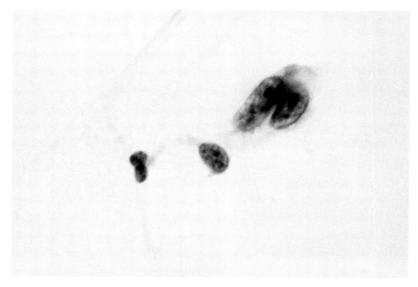

Fig. 7-116. Diagnostic Reed–Sternberg cell from mediastinal aspiration. (P stain, ×800)

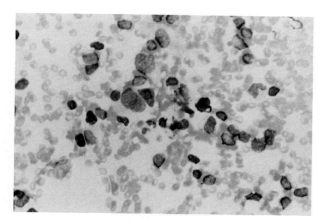

Fig. 7-117. Reed–Sternberg cells from mediastinal aspirate. (Wright–Giemsa stain, ×625)

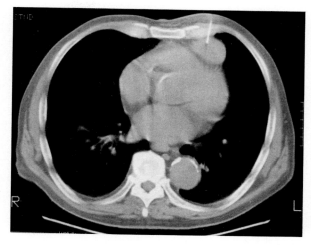

Fig. 7-119. CAT scan of a 74-year-old male demonstrating bosselated mass in the anterior mediastinum with a single needle in position for the aspiration biopsy.

tern with prominent nucleoli suggests the typically cliched "owl's eye" configuration of pattern recognition. Air-dried, Wright–Giemsa stained smears recapitulate the pattern but with less clarity (Fig. 7-117). A diagnosis of Hodgkin's lymphoma was offered, with the suggestion of lymphocyte predominance. The skeptical oncologist required a histological diagnosis for subclassification prior to the initiation of chemotherapy, and this was accomplished through mediastinoscopy. The node is effaced by the process that is primarily lymphoid with a classical Reed–Sternberg cell at the center (Fig. 7-118), compatible with lymphocyte predominant Hodgkin's lymphoma.

A 74-year-old man complained of a pressure sensation in his chest and was discovered to have mediastinal widening. Computed tomography demonstrated a bosselated mass in the anterior mediastinum, which was investigated pre-operatively by directed aspiration biopsy (Fig. 7-119). The needle extracted its harvest from the periphery of the lesion. The cells are singly dispersed and vary from oval and polygonal to stellate or fusiform (Fig. 7-120). Nuclei present microvacuolization and cytoplasm is ambiguously demarcated, often merging with the fibrillar, elusive background. A diagnosis of low-grade myxoid liposarcoma preceded histologic confirmation by

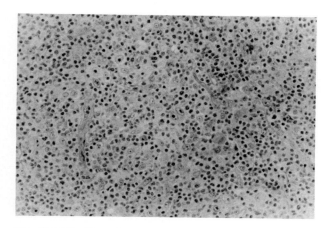

Fig. 7-118. Histologic section. Tissue confirmation of Hodgkin's lymphoma of mediastinum illustrating Reed–Sternberg cell at the center.

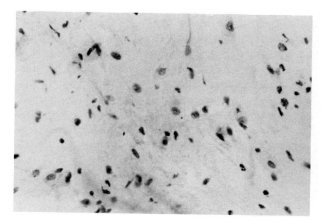

Fig. 7-120. Polygonal and stellate cells consistent with low-grade myxoid liposarcoma. (P stain, ×625)

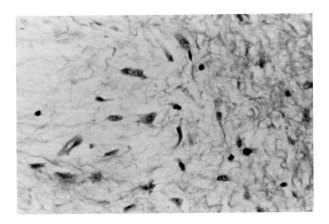

Fig. 7-121. Histologic section. Tissue correlate of myxoid liposarcoma. (×625)

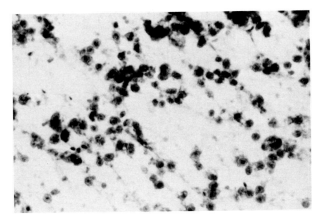

Fig. 7-123. Panoramic view, metastatic seminoma. (P stain, ×500)

open biopsy (Fig. 7-121). Histologic sections demonstrate a myxoid stroma with areas of compression and immature fibrosis supporting a cellular network of spindled and stellate elements with enlarged, atypical nuclei offering prominent nucleoli. These are essentially immature mesenchymal cells, some of which have properties of lipoblasts. There is no true capsule (not illustrated). Oil red O stains performed on frozen tissue separately demonstrated fine cytoplasmic lipid droplets.

A 52-year-old man sustained a radical orchiectomy for treatment of anaplastic seminoma of the testis 2 years prior to hospitalization for evaluation and management of dysphagia. A barium swallow (Fig. 7-122) demonstrated deflection of the espha-

gus by a mediastinal mass clinically consistent with metastatic seminoma. Aspiration biopsy was performed with fluoroscopic control and barium enhancement of the deviated esophagus. The panoramic view of the smear presents cellular, but minimally cohesive aggregates of essentially uniform polygonal cells with fragile, fragmented cytoplasm clinging to round or oval nuclei with prominent central nucleoli (Fig. 7-123). Background debris includes ghost-like remnants of cells and small, dark, round elements suggesting lymphocytes. High-power focus on a cohesive cluster illustrates the characteristic nuclei (Fig. 7-124) that are round to oval with discrete membranes; delicate, translucent chromatin; and strikingly prominent nucleoli. Peripheral cytoplas-

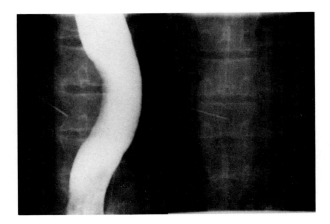

Fig. 7-122. Barium swallow for augmentation of needle positioning into a mediastinal mass in a 52-year-old male.

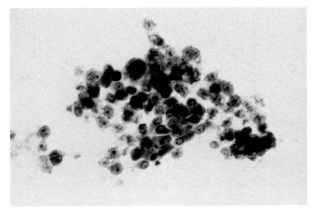

Fig. 7-124. High-power view illustrating characteristic cells of metastatic seminoma. (P stain, ×625)

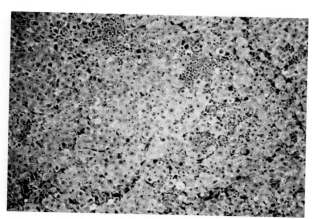

Fig. 7-125. Histologic section. Low-power view demonstrating primary seminoma of testis as the source for the mediastinal metastasis.

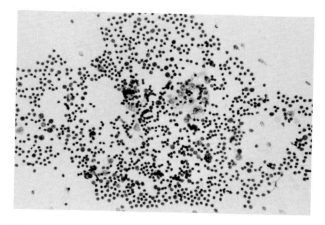

Fig. 7-127. Structure of normal thymus. (P stain, ×400)

mic boundaries are indistinct. Cohesion is tenuous, and at the periphery the cells seem precariously fragile. A diagnosis of metastatic seminoma was based on these features. Review of the primary testicular neoplasm confirmed this impression. The solid tumor is composed of polyhedral cells in mosaic sheets, intimately intermingled with lymphocytes that intrude into the interstitial stroma (Fig. 7-125). Nuclei are uniform and monotonously stereotyped by their similarities of delicate to translucent chromatinic texture surrounding prominent dark, round, regular nucleoli in dominance at the centers. Cytoplasm is vacuolated, lacey, and pale. Magnified cells seem somewhat less uniform, chromocenters become legible,

and nucleoli are often duplicate (Fig. 7-126). Fibrous septa conduct minute capillaries and lymphoid congeries. The composite architecture of typical seminoma emerges and predicts its own appearance in the clinical smears from the mediastinum.

Cells derived from thymus can present treacherous identity challenges in aspirate smears because normal, persistent thymus and thymomas can contribute a uniform lymphoid population. Clinical and radiographic correlation are essential to appropriate interpretation. Epithelioid thymomas can contribute more succulent, oval, or fusiform cells, making the task of distinction somewhat simpler. Aspirates prepared from normal residual thymus in a 14-year-old boy demonstrate a dominant population of lymphocytes that vary subtly in size but are not otherwise polymorphic and coexist with endothelial-like cells

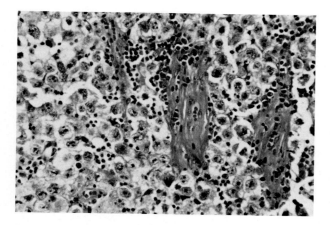

Fig. 7-126. Histologic section. High-power view of primary seminoma of testis representing the source of the mediastinal metastasis. (×625)

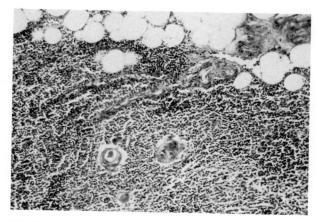

Fig. 7-128. Histologic section. Tissue correlate of normal thymic structure. (×200)

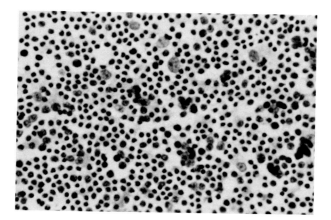

Fig. 7-129. Cells from enlarged thymus of a 17-year-old female with clinical myasthenia gravis. (P stain, ×625)

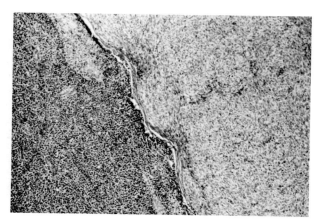

Fig. 7-131. Histologic section. Tissue correlate demonstrating the juxtaposition of lymphoid and epithelial components of thymoma. (×160)

with pale nuclei and delicate cytoplasm (Fig. 7-127). Correlation is noted with tissue that organizes a lymphoid population around vascular structures and Hassall's corpuscles (Fig. 7-128). Compare this with an aspirate from a 17-year-old girl with a thymic mass and clinical myasthenia gravis that shows a polymorphous population of mature and transformed lymphocytes, occasional histiocytes, and rosette-like clusters of small round cells (Fig. 7-129). The tissue correlate is a solid, parenchymatous lymphoid process with a tendency to germinal center organization (Fig. 7-130). The true thymoma may be dominantly lymphoid or a mixture of lymphoid and epithelial elements (Fig. 7-131) and yet donate only

one component to the needle probe (Figure 7-132). Note that the lymphoid population from this verified thymoma mimics the cells aspirated from the normal, persistent thymus (Fig. 7-127). The distinction between benign and malignant thymomas in the absence of anaplasia is generally made by the surgeon who verifies invasion of mediastinal structures and stroma by the malignant thymic lesions.

Although thymic carcinoid (Plate XX) was originally regarded as a peculiar type of thymoma, it is now clearly recognized as a neoplasm separate from thymic derivation and is an expression of carcinoid apud derivatives that may occur throughout the body. A 61-year-old man presented with uncon-

Fig. 7-130. Histologic section. Tissue correlate, enlarged thymus from a 17-year-old female with clinical myasthenia gravis. (×160)

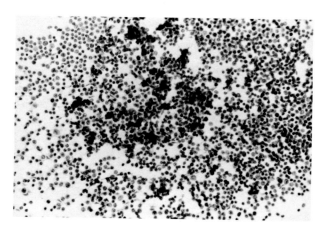

Fig. 7-132. Lymphoid component of a mixed lymphoid and epithelial thymoma. (P stain, ×400)

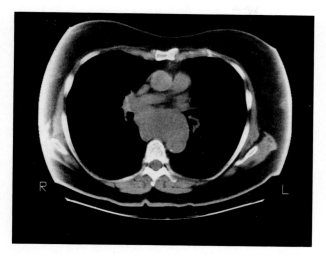

Fig. 7-133. CAT scan of a 61-year-old male demonstrating anterior-superior mediastinal mass.

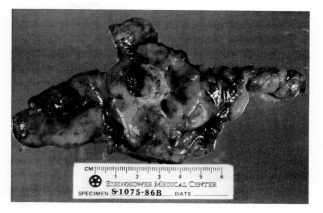

Fig. 7-135. Gross enlarged thymus specimen incorporating thymic carcinoid of a 61-year-old male.

trolled diabetes mellitus and hypercorticism with high serum cortisol and adrenocorticotropic hormone levels. Computed tomography demonstrated a left pulmonary infiltrate, left pleural effusion, ascites, and an anterior-superior mediastinal mass (Fig. 7-133). An aspiration biopsy was fixed in ethanol and stained by the P technique. A uniform population of small round cells with central or slightly eccentric nucleoli is arranged in anastomosing ribbons and cords with acinar formations (Fig. 7-134). Nuclei dominate the cytoplasm and are conspicuous in their conformity of pattern and structure. This is typical of carcinoid lesions and is a quality that often precludes

accurate prediction of malignant potential. Exploration of the mediastinum revealed a thymic origin and resulted in excision of a partially encapsulated, indurated, multilobulated thymus measuring $12 \times 8 \times 3$ cm, incorporating at its center a well-delimited neoplasm with a central area of stellate, granular necrosis, measuring $7 \times 6 \times 3$ cm (Fig. 7-135). Histological sections demonstrate a solid, lobulated neoplasm subdivided by fibrous septal extensions that circumscribe packets of uniform epithelial cells creating lobules with occasional central necrosis (Fig. 7-136). Nuclei are uniform in size, contour, chromatinic texture, distribution, and in the presence of single or multiple nucleoli. Two to three mitoses are present per high-power field (hpf). Circumscribing eosinophilic cytoplasm is granular and stains posi-

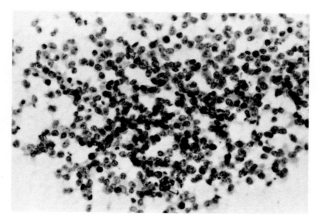

Fig. 7-134. A uniform population of small round cells in anastomosing ribbons consistent with thymic carcinoid. (P stain, ×500)

Fig. 7-136. Histologic section. Tissue correlate of atypical thymic carcinoid. (×160)

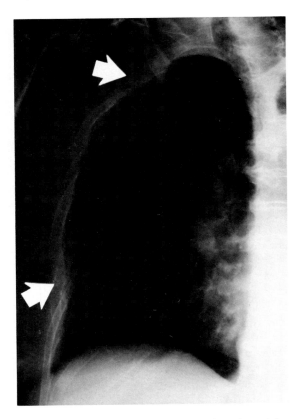

Fig. 7-137. Chest film of a 69-year-old female with bone pain and lytic rib lesions.

drome and ACTH may be identified in elevated quantities in the blood or by immunoperoxidase in the tumor itself.

Lesions of the Chest Wall

Involvement of the chest wall by tumor or infection generally results in a visible or palpable anatomic or cosmetic disturbance that permits aspiration biopsy under direct visualization. The procedure is standard, and the information generally invaluable. When the skeleton of the thorax is the focus of attention, it may be necessary to utilize image intensification for directing the puncture. A 69-year-old woman complained of back pain and tenderness in the anterior chest with discomfort in the right shoulder. Radiographic examination demonstrated lytic rib lesions (Fig. 7-137) that were evaluated by aspiration biopsy under fluoroscopic guidance (Fig. 7-138). The harvest is proliferating atypical plasma cells with eccentric nuclei and abundant, well-defined ovoid cytoplasm with crisply delimited cytoplasmic boundaries (Fig. 7-139). A diagnosis of plasma cell myeloma was confirmed by conventional bone marrow examina-

tively for silver argyrophilic granules and sporadically by immunoperoxidase for ACTH production. Ultrastructure confirmed the presence of numerous ovoid dense-core neuroendocrine granules measuring 100 to 300 nm in diameter. There are focal areas of capsular invasion with extension into fat and blood vessels. Silverberg[38] references Rosai's grade II thymic carcinoid as an atypical variant in which mitoses are frequent, pleomorphism is present, and lobulated nests of cells are centrally necrotic and mineralized. The latter feature is characteristic of this tumor. Thymic carcinoid tumors are more aggressive than conventional thymomas and in about two thirds of cases, invasion may be present initially or lesions recur. Distant metastases may be expected in 30% of cases, despite indolent local growth that may ultimately involve critical mediastinal structures. About 34% of patients with thymic carcinoids have Cushing's syn-

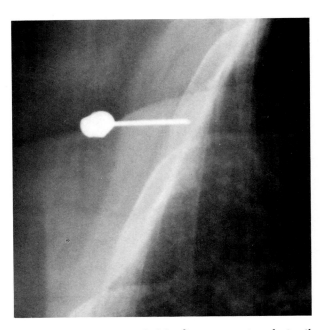

Fig. 7-138. Needle guided by fluoroscopy into lytic rib lesion preparatory to aspiration biopsy.

tion and by serum electrophoresis (Fig. 7-140), which disclosed an IgA-lambda monoclonal gammopathy with beta mobility.

CONCLUSIONS

The description of cell morphology and patterns from aspiration biopsies is, of necessity, a work in progress, modulated by nuances of the critical eye as it witnesses a novel format, an appealing new expression of a remembered pattern, or simply a lesion defined by the technique for the first time in its own visual vocabulary. It is a discipline in motion that retrains the inquisitive analytical mind to see tissues in the abstract presented as the sum of their cellular parts in a scientific, yet artistic format reliable for diagnosis universally. It is a process of learning, definition, classification, and intellectual challenge that liberates the diagnostic effort from traditional labor-intensive and expensive approaches that sometimes ignore the propriety for the patient's convenience, dignity, and financial or psychological resources. And finally, it is a dynamic concept of unification that integrates medical specialists with technology, promotes expedient cooperation among physicians, and most importantly recalls the humanizing influences to the patient-doctor relationship. Descriptions may change as sophisticated information accrues, but the intention is for the procedure to advance its simplicity to protect its universality.

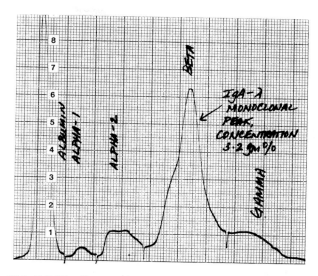

Fig. 7-140. Serum electrophoretogram demonstrating IgA lambda monoclonal gammopathy with gamma mobility.

ACKNOWLEDGMENTS

We wish to address the expertise and enthusiastic contributions of Dr. Richard Lynch, quintessential interventional radiologist, Department of Radiology, Eisenhower Memorial Hospital, for understanding the irrevocable value of aspiration biopsy and for his abilities to acquire the cellular sample with precision and care; our clinical colleagues who believe in the procedure and in our abilities to use it well and who allowed us involvement with their patients; the photographic talents of Kelly Fisher, Annenberg Center for Health Sciences, for his assistance with the illustrations; and our families, who graciously spent time without us, so that we could be together in our work in progress.

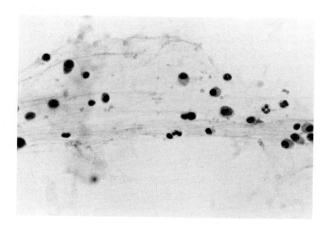

Fig. 7-139. Plasma cell myeloma. (P stain, ×625)

REFERENCES

1. Adamson JS, Bates JH: Percutaneous needle biopsy of the lung. Arch Intern Med 119:164–169, 1967
2. Berg JW, Robbins GF: A late look at the safety of aspiration biopsy. Cancer 15:826–827, 1962
3. Berquist TH et al: Transthoracic needle biopsy, accuracy and complications in relation to location and type of lesion. Mayo Clin Proc 55:475–481, 1980

4. Buchanan AJ, Gupta RK: Cytomegalovirus infection of the lung: Cytomorphologic diagnosis by fine needle aspiration cytology. Diag Cytopathol 2(4):341–342, 1986

5. Castellino RA, Blank N: Etiologic diagnosis of focal pulmonary infection in immunocompromised patients by fluoroscopically guided percutaneous needle aspiration. Radiology 132:563–567, 1979

6. Chaudhary S et al: Percutaneous transthoracic needle aspiration of the lung diagnosing *Pneumocystis carinii* pneumonitis. Am J Dis Child 131:902–907, 1977

7. Cormier Y et al: Prevention of pneumothorax in needle lung biopsy by breathing 100% oxygen. Thorax 35:37–41, 1980

8. Dahlgren SE, Lind B: Comparison between diagnostic results obtained by transthoracic needle biopsy and by sputum cytology. Acta Cytol 16(1):53–58, 1972

9. Dahlgren SE, Nordenstrom B: Transthoracic Needle Biopsy. Stockholm, Almqvist & Wiksells, 1966

10. Dalton HP, Nottebart HC: Interpretative medical microbiology. New York, Churchill Livingston, 1986

11. Davies DP: Percutaneous lung biopsy in the investigation of intrathoracic lesions. Radiography 51:179–185, 1985

12. DeVita VT et al (eds) AIDS. New York, JB Lippincott, 1985

13. Dick R et al: Aspiration needle biopsy of thoracic lesions: an assessment of 227 biopsies. Brit J Dis Chest 68:86–94, 1974

14. Dutra FR, Geraci CL: Needle biopsy of the lung. JAMA 155(1):14–21, 1954

15. Engzell U et al: Investigation on tumor spread in connection with aspiration biopsy. Acta Radiol 10(4):385–398, 1971

16. Fauci AS et al: The acquired immune deficiency syndrome: an update. Ann Int Med 102:800–813, 1985

17. Fontana RS et al: Transthoracic needle aspiration of discrete pulmonary lesions: Experience in 100 cases. Med Clin North Am 54(4):961–971, 1970

18. Frable WJ: Thin-needle aspiration biopsy. Philadelphia, WB Saunders, 1983

19. Gobien RP et al: CT-assisted fluoroscopically guided aspiration biopsy of central hilar and mediastinal masses. Radiology 141:443–447, 1981

20. Johnston WW, Frable WJ: The cytopathology of the respiratory tract. A review. Am J Pathol 84:317, 1976

20a. Johnston WW, Frable WJ: Diagnostic Respiratory Cytopathology. New York, Masson Publishing, USA, 1979

21. Kaminsky DB: Aspiration Biopsy for the Community Hospital. New York, Masson Publishing USA, 1981

22. Kaminsky DB: Aspiration biopsy in the context of the new medicare fiscal policy. Acta Cytol 28(3):333–336, 1984

23. Kato H et al: Percutaneous fine-needle cytology for lung cancer diagnosis. Diag Cytopathol 2(4):277–283, 1986

24. Kini SR et al: Cytopathology of papillary carcinoma of thyroid. Acta Cytol 24:511–521, 1980

25. Klein DL, Gamsu G: Pneumomediastinum—an unusual complication of needle biopsy of the lung. Chest 74(2):229–230, 1978

26. Koss L: Diagnostic Cytology. Philadelphia, JB Lippincott, 1979

26a. Koss LG et al: Aspiration Biopsy: Cytologic Interpretation and Histologic Bases. New York, Igaku–Shoin, 1984

27. Lalli AF et al: Aspiration biopsies of chest lesions. Radiology 127:35–40, 1978

28. Landman S et al: Comparison of bronchial brushing and percutaneous needle aspiration biopsy in diagnosis of malignant lung lesions. Radiology 115:275–278, 1975

29. Linder J et al: Fine needle aspiration biopsy of the mediastinum. Am J Med 81:1005–1008, 1986

30. Matthay RA, Mortiz ED: Invasive procedures for diagnosing pulmonary infection—a critical review. Clin in Chest Med 2(1):3–18, 1981

31. Moloo Z et al: Possible spread of bronchogenic carcinoma to the chest wall after a transthoracic fine needle aspiration biopsy, a case report. Acta Cytol 29(2):168–169, 1985

32. Norenberg R et al: Percutaneous needle biopsy of the lung: Report of two fatal complications. Chest 66(2):216–218, 1974

33. Pearce JG, Patt NL: Fatal pulmonary hemorrhage after percutaneous aspiration lung biopsy. Amer Rev Resp Dis 110:346–349, 1974

34. Pintozzi RL et al: The morphological identification of *Pneumocystis carinii*. Acta Cytol 23(1):35–39, 1977

35. Ranchod M, Levine GD: Spindle cell carcinoid tumors of the lung: a clinicopathological study of 35 cases. Am J Surg Pathol 4:315–330, 1980

36. Roggli VL et al: Asbestos bodies in fine needle aspirates of the lung. Acta Cytol 28(4):493–498, 1983

37. Schreeb T Von et al: Renal adenocarcinoma—Is there a risk of spreading tumor cells in diagnostic puncture? Scan J Urol Nephrol 1:270–276, 1967

38. Silverberg SG (ed): Principles and Practice of Surgical Pathology, vol 1, pp 745–748. New York, John Wiley and Sons, 1983

39. Sinner WN: Transthoracic needle biopsy of small peripheral malignant lung lesions. Invest Radiol 8(5):305–314, 1973

40. Sinner WN: Complications of percutaneous transthoracic needle aspiration biopsy. Acta Radiol 17(6):813–828, 1976

41. Sinner WN, Zajicek J: Implantation metastasis after percutaneous transthoracic needle aspiration biopsy. Acta Radiol 17(4):473–480, 1976

42. Sonnenberg E Van et al: Percutaneous biopsy of difficult mediastinal, hilar, and pulmonary lesions by computed-tomographic guidance and a modified coaxial technique. Radiology 148[1]:300–302, 1983

43. Struve–Christensen E: Iatrogenic dissemination of tumour cells. Danish Med Bull 25(2):82–87, 1978

44. Tao LC et al: Cytologic diagnosis of bronchioloalveolar carcinoma by fine needle aspiration biopsy. Cancer 57:1565–1570, 1986

45. Tao LC et al: Value and limitations of transthoracic and transabdominal fine needle aspiration cytology in clinical practice. Diag Cytopathol 2(4):271–276, 1986

46. Wallace JM et al: Percutaneous needle lung aspiration for diagnosing pneumonitis in the patient with AIDS. Am Rev Resp Dis 131:389–392, 1985

47. Walsh HA: Medical microbiology and infectious diseases, chap 72, pp 658–664. Philadelphia, WB Saunders, 1981

48. Watts JC, Chandler, FW: *Pneumocystis carinii* pneumonitis. Am J Surg Pathol 9(10):744–751, 1985

49. Weber WR et al: Lung biopsy in *Pneumocystis carinii* pneumonitis: a histopathologic study of typical and atypical features. Am J Clin Path 67:11–19, 1977

50. Westcott JL: Air embolism complicating percutaneous needle biopsy of the lung. Chest 63(1):108–110, 1973

51. Wolinsky H, Lischner, MW: Needle track implantation of tumor after percutaneous lung biopsy. Ann Int Med 71(2):359–362, 1969

52. Yamashina M: An 18-year history of a corticotropin-secreting spindle cell carcinoid in the lung. Arch Pathol Lab Med 109:673–675, 1985

52a. Zajicek: Aspiration Biopsy Cytology, Part I. Basel, S Karger, 1974

53. Zelch JV, Lalli AF: Diagnostic percutaneous opacification of benign pulmonary lesions. Radiology 108:559–561, 1973

54. Zornoza J et al: Aspiration biopsy of discrete pulmonary lesions using a new thin needle. Work Prog 123:519–520, 1977

CHAPTER 8

Abdominal Aspiration

JOSEPH A. LINSK

SIXTEN FRANZEN

The scope of abdominal aspirations includes (1) palpable masses involving the abdominal wall and peritoneum, (2) palpable intra-abdominal masses, (3) retroperitoneal disease, and (4) aspiration during surgery (particularly of the pancreas). The spleen is considered in Chapter 14.

Direct transabdominal aspiration continues to be an underused clinical tool. The bulk of the literature on the subject deals with technologically guided needle puncture of mainly nonpalpable masses.[13,23,26,28,35,36,38,40,41,44,61,62,81,83,92,97–99,105] It is curious that the simplicity of direct transabdominal puncture of a palpable mass, regardless of organ or location, has received so little attention. Surgical biopsy of abdominal masses usually requires major anesthesia and laparotomy. In our experience, direct puncture of a palpable abdominal mass, coupled with radiographic data where necessary, has almost invariably eliminated the need for diagnostic surgery.

METHOD: PALPABLE MASSES

Fine needles, 25 gauge to 22 gauge and 2 cm to 8 cm (1 in – 4 in) long (see Chap. 1), offer maximum adaptability. The limiting factor in using long fine needles is the physical flexibility of the instrument. For abdominal wall and easily palpable intra-abdominal masses, a 2-cm or 3-cm 22-gauge needle is quite satisfactory. Syringes have been discussed in Chapter 1.

The procedure should be explained to the patient to allay anxiety and gain cooperation. It is quite accurate to state that ill effects are rare. A consent form may be signed. To proceed, after careful palpation, clean the skin with alcohol or any generally used surface disinfectant. Using this simple method, we have never seen infection introduced into the abdomen, although many of the patients are immunosuppressed. Local anesthesia is optional; it has been recommended for retroperitoneal punctures. Infiltration may be carried out to the peritoneum, but we have usually found it to be unnecessary. Carry out the puncture by direct thrust into the mass after warning the patient or by initial push through the skin followed by entrance into the mass. For intra-abdominal masses, depress the abdominal wall so that the mass is immediately palpable under the fingers. For mobile masses, including the liver, instruct the patient to hold his breath. The patient may feel pain briefly when the needle enters the skin and passes through the peritoneum. Carry out aspiration and smearing as discussed in Chapter 1.

Retroperitoneal (nonpalpable) targets are discussed below.

A diagnostic scheme illustrating the approach to abdominal masses utilizing fine needle aspiration (FNA) is shown in Figure 8-1.

TARGETS FOR ASPIRATION

The following palpable and nonpalpable structures make up the targets for abdominal punctures:

Abdominal wall masses
Umbilicus
Subperitoneal masses

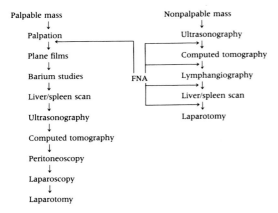

Fig. 8-1. Fine needle aspiration and the diagnostic profile. Abdominal masses. (Linsk JA, Franzen S: Fine Needle Aspiration for the Clinician, p 6. Philadelphia, JB Lippincott, 1986)

Palpable intra-abdominal masses
  Tumors obscured by ascites
  Palpable intra-abdominal masses due to primary or recurrent tumors of the gastrointestinal (GI) tract
  Mucinous tumors
  Miscellaneous metastatic tumors
  Small cell tumors
  Liver
  Retroperitoneal masses and nodes, including the pancreas, kidney, and adrenal

Abdominal Wall Masses

The abdominal wall is a frequent site for metastases spread by hematogenous, lymphatic, and direct routes of extension. Dermal and subcutaneous nodules, although they may result from intra-abdominal cancer, are considered elsewhere (see Chap. 13). Full-thickness masses, often involving scars postoperatively, yield diagnostic cells to distinguish tumor, granuloma, and keloid. Keloid is diagnosed by exclusion because there is no cell yield.

Umbilicus

Secretions and crusting from so-called granulomas of the umbilicus may mask metastases. Metastases may also present with a periumbilical firm nodular mass with intact smooth or puckered skin. It may be mistaken for the normal fibrotic nodule of the umbilicus.

The skin immediately overlying the implant may be taut, and a better approach is achieved through intact periumbilical skin with angling into the mass.

Subperitoneal Masses

Particularly during postcancer resection follow-up study, careful palpation may reveal subperitoneal abdominal masses. These smooth, ill-defined swellings move with the abdominal wall on pressure and are easily distinguished from body wall masses and infiltrates, which usually elevate the skin. Direct puncture and aspiration will yield cytologic smears that are usually diagnostic in the clinical context of a known primary tumor, for instance, an ovarian, GI, or renal tumor.

Palpable Intra-abdominal Masses

X-ray films and scans may provide useful information in selected cases (see Chap. 2). However, most masses require no adjuvant studies. Without a bleeding history or symptoms, coagulation studies are not required. Routine blood count should be observed for granulocyte numbers and platelet adequacy. Severe granulocytopenia is a relative contraindication. The platelet count should be above 100,000/mm³ for puncture of metastatic liver masses, since severe bleeding has been seen. A platelet count above 70,000 is adequate for the usual abdominal mass.

Clinical judgment and testing may allow abdominal puncture with fewer than 70,000 or 100,000 platelets, depending on the patient.

Tumors Obscured by Ascites

The search for cancer cells in ascitic fluid is sometimes frustrating; and removal of ascites in suspected tumor cases may reveal a palpable intra-abdominal mass, as shown in the following case report:

■   *Case 8-1*
A 58-year-old woman was referred to the oncology clinic for carcinoma of the pancreas with recurrent chylous ascites. On review, it was found that diagnosis was established at laparotomy, during which a large mass occupying the pancreas was seen. An enlarged lymph node was selected for biopsy to avoid violating the pancreas. Postoperatively, the node was reported as *reactive hyperplasia,* **but rapidly developing chylous ascites appeared to confirm the diagnosis of pancreatic carcinoma. In the clinic, paracentesis uncovered a fixed, intra-abdominal mass in the left epigastrium, which, on aspiration, yielded a smear compatible with lymphocytic–histiocytic non-Hodgkin's lymphoma. The patient improved with treatment but subsequently died. Lymphoma was confirmed at autopsy.**

Palpable Intra-abdominal Masses Due to Primary or Recurrent Tumors of the Gastrointestinal Tract

The patient with a palpable intra-abdominal mass may present with abdominal complaints varying from vague discomfort to severe pain and including anorexia, weight loss, and "functional" GI symptoms. The mass may also be an unexpected finding, unaccompanied by symptoms. On physical examination of the relaxed abdomen, ill-defined or clear-cut, small or large, fixed or mobile masses may be detected or suspected. Direct puncture at the time of palpation, will often give immediate clarifying diagnostic data.

The most common site of origin in any quadrant is the colon. Recurrence is often outside the colon and not revealed by barium enema or endoscopy. The concept of intraluminal and extraluminal tumors is discussed in *Fine Needle Aspiration for the Clinician.*[53] A primary tumor may also become a target for aspiration.

Gastric carcinoma appears and usually recurs in the epigastrium and upper abdomen. Differential diagnosis includes palpable masses in the pancreas (including cysts), biliary tract, liver, kidneys (see below), and spleen (see Chap. 14). Omental, mesenteric, and large lymph node metastases from multiple sources, including lymphoma and pelvic tumors, are common presenting targets.

Ovarian tumors, which are considered in detail in Chapter 10, not infrequently present as abdominal masses and may provide a massive cell yield, as illustrated in Figure 8-2 (Color Plate XXI). Vague abdominal masses may sometimes by clarified by deep palpation and puncture through a colostomy orifice with the Franzen needle guide (see Chap. 11). In addition, submucosal infiltration of a colostomy orifice is easily diagnosed by transmucosal puncture.

While primary gastric carcinoma becomes a target for fine needle aspiration (FNA) only in late stages, nonpalpable small and large gastric carcinomas are now regularly detected and surgically biopsied endoscopically. Diffusely infiltrating carcinoma may present a thickened wall that does not exfoliate cells and gives poor yield on forceps biopsy. A technique of FNA using an endoscopic unit has

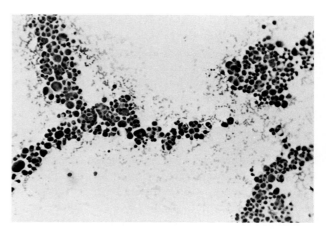

Fig. 8-2. Papillary carcinoma, ovary, poorly differentiated. This is a large cell yield of polymorphous cells forming papillary strands aspirated from a palpable mid-abdominal mass. (R stain, ×90)

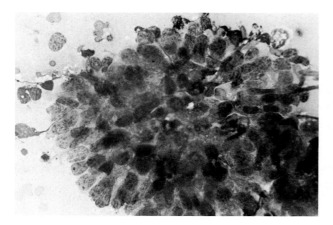

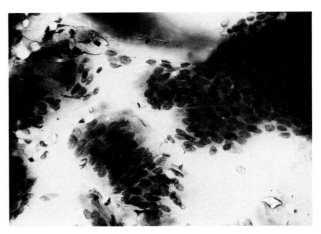

Fig. 8-3. Adenocarcinoma, colon, intra-abdominal mass. An intact papillary mass of tumor cells has been extracted. The peripheral columnar arrangement is evident, with deeper cells in haphazard arrangement. (R stain, ×600)

Fig. 8-4. Colonic mucosa. A more orderly arrangement of columnar cells is evident, with normal colonic mucin (staining purple) in the background. (R stain, ×300)

been presented by Iishi and coworkers.[43] It yielded positive diagnoses in all of 11 cases not diagnosable by routine methods.

Cytology. Recognition of adenocarcinoma in smears is not difficult. The most characteristic feature is a rather dense aggregation of cells with overlapping, often present within a necrotic background. Necrosis is a positive finding, and almost invariably, a few intact cells or convincing cell shadows will be found. These findings must be considered within the clinical context of a suspected GI carcinoma. Necrosis must be distinguished from fecal material, which is identified by color, odor, and microscopic features.

An acute inflammatory process may be present, and the aspirate may consist of an acute inflammatory exudate intermixed with tumor cells. In aspirates from colon carcinoma, the cells may have a columnar arrangement (Fig. 8-3). This must be distinguished from columnar cells in benign groups (Fig. 8-4). The cytologic findings vary with the degree of differentiation, and acinar structures may be identified in well and moderately differentiated tumors (Figs. 8-5 and 8-6). With Papanicolaou (P) stain, colon carcinoma often presents in clusters and larger tissue fragments. Nuclei are distinctly elongated and irregular. Nucleoli vary from inconspicuous small structures to large, irregular, multiple structures. There are many stripped nuclei. The cytoplasm, in general, is not well defined (Fig. 8-7).

In gastric carcinoma, cells are rounded rather than columnar. They are more likely to be poorly differentiated and less well preserved. They form rounded clusters (Fig. 8-8) and, less often, partially intact acinar structures (Fig. 8-9).

Cytologic findings vary with the degree of differentiation. Poorly differentiated adenocarcinoma may yield dissociated primitive cells and clusters composed of immature cells with granular nuclei (Fig. 8-10). Individual cells are recognized as cancer cells by their enlarged irregular nuclei, high nuclear

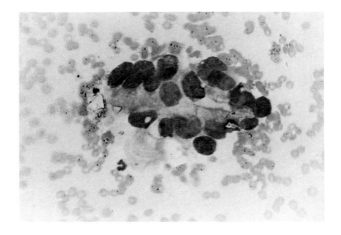

Fig. 8-5. Adenocarcinoma, colon. This is an intact alveolar structure together with some foreign debris, part of a smear aspirated from a colonic mass. (R stain, ×300)

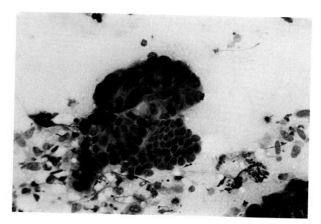

Fig. 8-6. Adenocarcinoma, colon. Complex cellular aggregate with well, intermediate, and frankly malignant differentiation. Extracted from a late recurring mass at anastomotic site. (R stain, ×600)

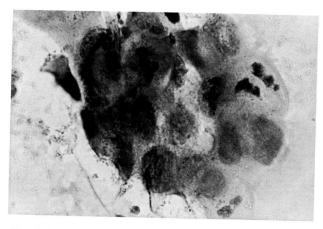

Fig. 8-8. Adenocarcinoma, gastric, omental mass. This is a rounded cluster of somewhat poorly defined cancer cells. There is cellular molding. One prominent mitosis is evident. (P stain, ×400)

cytoplasmic ratio, nucleoli, and poorly defined cytoplasm (Fig. 8-10).

A small dispersed cell type occurs that includes gastric carcinoma in the differential diagnosis of small cell tumors (see below) (Fig. 8-11). A small lymphocyte-like cell with dark nucleus and poorly defined cytoplasm (Fig. 8-12) is characteristic of linitis plastica. Such cells should not be confused with lymphocytes that infiltrate these lesions.

Signet ring cells (see below) (Fig. 8-13) are highly characteristic of gastric carcinoma within the appropriate clinical context.

Occasionally, cells may appear bizarre, anaplastic, and entirely unidentifiable as being of gastric origin (Fig. 8-14).

Clinical data are important. Cytologic distinction of a gastric from a pancreatic carcinoma (see below) in an epigastric mass may be difficult. Dis-

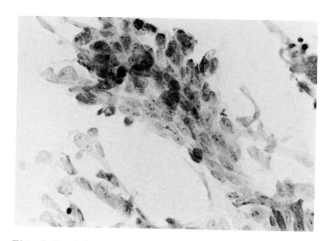

Fig. 8-7. Adenocarcinoma, colon. This smear shows clusters of irregular elongated cells, many with nucleoli. A columnar arrangement can be detected. Cytoplasm is poorly preserved. (P stain, ×200)

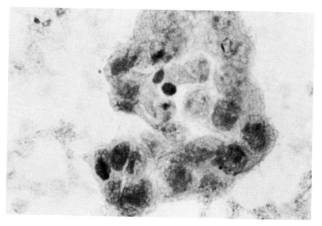

Fig. 8-9. Adenocarcinoma, gastric, moderately differentiated. This is a somewhat poorly preserved cluster of adenocarcinoma cells forming an acinar structure. (P stain, ×360)

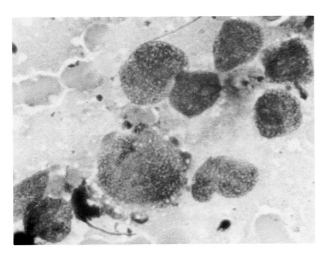

Fig. 8-10. Adenocarcinoma, gastric, poorly differentiated. This is a cluster of poorly differentiated cells with granular nuclei. (R stain, ×500)

tinction from colon carcinoma in which cells are better preserved and columnar (see below) may be more evident.

Mucinous Tumors

The presence of carcinoma cells, sometimes scanty, in a thick mucinous stroma is diagnostic of mucinous adenocarcinoma (Fig. 8-15 and Color Plate XI; also see Fig. 8-4). The mucin stains violet to purple to dark

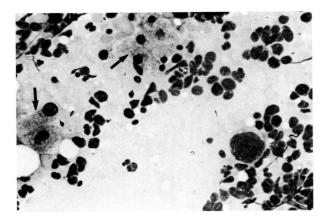

Fig. 8-11. Gastric carcinoma, metastatic to liver. This is a dispersed, somewhat polymorphous small cell pattern. One giant tumor cell is seen in contrast. There are scattered hepatocytes with granular poorly defined cytoplasm *(arrows)*. (R stain, × 400)

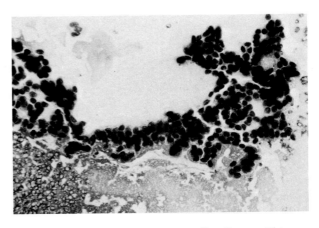

Fig. 8-12. Gastric carcinoma, small cell type. This pattern is seen in linitus plastica. (R stain, ×600)

blue with Romanowsky (R) stain (Fig. 8-16) and pale tan-orange with P. Cell groups appear as irregular islands within the strands of mucus. The individual cell cytology is similar to that described for other GI tumors and ovarian and metastatic breast (colloid) carcinoma. The presence of mucin may also be helpful in clarifying a clinical syndrome caused by mucinemia that results in multiple thromboses.[64,87]

Signet ring cells are not usually identified by R stain. Intracytoplasmic pink-staining acid mucopolysaccharide is evident in some adenocarcinomas of the salivary gland with R stain (see Chap. 5). This, however, is not mucin. Signet ring cells of breast carci-

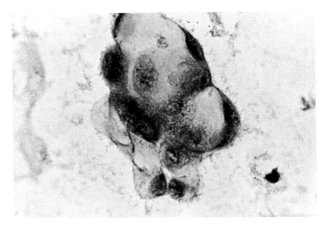

Fig. 8-13. Adenocarcinoma, gastric, signet ring type. This is a molded cluster of gastric carcinoma cells with intracytoplasmic vacuoles. (P stain, ×600)

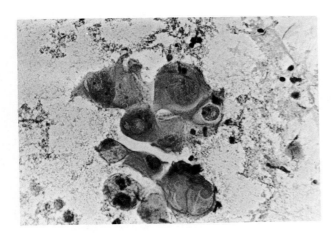

Fig. 8-14. Adenocarcinoma, gastric, poorly differentiated. This is a cluster of rather bizarre gastric carcinoma cells with giant nucleoli on a necrotic background. (P stain, ×400)

noma (see Chap. 6) can be identified (see Fig. 6-44 and 6-46), but an intracytoplasmic structure rather than a vacuole is evident (Fig. 6-45). Vacuoles of signet ring adenocarcinoma of the GI tract can be identified by P stain (see Fig. 8-13 and Color Plate XXI; also see Fig. 10-26).

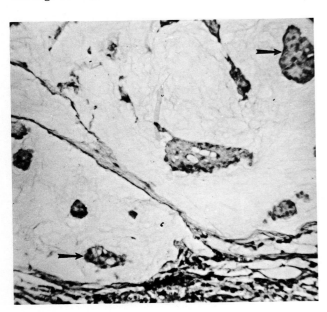

Fig. 8-15. Adenocarcinoma, gastric, mucinous type. Tissue section. There are dense pools of mucin, within which are small acinar tumor cell clusters (*arrows*). (H & E, ×200)

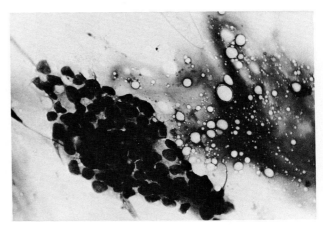

Fig. 8-16. Adenocarcinoma, gastric, mucinous type. This is a dense cluster of rounded carcinoma cells within a thick mucinous stroma staining purple and violet. There is no intracellular mucin. (R stain, ×300)

Pseudomyxoma peritonei, producing abdominal distention by gross production of gelatinous mucin, can be identified by FNA (Fig. 8-17).

Small Cell Tumors

Small cell tumors are listed in Figure 8-18. They may be anticipated or unexpectedly present in smears obtained from any quandrant of the abdomen or pelvis. Cytologically, they contrast clearly with the varieties of adenocarcinoma harvested from the GI tract.

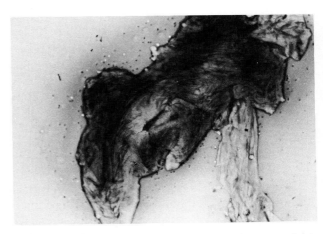

Fig. 8-17. Pseudomyxoma peritonei. This is a thick mass of mucin extracted from a distended abdomen. Cellular content was absent, with the exception of a few atypical mesothelial cells. Smear stains purple with R. (R stain, ×200)

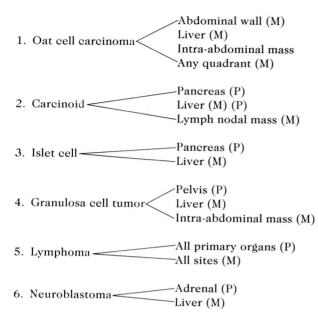

1. Oat cell carcinoma
   - Abdominal wall (M)
   - Liver (M)
   - Intra-abdominal mass
   - Any quadrant (M)

2. Carcinoid
   - Pancreas (P)
   - Liver (M) (P)
   - Lymph nodal mass (M)

3. Islet cell
   - Pancreas (P)
   - Liver (M)

4. Granulosa cell tumor
   - Pelvis (P)
   - Liver (M)
   - Intra-abdominal mass (M)

5. Lymphoma
   - All primary organs (P)
   - All sites (M)

6. Neuroblastoma
   - Adrenal (P)
   - Liver (M)

Fig. 8-18. Diagnosis of small cell tumors by FNA of primary (P) or metastatic (M) site.

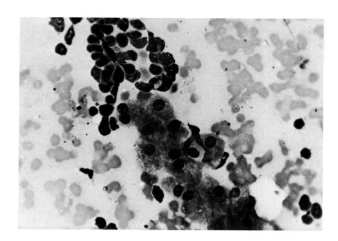

Fig. 8-19. Oat cell carcinoma, metastatic to liver. This figure contrasts a double-layer group of hepatocytes with a cluster of small, dark, elongated, naked nuclei (oat cells) (see Fig. 8-20 for comparison). (R stain, ×200)

Small cell gastric carcinoma (see above) is an exception. Hepatic and renal cytologic findings (see below) and metastatic squamous carcinoma should also be readily distinguished.

Apudomas

The amine precursor uptake and decarboxylation (APUD) tumors include oat cell (see Chap. 7), islet cell, carcinoid, medullary carcinomas of thyroid (see Chap. 4), pheochromocytomas (see below), and paragangliomas (see Chap. 3). Of this group, carcinoid, oat cell, and islet cell tumors may present cytologically as small cell tumors in the abdomen. These tumors share the presence of neurosecretory granules identified by Gremilius stain[55] or electron microscopy. Such granules help to distinguish apud tumors from other small cell tumors (lymphomas [see Chap. 14], neuroblastomas [see Fig. 8-28], granulosa cell tumors [see Figs. 8-22 and 8-23]) and from tumors that are occasionally small cell types (nasopharyngeal carcinoma [see Fig. 3-23], gastric carcinoma [see Figs. 8-11 and 8-12], and rhabdomyosarcoma [see Fig. 15-27]).

APUD tumors are more specifically identified by clinical application of markers, including serotonin and 5-hydroxyindoleacetic acid (carcinoid), in-

sulin (islet cell carcinoma), and ACTH (oat cell carcinoma). Markers are useful in cytologically puzzling problems, but the clinical syndrome combined with cytologic findings will often determine the diagnosis without marker studies.

Oat Cell Carcinoma

Oat cell carcinoma may present in the abdominal wall, the liver, and as an intra-abdominal (metastatic) mass. Contrast the small molded oat cells in Figure 8-19 with the large irregular randomly dispersed metastatic adenocarcinoma cells in Figure 8-20.

Cytology. The cytology of oat cell carcinoma is considered in Chapter 7. Metastases of oat cell carcinoma to the liver may be accompanied by considerable necrosis. Figure 8-21 illustrates dispersed primitive cells with granular nuclei, surrounded by multiple degenerated nuclei of hepatocytes simulating lymphocytes. The cytopathologist must be prepared to interpret smears in which cellular alterations have occurred as a result of partial necrosis. Smears of primitive cells may not conform to the classic cytologic pattern of oat cell carcinoma, but a diagnosis can be made with supporting clinical data.

Granulosa Cell Carcinoma

Granulosa cell carcinoma occurs in the pelvis (see Chap. 10), as an intraabdominal metastatic mass, and in the liver. Indolent metastases may be present for

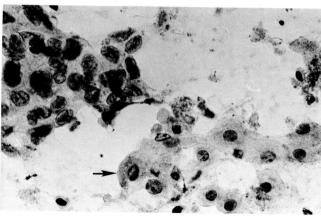

Fig. 8-20. Adenocarcinoma, metastatic to liver from colon. Hepatocytes *(arrow)* with small nuclear : cytoplasmic ratio and orderly arrangement are seen in contrast with cluster of adenocarcinoma cells. Nuclei are irregular, nucleoli are prominent, cells are in haphazard arrangement, and cytoplasmic borders are indistinct. (P stain, ×280)

years in the liver, simulating a carcinoid (see below) both clinically and cytologically (Fig. 8-22, and Color Plate XXI C).

Cytology. Single cells as well as clusters, lined up in rounded groups and in monolayer, are present. Nu-

clei are quite variable in size and have a reticulogranular texture. Nucleoli are inconspicuous with R stain. Although blue-gray cytoplasm is often stripped, it is occasionally abundant. Cells are about two times the size of a polymorphonuclear leukocyte (see Fig. 8-22). Cytology with P stain is further considered in Chapter 10. Nuclear membrane indentations and longitudinal grooves may be distinctive findings in the usual clinical setting.[18]

Granulosa cell carcinoma metastatic to bone marrow is seen in Figure 8-23, which illustrates cell size in relation to hematopoietic cells and megakaryocyte.

Islet Cell Carcinoma

Malignant islet cell tumors may present suitable palpable targets in a metastatic liver, in the pancreas, and as an intra-abdominal metastatic mass.

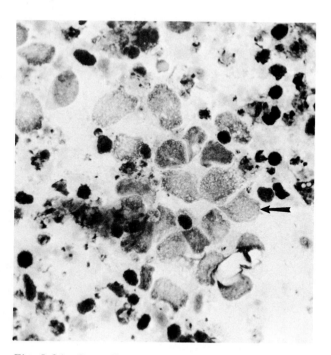

Fig. 8-21. Oat cell carcinoma, metastatic to liver. This smear was obtained from a massively enlarged liver. There was extensive necrosis. The small cells are pyknotic nuclei of hepatocytes. Tumor cells consist of stripped, somewhat swollen, primitive nuclei *(arrow)*. (R stain, ×300)

Cytology. In islet cell carcinoma the cells appear singly and in irregular groups. There may be some larger cell masses. Nuclei are round to oval and occasionally double. The nuclear chromatin varies from clumped to finely granular, and there are occasional large nucleoli. The overall cell size is about three times that of a polymorphonuclear leukocyte. The nuclei lie both centrally and eccentrically within the cytoplasm, which is gray-blue and abundant (Fig. 8-24). There are, however, many stripped nuclei. The cytologic findings mirror those in tissue section (Fig. 8-25). Distinction from carcinoid on light microscopy may be difficult. Red granulation in carcinoid is a distinctive finding (see below). (Also see Color Plate XXI D.)

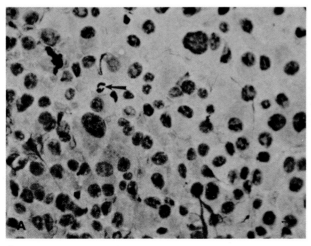

A

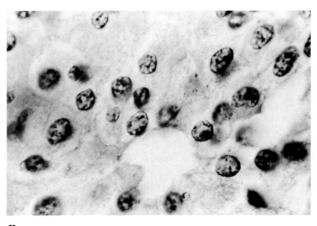

B

Fig. 8-22. A. Granulosa cell carcinoma, subperitoneal mass. There are occasional large nuclei. The chromatin is granular, and there is a tendency to form nodular groups. Note polymorphonuclear leukocyte for size comparison *(arrow)*. (R stain, ×300) B. Granulosa cell carcinoma, hepatic nodule. Cells maintain a fairly regular arrangement without overlapping. Nuclei are oval and moderately pleomorphic with inconsistent presence of nucleoli. An incomplete acinus, possible site of a Cell-Exner body, is evident. (P stain, ×400)

Carcinoid

Carcinoid appears as a cytologic target in the thorax (see Chap. 7), liver, and lymph nodes. Primary carcinoid will rarely present as a clinically palpable or identifiable abdominal mass.[67] It may present in the pancreas. Metastatic carcinoid in the liver presents with distinct masses that are palpable or observed with imaging procedures. It is indistinguishable from hepatic metastases of other small cell tumors and various non-small-cell metastatic tumors.

Clinical problems associated with metastatic carcinoid are illustrated in the following case reports:

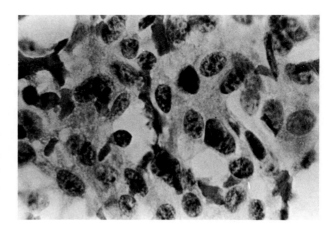

Fig. 8-24. Islet cell carcinoma, metastatic to omentum. This is a dense aggregate of tumor cells; the cells have maintained some tissue integrity. Nuclear chromatin is more clumped, with occasional nucleoli. This tumor had metastasized widely. (R stain, ×600)

Fig. 8-23. Bone marrow smear illustrates cluster of cells from granulosa cell carcinoma adjacent to a megakaryocyte. (R stain, ×1000)

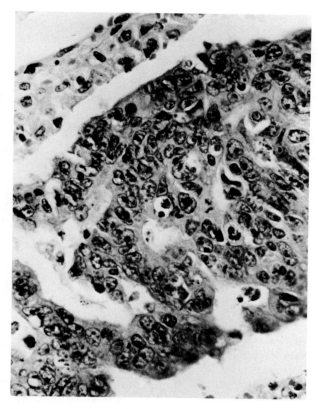

Fig. 8-25. Islet cell carcinoma. Tissue section. This is a pleomorphic area of the tumor in Fig. 8-25, with obvious malignant characteristics. (H & E stain, ×400)

■ *Case 8-2*

A 55-year-old obese woman underwent surgery to define an upper abdominal mass. A metastatic liver was identified and on wedge biopsy was interpreted as metastatic adenocarcinoma. On palpation, the pancreas was felt to be the source of the tumor. Two years later, the patient was more obese and suffered from continuous diarrhea. Aspiration of a palpable epigastric mass yielded smears interpreted as carcinoid (Fig. 8-26A). On review of the original biopsy, the diagnosis was revised to trabecular carcinoid in a fatty liver (Fig. 8-26B). At subsequent autopsy, no primary source was found outside of the liver.[67]

■ *Case 8-3*

A 78-year-old man was found to have a right hilar mass on routine chest x-ray. An enlarged liver was palpated and aspirated (Fig. 8-27). The diagnosis was small cell tumor, compatible with carcinoid. Neuroendocrine granules were identified on electron microscopy. Smears were considered not compatible with oat cell carcinoma. There was marked regression of the tumor masses on a regimen of streptozocin and fluorouracil, fairly specific therapy for carcinoid and endocrine tumors. The tumor ultimately ran a malignant course.

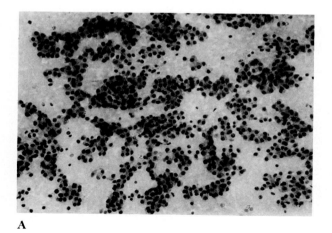

A

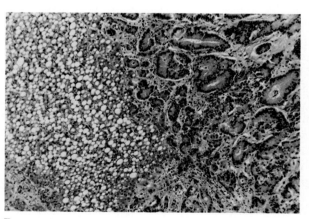

B

Fig. 8-26. A. Metastatic carcinoid. Abundant cell yield. Cells have been stripped away from the tumor stroma. (R stain, ×330) **B.** Trabecular carcinoid metastatic to a fatty liver. Tissue section. Cells with abundant cytoplasm and eccentric nuclei are seen in smears aspirated from such a tumor. (H & E stain, ×600)

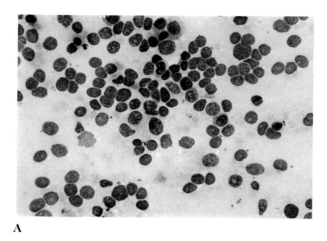
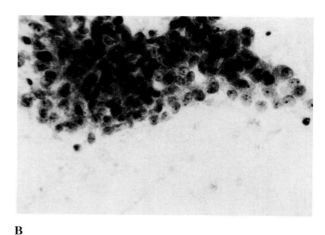

A   B

Fig. 8-27. A. Metastatic carcinoid. There are dispersed nuclei with sharply etched nuclear membranes and poorly defined cytoplasm. Nucleoli are prominent. (R stain, ×600) **B.** Agglutinated cluster of small cells with one or more nucleoli. Same tumor. (P stain, ×600)

Alertness to the possibility of metastatic carcinoid is important in three clinical contexts: (1) a pulmonary mass evident at the same time as hepatic metastases, ordinarily suggestive of a bronchogenic carcinoma with metastasis; (2) metastatic liver on scan and examination in a patient with a previous carcinoma; and (3) metastatic liver of unknown origin.

Prompt FNA of all liver metastases will yield a number of small cell tumors that will alter diagnostic and therapeutic programs.

FNA diagnosis of carcinoid appears to be an underutilized method. In a series of 64 patients with carcinoid observed over 8 years at Memorial Hospital, New York, 2 were diagnosed by aspiration.[55]

The cytologic findings of carcinoid may present unexpectedly in aspiration of metastatic livers identified by palpation or scan (Color Plate XXI E). The usual expected yield is large cell adenocarcinoma with necrosis, which has been our most common experience.

Wasastjerna published a case of J. Zajicek's in which small cells were aggregated in alveolar structures. Abundant cytoplasm containing striking red granules was evident on May–Grünwald–Giemsa (MGG) stain.[103]

With R stain, well-differentiated carcinoids will yield densely cellular smears (see Fig. 8-26). Nuclei are irregularly oval, dispersed, and eccentric with fairly abundant pale blue cytoplasm that is sometimes vacuolated. When there is cell crowding, cell borders are not evident, and there is an appearance of a syncytium. With P stain, cytoplasm varies from pink to pale green. Nucleoli are not evident in well-differentiated tumors.

In tumors that are not quite as well differentiated, there is greater variation in nuclear size, but dispersion is still prominent. Single central nucleoli are evident (see Fig. 8-27).

In poorly differentiated tumors, nuclei are largely stripped of cytoplasm. There is occasional molding, and there may be confusion with oat cell carcinoma (see Fig. 8-27A).

Red granulation in the cytoplasm is more evident with MGG than with Wright–Giemsa or Diff-Quik staining.

Well-differentiated lymphocytic lymphoma, palpable and nonpalpable, will yield small cells that should be recognizable as of lymphoid origin (see Chap. 14). Finally, smears of neuroblastoma must be considered in the differential diagnosis (see Chap. 17). An example is seen in Figure 8-28.

Liver

Since the first edition of *Clinical Aspiration Cytology*, there has been increasing interest in the liver as a target for FNA.[8,11,32,34,75,96,104] Much has been due to the explosive growth and widespread use of imaging, which has disclosed often unsuspected lesions, leading to aspiration. The use of FNA as a primary tool for diagnosis even for palpable disease is still undeveloped.

The largest group of liver aspirations, presented by Lundquist,[60] consisted of a consecutive series of 6357 biopsies performed on 6050 patients from 1964 through 1985. Of this group, 1000 consecutive liver biopsies (1976–1981) were selected to illustrate the spectrum of liver disease diagnosable by FNA. There were 123 metastatic cancers, 36 hepatomas, and 3 lymphomas. The bulk of the remainder was inflammatory diseases, many specifically diagnosable. (It is evident that the percentage of hepatomas would be sharply increased in series collected in Africa or Southeast Asia, where the incidence is 20 to 30 times higher.[69] The establishment of a specific diagnosis obviates the need for expensive imaging in most cases.

Palpable Disease

In Western Europe and North America the most common abdominal target for puncture is still the metastatic liver; however, it is evident from the Lundquist study[60] (see above) that nonneoplastic palpable diseases may become important targets, depending on the interest and orientation of the cytopathologist. Abnormal enlargements are seen in cirrhosis, particularly the postnecrotic type, chronic

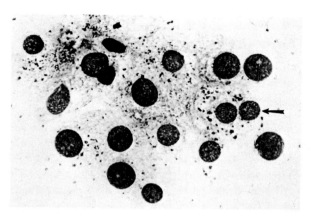

Fig. 8-29. Normal liver. Typical lysosomal granulation. Note binucleated cell (*arrow*). (MGG stain, ×800)

active hepatitis, and parasitic, cystic, and metabolic diseases.[89] The six most common mass lesions in the Mayo Clinic experience[1] are cysts, cavernous hemangiomas, focal nodular hyperplasias, adenomas, hepatomas, and metastatic lesions. With the exception of cavernous hemangiomas—found almost always as an incidental lesion on imaging—all of these are potential targets for FNA.

Extension of the left lobe may form a left upper quadrant mass that must be differentiated from spleen, kidney, and pancreatic abnormalities (see below). Epigastric projection of the liver profile may suggest pancreatic, omental, and gastric masses. The usual right upper quadrant site may be altered by an extremely irregular contour. Discrete nodular masses or massive lobar or diffuse enlargement can also be appreciated. The texture is hard. Without a discrete mass, the initial examination of the patient with a history of a primary tumor and suspected metastasis is critical. With complete relaxation of the patient as well as the examiner, the irregular firm edge of a metastatic liver will often be readily palpable, allowing direct puncture.

Although FNA of palpable lesions has regularly been carried out by us and recorded in publications by other authors,[58,59,103] it appears that the greatest number of liver aspirations currently being performed have been carried out under guidance of an imaging procedure. In several recently reported series, all aspirations were done under guidance.[8,11,32,104] Of the 520 cases reported, there can be no question that many of these lesions were palpable. These results are a product of procedural arrangements in which radiology departments do all aspira-

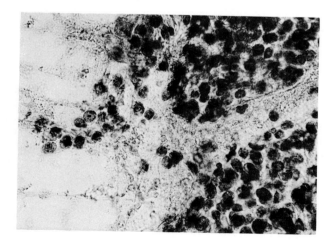

Fig. 8-28. Neuroblastoma. Small cells are in compact masses. Rosettes are not evident in this smear but may be present. There is no cell molding. Cytoplasmic borders are indistinct, and small nucleoli are present. (P stain, ×190)

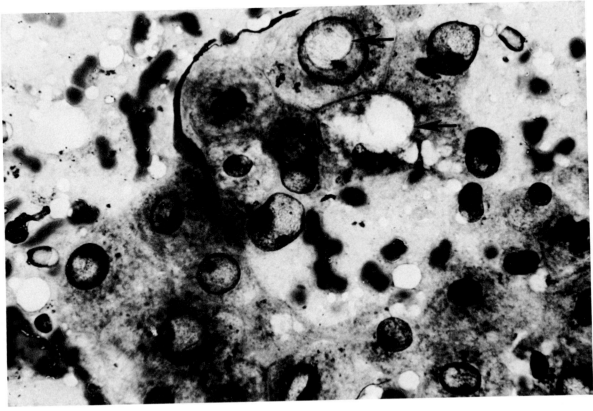

A

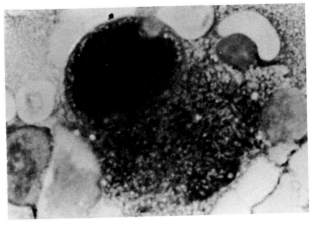

B

Fig. 8-30. A. Liver. Congestive heart failure. Hepatocytes with apparently empty ballooning nuclei *(arrows)* due to the presence of giant PAS–positive inclusion bodies. This is a normality specific for hepatocytes and is especially common in diabetes and congestive liver injury. (MGG stain, ×1200) B. Hepatocytes from A. (PAS, ×2000)

tions and submit slides to the cytopathologist. Puncture of palpable liver masses is within the range of expertise of the interventional cytopathologist and clinician and carries with it the advantage of hands-on cytology.

Nonpalpable Disease

Nonpalpable disease is detected, of course, only with some imaging procedure. Such abnormalities detected on scan (see Chap. 2) can be aspirated with a high percentage of accuracy on the basis of films, without CT or ultrasound guidance. With this technique, a correct diagnosis of malignancy in the liver was made in 51 of 54 patients.[40] Small solitary nodules and nodules in posterior or diaphragmatic positions will require active guidance with CT or ultrasound.[44]

As noted above, there now are numerous large series of cases aspirated under guidance with high degrees of diagnostic accuracy. With a rapid stain method and immediate slide review, the number of punctures can be sharply curtailed because the diagnosis will often be apparent on the initial puncture. False-positive liver scans are not rare,[45] and the failure of multiple FNAs to yield positive cells is strong evidence that the radiographic changes do not represent cancer. Random biopsies of the liver for malignancy with a Menghini needle produce a low yield, in contrast with scan-directed FNA. Ultrasonically guided needle puncture will also produce a higher percentage of positive biopsies than will the Menghini core biopsies.

Cytology of The Normal Liver

The interpretation of liver cytology is aided considerably by an appreciation of "normal" cytology and the variations induced by general pathologic disorders. Descriptive analysis by Söderstrom from the first edition of *Clinical Aspiration Cytology* is herein included.

The normal liver is represented in aspiration biopsy cytology (ABC) by glandular epithelium cells. Kupffer's cells and cuboid duct cells are present but contribute little to identification. Identification must be based on the typical gland cells, which in the smears are seen in clusters, representing two-dimensional fragments of hepatic muralia, normally without quite distinct borders between individual cells.

Hepatic gland cells have sharply demarcated, spherical nuclei. Distinct nucleoli are always present; there is often one larger and one smaller. A typical feature is the presence of two equal-sized nuclei in some of the cells, a finding that increases with the patient's age. The cytoplasm has the same floccular opacity and staining reactions as thyroid epithelium and contains similar small lysosomal (paravacuolar) granules (Fig. 8-29). Benign hepatocytes are seen in Figures 8-11, 8-19, and 8-20.

Some typical pathologic findings are common enough to be valuable identification marks, such as

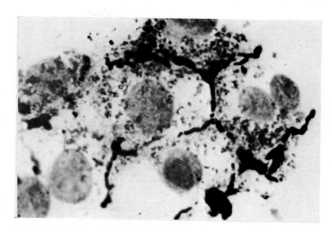

Fig. 8-31. Liver. Hepatocytes in cholestasis; bile capillaries are filled out with bile "thrombi." This picture may be seen also in metastases of hepatocellular carcinoma. (MGG stain, ×1000)

the giant periodic acid-Schiff (PAS)-positive intranuclear inclusion bodies, which make the nuclei appear as empty rings with other staining methods (Fig. 8-30). They are especially common in diabetic patients. Other typical pictures are presented by the curious type of mitochondrial swelling called cloudy swelling and by visible bile capillaries during and after the bile stasis. With naphthylamidase they may be visualized normally (Figs. 8-31 and 8-32).

The liver is a giant organ, and a mixture of liver cells in aspirates from other targets is not exceptional. It should be possible to interpret such contaminations correctly without much hesitation.

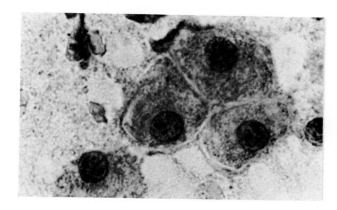

Fig. 8-32. Liver. Hepatocytes in the recovery stage after extraheptic cholestasis (choledocholithiasis). Bile capillaries are still visible, but the pigment has disappeared. (MGG stain, ×800)

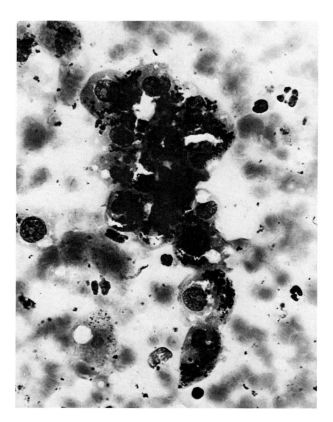

Fig. 8-33. Liver, acute hepatitis. There is extensive disruption of liver plates with cell destruction, clumping, and release of bile (staining black), inflammatory cells, and cellular debris. (R stain, ×300)

Cytology of the Nonneoplastic Liver

Nonneoplastic disorders may yield fairly characteristic cell patterns.[58,59,94,103] The normal cytologic findings, noted above, blend with a series of pathologic changes. Acute hepatitis, cirrhosis, and granulomatous changes can be demonstrated cytologically.[94,95,106]

In acute hepatitis, the cytologic yield is generally representative of the histopathologic alterations, but classification of subtypes of the acute processes comparable to histology requires considerable experience and should be undertaken only with clinical correlation (see below). Swollen hepatocytes with cytoplasm clearing toward the cell membrane and pyknotic cells with gray-blue or green structural cytoplasm and shrunken nuclei have been described. There is marked disruption of liver plates, distortion and pleomorphism of cells, and intranuclear inclu-

sions (Figs. 8-33 and 8-34). There are increased numbers of granulocytes mixed with hepatocytes as well as Kupffer's cells and mononuclear inflammatory cells. Fatty vacuolization is absent. There may be intracellular and intercellular bile stasis.

In cirrhosis, there are increased numbers of ductular cells that are small and regular with well-defined bland nuclei and no pleomorphism. They must be differentiated from tumor cells (Fig. 8-35). Hepatocytes are pleomorphic, and fatty vacuolization is often present and is seen easily under low power (Figs. 8-36 and 8-37). Finding fibrous tissue elements is usually confirmatory. Although the clinical context is important, problems of differential diagnosis between cirrhosis and tumor may occur.

Puncture of the liver capsule may yield mesothelial cells mixed with liver parenchyma. Mesothelial cells must be distinguished from bile duct epithelium. Cells are small and regular with round nuclei and often form a honeycomb pattern (Fig. 8-38). They may assume cellular configurations that simulate tumors (Fig. 8-39).

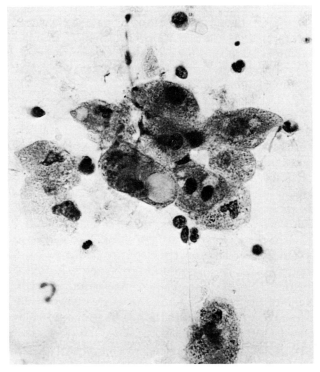

Fig. 8-34. Liver, acute hepatitis. This is a disrupted liver plate. Cytoplasmic degeneration with formation of intracytoplasmic vacuoles is evident. (P stain, ×600)

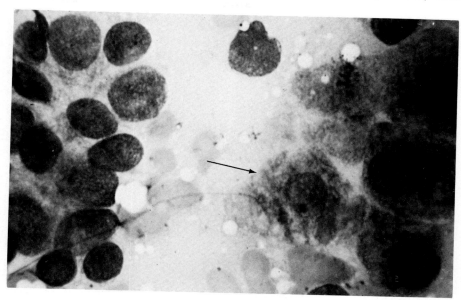

Fig. 8-35. Liver, metastasis. Granular and mottled hepatocytes *(arrow)* are seen in contrast to a cluster of tumor cells that are finely granular, moderately pleomorphic, and forming acini. (R stain, ×600)

Granulomatous changes, although infrequent, provide fairly specific cytologic findings.[95] They may be a surprise finding or anticipated by the presence of a number of clinical states, particularly tuberculosis, sarcoidosis, and Hodgkin's disease.[33] In the 1000 liver aspirations reported by Lundquist, 7 were granulomatous.[60] The granuloma is composed of aggregates of epithelioid cells surrounded by fibroblasts, lymphocytes, and plasma cells. There may be central necrosis and Langhans' (giant) cells. Aspiration will dislodge clusters of epithelioid cells, which are the

hallmark of the lesion and diagnostic in the absence of other elements. With R stain, these cells are enlarged and oval with eccentric elongated and curved nuclei and prominent blue cytoplasm. Nuclear chromatin is finely granular, and nucleoli are absent or small and not well defined (see Chap. 14). The cells are in distinctive groups and have a background of hepatocytes and mononuclear cells. Histiocytes, from which epithelioid cells are formed, have rounder nuclei with a filmy, less distinct cell outline.

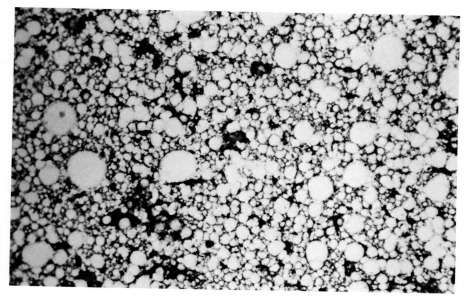

Fig. 8-36. Liver, cirrhosis. Gross, fatty engorgement of hepatocytes is evident. Compare to tissue section (see Fig. 8-37). (R stain, ×90)

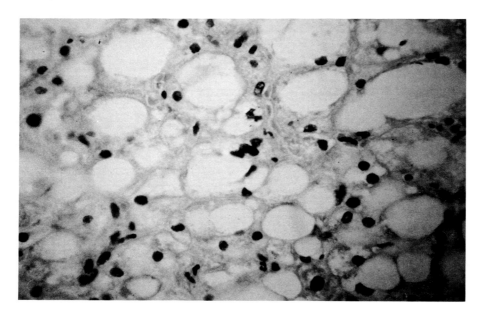

Fig. 8-37. Liver, cirrhosis. Tissue section. Note fat infiltration. (H & E stain, ×190)

In a cytologic study of nonneoplastic liver disease by Perry and Johnson,[78] hepatocyte morphology was found useful in distinguishing normal from pathologic specimens. The technique was deemed limited, however, in classifying liver disease. The study was based on cell washings from a core biopsy needle; this is not entirely comparable to an aspiration smear, which yields greater cellularity and a variety of fairly consistent diagnostic or suggestive patterns. Aspiration smears obtained with accurately placed needles will correctly determine or at least suggest the diagnosis in most cases. Considerable experience is needed to subclassify benign hepatic disease on cytologic bases. However, with clinical correlation including history and current state of health, physical findings (particularly hepatic enzyme levels), the addition of hepatic cytology may obviate the need for core biopsy in selected cases. Equivocal imaging reports, both mottled patterns on isotopic scan and insufficient resolution for multiple small

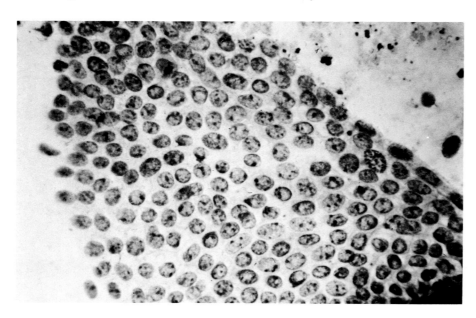

Fig. 8-38. Liver, mesothelium. Mesothelium has stripped and forms a cell ball that may simulate a small cell tumor. (R stain, ×280)

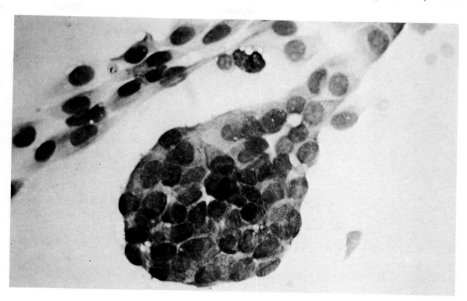

Fig. 8-39. Liver, mesothelium. This is a honeycomb pattern characteristic of benign mesothelium. (R stain, ×300)

lesions with CT are not infrequent. In a patient with an enlarged liver, weight loss, and a mottled imaging pattern, FNA may identify a tumor. Nodules less than 2 cm in diameter may not be revealed by current imaging modalities.

Cytology of the Neoplastic Liver

The interpretation of liver aspiration smears from suspected tumors is complicated by a number of factors not present in other targets. The liver is a secondary tumor site for every metastasizing tumor in the body. The cell yield can therefore be keratinizing and nonkeratinizing squamous cells from head and neck, esophageal, cutaneous, pelvic, and anal primary tumors. Adenocarcinoma from breast, lung, prostate, and glandular and gastrointestinal mucinous and nonmucinous primary tumors may be sources. Melanoma, germinal tumors, sarcomas, small cell tumors (oat cell, endocrine, lymphomas, carcinoid), and genito-urinary primary tumors also seed the liver and are clinically identified by palpation or imaging studies.

There is a high incidence of primary liver tumors in Asia and Africa,[25,69] and this neoplasm is being recognized with increasing frequency in Western Europe and North America. Problems in distinguishing hepatocytes from well-differentiated hepatoma cells (see Fig. 8-40) and from metastatic tumor cells must be addressed.

The difficulties are reduced by careful attention to clinical history, physical examination, which may indicate an extrahepatic primary tumor, and history of a primary liver disorder (see below).

Hepatocellular Carcinoma. Description of the cytology of hepatocellular carcinoma has appeared in several sources.[28,34,75,77,96,106] Cytologic features of well-differentiated tumors resemble normal hepatocytes primarily because the cytoplasm is flocculent and granular and may contain fat, bile, or both (Figs.

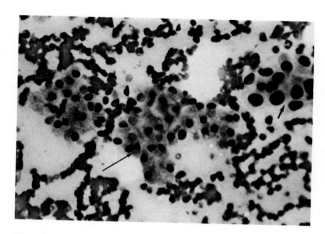

Fig. 8-40. Hepatoma, well differentiated. Hepatocytes *(long arrow)* blending with tumor cells *(short arrow).* Cytoplasm is similar. (R stain, ×600)

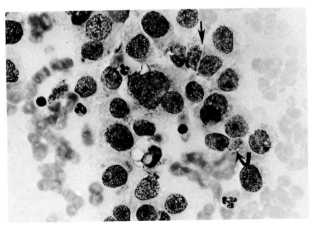

Fig. 8-41. Liver, hepatoma. Polymorphous cells are present in a disorganized cluster that still maintains a tendency to layer. Oblong cells with flattening of one side are seen *(arrow)*. Nuclei are reticulogranular and contain large nucleoli. Some bile pigment is present *(curved arrow)* but is seen better with R. (P stain, ×400)

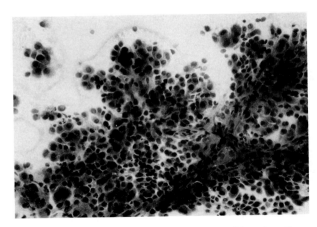

Fig. 8-43. Hepatoma. Abundant cell yield with cells attached to capillaries. Flattened cell borders and indented nuclei evident at low power. (P stain, ×200)

8-40 and 8-41). However, the nuclear cytoplasmic ratio of the tumor cells is higher (Fig. 8-41). Nuclei are larger and may be indented (Fig. 8-42). Large nucleoli are highly suggestive of malignancy but are not necessarily present. Large nucleoli may also be seen in benign hyperplasias. Benign hepatocytes are cohesively found in single and double layers (see above), whereas in hepatocellular carcinoma the cells tend to dissociate and form small aggregates. Cells may, however, retain the flattened cell borders of hepatocytes (Figs. 8-41 and 8-43). Cell patterns with crowd-ing and overlapping more strongly suggest malignancy (Fig. 8-44). A contrast with normal cells in the same field is the most useful diagnostic finding (see Fig. 8-40).

Well-differentiated hepatocellular carcinoma is difficult to distinguish histologically from adenoma and focal nodular hyperplasia. Using immunohistochemical parameters, studies for alpha-fetoprotein, antitrypsin, and HBsAg were not found to be useful in making the separation. Strict histologic criteria were used.[49]

Attempts at cytologic differentiation must therefore be undertaken with caution, and a diag-

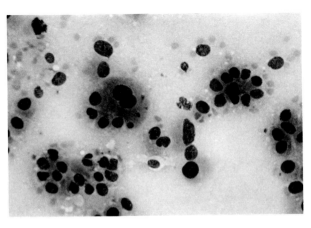

Fig. 8-42. Hepatoma, moderately differentiated. Note indented nuclei and intracytoplasmic inclusions. (R stain, ×600)

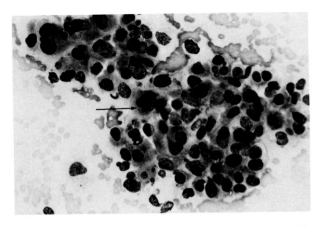

Fig. 8-44. Hepatoma. Overlapping polymorphic cells. Indented nuclei *(arrow)*, intracytoplasmic inclusions, and occasional cell border flattening are diagnostic features. (R stain, ×600)

nosis should be heavily supported by clinical data, particularly the presence of postnecrotic and alcoholic or nutritional cirrhosis, which predisposes to hepatoma.[68] Alpha-fetoprotein is elevated above 20 mg/ml to 40 mg/ml in the serum in 75% to 90% of patients.[102]

There are degrees of differentiation leading to poorly differentiated hepatoma. It may be difficult to distinguish this tumor from a number of poorly differentiated metastatic adenocarcinomas. Careful attention to and comparison with adjacent and intermixed hepatocytes may aid in the identification (see Fig. 8-40).

Intracytoplasmic inclusions have been noted by Orell and colleagues,[77] and in their presence metastatic breast carcinoma (see Chap. 6) must be considered in the differential diagnosis (Fig. 8-44). In addition, intranuclear inclusions may occur, and the differential diagnosis will include melanoma, adrenal carcinoma, and hypernephroma (Fig. 8-45). In contrast, well-differentiated bile duct carcinoma simulates bile duct structures (Fig. 8-46) or adenocarcinoma from other sites in the GI tract.

Hepatoblastoma is composed of primitive cells with large, round to ovoid, vesicular nuclei and fairly prominent nucleoli. Cytoplasm is fragile and ill defined. Cells may be dispersed and resemble nephroblastoma or even acute leukemia (Fig. 8-47). In one reported case of hepatoblastoma that was stained with Giemsa, abundant eosinophilic cytoplasm and a background of immature hematopoietic cells were noted.[14]

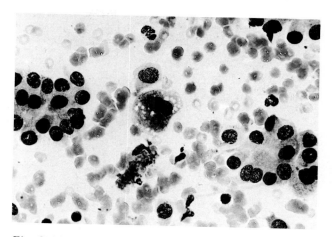

Fig. 8-46. Liver, bile duct carcinoma. Structures reproducing bile duct formations are seen in the well-differentiated tumor. (R stain, ×280)

Metastatic Tumors. Aspiration of metastatic nodules will produce several patterns.[8,40,42,54] Tumor cells may be scattered singly and in groups among clearly benign hepatocytes. There may be masses of tumor cells with no identifying landmarks of the liver bed (Fig. 8-48). Necrosis is a highly significant finding and should be distinguished from the necrosis associated with the inflammatory reaction of a nonmalignant disorder. There are often groups of ghost cells

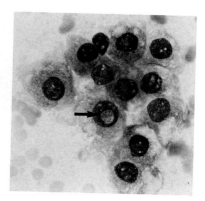

Fig. 8-45. Hepatoma. There is an intranuclear inclusion (*arrow*) also seen in hypernephroma and melanoma. Clinical correlation is important. (R stain, ×300) (Linsk JA, Franzen S: *Fine Needle Aspiration for the Clinician,* p. 185. Philadelphia, JB Lippincott, 1986)

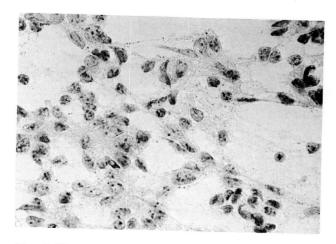

Fig. 8-47. Liver, hepatoblastoma. Primitive cells are haphazardly dispersed. Nuclei are round to oval and vesicular with two or more nucleoli. Cytoplasm is granular and poorly delineated. (P stain, ×280)

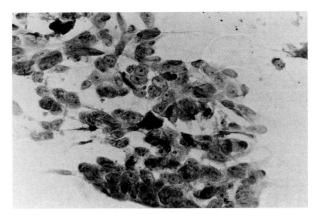

Fig. 8-48. Metastatic colon carcinoma in liver with no identifying liver cells in background. Note columnar formation. (P stain, ×800)

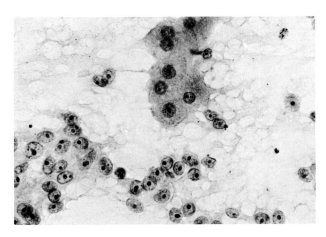

Fig. 8-50. Liver, metastatic melanoma. Rounded melanoma cells with well-defined enlarged nucleoli stand in contrast to the cluster of hepatocytes with granular cytoplasm and low nuclear:cytoplasmic ratio. (P stain, ×250)

intermixed with the necrotic material as well as occasionally clearly defined tumor groups (Fig. 8-49).

The most common cell yield is derived from metastatic carcinoma of GI tumors. However, cytology of melanoma (Fig. 8-50; see Chap. 13), monolayer pattern of breast carcinoma (see Chap. 6), squamous carcinoma of head and neck and thoracic tumors (Color Plate XXI F; see Chaps. 3 and 7); and renal tumors (see below) may be identified, as may the series of small cell tumors described above.

Diffuse infiltration of the liver with enlargement but no focal defects may occur in undifferentiated carcinomas and in lymphoma. Lymphoid smears are diagnostic if they consist of poorly differentiated cells (Fig. 8-51) or if they occur in the context of a known non-Hodgkin's lymphoma. Random puncture is satisfactory. Failure to identify the primary tumor from the cytologic findings in metastatic livers results in a classification of cancer of unknown primary site.[6]

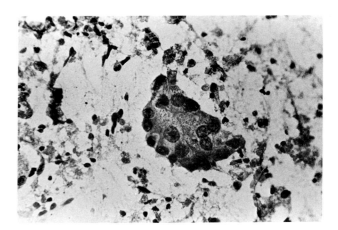

Fig. 8-49. Liver, metastatic pancreatic carcinoma. The background of this smear was primarily necrotic with an infrequent intact cluster of tumor cells. This is an organoid complex which could be of ductal or acinar origin. (P stain, ×300)

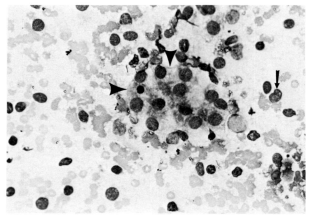

Fig. 8-51. Liver, lymphomatous infiltration. This is a cluster of hepatocytes (arrowheads) surrounded by scattered atypical lymphoid cells. Note cleaved cell (arrow). (R stain, ×280)

Differential Diagnosis of
Left Upper Quadrant Masses

In puncture of a left upper quadrant mass, it is important to keep in mind the variety of potential targets from which cells may be extracted. The left lobe of the liver (see above), the spleen (see Chap. 14), the kidney (see below), recurrent GI tumors (see above), retroperitoneal nodes (see below), and pancreatic cysts must all be considered. Fine needle perforation of a pancreatic cyst has produced no complications in our experience. However, a case of fatal septic shock was reported by Ulich and Layfield.[100]

Markedly elevated amylase levels in the fluid are diagnostic. There is no significant cytologic yield.

The following case report illustrates the problems encountered in left upper quadrant masses.

■   *Case 8-4*

A 55-year-old man had valvular heart disease, fever, and a palpable left upper quadrant mass considered to be a spleen. He was treated for subacute bacterial endocarditis but did not improve. Multiple x-ray, ultrasound, and scan studies, including CT scan, led physicians to conclude that a retroperitoneal mass appeared to be invading the kidney and possibly the spleen. Finally, 4 months after onset of the syndrome, with multiple studies and prolonged hospitalization, fine needle puncture yielded malignant spindle cells consistent with a retroperitoneal sarcoma (Fig. 8-52).

Retroperitoneal Disease

Retroperitoneal targets yielding clinically useful cytologic smears include the pancreas, duodenum and ampullar structures, paravertebral and para-aortic lymph nodes (see below), renal and adrenal masses, ureteral and para-ureteral structures, and retroperitoneal sarcomas (see Case 8-4) originating in the somatic tissues of the posterior coelomic wall.[3,10,13,23,29,38,62,105,106]

Although special techniques may be required for target localization, extremely careful palpation of the relaxed abdomen will often reveal adequate palpatory landmarks for direct puncture (see above). This is particularly useful in the clinical context of a known primary tumor, for instance, a cervical or

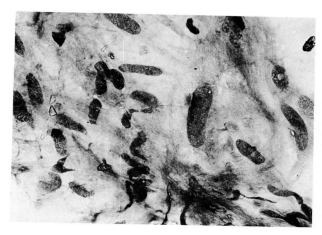

Fig. 8-52. Spindle cell sarcoma, retroperitoneal. There are numerous polymorphous spindle cells with sausage-like nuclei and poorly defined cytoplasm. (R stain, ×600)

prostate carcinoma, with clinical suspicion of lymph node metastases. Increasing the tumor stage by this maneuver will sharply reduce the number of diagnostic and operative procedures and eliminate hospitalization. Most retroperitoneal punctures, however, will require guidance. Localization and puncture methods are discussed in Chapter 2.

Retroperitoneal Nodes

Puncture of nodes opacified by lymphangiography can be carried out under fluoroscopy.[28,81,106] Similarly, CT scanning demonstrates enlarged retroperitoneal nodes. Lymphoma and metastatic carcinoma, particularly squamous carcinoma from the cervix, transitional cell carcinoma from the bladder, and prostatic carcinoma, can be identified and punctured.

Germinal Tumors

Considerable attention has been given to poorly differentiated carcinomas of unknown primary site. Many of these tumors are found in retroperitoneal nodes.[30,31]

Cytologic findings of these nodes under light microscopy will often, but not always, provide a diagnosis or direct further diagnostic measure. Ultrasound and immunocytochemical studies are not always readily available.[9] Large cell lymphoma (see Fig. 14-6A), amelanotic melanoma (see Figs. 13-2 and 13-10), and germinal tumors (see Chap. 12) are often identifiable on light microscopy. In a recent

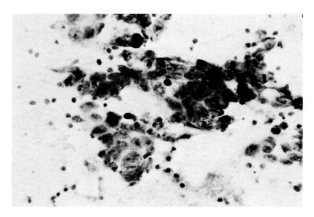

Fig. 8-53. Germinal tumor, metastatic mass, left supra-clavicular cluster of primitive cells with prominent nucleoli and necrotic background. (P stain, ×800)

study, Greco and colleagues established that cases that were diagnosed as poorly differentiated carcinoma of unknown primary site responded to a platinum-based combined chemotherapy designed for metastatic gonadal tumors. Of 66 patients measured, 13 had elevation of either human chorionic gonadotrophin or alpha-fetoprotein and 3 had elevations of both. Responses were best in patients with thoracic, retroperitoneal, or lymph node disease. Questions remained as to whether the responding patients all had germ cell tumors.[30] Tumor markers are mandatory in all young males with metastatic carcinoma of unknown origin.[84] Refinement of the diagnosis by cytologic criteria will contribute to solutions of these problems.

■ *Case 8-5*
A 34-year-old, well-developed, well-nourished man in no distress presented with a left supraclavicular mass measuring 3 cm × 6 cm. It had gradually appeared and increased in size, prompting investigation. Chest x-ray findings were negative, and a surgical biopsy was carried out that was interpreted as poorly differentiated large cell carcinoma, probably of bronchogenic origin. The patient was referred for oncologic opinion. FNA of residual mass was carried out and yielded a cytologic pattern strongly suggestive of a germinal tumor (Fig. 8-53). (Compare to Fig. 12-17, illustrating a testicular aspirate.) The histopathology was again reviewed by the

pathologist, who agreed that this was a possible diagnosis. Testicles were negative for tumor by both palpation and ultrasound. CT of the abdomen revealed enlarged para-aortic nodes. Beta-subunit of human chorionic gonadotrophin titer was elevated to 1017. The patient responded very rapidly to a platinum-based combined chemotherapy program.

Ureteral Compression

Diagnosis of lesions compressing one or both ureters becomes necessary during the course of a number of pelvic, intra-abdominal, and, occasionally, extra-abdominal primary tumors. The common primary tumors are those of the cervix and prostate, but no metastasizing primary tumor is excluded.

With intravenous (IV) urography as a guide (see Chap. 2), aspiration yields diagnostic material that may distinguish tumor from radiation fibrosis and that helps guide therapy.[23] Cytologic descriptions are discussed in the sections on specific primary tumors and tissue types.

Abdominal Lymphoma

Diagnosis of abdominal lymphoma by FNA is a shortcut to treatment in many patients, particularly those in whom the diagnosis was previously established by surgical biopsy of an accessible node. Such targets are often palpable or may require one of the methods used in retroperitoneal aspiration. Puncture of abdominal masses in patients with known regional node lymphoma is important for other reasons: second primary tumors, such as colon carcinoma, are not rare; and cytologic transformation of the lymphoma to more primitive cell types may occur and can easily be detected.[85]

Primary retroperitoneal lymphoma can also be diagnosed and treated with radiation and chemotherapy without surgery, particularly in older patients in whom laparotomy may not be medically indicated. The cytology of lymphoma is considered in Chapter 14.

Retroperitoneal Sarcoma

Retroperitoneal sarcomas are often palpable, as illustrated in Case 8-4. Nonpalpable sarcomas are detected and aspirated under imaging guidance (see Chap. 2). The cytology of sarcomas is described in Chapter 15.

Pancreas

The clinical aspects of diagnosis of pancreatic disease have been considered in detail in *Fine Needle Aspiration for the Clinician.*[53] The diagnosis of acute pancreatitis and chronic relapsing pancreatitis is clinically straightforward. Problems arise in distinguishing an unsuspected chronic benign induration and enlargement of the head of the pancreas from a tumor. The diagnosis of tumors is obscure, late, and incurable in 90% of cases.[53]

Noninvasive modalities play a major diagnostic role in patients with unexplained anorexia, weight loss, epigastric and back pain, and atypical thrombophlebitis who have had negative gastro-enterologic studies. Transabdominal aspiration yields the most rapid diagnosis, and clinical judgment ripened by experience with FNA will determine whether CT should precede or follow GI studies. Unfortunately, identifying pancreatic carcinoma by FNA rarely benefits the patient.[82] Where FNA has established a diagnosis of carcinoma, angiographic studies may determine whether surgery can be attempted. (The Technique of aspiration is presented in Chapter 2.) Of greater importance is the determination that the lesion is not a pancreatic adenocarcinoma but a lymphoma, carcinoid, islet cell tumor, cystadenoma, or inflammatory lesion, illustrated in the following case.

■ *Case 8-6*
A husky 20-year-old man presented with painless obstructive jaundice and no history of acute illness. He underwent multiple chemical, imaging, and endoscopic studies by the gastroenterologic service. The oncology service was asked to carry out FNA of an abnormal mass in the region of the pancreatic head (Fig. 8-54). This was done with ultrasound yielding tissue fluid, few inflammatory cells, and fibroblasts. At laparotomy the surgeon identified a poorly delimited mass obscuring the pancreas and compressing the bile duct. Wedge biopsy and multiple FNAs were carried out, again yielding only fibrous tissue. Percutaneous transhepatic catheterization was carried out, and exfoliative cells were described as "suspicious for cancer."

The patient is clinically well 3 years later with a normal biliary channel. Presumptive diagnosis is inflammatory lesion of unknown origin.

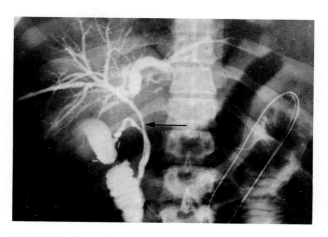

Fig. 8-54. Visualization of the biliary tract. There is irregular narrowing of the distal common bile duct.

In this case, the clinical evidence for tumor was overwhelming. On the other hand, multiple well-placed cytologic aspirations would rarely produce a negative diagnosis. The interrelationship of clinical and cytologic findings and clinical judgment is well demonstrated.

The cytology of pancreatic disease has been studied in specimens obtained by a transabdominal guided puncture[36,38,39,48,62,92,97,98] and by direct aspiration at laparotomy (see perioperative aspirations below). In transabdominal aspirations, smears may contain an admixture of cells from surrounding organs, including liver, stomach, colon, duodenum, and bile ducts. Although benign hepatic cytology is distinctive, there is considerable overlap in morphology of gastrointestinal, biliary, and pancreatic aspirates. Aspiration under direct visualization will yield smears entirely of pancreatic origin, and such smears, when available, are worthy of study.

Histologic Structure

The basic structure of the pancreas consists of large ducts lined by columnar epithelium with smaller branches lined by cuboidal epithelium. The ducts carry the secretion of the mass of surrounding glandular tissue. Imbedded in the entire gland are multiple islets. Although tumors originate in all of these tissue types, by far most carcinomas are of ductal origin. Acinar carcinoma is rare, difficult to clearly distinguish from ductal carcinoma histologically,[20] and probably rarely diagnosed cytologically (see Fig. 8-49).

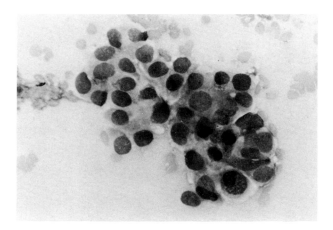

Fig. 8-55. Pancreas, benign cytology. This is a cluster of atypical ductal epithelium extracted from the same lesion that yielded frankly malignant cytology (see Fig. 8-58). Nuclei are bland and well defined with moderate variation in size and shape. Nucleoli are not evident. (R stain, ×600)

In well-differentiated carcinoma examined histologically, atypical glands and tubules, small and large, lined by columnar or cuboidal epithelium, or both, invade the pancreatic stroma and surrounding lymph nodes and organs.[20] Problems arise in distinguishing well-differentiated cell clusters and glandular complexes from atypical benign duct structures. Identification of poorly differentiated carcinoma poses no problem.

Benign Cytology

It is crucial to distinguish benign glandular, ductal, mesothelial, and endothelial cells from cancer. *Glandular or acinar* cells appear in clusters or groups forming acini in some samples. Nuclei are peripheral, uniform, and round or oval with benign, finely granular chromatin. The nuclear membranes are smooth and thin. The cytoplasm is granular, secretory, and fairly abundant, although naked nuclei may be seen. A well-preserved acinus seen in a metastatic site in Figure 8-49 might be construed as an atypical benign acinus at a primary site.

Ductal cells are both cuboidal and cylindric and are usually in tight groups, monolayer sheets, and sometimes a linear arrangement (Fig. 8-55). Nuclei are well defined, uniform, and round to oval. Nucleoli, if present, are small. The chromatin is coarsely granular. Multinucleation is not a feature. There may

be mild to moderate atypia (Fig. 8-55). This atypia may be a pitfall in diagnosis because it blends with well-differentiated ductal carcinoma.[21] Mesothelial cells appear in sheets (see Fig. 8-38). Such sheets may be interpreted as benign ductal epithelium when, in fact, the needle did not penetrate the pancreas. The dominant feature of benign cells is regularity of arrangements in cell groups with a pavement-like appearance.

Nonpancreatic benign cellular elements may be recruited from surrounding structures, including the liver (see above) and the lining epithelium of viscera (see Fig. 8-4).

The yield in acute pancreatitis is cellular debris, inflammatory cells (including polymorphonuclear leukocytes), macrophages, and lipophages. There may be tightly packed clusters of acinar cells with somewhat pyknotic and irregular nuclei. The background consists of necrotic debris with a finely and coarsely granular appearance mixed with polymorphonuclear leukocytes (Fig. 8-56). Subacute and chronic pancreatitis yield lymphocytes and plasma cells as well as scattered large histiocytes, occasional fibroblasts, and a few epithelial clusters.

Malignant Cytology

Carcinoma (of Ductal Origin). Smears from carcinoma are usually quite cellular. Carcinoma cells in groups have the characteristics described for adenocarcinoma in other GI neoplasms (Fig. 8-57). There is

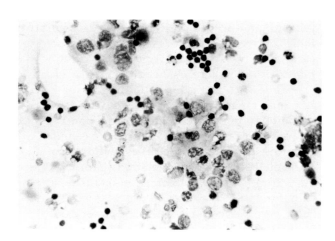

Fig. 8-56. Pancreatitis. Degenerated, disorganized benign cells are surrounded by inflammatory cells. (R stain, ×190)

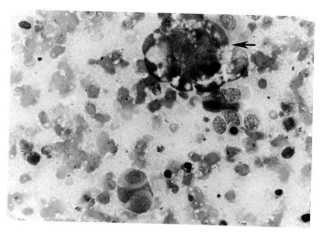

Fig. 8-57. Pancreas, carcinoma. There are two carcinomatous clusters with overlapping and cell molding. The morphology is not unlike that of gastric carcinoma *(arrow)*. The granular background arises from necrotic cell material. (R stain, ×600)

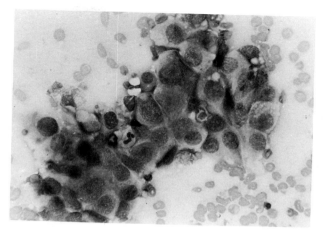

Fig. 8-59. Pancreas, carcinoma, well differentiated. This is a dense cluster with some maintenance of cell boundaries and polarity, suggesting a benign diagnosis. Cell variation is marked, however, and there are prominent nucleoli. (R stain, ×300)

crowding and overlapping with irregular nuclei and varying pleomorphism. Nuclei are granular with R stain, and occasional enlarged nucleoli are stained (Fig. 8-58).

Well-differentiated carcinomas may yield groups difficult to differentiate from atypical nonneoplastic cells[65,98] (Fig. 8-59). Prominent nucleoli (with P stain) and occasional mitoses will indicate a diagnosis of cancer. Cells appear in irregular clusters and microglandular structures. Nuclear pleomorphism can

usually be detected. There is a disorderly pattern in chromatin content and distribution, and multiple nucleoli are frequently seen.

In *moderately differentiated* carcinoma, smears consist of single cells, clusters, and tumor fragments (Figs. 8-60 and 8-61), with increasing pleomorphism. There may be some retention of microglandular structures. Disorientation and crowding of cells as reported by Mitchell and Carney[65] are the major diagnostic findings in our experience.

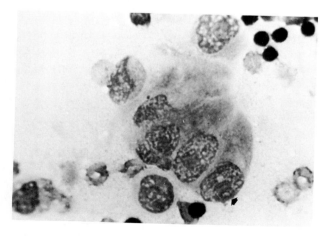

Fig. 8-58. Pancreas, carcinoma. Somewhat elongated cells suggesting columnar arrangement are not usually seen in pancreatic tumors. Nuclei are granular, and nucleoli *(arrow)* are large. (R stain, ×600)

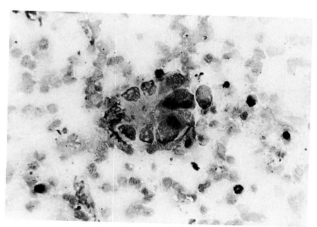

Fig. 8-60. Pancreas, carcinoma, moderately differentiated. Acinus is composed of bizarre cells with some artefactual change. There is a dirty background with cell fragments and granular necrosis. (R stain, ×300)

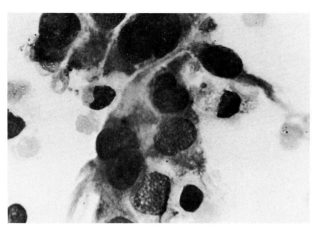

Fig. 8-61. Pancreas, carcinoma. This is a cluster of tumor cells with irregular granular nuclei. Cell borders are well demarcated, suggesting better differentiation. (R stain, ×600)

In *poorly differentiated* carcinoma, anaplastic changes occur. There is little cellular cohesion and many single cells are seen. Loose clusters are also present. The cells may be quite bizarre, showing a primitive chromatin pattern with R stain (Fig. 8-62). The cytologic findings of pleomorphic giant cell carcinoma of the pancreas were reported by Pinto and associates.[80] Bizarre tumor giant cells and osteoclast-like giant cells were present. Such a yield from a metastatic site must include giant cell tumors of the lung (see Chap. 7) and thyroid (see Chap. 4) in the differential diagnosis. In an ultrastructural study of osteoclast-type giant cell tumors of the pancreas, epithelial histogenesis was established, but ductal versus acinar origin could not be determined.[12]

Columnar formations, although infrequent in any of the tumor types, do occur (see Fig. 8-58). Tumor necrosis must be differentiated from an inflammatory exudate. It consists of granular debris with tumor cell fragments and shadows (see Fig. 8-57) in contrast to cell debris mixed with acute inflammatory cells (granulocytes) in acute pancreatitis (see Fig. 8-56). Islet cell tumors have been considered (see above).

False-Negative and False-Positive Pancreatic Cytology. As with all abdominal punctures, false-negative results will occur almost entirely from missing the lesion. A false-positive diagnosis in pancreatic aspiration after a reasonable amount of experience is

rare.[10,19,65,106] However, atypical ductal cells must be considered in a differential diagnosis, as noted above. In contrast, false-positive results with exfoliated cells obtained from duodenal drainage have been reported to be as high as 12%.[74] False-positive exfoliated cells were reported in Case 8-6.

Kidney Aspiration

The most common palpable renal masses that may become appropriate targets for ABC are solitary cysts. Hydronephrosis, polycystic kidneys, and displaced kidneys caused by congenital malformations or the pressure of an enlarged spleen may also present as palpable upper- or lower-quadrant masses. Usually, the clinical context and radiographic study will promptly clarify the diagnosis.

Adenomas of the kidney cannot be distinguished histologically from well-differentiated carcinoma. Such lesions under 5 cm are deemed adenomas or, if larger, carcinomas.[20] However, the hallmark of malignancy is metastasis, and, with no evidence of dissemination, such well-differentiated tumors (adenomas, papillary cystadenomas) may be considered benign. The cytologic findings will not differ from those of well-differentiated carcinoma.

Renal cell carcinomas are the most common malignant tumors of the kidney. They occur more often in middle-aged or elderly men. The tumor is usually silent, with pain, hematuria, and mass occurring late in its course.

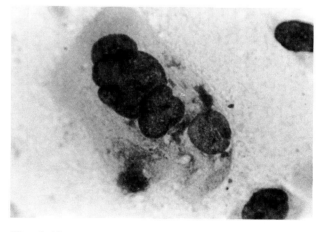

Fig. 8-62. Pancreas, carcinoma, poorly differentiated. This is a bizarre, multinucleated, giant tumor cell with immature nuclear chromatin. (R stain, ×800)

Nonpalpable tumors detected radiographically are suspected because of various symptoms and signs, including fever, weight loss, left varicocele, microscopic hematuria, and metastases of unknown origin. It has been noted that fever is more common in clear cell than in granular cell types (see below).[15]

Transitional cell carcinomas, arising in the renal pelvis, are less common; they account for about 10% of renal neoplasms.[56] Hematuria is the most frequent presenting sign. Urinary cytology is a valuable adjuvant study, with cytologic findings identical to those in transitional cell carcinoma of the bladder.[88]

Nephroblastoma (Wilms' tumor) is an embryonic tumor presenting clinically in infancy or early childhood.[20] It constitutes 6% of renal cancers.[56] Although metastatic spread, particularly to the lungs, may be the initial finding, an abdominal mass is present in the great majority of patients and, coupled with roentgenogram findings, presents no major clinical diagnostic difficulties. It expands within the renal parenchyma, ultimately breaking through the capsule to invade veins and surrounding tissues. Because it presents as an abdominal tumor, it is a prime target for aspiration biopsy. The differential diagnosis includes neuroblastoma and lymphoma (see Fig. 8-28).

Palpable Masses

Palpable masses of the kidney are targets for direct puncture and require no special preparation or studies. Problems of differential diagnosis with other left upper quadrant masses are considered above.

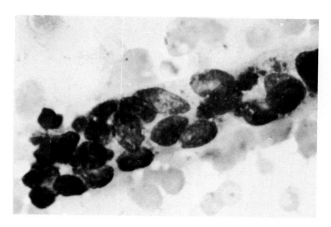

Fig. 8-64. Kidney. Fragment of the loop of Henle. (MGG stain, ×1000)

Nonpalpable Masses

Puncture of nonpalpable masses with radiographic techniques is described in Chapter 2.

Cytology of the Normal Kidney

Benign renal enlargements or displacements will occasionally result in puncture of a histologically normal kidney. Normal cytologic findings as described by N. Söderstrom in the first edition of *Clinical Aspiration Cytology* are presented herein.

The cell yield of the kidney has a broader cytologic spectrum than that of the liver. Only epithelial cells are seen, and sometimes tubular fragments are extracted (Fig. 8-63). Typically, clear, ungranulated cells from the collecting tubules and often heavily granulated cells from Henle's loop can be identified (Figs. 8-64 and 8-65). Larger tubular cells may approach the size and have an opaque cytoplasm similar to that of liver epithelium. Such cells when granulated are considered by Söderstrom to arise from proximal convoluted tubules, whereas cells of the same size with a clear structureless cytoplasm originate from distal tubules (Fig. 8-66). Occasionally, a single lysosomal body closely adherent to the nuclear membrane is seen in tubular epithelium cells. It stains turquoise green with MGG stain and may be used as an identifying feature of its renal origin (see Figs. 8-65 and 8-66). Identification of benign renal epithelium cells is important because they may appear in liver aspirates and be confused with metastatic cells.

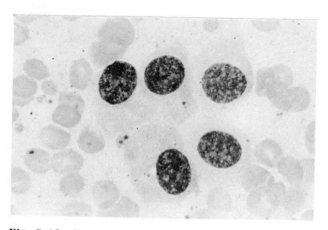

Fig. 8-63. Kidney. Epithelium from collecting tubules (same specimen as Fig. 8-65). (MGG stain, ×1000)

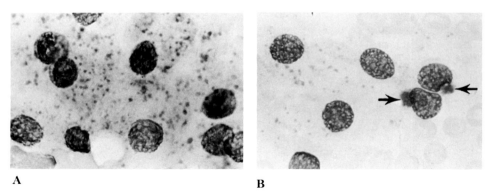

A   **B**

Fig. 8-65. Kidney. Granulated epithelium from convoluted tubules. Note in **B**, paranuclear bodies *(arrows)*, a specific identifying feature of renal tubular epithelium, usually seen in only some of the cells. (MGG stain, ×1000)

Cysts. Cysts may be punctured, yielding clear fluid. Cytologic examination is not usually carried out because of the rarity of carcinoma in a cyst.[47] Cytologic findings have been reported by Orell and coworkers to consist of degenerating epithelial cells and cyst macrophagias. With clinical suspicion of a tumor or if the fluid is cloudy or sanguineous, a smear is submitted for examination.[77]

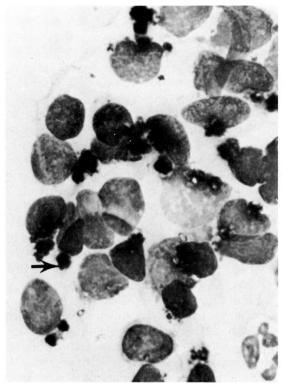

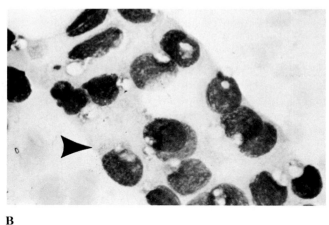

B

Fig. 8-66. Kidney. "Clear" epithelium cells from (probably distal) convulated tubules. **(A,** *triangular arrow)* Note numerous paranuclear bodies. **(B,** *arrows)*. (MGG stain, ×1200)

A

Abnormal Findings

The most common cell yield is from renal cell adeno-carcinoma. Drops of thin fluid or a voluminous fluid mixed with blood are obtained. Smears can be classified cytologically into well, moderately, and poorly differentiated carcinomas.

Well-Differentiated Carcinoma. Well-differentiated carcinoma shows fairly strong cohesion of cells (see Fig. 8-70). Two cytologic types can be identified: clear and granular cell or oncocytic.

Clear cell tumors are highly characteristic and are one of the few adenocarcinomas readily identifiable in metastatic lesions. Nuclei are well defined, round to slightly irregular, and inconsistently eccentric, with finely granular chromatin. The nuclear:cytoplasmic ratio is low. Nucleoli are difficult to see. Cytoplasm is abundant and finely or grossly vacuo-

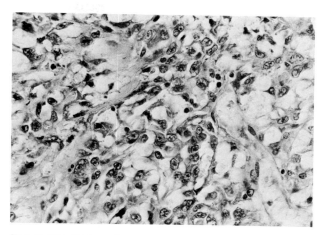

Fig. 8-68. Kidney, hypernephroma. Tissue section. The majority of the cells are clear, with a few granular cells intermixed. Both types may appear in one smear. (H & E stain, ×190)

lated with both R and P stains. Vacuoles at times virtually replace the cytoplasm (Fig. 8-67). The findings correspond to the clear cell tissue section (Fig. 8-68). These vacuolizations, a result of lipid dissolution, are not seen in normal renal epithelium.

Cells may be spread in monolayers (Fig. 8-69) or in microfollicular tubules or papillary structures (Fig. 8-70) or be dissociated. Cells may also be distributed along capillaries. Solitary vacuolated cells must be distinguished from histiocytes. The presence of

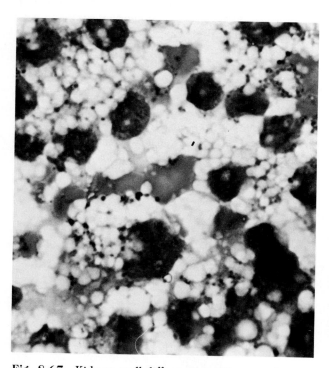

Fig. 8-67. Kidney, well-differentiated hypernephroma, clear cell type. This is a highly characteristic pattern of clear cell carcinoma with lucent globules producing identations into the nuclear membrane. The nuclear:cytoplasmic ratio is low. This should not be confused with benign renal epithelium, which is not vacuolated. (R stain, ×600)

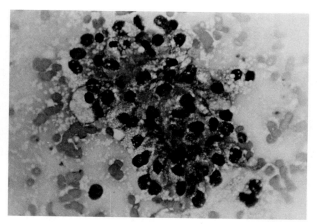

Fig. 8-69. Hypernephroma, well differentiated. This is a monolayer of mixed vacuolated and opaque cells. (P stain, ×400)

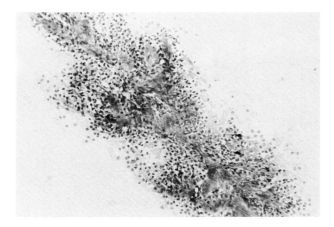

Fig. 8-70. Kidney, well-differentiated hypernephroma, clear cell type. Needle aspirate. This is a low-power view illustrating clear cells adherent to a connective tissue stroma and forming papillary fronds. (P stain, ×33)

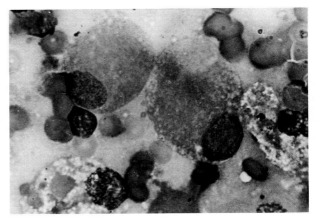

Fig. 8-72. Hypernephroma, well differentiated, oncocytic type. There is abundant opaque cytoplasm staining violet-gray. (R stain, ×1000)

groups and clusters composed of the same cell places the solitary cells in perspective.

The granular or oncocytic cells also have well-defined round eccentric nuclei with insignificant nucleoli and bland chromatin. The nuclear cytoplasmic ratio may be higher than in the clear cell type (Fig. 8-71). Cytoplasm is large, dense, and violet-gray with R stain (Fig. 8-72). Tinctorial characteristics with R stain provide important cytologic criteria. Cytoplasm appears granular, pink granules sometimes appearing with P stain. Cell borders are well defined in the

oncocytic cells (Fig. 8-72). The cellular arrangement is similar to that of the clear cell type. The well-differentiated opaque cell type may be confused with normal cells (see Figs. 8-66 and 8-71). More commonly, both clear and granular cells are mixed in varying proportions (see Fig. 8-76).

Moderately Differentiated Carcinomas. As tumors dedifferentiate there are larger cell yields; and hemorrhage and necrosis, which are highly suspicious of underlying cancer, may be encountered.

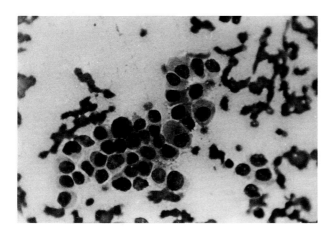

Fig. 8-71. Hypernephroma, well differentiated, opaque (granular) type. Nuclear:cytoplasmic ratio is larger. Such a smear can be confused with benign cells. (R stain, ×600)

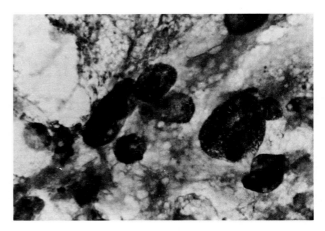

Fig. 8-73. Hypernephroma, moderately differentiated. This is an obviously malignant smear with pleomorphic nuclei. Vacuolization is still evident. (R stain, ×1000)

Tumors may be mixed, with varying differentiation, and the cytologic picture may depend on sampling.

In the cytologic study of moderately differentiated tumors, clear-cut cancer is evident. Although some of the clear cell vacuolization may still be seen as an identifying marker (Fig. 8-73), cells are usually granular. The opaque cytoplasm may, however, have fine peripheral vacuoles (Fig. 8-74). An extruded nucleus and intranuclear cytoplasmic inclusions have been noted in the study of metastatic hypernephroma.[52] However, diagnosis very much depends on enlarged and irregular nuclei and enlarged and usually single nucleoli. A relatively low nuclear:cytoplasmic ratio may still be evident (Fig. 8-75). There is further cellular dissociation, but cells may still be strung out on capillaries or appear as masses in syncytium (Fig. 8-76). In these formations the nuclei are fairly crowded and round to oval, and the cytoplasm is not well defined. (There may be mixed masses of vacuolated and opaque tumor cells (Fig. 8-76). With P stain, reddish cytoplasmic granules may be present, and a malignant chromatin pattern may not be readily evident (see Fig. 8-75).

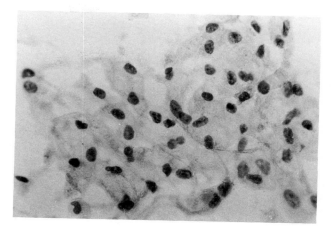

Fig. 8-75. Kidney, hypernephroma, moderately differentiated. Cells are in a syncytium. There is a tendency to nuclear elongation. (P stain, ×190)

Poorly Differentiated Carcinoma. There is increased atypia in poorly differentiated carcinoma with occasional multiple nuclei and malignant giant cells (Fig. 8-77). Necrosis and hemorrhage are frequently present. The nuclei are obviously malignant with hyperchromatism and pleomorphism. However, the nuclear:cytoplasmic ratio may still be somewhat low even though the nucleus as well as the cell is enlarged (Fig. 8-78). The cytoplasm is poorly defined and violet-gray and may still have many fine vacuoles and occasional linear lines and streaks with R stain. With further anaplasia, sarcomas are simu-

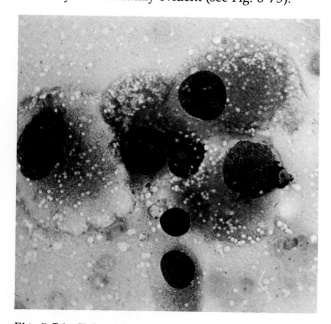

Fig. 8-74. Kidney, hypernephroma, moderately differentiated. Nuclei are well defined and variable in size and shape. Single, large nucleoli are present. Cytoplasm is abundant and still finely vacuolated. The nuclear:cytoplasmic ratio is still low and is a useful identifying marker of the primary tumor in metastatic sites. (R stain, ×1000)

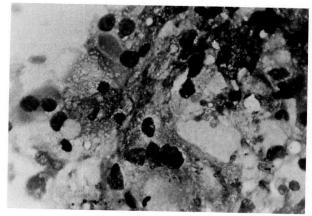

Fig. 8-76. Hypernephroma, moderately differentiated. This is a mixed mass of tumor cells, some vacuolated, some opaque. Tumor nuclei are evident. (R stain, ×400)

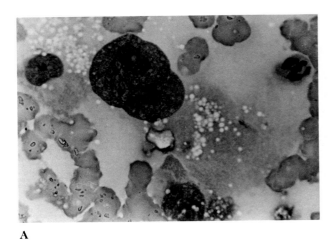
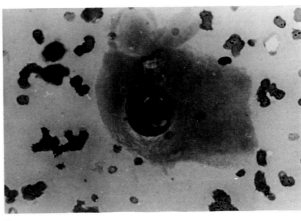

A   B

Fig. 8-77. A. Hypernephroma, poorly differentiated. There is a lobulated nucleus with massive nucleolus, partially extruded from opaque cytoplasm which still has some fine vacuoles. (R stain, ×1000) **B.** Tumor giant cell with massive intranuclear inclusion. (R stain, ×800)

lated (Fig. 8-79). Cells are largely dissociated at this time unless there is a mixture of well-differentiated material from another part of the mass.

Carcinoma of the Renal Pelvis.

Puncture of carcinoma of the renal pelvis is carried out as described above for other renal tumors. Cytologic diagnosis can often easily be made by aspira-

tion. These are transitional cell carcinomas, and the morphology does not differ from that of papillary lesions originating in the bladder.

The yield is fairly cellular and often consists of flattened, elongated, pointed urothelial cells (see Chap. 9). The cells are crowded into papillary masses seen best under low power (Fig. 8-80). Nuclei are fairly hyperchromatic and vary in size and shape de-

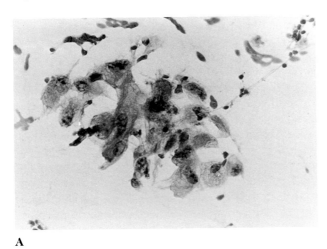
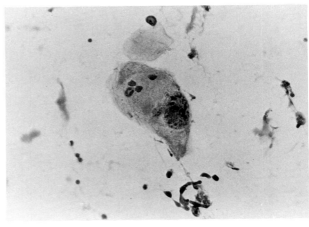

A   B

Fig. 8-78. A. Kidney, hypernephroma, poorly differentiated. This is a bizarre, irregular cluster of granular tumor cells that still retains low nuclear:cytoplasmic ratios. (P stain, ×600) **B.** Kidney, hypernephroma, poorly differentiated. This is a markedly enlarged granular tumor cell with low nuclear:cytoplasmic ratio. A polymorphonuclear cell has been phagocytized. (P stain, ×800)

pending on differentiation. As the tumors dedifferentiate, their identification as being of transitional cell origin becomes more difficult. Squamous carcinoma may be seen.

Tumor Grading as an Aid in Patient Management

The cytologic distinction between well and poorly differentiated carcinoma is clear-cut. Some moderately differentiated tumors, however, may fall into one of the other groups depending on the experience and impressions of the pathologist and the strictness of the criteria used. A more important factor is that the true grade of a tumor depends on its least differentiated portion, and sampling may fail to yield that portion of the target in a complex tumor. Multiple punctures, however, tend to reduce that variable.

The correlation between cytologic and histologic grading has been studied, and, although there is modest agreement, it is still imperfect.[76] However, correlation between cytologic and histologic grading with survival was generally good and even excellent in the poorly differentiated tumors.

Pre-operative diagnosis and grading have been used to mandate pre-operative radiation in poorly differentiated cases. Ultimately, the important potential clinical applications of grading may be in guiding the use of chemotherapy, interferon, and immunologic and biologic modifier therapy currently being investigated in many centers.

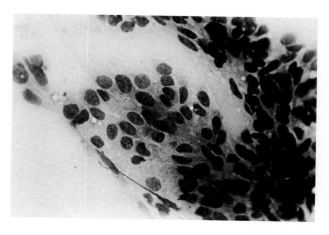

Fig. 8-80. Kidney, transitional cell carcinoma. This is a papillary frond composed of elongated, sometimes pointed epithelial cells. Nuclei are fairly bland in this well-differentiated variety (see Chap. 9). (R stain, ×400)

Sarcoma

Renal cell carcinoma may dedifferentiate to a sarcomatoid picture that includes giant tumor cells and spindle varieties. All sarcoma types can potentially be represented in the kidney (see Chap. 15). Cytologic diagnostic reporting of undifferentiated sarcomatoid lesions should indicate only the presence of a poorly differentiated malignant neoplasm.

Nephroblastoma

Nephroblastoma may be the site of central necrosis, hemorrhage, and cyst formation and will yield non-diagnostic hemorrhagic and necrotic cell masses. The general cellular content consists of two cell types: round to oval embryonal cells that occur in a loose and compact arrangement and form primitive tubules and sarcomatous spindle cells similar to those observed in soft-tissue tumors (see Chap. 15). The two types vary in number. The basic stroma consists of myxoid material (see Fig. 8-81) as well as stellate and fibroblastic cells. Metaplastic squamous cells may also be present.

In cytologic smears, cells occur in clusters and sheets and may form incomplete tubules (Fig. 8-82). With R stain, the nuclear chromatin is reticulogranular and the cytoplasm forms a thin rim, giving cells the appearance of leukemic cells (see Fig. 8-81). Masses of cells sometimes appear to form woven col-

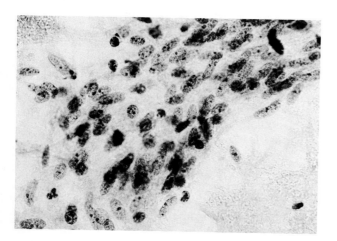

Fig. 8-79. Kidney, transformation to sarcomatoid pattern. These are dispersed, elongated, primitive nuclei. With further pleomorphism, distinction from a spindle cell sarcoma becomes difficult. (P stain, ×300)

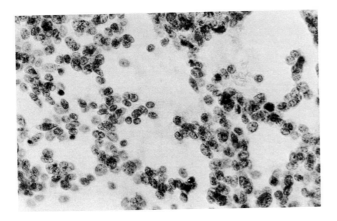

Fig. 8-81. Nephroblastoma. This is a cellular smear with uniform round to oval cells forming *incomplete tubules.* Nucleoli are inconspicuous, and the nuclear:cytoplasmic ratios are high, although cytoplasm is generally pale and poorly visualized. (P stain, ×190)

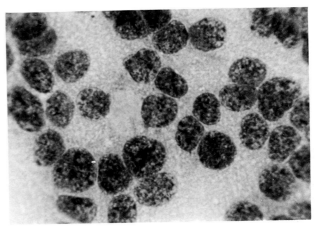

Fig. 8-82. Nephroblastoma. This is from the same aspirate as seen in Fig. 8-82. Note finely granular background similar to the background described for a seminoma (see Chap. 12; Color Plate XVII). (R stain, ×400)

umns, and cells may appear elongated. The fibrous stroma is normally left behind in the aspiration, but the loose myxoid material may be seen.

Angiomyolipoma

Angiomyolipoma is a rare hamartomatous lesion that may be a symptomatic mass and become a target for FNA. Accurate cytologic diagnosis of this lesion will limit surgical intervention. Diagnosis depends on the presence of smooth muscle cells and fat cells. Nguyen cautions against interpreting bizarre mesenchymal elements as malignant.[71] Thick vessels, cigar-shaped nuclei, smooth muscle cells in bundles, fat, and the absence of epithelial cells are characteristic features.[26]

Renal Oncocytoma

Renal oncocytoma is an uncommon tumor that must be distinguished from well-differentiated granular cell hypernephroma. Few cases have been studied cytologically.[2,22,72,86] In cases reported by Alanen,[2] there were abundant tumor cells singly and in clusters with large, well-demarcated, eosinophilic fine granular cytoplasm. The nuclear:cytoplasmic ratio was small. Nuclei were hyperchromatic and nucleoli were prominent. On review of reported material, the oncocytic cells have much in common with those of salivary and thyroid oncocytic cells (see Chaps. 4 and 5).

Indications for Puncture

Angiography is highly diagnostic of renal tumors, and this raises the question of when to use FNA. The partial list below illustrates the usefulness of presurgical aspiration.

1. The presurgical diagnosis of cell type, by defining poorly differentiated and anaplastic tumors, allows for preoperative radiation.
2. Infarction therapy by renal artery occlusion should be preceded by cytomorphologic diagnosis.[5,27,50]
3. Cysts, lipomas, and arteriovenous malformations may simulate tumors, leading to possibly unnecessary nephrectomy without presurgical diagnosis.
4. Invasion of the kidney by colon carcinoma may simulate primary renal tumor radiographically and lead to the wrong operative procedure. Presurgical cytologic diagnosis offers a safeguard because of the fairly characteristic cytology of renal tumors.
5. Metastases to the kidney presenting as primary tumors are uncommon but should not be discounted.[17] Bilateral renal tumors are often metastatic.[101]
6. It facilitates the study of human renal allografts.[37]

No significant local or remote complications from puncture have been reported, nor have we encountered any.[3,106] No evidence of spread of tumor was found in a controlled study by von Schreeb.[90]

Adrenal Disorders

The adrenal mass revealed by an imaging method is a generic tumor[53] until identified by morphologic examination or a corroborating clinicopathologic study such as catecholamines (pheochromocytoma). The following lesions make up the bulk of pathologic adrenal disorders that might become targets for FNA.

> Adrenal cortical nodules
> Adrenal hyperplasia
> Pheochromocytoma
> Myelolipoma
> Neuroblastoma
> Adrenal cysts
> Metastatic carcinoma
> Adrenal cortical carcinoma

Benign *adrenal cortical* nodules may vary from 2 g to 200 g.[20] This variability must be considered in evaluating apparently benign cytologic smears (Fig. 8-83). Adrenal cortical carcinomas are rarely less than 100 g at diagnosis.[24]

Adrenal hyperplasia (up to 12 g) would be an unusual target for FNA because associated endocrine disorders are ordinarily diagnosed by clinical methods.

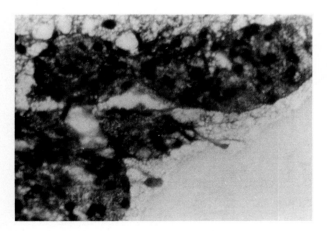

Fig. 8-83. Benign adrenal cortical nodule. Clusters of epithelial cells with round regular nuclear and vacuolated cytoplasm. (P stain, × 400)

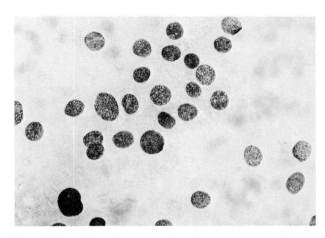

Fig. 8-84. Adrenal tumor, pheochromocytoma. Smear is obtained from a metastatic node. Cells are oval and characteristically variable in size. Chromatin is granular, nucleoli inconspicuous, and cytoplasm is not evident. There is a vague organoid pattern. (R stain, ×300)

Pheochromocytoma yields a cytologic pattern similar to that of carotid body tumors (see Chap. 3).[70] Figures 8-84 and 8-85 depict cellular patterns obtained from aspiration of metastatic pheochromocytoma to the left supraclavicular node. Puncture of primary pheochromocytoma is potentially hazardous.

Myelolipoma should be easily identified by the presence of fat and hematopoietic cells. Cases have been reported by Katz and colleagues and Pinto.[46,79]

Adult *neuroblastoma* is a rare tumor, and not all adult neuroblastomas originate in the adrenal gland.[4] The cytology of four cases of neuroblastoma presenting in childhood, of which one originated in the adrenal gland, was reported by Miller and associates.[63] The cytology of neuroblastoma is presented in Chapter 17 and Figure 8-28.

Four *adrenal cysts* were seen in a group of 22 patients with adrenal masses reported by Katz and coworkers.[46] As with most cystic lesions, the cytologic content was sparse and nonspecific. Fluid was clear, turbid, or bloody and contained foamy or hemosiderin laden macrophages.

Metastatic carcinoma was present in 7 of the 12 adrenal masses in the Katz series. With increasing use of CT for staging purposes, particularly in bronchogenic carcinoma, cytopathologists can expect to see a larger volume of material originating in adrenal

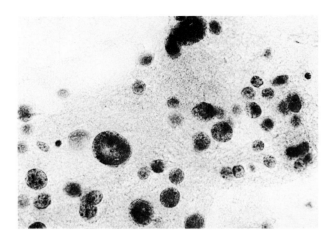

Fig. 8-85. Adrenal tumor, pheochromocytoma. Same tumor as in Fig. 8-83. With this stain, nucleoli are more prominent. (P stain, ×300)

masses. Even with a known primary tumor, however, a significant number of adrenal masses are due to unrelated benign lesions. In two studies with a total of 38 cases,[13,46] 12 patients with known extra-adrenal primary tumors had benign masses.

Interpretation of adrenal cytology from possible *adrenal cortical carcinoma* should not be attempted without full clinical disclosure. Pitfalls remain. Several primary tumors can mimic adrenal epithelium and adrenal tumors. Pulmonary adenocarcinoma metastatic to the adrenal gland simulated normal epithelium in a case reported by Mitchell and associates.[66] Metastatic hypernephroma, hepatoma, and melanoma should all be included in the differential diagnosis. In a patient with a scalp metastasis, the cytologic smear was highly suggestive of clear cell hypernephroma (see Fig. 8-45). An adrenal mass was demonstrated on CT. Although the origin of a metastatic mass can often be cytologically identified, a significant number will be listed as metastases with unknown primary tumor.[52]

Experience with cytologic patterns of adrenal cortical carcinoma is limited. In the series reported by Katz and colleagues,[46] tumor cells identifiable as being of adrenal origin had eccentric nuclei and vacuolated cytoplasm reminiscent of hypernephroma (see Figs. 8-45 and 8-83). Poorly differentiated carcinomas probably cannot be specifically identified as being of adrenal origin.

Perioperative and Postoperative Punctures of Abdominal Masses

Perioperative Puncture

FNA of selected targets during abdominal surgery is a useful and sometimes indispensable procedure.[7,16,48,73,93] Clinical observations by the surgeon may give erroneous impressions. Both wedge and core biopsies of the pancreas are attended by significant morbidity and mortality.[10,51,57,93] In addition, such biopsies lack flexibility and may provide poor specimens for frozen section. FNA is adaptable to operative findings and will, for the most part, distinguish tumors, inflammations, and fibrotic indurations (see Case 8-5). Several deep passes may be made into the target area to assure adequate sampling. The result is that complications have been virtually eliminated, false-negative results are infrequent, and false-positive results are rare.[93]

Although lesions of the pancreas are the most useful for this application, of FNA, lesions adjacent or fixed to major vascular channels are also desirable targets.

Suspicious but not easily biopsied nodes can be sampled when staging laparotomies for lymphoma and cervical, testicular, and prostatic carcinoma are carried out. Failure to obtain easily available cytologic material may result in puzzling and often frustrating postoperative clinical problems. A common problem in oncologic practice is a surgical report indicating probable liver metastasis based on palpatory findings. Definitive therapy is clearly inhibited by such a report. Nothing is lost by needling lesions, even those the diagnosis for which appear most obvious.

Material obtained can be examined at once with a quick-stain technique analogous to a frozen section. A trained health care employee will receive the syringe or detached needle, expel the aspirate on slides, make and fix appropriate smears for later examination with P stain, and transmit air-dried smears for immediate staining and reading. Nursing personnel can easily be trained to do this.

Postoperative Puncture

Failure to obtain a satisfactory surgical biopsy occurs because necrosis and inflammation can simulate identifiable tumor tissue. Superficial biopsies may be obtained because of reluctance to incise tumors adjacent to vital structures, as noted above. In the presence of an obvious large, nonresectable tumor, the surgeon may obtain an inadequate surgical biopsy without frozen section and find himself with a closed abdomen and no tissue diagnosis. Transabdominal penetration into the mass with a fine needle aided by surgical description or by an imaging procedure will usually obtain adequate diagnostic material.

Hazards and Side-Effects

The finer the needle, the less trauma there is. In contrast with transthoracic puncture, it is rare to experience untoward effects from transabdominal puncture of any organ with a fine needle. Puncture of an intestinal loop may occur, particularly in patients who have undergone previous abdominal surgery or radiation and now have adhesions or misplaced or dilated loops. Aspiration of feces is described above. Even in leukopenic and immunosuppressed patients, puncture of the bowel is not followed by peritoneal contamination in our experience. Complications of pancreatic needling are rare.

Penetration of a needle through the stomach or through small bowel loops into a retroperitoneal structure fails to produce contamination.[97] Penetration through the colon, with entrance of the needle into a solid structure, is described in Chapter 19. This maneuver might produce infection, but, by analogy, transrectal punctures of the prostate or pelvic organs extremely rarely results in infection (see Chaps. 10 and 11). In addition, no significant tissue trauma, such as a free perforation, occurs. These are self-healing punctures, as illustrated in the following case report.

■ *Case 8-7*
A 59-year-old man presented with anemia (hemoglobin 8 g/dl), ascites, and suggestive evidence of cirrhosis but no conclusive chemical findings and no available tissue. Alkaline phosphatase was 139 IU, and bilirubin total was 1.2 mg/dl. A right upper quadrant firm mass was palpable. Barium enema and intravenous pyelogram (IVP) findings were negative. Ultrasound was not available. Puncture yielded 5 ml of bile, indicating a distended gallbladder. Subsequent studies identified an obstructive biliary lesion.

This procedure obviously is not recommended but is cited to demonstrate the self-sealing quality of a muscular viscus. The following case report is included in contrast.

■ *Case 8-8*
A 71-year-old man was admitted to the hospital in an alert state with marked jaundice (bilirubin total of 14 mg/dl). History indicated a slow, silent onset suggesting biliary obstruction by tumor, but x-ray and other studies available at that time were insufficient to exclude cholestatic hepatitis. The attending clinician requested a core biopsy, which was deferred; and a fine needle puncture of a rather markedly enlarged liver was carried out. Immediately after removing the needle, there was intense peritoneal irritation, indicating chemical peritonitis due to bile. The patient responded to supportive measures. The cytologic material showed only bile stasis.

In this situation, a distended bile ductule probably leaked bile. Obstructive juandice is a relative contraindication to the procedure. This hazard is discussed further by Schulz.[91]

REFERENCES

1. Adson MA: Mass lesions of the liver. Mayo Clin Proc 61:362, 1986
2. Alanen KA, Tyrkkö JES, Nurmi NJ: Aspiration biopsy cytology of renal oncocytoma. Acta Cytol 29:859, 1985
3. Alfthan O, Koivuniemi A: Transcutaneous cytological aspiration biopsy in the diagnosis of renal tumours. Ann Chir Gynecol 92:1, 1969
4. Allan SG, Cornbleet MA, Carmichael J et al: Adult neuroblastoma. Cancer 57:2419, 1986
5. Almgård LE, Fernstrom HM, Ljungquist A: Treatment of renal adenocarcinoma by embolic occlusion of the renal circulation. Br J Urol 45:474, 1973
6. Altman E, Cadman E: An analysis of 1539 patients with cancer of unknown primary site. Cancer 57:120, 1986
7. Arensjö B, Stormby N, Akerman M: Cytodiag-

nosis of pancreatic lesions by means of fine-needle biopsy during operation. Acta Chir Scand 138:363, 1972

8. Axe SR, Erozan YS, Ermatinger SV: Fine needle aspiration of the liver. Am J Clin Pathol 86:281, 1986

9. Azar HA, Espinoza OG, Richman AV et al: "Undifferentiated" large cell malignancies: An ultrastructural and immunocytochemical study. Hum Pathol 13:323, 1982

10. Beazley RM: Needle biopsy diagnosis of pancreatic cancer. Cancer 47:1685, 1981

11. Bell DA, Carr CP, Szyfelbern WM: Fine needle aspiration cytology of focal liver lesions. Acta Cytol 30:397, 1986

12. Berendt RG, Schnitka T, Wiens E et al: The osteoclast-type giant cell tumor of pancreas. Arch Pathol Lab Med 111:43, 1987

13. Berkman WA, Bernardino ME, Sewell CW et al: The computed tomography guided adrenal biopsy? An alternative to surgery in adrenal mass diagnosis. Cancer 53:2098, 1984

14. Bhatia A, Mehrotic P: Fine needle aspiration cytology in a case of hepatoblastoma. Acta Cytol 30:439, 1986

15. Böttiger LE, Ivemark BI: The structure of renal carcinoma correlated to its clinical behavior. J Urol 81:512, 1959

16. Christoffersen P, Poll P: Peroperative pancreas aspiration biopsies. Acta Pathol Scand [suppl]. 212:28, 1970

17. Davis RI, Corson JM: Renal metastases from well differentiated follicular thyroid carcinoma. Cancer 43:265, 1979

18. Ehya H, Lang WR: Cytology of granulosa cell tumor of the ovary. Am J Clin Pathol 85:402, 1986

19. Evander A, Ihse I, Lunderguist A et al: Percutaneous cytodiagnosis of carcinomas of the pancreas and bile duct. Ann Surg 188:90, 1978

20. Evans RW: Histological appearances of tumours. Edinburgh, E & S Livingston, 1968

21. Fekete PS, Nunez C, Pitlik DA: Fine-needle aspiration biopsy of the pancreas: A study of 61 cases. Diag Cytol 2:301, 1986

22. Franzen S, Brehmer–Andersson E: Cytologic diagnosis of renal cell carcinoma. In Kuss R, Murphy G, Khoury S et al (eds): Renal Tumors: Proceedings of the First International Symposium on Kidney Tumors, pp 425–432. New York, Alan Liss, 1981

23. Freiman DB, Ring EJ, Oleaga JA et al: Thin needle biopsy in the diagnosis of ureteral obstruction with malignancy. Cancer 42:714, 1978

24. Gandour MJ, Grizzle WE: A small adrenocortical carcinoma with aggressive behavior. Arch Pathol Lab Med 110:1076, 1986

25. Geddes EW, Faulkson G: Differential diagnosis of primary malignant hepatoma in 569 Bantu mineworkers. Cancer 31:1216, 1973

26. Glentøj A, Partoft S: Ultrasound-guided percutaneous aspiration of renal angiomyolipoma. Acta Cytol 28:265, 1984

27. Goldstein HM, Medellin H, Beydoun MT et al: Transcatheter embolization of renal cell carcinoma. Am J Roentgen 123:557, 1975

28. Gondos B, Faripour F: Fine needle aspiration cytology of liver tumors. Ann Clin Lab Sci 14:155, 1984

29. Göthlin JH: Post-lymphographic percutaneous fine needle biopsy of lymph nodes guided by fluoroscopy. Radiology 120:205, 1976

30. Greco FA, Oldham RK, Fer MF: The extragonadal germ cell cancer syndrome. Semin Oncol 9:448, 1982

31. Greco FA, Vaughn WK, Hainsworth JD: Advanced poorly differentiated carcinoma of unknown primary site: Recognition of a treatable syndrome. Ann Intern Med 104:547, 1986

32. Greene C–A, Suen KC: Some cytologic features of hepatocellular carcinoma as seen in fine needle aspirates. Acta Cytol 28:713, 1984

33. Guckian JC, Perry JE: Granulomatous hepatitis. Ann Intern Med 65:1081, 1966

34. Gupta SK, Das DK, Rajwanshi A et al: Cytology of hepatocellular carcinoma. Diag Cytol 2:290, 1986

35. Haaga JR, Alfidi RJ: Precise biopsy localization by computed tomography. Radiology 118:603, 1976

36. Hancke S, Holm HH, Koch F: Ultrasonically guided percutaneous fine needle biopsy of the pancreas. Surg Gynecol Obstet 140:361, 1975

37. Häyry P, Willibrand E von: Practical guidelines for fine needle aspiration biopsy of human renal allografts. Ann Clin Res 13:288, 1981

38. Hidvegi D, Nieman HL, DeMay RM et al: Percutaneous transperitoneal aspiration of pancreas guided by ultrasound; morphologic and cytochemical appearance of normal and malignant cells. Acta Cytol 23:181, 1979

39. Ho CS, McLoughlin MJ, McHattie J et al: Percutaneous fine needle aspiration biopsy of the pancreas following endoscopic retrograde cholangio-pancreatography. Radiology 125:351, 1977

40. Ho CS, McLoughlin MJ, Tao LC et al: Guided percutaneous fine-needle aspiration biopsy of the liver. Cancer 47:1781, 1981

41. Holm HH, Rasmussen SN, Kristensen JK: Ultrasonically guided percutaneous puncture technique. J Clin Ultrasound 1:27, 1973
42. Hurwitz AL, Gueller R, Pugay P: Fine-needle aspiration of malignant hepatic nodules for cytodiagnosis. JAMA 229:814, 1974
43. Iishi H, Jamamoto T, Tatsuta M et al: Evaluation of fine-needle aspiration biopsy under direct vision gastro-fibroscopy in the diagnosis of diffusely infiltrative carcinoma of the stomach. Cancer 57:1365, 1986
44. Johansen P, Svendsen K: Scan guided fine needle aspiration biopsy in malignant hepatic disease. Acta Cytol 22:292, 1978
45. Johnson PM, Sweeney WA: The false positive hepatic scan. J Nucl Med 8:451, 1967
46. Katz RL, Patel S, MacKay B et al: Fine needle aspiration of the adrenal gland. Acta Cytol 28:269, 1984
47. Khorsand D: Carcinoma within solitary renal cysts. J Urol 93:440, 1965
48. Kline TS, Neal HS: Needle aspiration biopsy—a safe diagnostic procedure for lesions of the pancreas. Am J Clin Pathol 63:16, 1975
49. Koelma IA, Nap M, Huitemas S et al: Hepatocellular carcinoma, adenoma and focal nodular hyperplasia—comparative histopathologic study with immunohistochemical parameters. Arch Pathol Lab Med 110:1035, 1986
50. Lalli AF, Peterson N, Bookstein JJ: Roentgenguided infarction of kidneys and lungs. Radiology 93:434, 1969
51. Lightwood R, Reber HA, Way LW: The risk and accuracy of pancreatic biopsy. Am J Surg 132:198, 1976
52. Linsk JA, Franzen S: Aspiration cytology of metastatic hypernephroma. Acta Cytol 28:250, 1984
53. Linsk JA, Franzen S: Fine Needle Aspiration for the Clinician. Philadelphia, JB Lippincott, 1986
54. Lopes–Cardozo P: Atlas of Clinical Cytology. Philadelphia, JB Lippincott, 1976
55. Lozowski W, Hadju SI, Melamed MR: Cytomorphology of carcinoid tumors. Acta Cytol 23:360, 1979
56. Lucke B, Schlumberger H: Tumors of the Kidney, Renal Pelvis and Ureters, sect VIII, fasc 30. Washington, DC, Armed Forces Institute of Pathology, 1957
57. Lund F: Carcinoma of the pancreas. Acta Chir Scand 135:515, 1969
58. Lundquist A: Fine needle aspiration biopsy for cytodiagnosis of malignant tumors in the liver. Acta Med Scand 188:465, 1970
59. Lundquist A: Fine needle aspiration biopsy of the liver. Acta Med Scand [suppl] p 520, 1971
60. Lundquist A: Cytodiagnosis of primary and secondary tumours of the liver. Presented at World Congress of Gastroenterology, São Paulo, Brazil, September, 1986
61. Lutz H, Weidenhiller S, Rettenmaier G: Ultrasonically guided fine-needle aspiration biopsy of the liver. Scweiz Med Wochenschr 103:1030, 1973
62. McLoughlin MJ, Ho CS, Langer B et al: Fine needle aspiration biopsy of malignant lesions in and around the pancreas. Cancer 41:2413, 1978
63. Miller TR, Bottles K, Abele JS et al: Neuroblastoma diagnosed by fine needle aspiration. Acta Cytol 29:461, 1985
64. Min KW, Gyorkey F, Sato CL: Mucin-producing adenocarcinomas and non-bacterial thrombotic endocarditis. Cancer 45:2374, 1980
65. Mitchell ML, Carney CN: Cytologic criteria for the diagnosis of pancreatic carcinoma. Am J Clin Pathol 83:171, 1985
66. Mitchell ML, Ryan FP, Sherner RW: Pulmonary adenocarcinoma metastatic to the adrenal gland mimicking normal epithelium on fine needle aspiration. Acta Cytol 29:994, 1985
67. Miura K, Shirasawa H: Primary carcinoid tumor of the liver. Am J Clin Pathol 89:561, 1988
68. Mori W: Cirrhosis and primary cancer of the liver. Cancer 20:627, 1967
69. Murray–Lyon IM: Liver tumors. In Weatherall DJ, Ledinham JGG, Warell DA (eds): Oxford Textbook of Medicine. Oxford, Oxford University Press, 1983
70. Nguyen G–K: Cytopathologic aspects of adrenal pheochromocytoma in a fine needle aspiration biopsy. Acta Cytol 26:354, 1982
71. Nguyen G–K: Aspiration biopsy cytology of renal angiomyolipoma. Acta Cytol 28:261, 1984
72. Nguyen G–K, Amy RW, Tsang S: Fine needle aspiration biopsy cytology of renal oncocytoma. Acta Cytol 29:33, 1985
73. Nieburg RK: Peroperative pancreatic aspirations. Ann Clin Lab Sci 9:11, 1979
74. Nieburgs HE, Dreiling DA, Rubio C et al: The morphology of cells in duodenal drainage smear; histologic origin and pathologic significance. Am J Dig Dis 7:489, 1972
75. Noguchi S, Yamamoto R, Tatsuta M et al: Cell features and patterns in fine-needle aspirated of hepatocellular carcinoma. Cancer 58:321, 1986
76. Nurmi M, Tyrkkö J, Puntala P et al: Reliability of aspiration cytology in the grading of renal ade-

nocarcinomas. Scand J Urol Nephrol 18:151, 1984

77. Orell SR, Sterrett GF, Walters MN–I et al: Manual and Atlas of Fine Needle Aspiration Cytology. Melbourne, Churchill Livingstone, 1986

78. Perry MD, Johnston WW: Needle biopsy of the liver for the diagnosis of non-neoplastic liver diseases. Acta Cytol 29:385, 1985

79. Pinto MM: Fine needle aspiration of myelolipoma of the adrenal gland. Acta Cytol 29:863, 1985

80. Pinto MM, Monteiro NL, Tizol DM: Fine needle aspiration of pleomorphic giant cell carcinoma of the pancreas. Acta Cytol 30:430, 1986

81. Prando A, Wallace S, Von Eschenbuch AC et al: Lymphangiography in staging carcinoma of the prostate. Radiology 131:641, 1979

82. Raskin HF, Mosely RD, Kirsner JB et al: Carcinoma of the pancreas, bilary tract and liver: II. Cancer 11:166, 1961

83. Rasmussen SN, Holm HH, Kristensen JK et al: Ultra-sonically guided liver biopsy. Br Med J 2:500, 1972

84. Richardson RL, Greco FA, Wolff S et al: Extragonadal germ cell malignancy: Value of tumor markers in metastatic carcinoma in young males (abstract). Proc Am Assoc Cancer Res 20:204, 1979

85. Richter MN: Generalized reticular cell carcoma of lymph nodes associated with lymphatic leukemia. Am J Pathol 4:285, 1928

86. Rodriquez CA, Buskop A, Johnson J et al: Renal oncocytoma: Preoperative diagnosis by aspiration biopsy. Acta Cytol 24:355, 1980

87. Rohner RF, Prior JT, Sipple JHL: Mucinous malignancies, venous thrombosis and terminal endocarditis with emboli. Cancer 19:1805, 1966

88. Sarnacker CT, McCormick LJ, Kiser WS et al: Urinary cytology and the clinical diagnosis of urinary tract malignancy. J Urol 106:761, 1971

89. Schiff L: Diseases of the Liver. Philadelphia, JB Lippincott, 1975

90. Schreeb T von, Arner O, Skovsted G et al: Renal adenocarcinoma. Is there a risk of spreading tumor cells in diagnostic puncture? Scand J Urol Nephrol 1:270, 1967

91. Schulz TB: Fine-needle biopsy of the liver complicated the bile peritonitis. Acta Med Scand 199:141, 1976

92. Smith EH, Royal JB, Young CC et al: Percutaneous aspiration biopsy of the pancreas under ultrasonic guidance. N Engl J Med 292:825, 1975

93. Smith RC, Lin BPC, Loughman NT: Operative fine needle aspiration cytology of pancreatic tumors. Aust NZ J Surg 55:145, 1985

94. Söderstrom N: Fine-Needle Aspiration Biopsy. New York, Grune & Stratton, 1966

95. Stormby N, Åkerman M: Aspiration cytology in the diagnosis of granulomatous liver lesions. Acta Cytol 17:200, 1973

96. Suen K: Diagnosis of primary hepatic neoplasms by fine-needle aspiration cytology. Diag Cytol 2:99, 1986

97. Tao LC, Ho CS, McLoughlin MJ et al: Percutaneous fine needle aspiration biopsy of the pancreas. Cyto-diagnosis of pancreatic carcinoma. Acta Cytol 22:215, 1978

98. Tatsuta A, Yamamoto R, Yamamurra H et al: Cytologic examination and CEA measurements in aspirated pancreatic material collected by percutaneous fine-needle aspiration biopsy under ultrasonic guidance for the diagnosis of pancreatic carcinoma. Acta Cytol 52:693, 1982

99. Tylen U, Arnesjö B, Lindberg LG et al: Percutaneous biopsy of carcinoma of the pancreas guided by angiography. Surg Gynecol Obstet 142:737, 1976

100. Ulich TR, Layfield LJ: Fatal septic shock after aspiration of a pancreatic pseudocyst. Acta Cytol 29:879, 1985

101. Wagle DG, Moore RH, Murphy GP: Secondary carcinomas of the kidney. J Urol 114:30, 1975

102. Waldmann TA, McIntire RR: The use of radioimmune assay for alpha-fetoprotein in the diagnosis of malignancy. Cancer 34:1510, 1974

103. Wasastjerna C: Liver. In Zajicek (ed): Aspiration Biopsy Cytology: Part 2. Cytology of Infradiaphragmatic Organs. Monographs in Clinical Cytology, vol 7. Basel, S Karger, 1979

104. Whitlatch S, Nûnez C, Pitlik DA: Fine needle aspiration biopsy of the liver. Acta Cytol 28:719, 1984

105. Willems JS, Löwhagen T: Aspiration biopsy cytologic of the pancreas. Schweiz Med Wochenschr 110:845, 1980

106. Zajicek J: Aspiration biopsy cytology: I. Cytology of infradiaphragmatic organs. In Wied GL (ed): Monographs in Clinical Cytology. Basel, S Karger, 1974

CHAPTER 9

Pelvis (Non-Gyn)

JOSEPH A. LINSK

SIXTEN FRANZEN

The pelvis consists of diverse organs and tissues but is here considered as a single region. This region is defined by a specific set of clinical target areas that provide a variety of clinical cytologic smears. The specific cytology of the gynecologic organs, the prostate, and the scrotal contents are presented in Chapters 10 to 12. The following anatomic sites are considered in this chapter:

> Suprapubic abdominal wall and bladder
> Inguinal and femoral areas
> Perineum
> Anus and peri-anal region
> Buttocks
> Rectum
> Presacral region
> Pelvic retroperitoneum
> External genitalia
> Bony pelvis and sacrum

LESIONS

Suprapubic Targets

Perhaps the most common palpable abnormality in the suprapubic region is seen in an oncologic setting. Dense tissue fibrosis appearing after radiation therapy for pelvic disease (tumors of the bladder, rectum, and gynecologic organs) can simulate a tumor. There may be a readily palpable superior ledge, and the mass cannot be clinically distinguished from infil-

trating carcinoma. Failure to appreciate this tissue alteration or to obtain a biopsy may result in misdiagnosis. Fine needle aspiration (FNA) yields tissue fluid with scarce cellular content in contrast with the yield from carcinomatous infiltration, which generally contains many tumor cells. Where invasion of the body wall has occurred, however, a few tumor cells mixed with striated muscle may be present (Fig. 9-1).

Primary bladder tumors are unlikely targets for direct suprapubic puncture, because diagnosis is usually determined at an earlier stage by urologic techniques. Bulky mesenchymal tumors may, however, become palpable. Myosarcoma of the bladder presented as a bulky suprapubic mass. In Figure 9-2A there are large sarcoma cells with macronucleoli *(arrow)* and numerous degenerated cells with changes due to chemotherapy. With serial aspirations effectiveness of chemotherapy can be evaluated (Fig. 9-2B). Leiomyosarcoma yields plump spindle cells, as described in Chapter 15.

Recurrent transitional cell carcinoma may produce pelvic masses easily reached by suprapubic palpation and aspiration. Knowledge of the original histologic diagnosis may be useful in assessing the cytologic smears obtained from such targets, but difficulties in grading tumors of the urinary tract have been stressed.[2] Well-differentiated tumors may recur, with increasing anaplasia and pleomorphism and loss of their papillary character. However, metastases with the clearly recognized pattern of papillary carcinoma may also occur locally as well as in remote sites.[17]

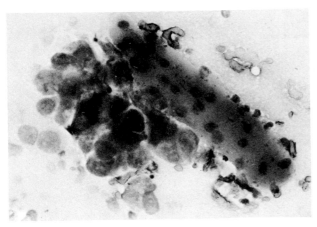

Fig. 9-1. Abdominal wall. This is a cluster of tumor cells adherent to a striated muscle bundle. (Romanowsky [R] stain, ×300)

Normal urothelium stained with Papanicolaou (P) is described in Chapter 10. With Romanowsky (R) stain, well-differentiated aspirated neoplastic transitional cells are distinguished by eccentric nuclei and fairly dense moderate to abundant cytoplasm (Fig. 9-3). In moderately differentiated tumors, the cell yield may be rich and the cells clustered into papillary groups (Fig. 9-4) or dispersed into monolayers (Fig. 9-5). The cells are fairly well defined with round to oval nuclei. Chromatin varies from irregularly clumped (Fig. 9-6) to an immature, finely granular

pattern in poorly differentiated cells (see Fig. 9-9). Nucleoli are evident. With P stain, cytoplasm stains yellowish tan and may be granular and not as well defined. With R stain, cytoplasm tends to be more dense, angular, and pointed (Fig. 9-7).

In poorly differentiated tumors, cells may usually be recognized by eccentric pleomorphic nuclei with retention of abundant cytoplasm in some cells. Chromatin varies from dense (Fig. 9-8) to finely divided (Fig. 9-9), and large irregular nucleoli are prominent in the latter group.

Inguinal and Femoral Targets

An inguinal bulge is clinically deceptive, and patients with bulky lymph nodes as well as cysts of the inguinal canal have been admitted to the hospital for hernia repair. All questionable masses should be subjected to FNA. Cysts yield fluid, usually clear yellow. Inadvertent puncture of a hernia is identified by its intestinal contents. The hazard in such an event is minimal. Enlarged lymph nodes provide the usual target in this area. Carcinomas that find their way to inguinal nodes by the normal lymphatic pathways are usually surface cancers (squamous carcinomas and melanomas) arising in the skin and mucosa of the lower extremities, external genitals (including the vagina), anus, buttocks, and lower abdomen. Cancers of the prostate, bladder, and female reproductive system migrate to the inguinal and femoral nodes

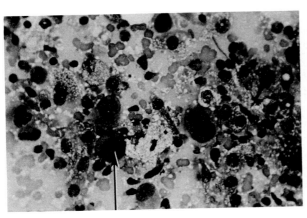

A

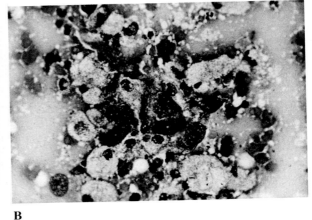

B

Fig. 9-2. A. Myosarcoma cells from palpable suprapubic mass. Prominent nucleoli (*arrow*). Some degenerated cells. **B.** Effects of chemotherapy. (R stain, ×800)

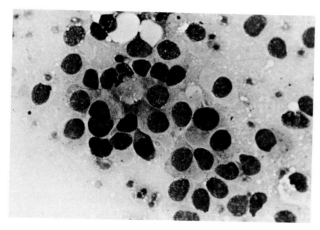

Fig. 9-3. Transitional cells, carcinoma, well differentiated. Cells appear singly and in clusters. Nuclei are round to oval, of uniform size, minimally irregular, and eccentric. Chromatin is finely granular, and occasional nucleoi are present. Cytoplasm is well defined and opaque. (R stain, ×300)

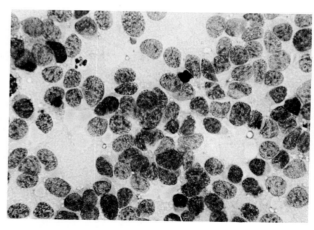

Fig. 9-5. Transitional cells, carcinoma, moderately differentiated. There is an abundant cell yield and moderate variation in nuclear size and shape. Chromatin has a looser structure and is less prominent. Cells have multiple nucleoli. (R stain, ×300)

only as a result of retrograde spread due to lymphatic blockage of an extensive tumor.

In addition, tumors of the head and neck (particularly nasopharyngeal carcinoma) (Fig. 9-10A and B), breast, thorax (Fig. 9-11), and abdominal organs may all reach the inguinal and femoral nodes by lymphatic, hematogenous, and transcoelomic routes.[17]

Such primary tumors must be kept in mind in cytologic interpretation.

The inguinal region, a site of lymph node dissection, may undergo nodular scarring as well as recurrent metastatic tumor nodules postoperatively. Fine needle puncture with a 25-gauge needle may reveal the earliest evidence of recurrent carcinoma in

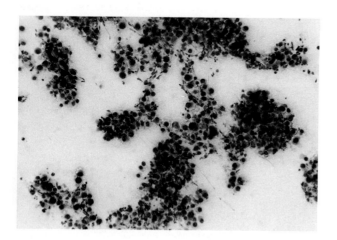

Fig. 9-4. Transitional cells, bladder carcinoma, suprapubic mass. This is a moderately to poorly differentiated tumor in which the smear continues to display a papillary configuration. (R stain, ×90)

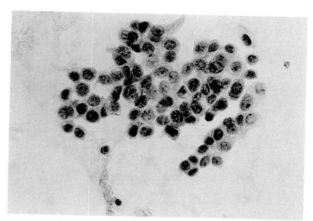

Fig. 9-6. Moderately differentiated transitional cells aspirated from suprapubic mass. Note clumped chromatin, prominent multiple nucleoli, and angular cytoplasm. (P stain, ×600)

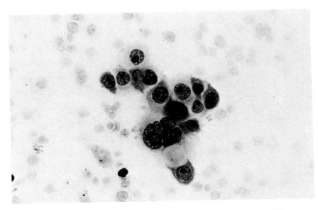

Fig. 9-7. Transitional cells. Bladder carcinoma. This is a cluster of characteristically angular and pointed carcinoma cells. (R stain, ×300)

nodules as small as 2 mm or 3 mm. Recurrent melanoma is easily detected by this method (see Chap. 13). Nodules may also occur at the site of high ligation of the vas deferens, and aspiration will distinguish essentially acellular scar tissue from recurrent tumor. Nodules also occur frequently after vasectomy for sterilization. Needle aspiration may be requested to relieve anxiety. The yield is essentially acellular.

Hodgkin's and non-Hodgkin's lymphomas may have their initial or secondary appearance in inguinal and femoral nodes. Interpretations are complicated by the frequency of hyperplasia secondary to inflammatory lesions in the extremities. Isolated, significantly enlarged femoral nodes are more likely to be neoplastic. Puncture in this region is a useful shortcut to staging of lymphoma (see Chap. 14).

Perineum

The perineum becomes a target for aspiration biopsy cytology (ABC) after abdominal–perineal resection if the patient complains of deep perineal pain or pain on sitting. On deep palpation, there is induration or localized nodularity that can be easily punctured. An asymptomatic mass may also be palpable and aspirated during routine follow-up examination. The cytologic yield consists of clusters of tumor cells mixed with necrosis, characteristic of adenocarcinoma. Squamous carcinoma, particularly the nonkeratiniz-

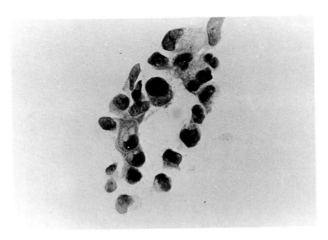

Fig. 9-8. Transitional cells, metastic inguinal mass. This cluster of polymorphous, poorly differentiated tumor cells has formed a pseudoacinus. Note persistent low nuclear:cytoplasmic ratio. Cytoplasm is vacuolated. (R stain, ×132)

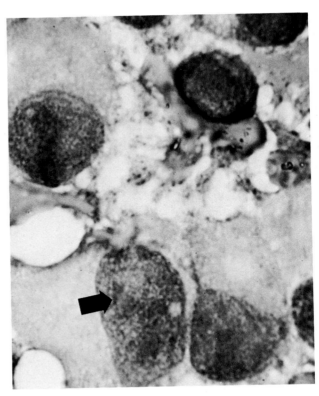

Fig. 9-9. Transitional cells, carcinoma, poorly differentiated. Note angular nucleolus (arrow). This is a high-powered view of individual tumor cells. (R stain, ×333)

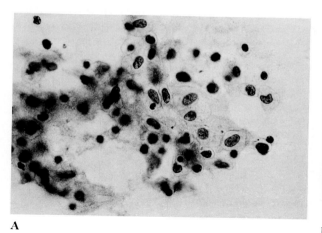
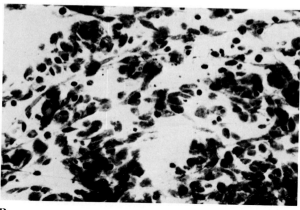

A   B

Fig. 9-10. A. Poorly differentiated tumor cells aspirated from 1-cm hard inguinal node in patient with radiation-treated nasopharyngeal cancer. (P stain, ×600) **B.** Same tumor with R stain. (R stain, ×600)

ing type, may be extracted when the primary tumor is anal carcinoma.

Palliative radiation can be monitored by serial punctures, which demonstrate the progressive disappearance of intact cells and their replacement by necrotic tissue (Fig. 9-12). The persistence of identifiable cells does not necessarily indicate treatment failure, since follow-up study of many months without tumor progression is not unusual.

Anal and Peri-anal Lesions

Anal cancers may present as ulcers, warty alterations, and ill-defined circumferential or nodular induration.[1] Targets may simulate external hemorrhoids and nonspecific inflammatory lesions. FNA will rapidly determine whether the patient should receive local anti-inflammatory treatment or be referred for oncologic evaluation. FNA should provide a rapid

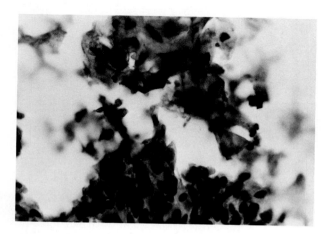

Fig. 9-11. Mixed keratinized and nonkeratinized squamous cells aspirated from large inguinal mass in patient with large pulmonary mass. (P stain, ×600)

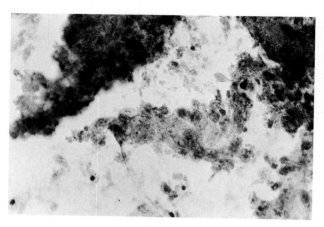

Fig. 9-12. Adenocarcinoma seen during radiation. Grossly necrotic tumor cells masses are adjacent to faded tumor groups. (P stain, ×250)

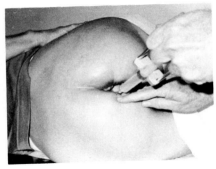

A   **B**

Fig. 9-13. A. Infiltrated granular tissue is visible at the anal verge. **B.** To avoid the necrotic portion of the tumor, the needle is inserted into intact skin where tumor has infiltrated. An abundant cell yield was obtained. The cytologic finding was carcinoma. (Linsk JA, Franzen S: Fine Needle Aspiration for the Clinician, p. 212. Philadelphia, JB Lippincott, 1986)

and adequate diagnosis for therapy of recurrent anal carcinomas. Because the prognosis is poor, particularly for anal tumors,[4] aspiration should be promptly done on minimal suspicion. Aspiration is demonstrated in Figure 9-13. Cytologically, the following three cell samples are encountered:

Squamous Carcinoma

Squamous carcinoma presents with varieties of differentiation (see Chap. 3). Lesions of the anal canal are usually less keratinized than those of the anal margin (Figs. 9-14 and 9-15).

Basaloid or Cloacogenic Carcinoma

Basaloid or cloacogenic carcinomas are infrequent tumors of the anus and canal that nevertheless may provide cytologic material. There are nonkeratinizing small cell tumors that cytologically may resemble basal cell and transitional cell carcinoma.

Inflammatory Exudates

Inflammatory exudates are more frequent than neoplasms and are thus more common cytologic specimens. Such specimens will be obtained or referred by

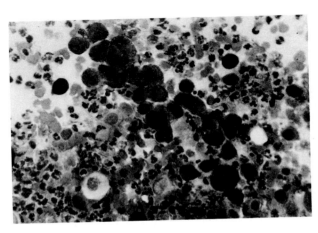

Fig. 9-14. Poorly differentiated nonkeratinizing squamous carcinoma of anal canal with overlying ulceration and inflammation. (R stain, ×600)

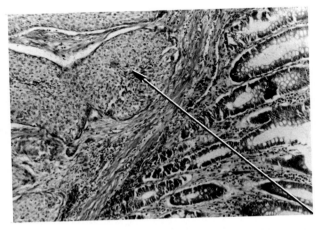

Fig. 9-15. Submucosal extension of small cell nonkeratinizing squamous carcinoma. Tissue section. Transrectal aspiration yields diagnostic smears. (H & E stain, ×400)

the examining physician depending on his level of clinical suspicion.

Melanoma, adenoid cystic carcinoma, and adenocarcinoma may also provide cytologic specimens with morphology, as described in Chapters 5, 8, and 13.

Buttocks

The buttocks are the site of both primary and metastatic masses. Unilateral swelling is suspicious for soft-tissue tumor, including chordoma. Recurrent chordoma presented as a large, firm, lobulated buttock mass in a personally observed case. Diagnosis by FNA was easily established. A similar case with recurrent extension into the buttock of a sacrococcygeal chordoma was reported by Rone.[14] The cytologic diagnosis was anaplastic chordoma. The cytology of chordoma is given in Chapter 16. Long FNA will easily obtain cytologic smears for the diagnosis of sarcomas, which are considered in Chapter 15. The buttocks are an uncommon area for metastatic lesions, but they must be considered in the differential diagnosis. Although some masses are within reach of relatively short 22-gauge needles, the depth of the mass may be deceptive on palpation, and 6-cm or 8-cm needles may be needed.

Rectal and Presacral Masses

Rapid diagnosis of palpable rectal cancers can be made by fine needle puncture and is of use in obviously inoperable patients. The usual diagnostic measure is surgical biopsy because of easy accessibility and the possibility of determining the extent of invasion. Sometimes a surgical biopsy may be too superficial, as illustrated in the following case report:

■ *Case 9-1*
A 63-year-old man had a partially necrotic mass on the anterior rectal wall. Only inflammatory necrotic tissue was obtained on surgical biopsy. Transrectal aspiration produced a characteristic picture of moderately differentiated prostate carcinoma (see Chap. 11), an unusual presentation through Denon villiers' fascia.

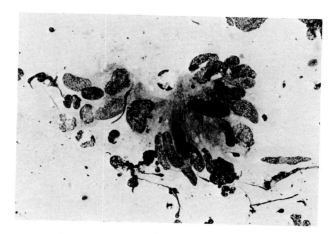

Fig. 9-16. Adenocarcinoma, rectum. Transrectal puncture. This is a cluster of elongated tumor cells with reticulogranular chromatin. Cytoplasmic borders are not defined. The background contains cells debris and inflammatory cells. (R stain, ×190)

Aspiration is carried out with the needle guide and technique described in Chapter 11. The cytology of rectal carcinoma is identical to colonic carcinoma. Figure 9-16 illustrates the yield from primary rectal carcinomas.

Transrectal puncture of presacral masses is the simplest method of rapid diagnosis.

Presacral Tumors

Presacral tumors in adults have been summarized by O'Dowd and colleagues.[11] A large variety of pathologic disorders was collected including inflammatory, developmental, germinal, neural, osseous, carcinomatous, and sarcomatous. Considerable diagnostic guidance will be offered by clinical history and physical findings. The sacrococcygeal concavity is often overlooked in routine examinations. Presacral chordoma was missed after extensive examinations in the case reported by O'Dowd and in a personally observed case reported in Fine Needle Aspiration for the Clinician.[9] Anaplastic chordoma may be quite soft.

The most common lesion, in our experience, is recurrent rectal and colon carcinoma. As with perineal screening, the cytology is usually straightforward and therapy can be monitored by serial punctures. However, cytologic smears may be compli-

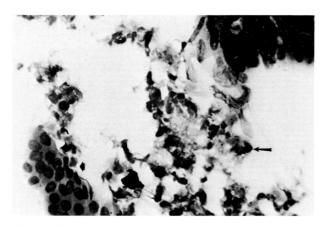

Fig. 9-17. Mixed aspirate, presacral mass, extension of colon carcinoma. Transrectal puncture. Note normal surface mucosa (*short arrow*) mingled with necrotic cells and partially degenerated tumor clusters (*long arrow*). (R stain, ×330)

cated by a mixture of benign, radiated, and malignant cells as well as necrosis and mucin (Fig. 9-17).

Diagnosis of childhood tumors is facilitated with a plastic tube as needle guide attached to the fifth finger.[18] Other infiltrations, such as Blumer's shelf, may yield diagnostic cells to clarify a diagnosis and facilitate staging.

Pelvic Masses

With the patient adequately relaxed, nodes and masses within the small pelvic volume can be palpated and punctured with deep finger pressure while control of the mass is maintained (see below). Such a puncture, together with rectal examination, may explain unilateral and bilateral leg edema as well as edema of the genitals.

Primary staging or diagnosis of recurrent lymphoma is often simply determined by an anterior puncture. Prostate carcinoma, moderately or poorly differentiated, may produce palpable iliac nodes. Cytology obtained in transabdominal pelvic puncture is typical of prostatic carcinoma without an inflammatory or "dirty" background.

Pelvic Retroperitoneum

Although the internal and external pelvic organs are easily palpable and visualized, most metastatic pelvic lymph nodes are not palpable; therefore clinical staging can only be surmised. The use of fluoroscopically guided FNA to puncture opacified pelvic nodes for tumor staging has increased extensively since the first edition of this book. Following lymphangiography, FNA and cytologic diagnoses are carried out with confidence and management is determined. This technique has been applied with significant yield in staging multiple gynecologic and nongynecologic primary tumors. Evaluation of pelvic nodes has been reported in patients with cancers of the bladder, testicle, penis, prostate, cervix, endometrium, ovary, vagina, and vulva.[3,10,12,13,15,16]

Finding more than two cytologically positive nodes in bladder carcinoma indicates systemic spread and prevents radical surgery.[12] Prognosis is sharply reduced by the presence of positive nodes. In a study of lymph node metastases in gynecologic disease, 20 of 24 patients with positive nodes died within a year, while 22 of 26 patients with negative nodes on FNA were alive at 6 to 36 months.[10] In staging genito-urinary cancer, in 40 patients with negative lymphangiography, 6 were found positive on FNA. In another series of 19 patients with visualization on normal lymph nodes, 7 were found positive on FNA.[3,16]

Because of the poor prognosis with positive nodes and the possibility of radiotherapeutic salvage, it is evident that the findings on lymphangiogram cannot be used as the exclusive determinant in staging. Where such staging is carried out for primary pelvic neoplasm, it appears to be a valid and valuable adjuvant procedure. After lymphangiography, enlarged pelvic nodes may yield oily contrast substance in FNA.[18] See Chapter 10 (Fig. 10-14).

Transvaginal and Transrectal Aspiration of Pelvic Masses

Nongynecologic pelvic masses not evident on abdominal and suprapubic palpation can often be reached by transvaginal and/or transrectal palpation and FNA using a prostate needle guide. A mass, either symptomatic or palpated in the course of a follow-up examination, usually occurs in the context of a known prior primary cancer. Recurrent colon carcinoma has been the most common tumor in our experience. Inflammatory disorders are also rarely diagnosed in selected cases. Actinomycosis was diagnosed on transvaginal aspiration in a recent report.[7]

The External Genitalia

Infiltration of the penis by contiguous, lymphatic, or hematogenous tumor spread results in diffusely firm or nodular tissue, often extremely painful, sometimes producing priapism. The most common primary tumors are prostate, colon, and bladder in our experience. FNA cytology is generally diagnostic with knowledge of the known primary. Palliative radiation can be administered on the basis of a cytologic diagnosis.

Invasion of the vagina by colon, bladder, or other carcinomas occurs; and transvaginal aspiration is easily carried out with a prostate needle guide (see Chapter 10). Nodules in the rectovaginal septum usually due to endometriosis are considered in Chapter 10. Two cases diagnosed by FNA were reported by Leiman and associates.[6] (See Figs. 10-13 and 13-31.)

Metastases to the vulva can presumably occur from a number of primaries. We have seen one patient with nodular masses infiltrating both labia as well as the vagina due to hypernephroma. Nephrectomy had been done 9 months previously. FNA was diagnostic.[8]

We have not seen metastases to the scrotum. Scrotal contents are discussed in Chapter 12.

Sacrum and Bony Pelvis

Localization of bony lesions in the pelvis and sacrum is easily achieved by palpation, x-ray study, and isotopic scans. Sizable defects or osteoblastic alterations with or without localized tenderness may be punctured in an outpatient setting with palpation and topographic landmarks as guidelines. This is particularly true of the posterior iliac crest, the sacrum, and the ischium. Most targets, however, are reached with the aid of fluoroscopy.

Lytic defects are easily penetrated with a fine needle (22-gauge), but even blastic or mixed alterations can be approached and sufficiently penetrated with a fine needle to obtain satisfactory material. Somewhat larger (20-gauge and 18-gauge) needles may be required and occasionally a formal bone marrow–aspirating needle (16-gauge or 17-gauge) is required. Core biopsy specimens are not required for diagnosis of myeloma, lymphoma, and metastatic carcinoma, although histologic structure in the specimen is sometimes useful.

It is not generally recognized that *fine needle aspiration of the periosteum without penetration of the cancellous bone at the site of the bony lesion identified by x-ray study or scan will yield diagnostic cells.* Radio-

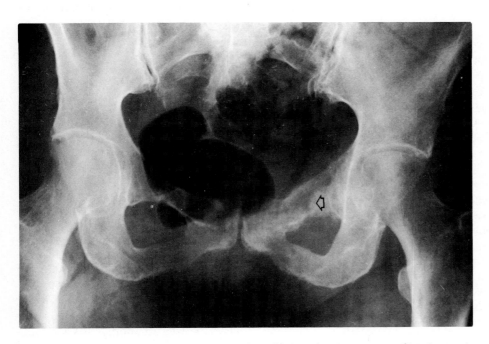

Fig. 9-18. Chondrosarcoma, pelvis. Note lytic lesion of pubis on x-ray film *(arrow).*

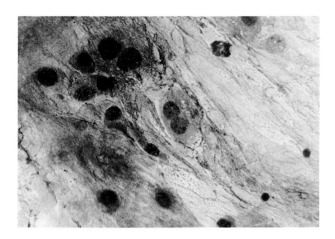

Fig. 9-19. Chondrosarcoma, pelvis. Penetration of the pubis was carried out with ease with a fine needle. Round nuclei are present. Note double nucleus in lacuna. The background is fibrillar and stained pink, violet, and blue. (R stain, ×600)

graphic changes are unlikely to be produced by a surface metastatic deposit; therefore, the malignant cells have penetrated to the surface from an endosteal site, presumably through vascular channels or, less likely, by way of the haversian system.[5] Infiltration with local anesthesia produces cellular distortion and should be avoided.

Metastatic carcinoma cells are the most common yield from bony lesions. Prostate, breast, and bronchogenic carcinomas are the usual sources, although no primary tumor is exempt. The dense osteoblastic lesions of prostate carcinoma often cannot be penetrated with less than a bone marrow–aspirating needle (16-gauge or 17-gauge), however. Myeloma and lymphoma yield cellular smears that are readily recognizable. The cytology of lymphomas is described in Chapter 14. The varieties of myeloma are best considered in hematopathologic sources. However, differential diagnosis with osteoblasts (see Fig. 16-8) and with melanoma (see Fig. 13-2) must be kept in mind.

Osteoblasts aspirated from a bony lesion indicate bone reparation. This may occur at the site of a pathologic destructive process, although osteoclasts are more likely to be found (see Chap. 14).

Primary bone tumors are considered in Chapter 16. Ewing's sarcoma, however, may be confused with myeloma, lymphoma, and metastatic neuro-

blastoma. Smears consist of groups of darkly staining (with R) or pyknotic (with P) nuclei that are irregularly round to oval with scant or unidentifiable cytoplasm. Chromatin is clumped, nucleoli are prominent, and the nuclear membrane is irregular. Cells may form acinar and linear structures.

Aspiration of lytic lesions with a fine needle may also yield unexpected primary lesions. Figure 9-18 is a radiograph of the pelvis of a woman aged 78 years. Chondrosarcoma is easily identified in smears obtained from the pubic bone (Fig. 9-19).

Chordoma also may produce destructive lesions of the sacrum. The cytology is presented in Chapter 16.

CLINICAL APPLICATION

The cytologic material obtained in the pelvic region is often from metastatic tumors and is, in general, considered in other chapters in this text.

Transitional epithelium is fairly distinctive. Nuclei are usually eccentric and may maintain this characteristic in less well differentiated lesions. Nucleoli are more prominent, with increasing dedifferentiation. Cytoplasm may be voluminous and well defined. In well-differentiated smears, cells are in clusters of papillary structures. In less-differentiated varieties, the cells show greater dispersion.

Targets in this region are usually palpable and easily punctured, although they may require radiographic guidance.

ABC should yield a diagnosis rather simply, can often be carried out in an office or clinic setting, and may provide shortcuts to diagnosis without the use of expensive technology.

REFERENCES

1. Coulson WF: Surgical Pathology. Philadelphia, JB Lippincott, 1978
2. Evans RW: Histological Appearances of Tumours. Edinburgh, E & S Livingston, 1968
3. Ewing TL, Buchler DA, Haagerland DL et al: Percutaneous lymph node aspiration in patients with gynecologic tumors. Am J Obstet Gynecol 143:824, 1982
4. Greenall MJ, Magill GB, Quan SHG et al: Recurrent epidermoid cancer of the anus. Cancer 57:1437, 1986

5. Ham AW, Cormack DH: Histology. Philadelphia, JB Lippincott, 1979

6. Leiman G, Markowitz S, Verga–Ferreira MM et al: Endometriosis of the rectovaginal septum: Diagnosis by fine needle aspiration cytology. Acta Cytol 30:313, 1986

7. Lininger JR, Frable WJ: Diagnosis of pelvic actinomycosis by fine needle aspiration. Acta Cytol 28:60, 1984

8. Linsk JA, Franzen S: Aspiration cytology in metastatic hypernephroma. Acta Cytol 28:250, 1984

9. Linsk JA, Franzen S: Fine Needle Aspiration for the Clinician. Philadelphia, JB Lippincott, 1985

10. McDonald TW, Morley GW, Chooy C et al: Fine needle aspiration of per aortic and pelvic lymph nodes showing lymphographic abnormalities. Obstet & Gynecol 61:383, 1983

11. O'Dowd GI, Schumann GB: Aspiration cytology and cytochemistry of coccygeal chordoma. Acta Cytol 27:178, 1983

12. Piscioli F, Pusiol T, Leonardi E et al: Role of percutaneous pelvic node aspiration cytology in the management of bladder carcinoma. Acta Cytol 29:37, 1985

13. Prando A, Wallace S, Van Eschenbuch AC et al: Lymphangiography in staging carcinoma of the prostate. Radiol 131:641, 1979

14. Rone R, Ramsey I, Duncan D: Anaplastic sacrococcygeal chordoma. Acta Cytol 30:183, 1986

15. Scappini P, Piscioli F, Pusiol T et al: Penile cancer — Aspiration biopsy cytology for staging. Cancer 58:1526, 1986

16. Wajsman Z, Gamarra M, Park JJ et al: Transabdominal fine needle aspiration of the retroperitoneal lymph nodes in staging of genitourinary tract cancer. J Urol 128:1238, 1982

17. Willis RA: The Spread of Tumours in the Human Body, 3rd ed. Scarborough, Ont, Butterworth & Co., 1973

18. Zajicek J: Aspiration Biopsy Cytology, Vol II, Cytology of infradiaphragmatic organs. In Wied GL (ed): Monographs in Clinical Cytology. Basel, S Karger, 1979

COLOR PLATE I.

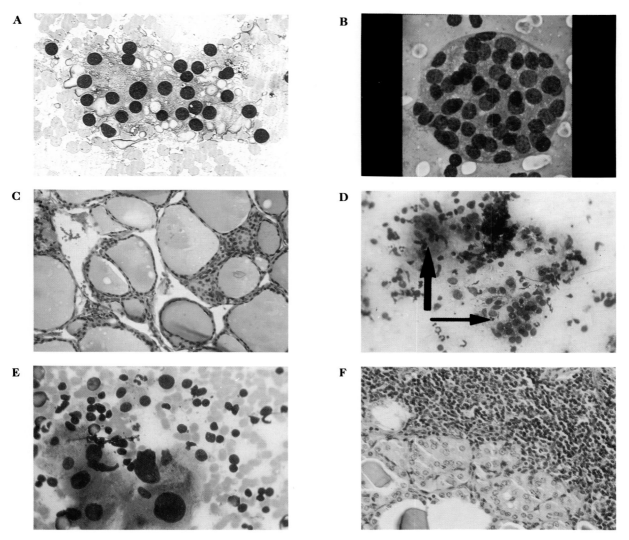

Color Plate I. **A.** Hyperthyroidism. This is a cluster of benign epithelium with vacuoles and "flaming" cytoplasm. This change may also be seen in Hashimoto's thyroiditis. Such changes are not evident in P stain. (MGG stain) **B.** Benign thyroid follicle aspirated intact from a benign colloid nodule. The flattened wall of such a follicle will yield sheets of small epithelial cells. (MGG stain) **C.** Tissue section, benign colloid nodule. Follicles are distended with colloid and become blue-purple with MGG stain and pale orange with P background in cytologic smears. **D.** Subacute thyroiditis. Multinucleated giant cells *(thick arrow)* are intermixed with regressive thyroid epithelium *(thin arrow)* and granulocytes. (MGG stain) **E.** Chronic lymphocytic (Hashimoto's) thyroiditis, 1983. Marked pleomorphism of nuclei suggests neoplasm. Abundant grey-blue cytoplasm identifies Hürthle's cells and the mixed lymphoid background confirms the diagnosis. (MGG stain) **F.** Tissue section taken in 1973 from the same patient. Source of the cytomorphology is evident in H & E stain. The surgical specimen obtained in 1973 was diagnosed as adenomatous goiter. After reviewing the cytologic smears of 1983 **(E)**, the original pathologic diagnosis was revised to lymphocytic thyroiditis.

COLOR PLATE II.

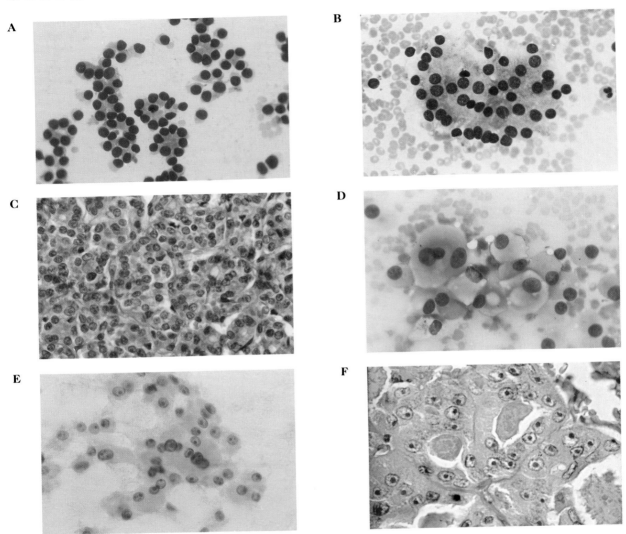

Color Plate II. **A.** Follicular adenoma, thyroid. Monotonous intact follicles with absent colloid identify a follicular neoplasm which could be adenoma or well-differentiated carcinoma. (MGG stain) **B.** Follicular adenoma. Cells present a trabecular as well as a follicular pattern. (MGG stain) **C.** Tissue section. Closely packed follicles as well as trabecular formation are the source of cytomorphology in **A** and **B**. Examination of vessels and capsule is necessary to distinguish well-differentiated carcinoma. (H & E stain) **D.** Hürthle cell. There are large polygonal cells with abundant cloudy blue cytoplasm and relatively small nuclei exhibiting less pleomorphism than in lymphocytic thryoiditis. Nucleoli, however, are prominent and there is no lymphoid background. (MGG stain) **E.** Hürthle cell carcinoma. Cytoplasm is not as well defined and nuclei are smaller but the nucleoli are more prominent. (P stain) **F.** Tissue section. Hürthle cell or oncocytic cytologic changes with a follicular structure. Nucleoli are prominent. Follicles are not usually reproduced in cytologic smears of Hürthle cell tumors.

COLOR PLATE III.

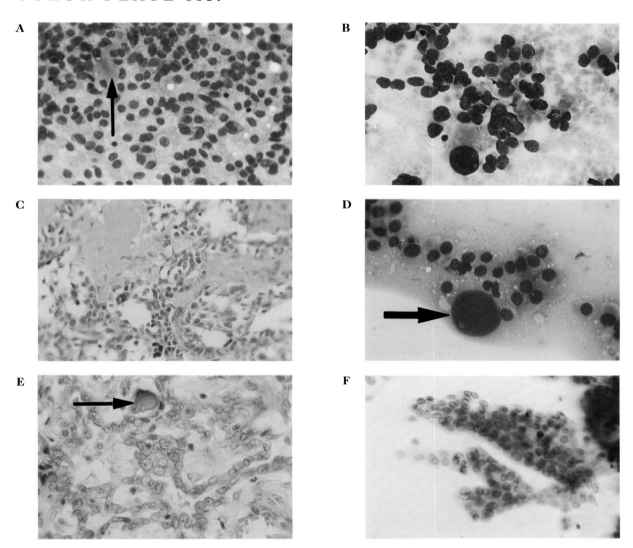

Color Plate III. **A.** Medullary carcinoma, thyroid. Dispersed round, oval, and some spindle cells on a colloid-free background are seen. An amyloid plaque is indicated *(arrow)*. (MGG stain) **B.** Medullary carcinoma. Cells are clustered with some large bizarre cells suggesting anaplastic change. See Color Plate IV. (MGG stain) **C.** Tissue section. Irregular cellular structures border and are imbedded in amyloid stroma. **D–F.** Papillary carcinoma **D.** Psammoma body. One of the important diagnostic features is noted *(arrow)*. A small intranuclear inclusion is also seen. (MGG stain) **E.** Psammoma body. Source of the finding on smear is noted on tissue section. (H & E stain) **F.** Papillary formation is clearly depicted. Nuclei have ground glass appearance. (P stain)

COLOR PLATE IV.

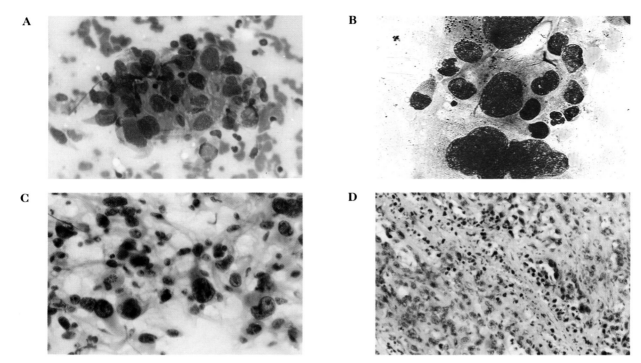

Color Plate IV. **A.** Anaplastic carcinoma, thyroid. Disorganized clusters of large tumor cells are seen. **B.** Anaplastic carcinoma. Marked variation in size and morphology of nuclei is evident. Fine reddish granulation is present. **C.** Anaplastic carcinoma. Haphazard distribution of large cancer cells with prominent nucleoli and bizarre cytoplasm formations are seen. (P stain) **D.** Tissue section of anaplastic carcinoma with pyknotic nuclei, necrosis, and intact undifferentiated carcinoma cells.

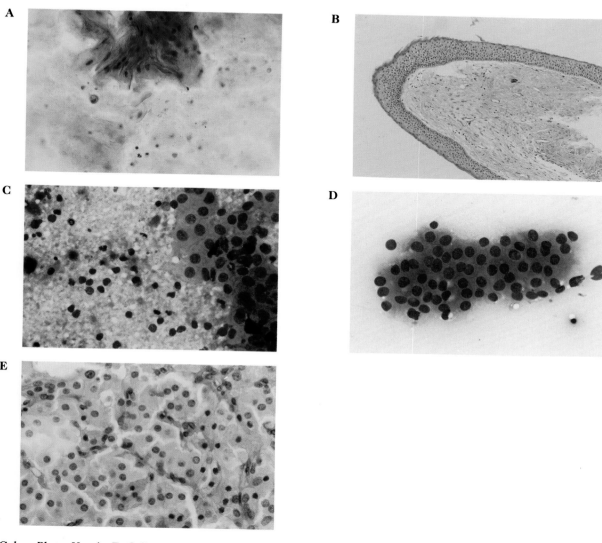

Color Plate V. **A–E.** Salivary gland. **A.** Cyst fluid. There is a turbid background composed of proteinaceous material and smudged cells, together with a cluster of keratinized squamous cells. Mucoepidermoid carcinoma must be considered even in the absence of a malignant component. (P stain) **B.** Tissue section. Benign simple squamous cyst with desquamated intracystic squamous cells. (H & E stain) **C.** Cystadenoma (Warthin's tumor). Smears contrast with squamous cyst. There is a cluster of oncocytic cells in a flocculent turbid background containing lymphocytes. (MGG stain) **D.** Oncocytoma. Cluster of oncocytic (Hürthle, Askenazi) cells on a clear background. Epithelial group is cytologically similar to the cells in **C**. (MGG stain) **E.** Tissue section. Histology clearly illustrates the source of the cytomorphology in **D**. (H & E stain)

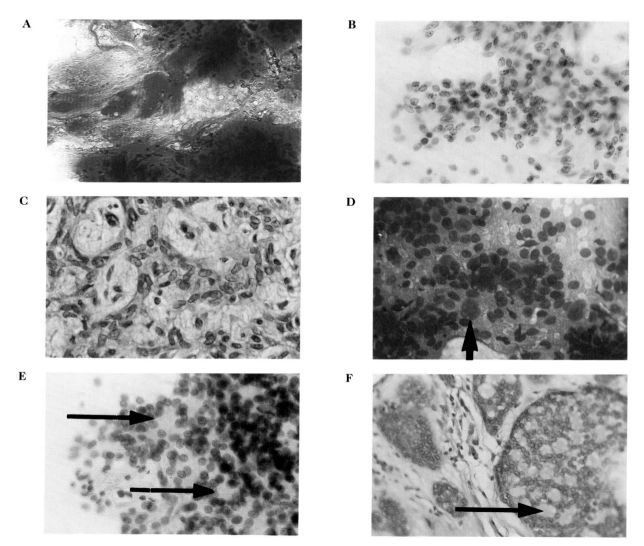

A

B

C

D

E

F

Color Plate VI. **A.** Pleomorphic adenoma. Pink stroma with imbedded blue nuclei is the classic cytomorphology of this tumor. In highly cellular lesions, stroma may be scant. (MGG stain) **B.** Pleomorphic adenoma. Randomly clustered bland nuclei with imperceptible stroma fails to provide the distinctive criteria seen in **A.** (P stain) **C.** Tissue section. Histology is variable depending on the amount of mucoid or chondroid stroma. Small oval nuclei are extracted with or without stroma. (H & E stain) **D.** Adenoid cystic carcinoma. Smear is composed of small, often bland appearing nuclei with minimal cytoplasm. The hallmark, but not pathognomonic (see Tubular Adenoma, p. 101 and Color Plate VII E) feature is the presence of mucous balls staining bright pink *(arrow)*. (MGG stain) **E.** Adenoid cystic carcinoma. Cells are round to oval and regular. Mucoid "balls" are unstained but may be surrounded by a spheroid mantle of cells *(arrows)*. (P stain) **F.** Tissue section. The tissue structure revealed the source of cell grouping surrounding the mucoid material extracted intact by the aspiration. (H & E stain)

COLOR PLATE VII.

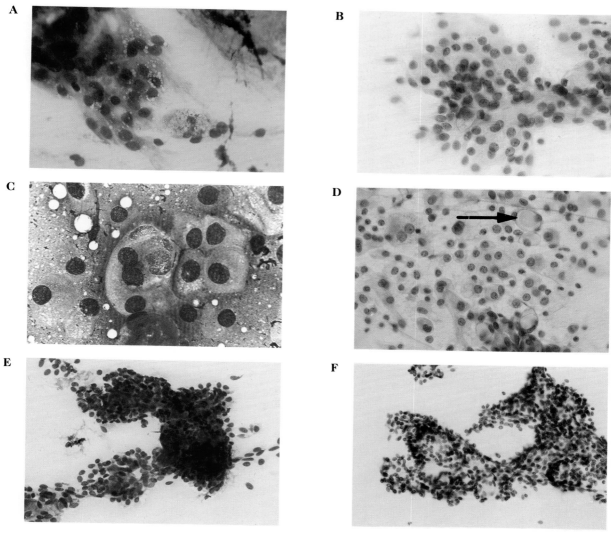

Color Plate VII. **A.** Mucoepidermoid carcinoma. Polygonal cells are loosely grouped. In the absence of frank keratinizing squamous cells, careful inspection for squamoid changes is necessary. Vacuolated cells suggest mucus formation. Purplish mucin is evident. (MGG stain) **B.** Mucoepidermoid carcinoma. Vacuolated cells with pink mucinous background are evident in the center of the cluster. In the absence of squamous differentiation, the surrounding cells are not distinctive. (P stain) **C.** Adenopapillary mucus-producing carcinoma, parotid. Carcinoma cells are bland without frank criteria of malignancy. Both intracellular and extracellular mucus balls are evident and characteristic. Intracellular inclusions are not seen in adenoid cystic carcinoma. (R stain) **D.** Adenopapillary carcinoma. Signet ring cells with well-defined intracytoplasmic globules are more prominent. Such a smear might be simulated by a metastatic gastric carcinoma. (P stain) **E.** Pleomorphic adenoma with globules. Globule formation can lead to overdiagnosis of adenoid cystic carcinoma and is a diagnostic pitfall. (MGG stain) **F.** Pleomorphic adenoma. Globules are not evident. (P stain)

COLOR PLATE VIII.

A
B

C
D

E
F

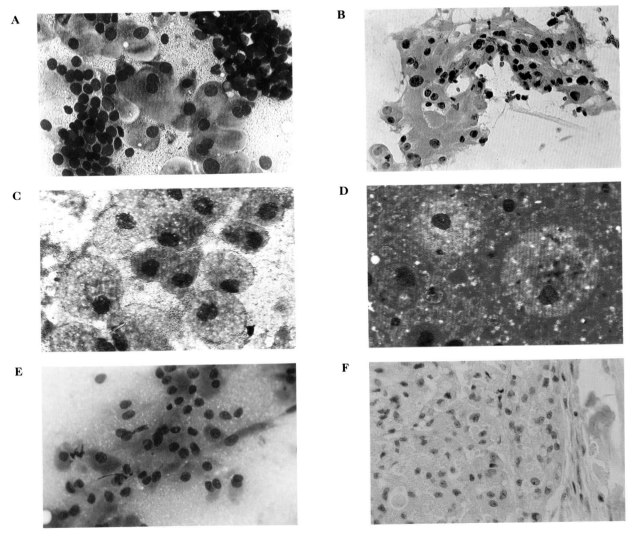

Color Plate VIII. **A.** Fibrocystic disease of the breast. Clusters of small epithelial cells with round-to-oval nuclei are present intermixed with apocrine cells. The latter have small nuclear cytoplasmic ratio. (MGG stain) **B.** Benign mammary dysplasia with atypia. This smear was obtained from the sediment of cystic fluid. Ductal carcinoma was found in the wall of the cyst. (P stain) **C.** Fat necrosis of the breast. A group of lipid histiocytes. The background is dirty. (MGG stain) **D.** Fat necrosis of the breast. The smear is dominated by an eosinophilic granular, necrotic background with scattered histiocytes and degenerated epithelial cells. (MGG stain) **E–F.** Granular cell tumor. **E.** Cells are dispersed or in loose clusters. Nuclei are small, round to oval, and eccentric with abundant purple cytoplasm. The cytology accurately reflects the histology. (MGG stain) **F.** Tissue section. Irregularly arranged nests of large cells with small eccentric hyperchromatic nuclei. Cytoplasm is abundant and granular. (H & E stain)

COLOR PLATE IX.

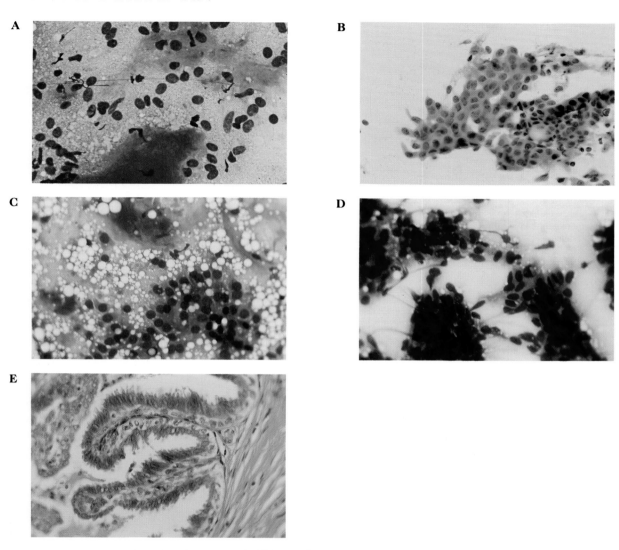

Color Plate IX. **A–C.** Fibroadenoma. **A.** Dispersed somewhat irregular oval nuclei are seen with typical amorphous pink stroma. (MGG stain) **B.** Cluster of small epithelial cells with adequate cytoplasm. Diagnosis is not confirmed without the presence of dispersed naked nuclei. Stroma is a secondary criterion. (P stain) **C.** Lactating adenoma. Round nuclei are surrounded by abundant purple cytoplasm. Extracellular fat globules and pink stroma are characteristic. Single central nucleoli are more evident in Fig. 6-19. (P stain) **D–E.** Papilloma. **D.** Fingerlike projections of elongated epithelial cells form solid clusters. (MGG stain) **E.** Tissue section of the same lesion displays projections with connective tissue cores covered with columnar epithelium overlying a basal cell layer. The aspiration strips epithelium from the core. (H & E stain)

COLOR PLATE X.

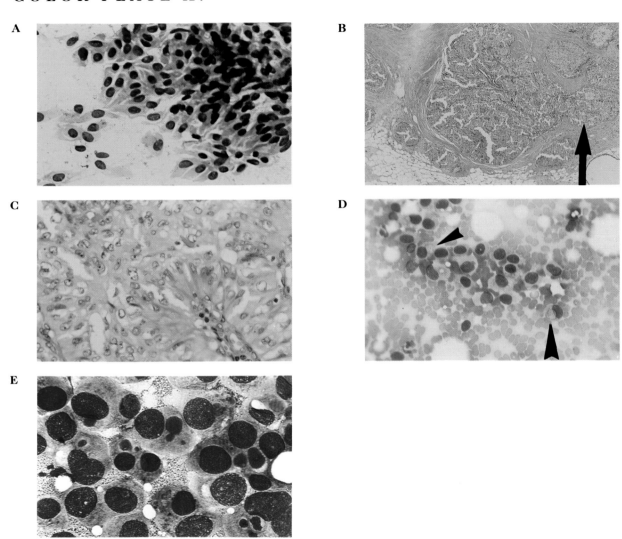

Color Plate X. **A–E.** Minimal carcinoma, breast. **A.** Movable mass less than 1 cm in a 70-year-old woman. Cellular smear is composed of oval and polygonal cells with intact, fairly abundant cytoplasm. Cancer cannot be definitively diagnosed, but excision is recommended. (MGG stain) **B–C.** Tissue section. Well-differentiated adenopapillary carcinoma 7 mm in diameter. *Arrow* suggests infiltration. (H & E stain) **D.** Stereotactic aspiration. Intracytoplasmic inclusions *(arrow)* confirm malignant disease. **E.** Ductal carcinoma of the breast. Smear is distinguished by the presence of intracytoplasmic magenta bodies. These rounded, stained inclusions are highly characteristic of breast carcinoma and can be identified in metastatic sites. (MGG stain)

COLOR PLATE XI.

A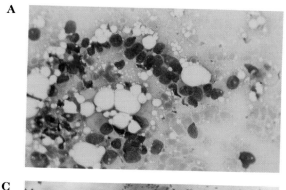

B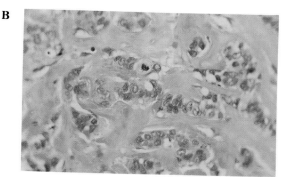

C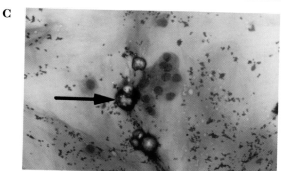

D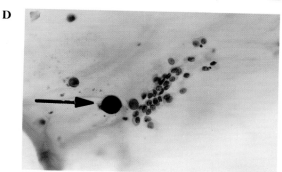

E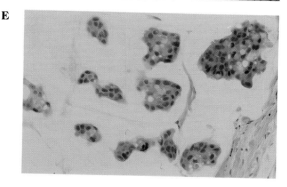

Color Plate XI. **A–E.** Breast carcinoma. **A.** Ductal carcinoma, NOS. Carcinoma cells are randomly arranged. Dispersed single cells, loose clusters, and cells in linear formation have been extracted from a scirrhous connective tissue stroma. (See **B.**) (MGG stain) **B.** Ductal carcinoma with desmoplastic stroma. Tissue section. Carcinoma cells, squeezed between thick collagenous cords are easily extracted. (H & E stain) **C.** Colloid carcinoma. Bland appearing carcinoma cells are sparsely spread through a violaceous background of mucin. Microcalcifications are evident *(arrow)*. (MGG stain) **D.** Colloid carcinoma. Background is pale greenish-violet. Calcifications are evident *(arrow)*. (P stain) **E.** Colloid carcinoma. Tissue section. Small clusters of carcinoma cells are surrounded by abundant clear mucin. (H & E stain)

COLOR PLATE XII.

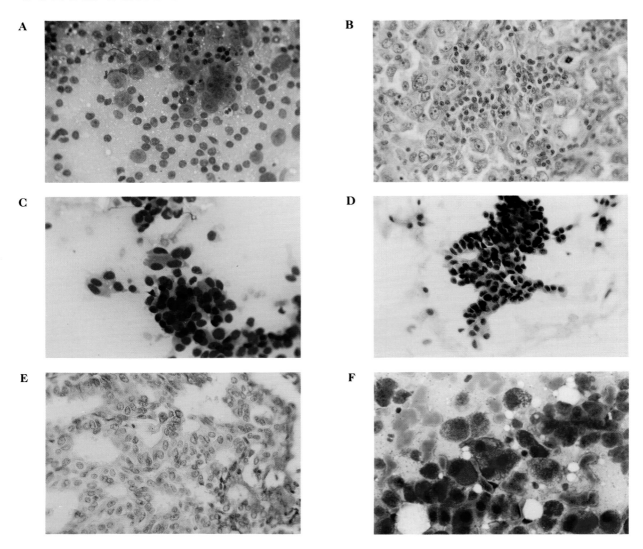

Color Plate XII. **A.** Medullary carcinoma, breast. Large rounded tumor cells with central macronucleoli and ill-defined cytoplasm are surrounded by lymphocytes. (MGG stain) **B.** Medullary carcinoma. Tissue section. Large poorly differentiated tumor cells with large nucleoli are infiltrated with small lymphocytes. On gross examination, the tissue is rubbery and simulates lymphoma. (H & E stain) **C–F.** Ductal carcinoma, well differentiated, papillary. **C.** Small crowded cells appear bland with minimal pleomorphism and a few elongated cells. (MGG stain) **D.** Small cells with dense nuclei and strands of columnar cells. (P stain) **E.** Tissue section. Solid strands of cells in papillary formation forming tubules and ducts. (H & E stain) **F.** Carcinoid. Small bland cells with eccentric nuclei and intensely staining red cytoplasmic granules identifiable in minimal lesion. (MGG stain)

COLOR PLATE XIII.

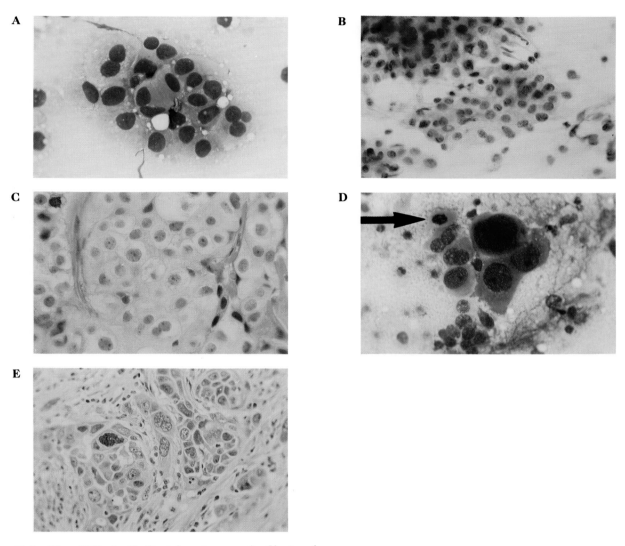

Color Plate XIII. **A–E.** Ductal carcinoma. **A.** Cluster of apocrine cells with enlarged mildly pleomorphic nuclei and abundant blue-grey cytoplasm. (MGG stain) **B.** Apocrine differentiation is not as prominent in P stain. (P stain) **C.** Source of extracted cells in **A** and **B** is clearly evident in tissue section. (H & E stain) **D.** Markedly pleomorphic tumor cells with finely divided chromatin and prominent nucleoli. Squamous differentiation is evident *(arrow)*. (MGG stain) **E.** Tissue section. Marked pleomorphism and epidermoid morphology are seen.

COLOR PLATE XIV.

A

B

C

D

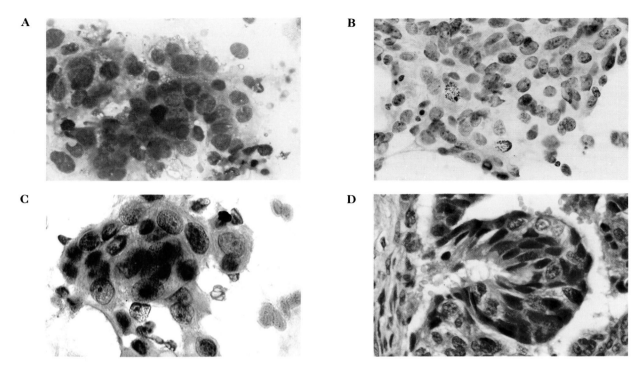

Color Plate XIV. **A–D.** Ductal carcinoma, poorly differentiated. **A.** There is crowding of large cells with finely divided chromatin and multiple nucleoli. Variation in cell size and shape is prominent. (MGG stain) **B.** Same case as **A.** Considerable crowding and overlapping of cells but nuclear chromatin is bland. Mitotic figures are evident. (P stain) **C.** Higher power reveals typical carcinoma cells with prominent nucleoli. (P stain) **D.** Same case as **A, B,** and **C.** Tissue section. Hyperchromatic cells with nucleoli. Elongated cells not reflected in aspiration smears are due to artifact.

COLOR PLATE XV.

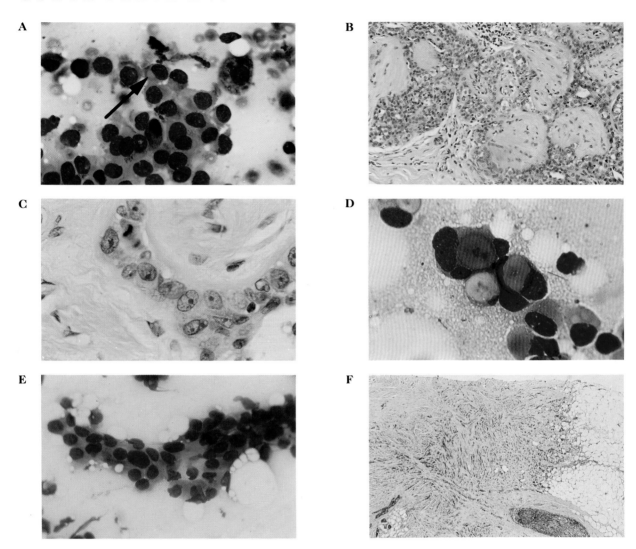

Color Plate XV. **A.** Lobular carcinoma. Small cells with indented nuclei and intracytoplasmic lumen *(arrow)*. (MGG stain) **B–C.** Aspiration in **A** extracted from lobular carcinoma in situ associated with fibroadenoma. Note carcinoma cells in higher power. **D.** Signet ring carcinoma. Inclusions are strikingly demonstrated. Inclusions identify lobular carcinoma. (MGG stain) **E.** Aspiration smear obtained by stereotactic puncture. Pleomorphism is sufficient to diagnose cancer. **F.** Tissue section. Invasive carcinoma 6 mm in greatest dimension. (H & E stain)

A

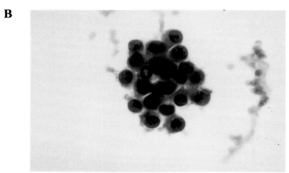

A

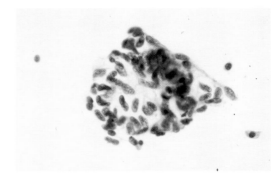

B

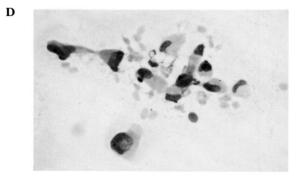

B

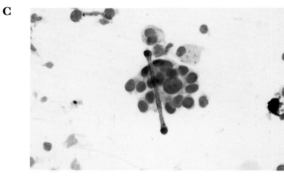

C

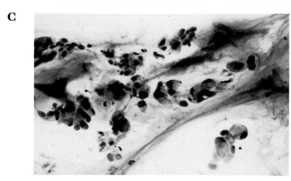

C

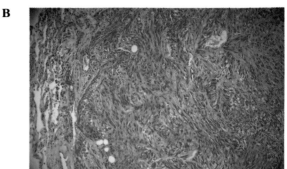

D

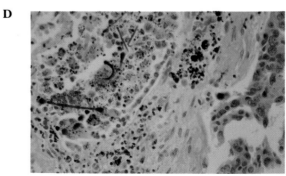

D

Color Plate XVI. **A–D.** Varieties of adenocarcinoma, lung. **A.** Papillary. (P stain) **B.** Bronchiolar. (P stain) **C.** Mucinous. (P stain) **D.** Poorly differentiated. (P stain)

Color Plate XVII. **A–D.** Unusual lesions, lung. **A.** Leiomyosarcoma. (P stain) **B.** Leiomyosarcoma. Tissue section. (H & E stain) **C.** Ferruginous body. Cytologic smear. (P stain) **D.** Ferruginous body. Histologic section. (H & E stain)

COLOR PLATE XVIII.

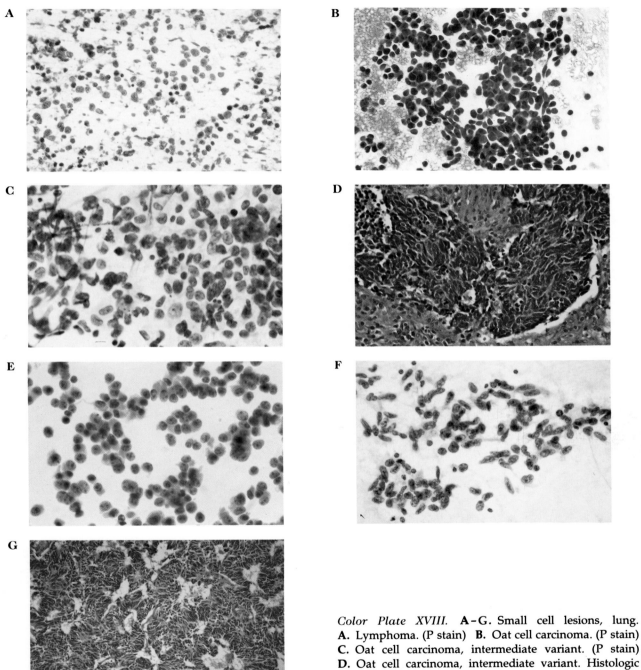

Color Plate XVIII. **A–G.** Small cell lesions, lung. **A.** Lymphoma. (P stain) **B.** Oat cell carcinoma. (P stain) **C.** Oat cell carcinoma, intermediate variant. (P stain) **D.** Oat cell carcinoma, intermediate variant. Histologic section. (H & E stain) **E–F.** Carcinoid round cell and spindle cell variants. (P stain) **G.** Spindle cell carcinoid. Tissue section. (H & E stain)

COLOR PLATE XIX.

A

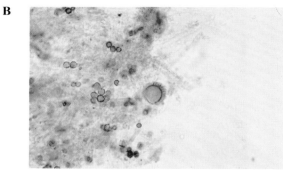

B

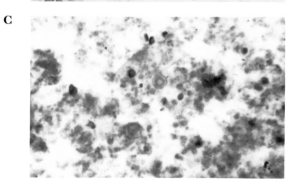

C

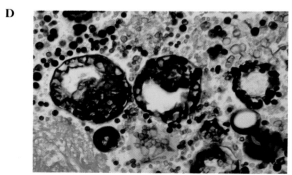

D

Color Plate XIX. **A–D.** Fungi, lung. **A.** Cryptococcus. (P stain) **B.** Aspergillosis. (P stain) **C.** Coccidiomycosis. (P stain) **D.** Coccidiomycosis. Tissue section. (H & E stain)

COLOR PLATE XX.

A

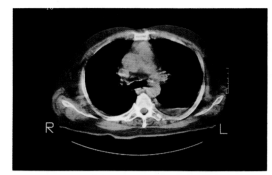

B

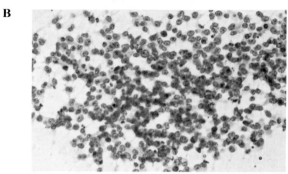

C

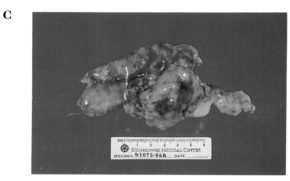

D

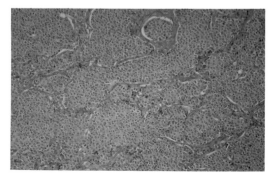

Color Plate XX. **A–D.** Thymic carcinoid. **A.** Computerized tomograph illustrating anterior mediastinal mass. **B.** Carcinoid. **C.** Thymus, gross specimen. **D.** Thymic carcinoid. Tissue section.

COLOR PLATE XXI.

A

B

C

D

E

F

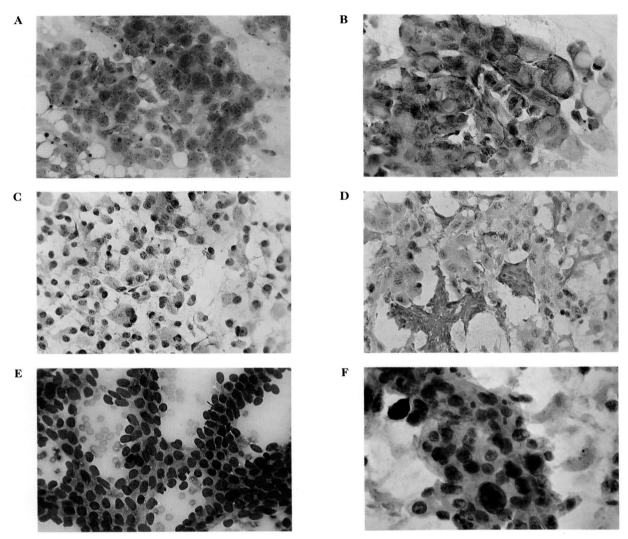

Color Plate XXI. **A–F.** Abdominal masses. **A.** Papillary cystadenocarcinoma of the ovary. Papillary masses of small cells aspirated transabdominally from a palpable metastasis. (P stain) **B.** Gastric carcinoma. Signet ring cells clearly identified in mass of tumor cells aspirated transabdominally from an epigastric mass. (P stain) **C.** Granulosa cell carcinoma. Transabdominal aspiration of palpable liver mass. Cells are dispersed with eccentric nuclei. Cytoplasm is abundant. (P stain) **D.** Islet cell tumor. Transabdominal aspiration of epigastric mass. Small cells are forming loose clusters with some alveolar formation. (P stain) **E.** Carcinoid transabdominal aspiration of enlarged liver. There are masses of small cells forming a trabecular pattern. (R stain) **F.** Metastatic squamous carcinoma in liver. Transabdominal aspiration yields masses of keratinized tumor cells. Primary is bronchogenic carcinoma.

COLOR PLATE XXII.

A

B

C

D

E

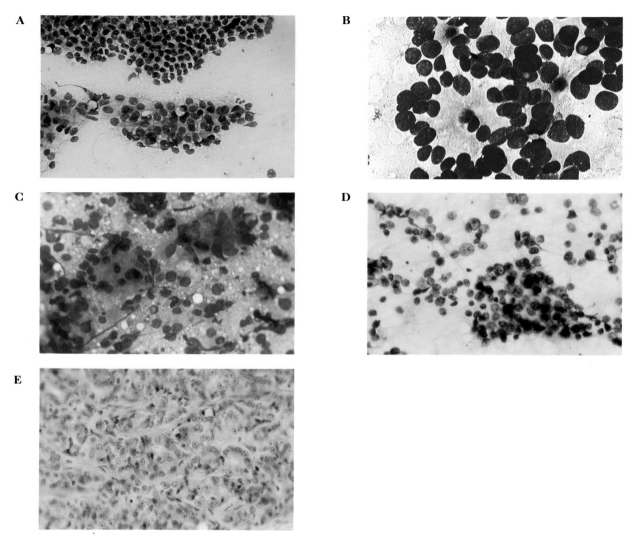

Color Plate XXII. **A–B.** Prostatic carcinoma, well differentiated. **A.** Normal epithelium is seen in contrast with large irregular cells with well-differentiated carcinoma. (MGG stain) **B.** Low-power view illustrating dense cellularity. On the gross slide, the material appears peppered. Cell masses may be connected by strands. (R stain) **C–E.** Prostatic carcinoma, moderate differentiation. **C.** Enlarged irregular cells forming ill-defined acini. Acinar structures are the usual pattern in well-differentiated tumors, but cytologic variabilities suggest less differentiated tumors here. (MGG stain) **D.** Polymorphic cell groups are present in loose clusters. There are hyperchromatism and nucleoli. (P stain) **E.** Tissue section. Compact cellular pattern. Some cells are forming small imperfect acini. Cells are typical carcinoma cells with nucleoli. (H & E stain)

COLOR PLATE XXIII.

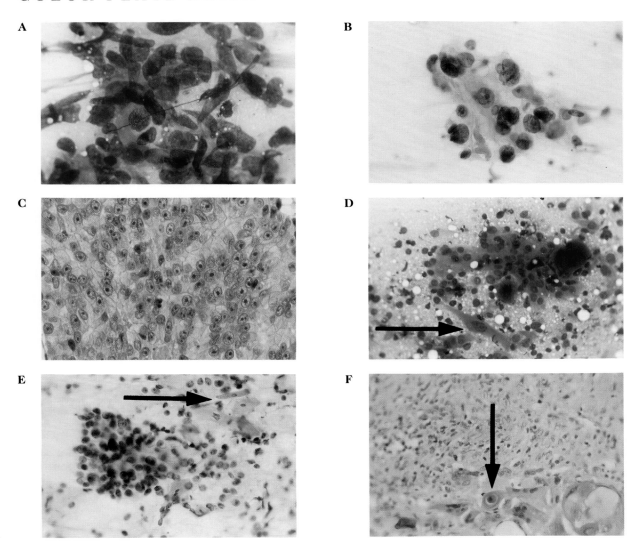

Color Plate XXIII. **A–C.** Prostatic carcinoma, poorly differentiated. **A.** Somewhat fragile large tumor cells are in haphazard arrangement. (MGG stain) **B.** Loose clusters of malignant cells with marked hyperchromatism. (P stain) **C.** Tissue section. Unstructured compact tumor composed of primitive cells with vesicular nuclei and primitive nucleoli. (H & E stain) **D–F.** Prostate carcinoma. Mixed adenocarcinoma and squamous carcinoma. **D.** Well-defined keratinizing squamous cells are intermixed with adenocarcinoma, moderately differentiated. (MGG stain) **E.** Same tumor as **D** with P stain. (P stain) **F.** Tissue section. Same tumor as **D** and **E** with H & E stain.

COLOR PLATE XXIV.

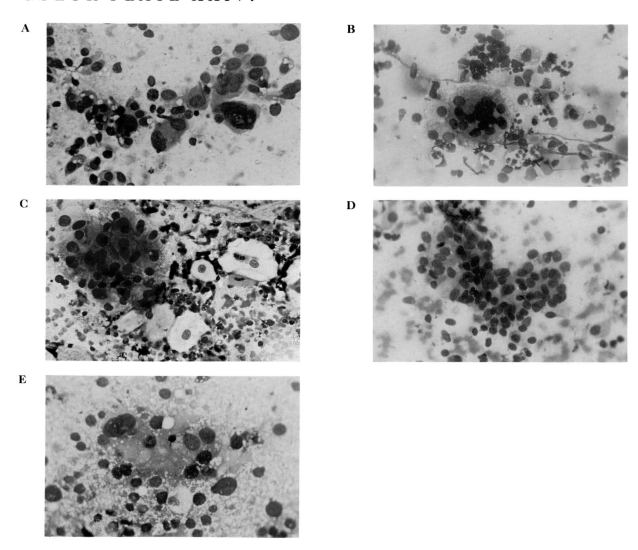

Color Plate XXIV. **A–B.** Giant cells, prostate. **A.** Seminal vessicle. Cells obtained on high lateral aspiration. There is marked variation in nuclear size. This simulates poorly differentiated tumor and is a pitfall in diagnosis. (MGG stain) **B.** Granulomatous prostatitis. Multinucleated giant cells are combined with epithelioid cells and granulocytes with considerable cell destruction. Multinucleated giant cells also occur in carcinoma. See Fig. 11-14. (MGG stain) **C–E.** Effects of treatment. **C.** Estrogen therapy induces specific glycogenic changes in the epithelial cells. (MGG stain) **D.** Orchiectomy. Removal of testosterone transforms tumor cells to a more normal morphology. **E.** Radiation. Cytologic response depends on dosage at the time of aspiration. Cells are stripped of cytoplasm and appear nonviable. In later stages, necrosis and fibrosis occurs.

A

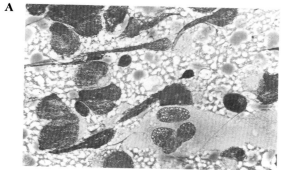

B

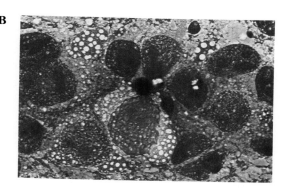

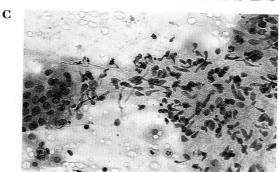

C

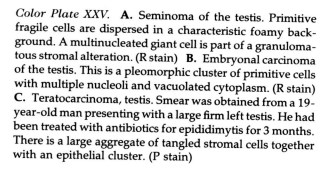

Color Plate XXV. **A.** Seminoma of the testis. Primitive fragile cells are dispersed in a characteristic foamy background. A multinucleated giant cell is part of a granulomatous stromal alteration. (R stain) **B.** Embryonal carcinoma of the testis. This is a pleomorphic cluster of primitive cells with multiple nucleoli and vacuolated cytoplasm. (R stain) **C.** Teratocarcinoma, testis. Smear was obtained from a 19-year-old man presenting with a large firm left testis. He had been treated with antibiotics for epididimytis for 3 months. There is a large aggregate of tangled stromal cells together with an epithelial cluster. (P stain)

COLOR PLATE XXVI.

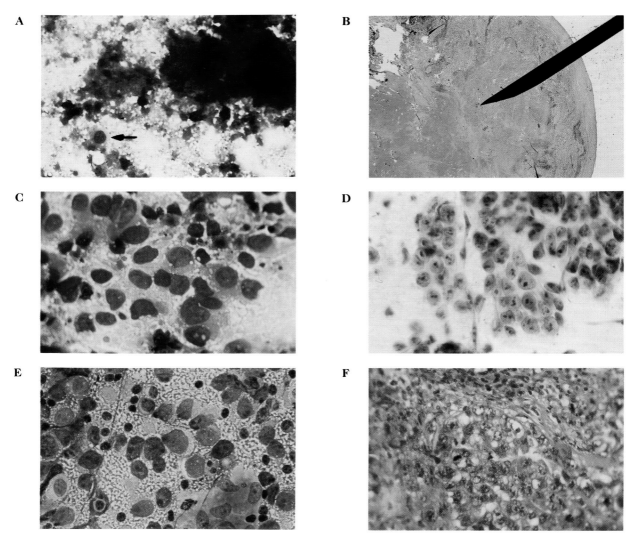

Color Plate XXVI. Testis. A 33-year-old man complained of pain in his right testicle. A hard nodule was palpated in the superior pole. Fine needle aspiration yielded necrosis and one intact primitive cell. The needle had penetrated to the necrotic center. On repeat aspiration, both embryonal and seminomatous elements were present. Alpha-fetoprotein and beta-subunit of human chorionic gonadotropin were normal. Necrosis even without tumor cells is a suspicious finding. **A.** Necrosis and a single intact cell *(arrow).* (MGG stain) **B.** Tissue section with needle entering necrotic center. **C.** Embryonal carcinoma. Polygonal tumor cells with immature nuclei and vacuolated cytoplasm on a necrotic background. (MGG stain) **D.** Embryonal carcinoma. Loose clusters of carcinoma cells with prominent nucleoli. (P stain) **E.** Seminoma. Dispersed "blastic" cells on a bubbly, lacy background. (MGG stain) **F.** Tissue section. Embryonal carcinoma simulating undifferentiated carcinoma. (H & E stain)

COLOR PLATE XXV.

A

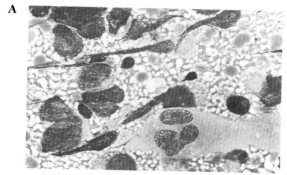

B

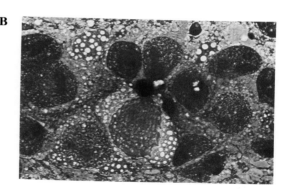

C

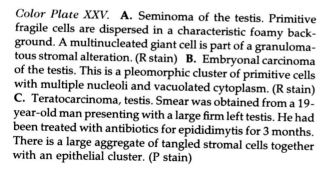

Color Plate XXV. **A.** Seminoma of the testis. Primitive fragile cells are dispersed in a characteristic foamy background. A multinucleated giant cell is part of a granulomatous stromal alteration. (R stain) **B.** Embryonal carcinoma of the testis. This is a pleomorphic cluster of primitive cells with multiple nucleoli and vacuolated cytoplasm. (R stain) **C.** Teratocarcinoma, testis. Smear was obtained from a 19-year-old man presenting with a large firm left testis. He had been treated with antibiotics for epididimytis for 3 months. There is a large aggregate of tangled stromal cells together with an epithelial cluster. (P stain)

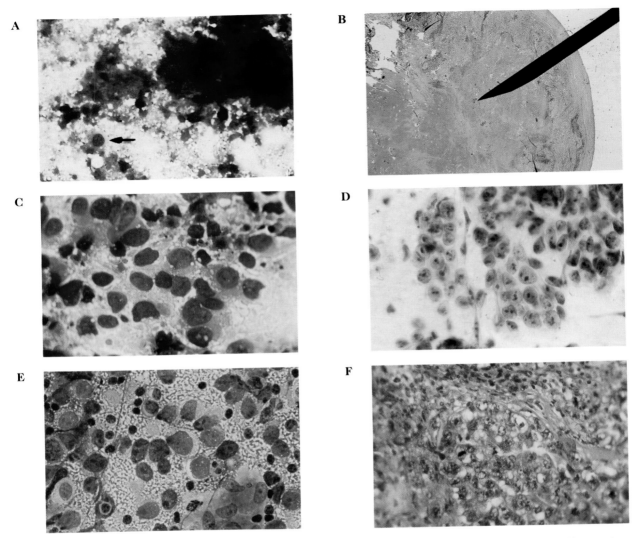

Color Plate XXVI. Testis. A 33-year-old man complained of pain in his right testicle. A hard nodule was palpated in the superior pole. Fine needle aspiration yielded necrosis and one intact primitive cell. The needle had penetrated to the necrotic center. On repeat aspiration, both embryonal and seminomatous elements were present. Alpha-fetoprotein and beta-subunit of human chorionic gonadotropin were normal. Necrosis even without tumor cells is a suspicious finding. **A.** Necrosis and a single intact cell *(arrow).* (MGG stain) **B.** Tissue section with needle entering necrotic center. **C.** Embryonal carcinoma. Polygonal tumor cells with immature nuclei and vacuolated cytoplasm on a necrotic background. (MGG stain) **D.** Embryonal carcinoma. Loose clusters of carcinoma cells with prominent nucleoli. (P stain) **E.** Seminoma. Dispersed "blastic" cells on a bubbly, lacy background. (MGG stain) **F.** Tissue section. Embryonal carcinoma simulating undifferentiated carcinoma. (H & E stain)

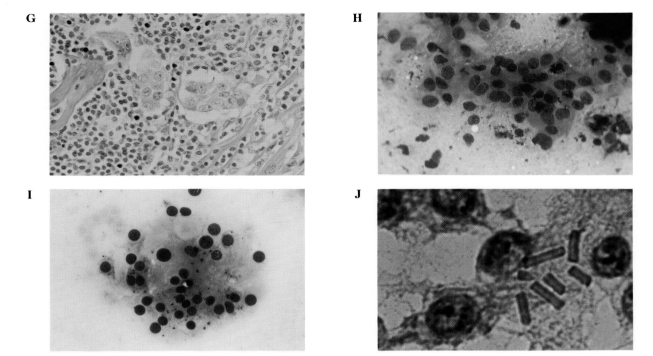

Color Plate XXVI (continued). **G.** Tissue section. Seminoma. Clusters of tumor cells surrounded by lymphoid stroma. (H & E stain) **H.** Adenomatoid tumor of epididymis. Cluster of mesothelial-like cells with royal blue cytoplasm and round to oval nuclei. (MGG stain) **I.** Cluster of Sertoli cells from patient with atrophic testis. Nuclei are round with mild anisokaryosis. Cytoplasm is geometric in shape. (MGG stain) **J.** Crystalloids of Reinke staining red. Aspirate from Leidig cell tumor. (P stain)

COLOR PLATE XXVII.

A
B

C
D

E

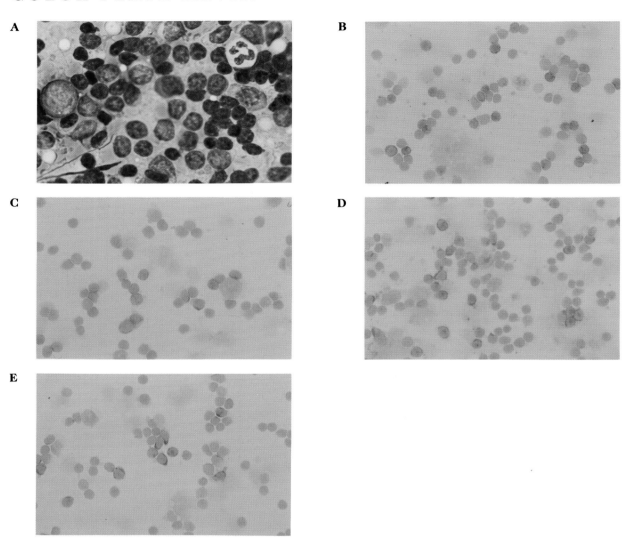

Color Plate XXVII. **A–E.** Reactive lymphadenitis. FNA of a lymph node in the right arm of a 17-year-old girl. **A.** Presence of variety of lymphoid cells: small lymphocytes, follicle center cells, neutrophil, and plasma cell. (MGG stain) **B.** Immunocytochemical detection of T cells by alkaline phosphatase technique. (APH) **C.** Immunocytochemical detection of B cells. (APH) **D.** Kappa light chains in cytoplasm of several cells. (APH) **E.** Lambda light chains in cytoplasm of some cells. (APH)

COLOR PLATE XXVIII.

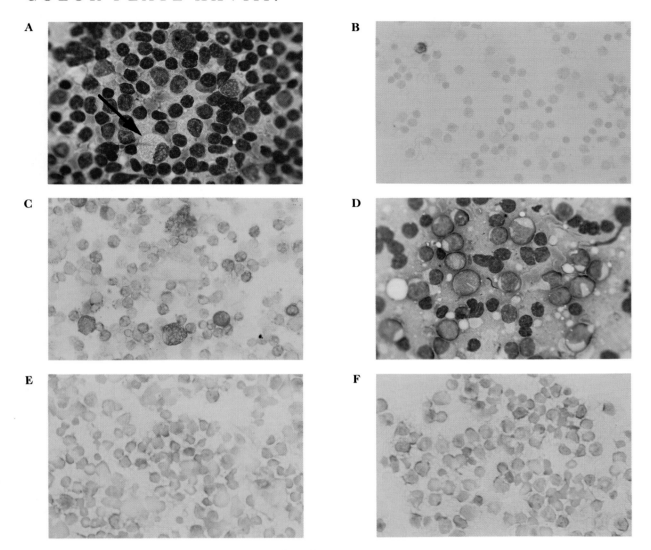

Color Plate XXVIII. **A–C.** Immunocytoma, lympho-plasmacytoid type. FNA of cervical lymph nodes of a 67-year-old man. **A.** Lymphocytes, lymphoplasmacytoid cells, plasma cells, and plasmacytoid cell with foamy cytoplasm *(arrow).* (MGG stain) **B.** Kappa light chains in cytoplasm of only one cell. (APH) **C.** Lambda light chains in cytoplasm of all cells revealing a monoclonal origin. (APH) **D–F.** Centroblastic/centrocytic lymphoma. FNA of thyroid tumor of a 69-year-old man. **D.** Mixed lymphoid cells: lymphocytes, centrocytes, large cleaved cells and centroblasts. (MGG stain) **E.** Immunocytochemical detection of B cells. (APH) **F.** Lambda light chains in the cytoplasm of tumor cells. (APH)

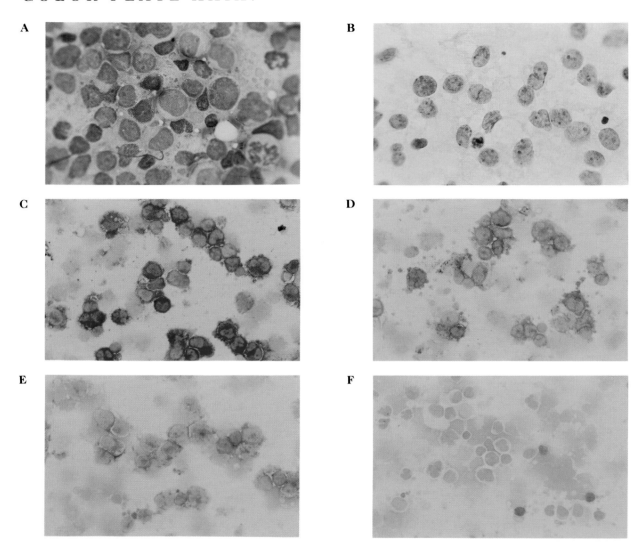

Color Plate XXIX. **A–F.** Centroblastic lymphoma. FNA of left tonsil of a 82-year-old man. **A.** Predominance of large blastic cells with prominent nucleoli at membrane and frequent mitosis. (MGG stain) **B.** Same aspirate as **A** stained with P. (P stain) **C.** Lambda light chains in the cytoplasm of blast cells. (APH) **D.** My-chains in the cytoplasm of tumor cells. (APH) **E.** Immunocytochemical detection of B cells demonstrating the phenotype of the tumor cells. (APH) **F.** Immunocytochemical detections of T cells showing few mature small lymphocytes of reactive origin. (APH)

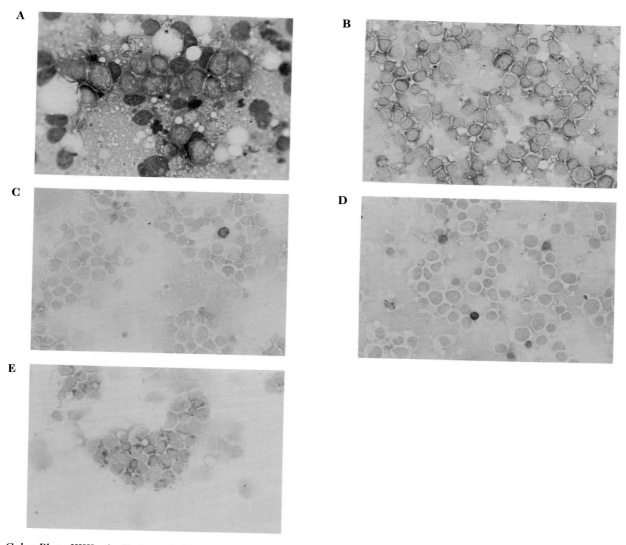

Color Plate XXX. **A – E.** Lymphoblastic lymphoma. FNA of left fossa iliaca tumor, taken through transrectal aspiration cytology from a 66-year-old man. **A.** Predominance of lymphoblasts with small vacuoles in the cytoplasm. (MGG stain) **B.** Kappa light chains in cytoplasm of the blastic cells. (APH) **C.** Lambda light chains in cytoplasm of only one small cell. (APH) **D.** Immunocytochemical detection of T cells showing few mature small lymphocytes of reactive origin. (APH) **E.** Immunocytochemical detection of B cells revealing the phenotype of the blastic cells. (APH)

COLOR PLATE XXXI.

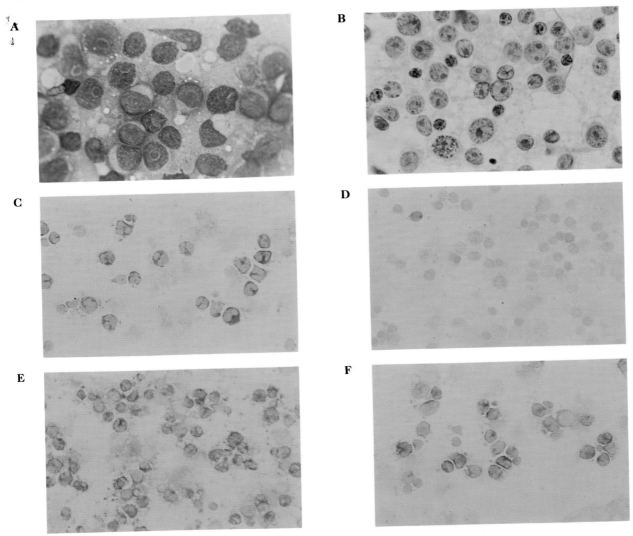

Color Plate XXXI. **A–F.** Immunoblastic lymphoma. FNA of axillary lymph node of 88-year-old man. **A.** Numerous large cells with eccentric nuclei and prominent single nucleolus. (MGG stain) **B.** Same aspirate as **A** stained with P stain. **C.** Kappa light chains in cytoplasm of the tumor cells. (APH) **D.** Lambda light chains in cytoplasm of only one cell. (APH) **E.** My-chains in cytoplasm of all tumor cells. (APH) **F.** Immunocytochemical detection of B cells showing the phenotype of the tumor cells. (APH)

COLOR PLATE XXXII.

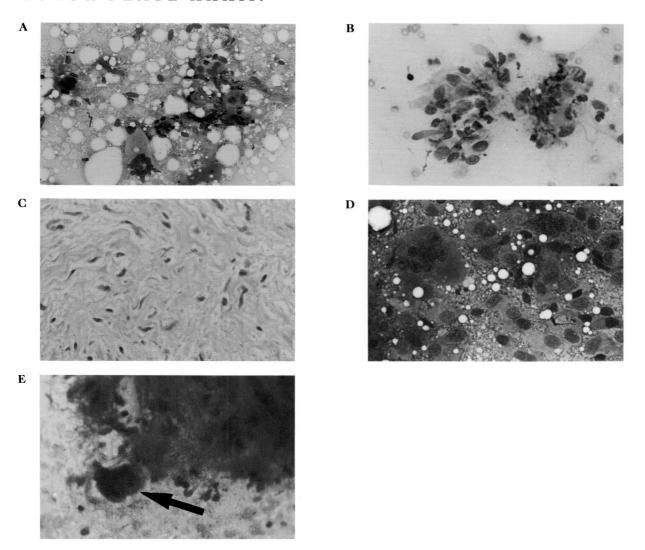

Color Plate XXXII. **A.** A 34-year-old woman presented 8 months after childbirth with a 2 cm-sized nodule in the rectus abdominis muscle. The aspirated material contained free-lying fibroblasts and atrophic striated muscle cells, some resembling multinucleated giant cells. Cytologically a desmoid tumor infiltrating the muscle was diagnosed. (The multinucleated muscle cells should not be mistaken for malignant cells.) **B–C.** A 31-year-old male presented with signs of urinary obstruction and an abdominal mass. Aspirated material **(B)** revealed cellular clusters of active fibroblasts with atypia. The diagnosis of mesenteric fibromatosis was suggested. A football-sized mass infiltrating the small intestine was surgically removed. Histopathologically, the preoperative diagnosis fibromatosis was con-firmed **(C)**. **D.** A 24-year-old female suddenly discovered a small subcutaneous nodule on her forearm. The aspirate showed large myofibroblasts against a pinkish stippled background of mesenchymal mucosubstances. Multinucleated giant cells of benign appearance were numerous. The cell picture demonstrated nodular fasciitis, which disappeared spontaneously within a month. **E.** A 65-year-old woman suddenly noticed a painful swelling of the right sternocleidomastoideus muscle. At FNA viscous material was obtained consisting of loose connective tissue and some very bizarre large-sized mononucleated and binucleated cells with basophilic cytoplasm and giant nucleoli *(arrow)*. A tentative diagnosis of proliferative myositis was made. Within a few weeks the mass had disappeared.

COLOR PLATE XXXIII.

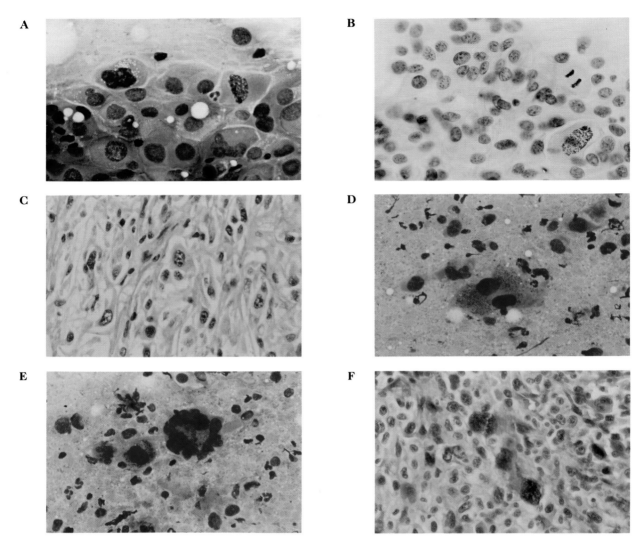

Color Plate XXXIII. **A–C.** A 60-year-old woman operated 1 year previously for malignant fibrous histiocytoma of the right popliteal fossa had observed a nodule in the surgical scar. FNA showed abundant mucous material containing histiocyte-like and multinucleated sarcoma cells. The pink-staining mesenchymal mucosubstances stand out clearly in MGG stain **(A)**, but the coarse chromatin pattern is easier to judge in P stain **(B)**. Frequent mitotic figures. The cytologic picture correlates well with previously diagnosed high grade malignant fibrous histiocy- toma of myxoid type **(C)**. **D–F.** A 76-year-old man was referred for aspiration of a walnut-sized tumor on the medial side of the thigh near the knee. A few drops of viscous fluid were obtained. The smears showed plump histiocyte-like and spindle-shaped, fibroblast-like pleomorphic sarcoma cells **(D)**. Note magenta-stained granula in the cytoplasm of histiocytic giant cells **(D–E)**. Paraffin sections of the resected tumor showed a pleomorphic malignant fibrous histiocytoma grade IV **(F)**.

COLOR PLATE XXXIV.

A

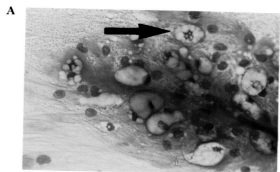

B

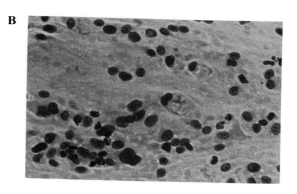

C

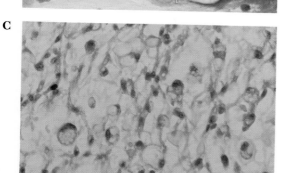

Color Plate XXXIV. **A.** A 46-year-old man had a soft tissue mass on the back of the right thigh. The smears revealed a myxoid tumor of monomorphic appearance containing numerous multivacuolated atypical lipoblasts, the hallmark of liposarcoma. This diagnosis of myxoid liposarcoma, grade II was confirmed histopathologically. **B–C.** A 75-year-old woman had a soft tissue mass in the distal and medial circumference of the left thigh. FNA showed a cellular round cell myxoid sarcoma containing atypical multivacuolated lipoblast. Note that the interstitial background mucin also covers obscure and to some extent large fat vacuoles in the lipoblasts **(B).** Histopathologic exam showed a round cell liposarcoma grade III **(C).**

(continued)

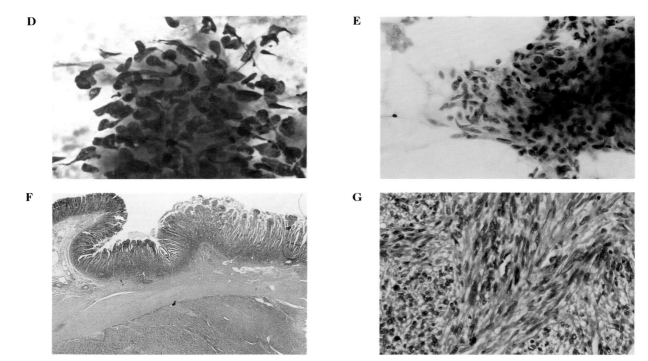

Color Plate XXXIV. (continued). **D–G.** A 73-year-old man with anemia presented with a left hypochondric mass. FNA showed cohesive clusters of spindle-shaped sarcoma cells with dense cytoplasm **(D–E).** Paraffin sections showed leiomyosarcoma of the small intestine, grade III **(F–G).**

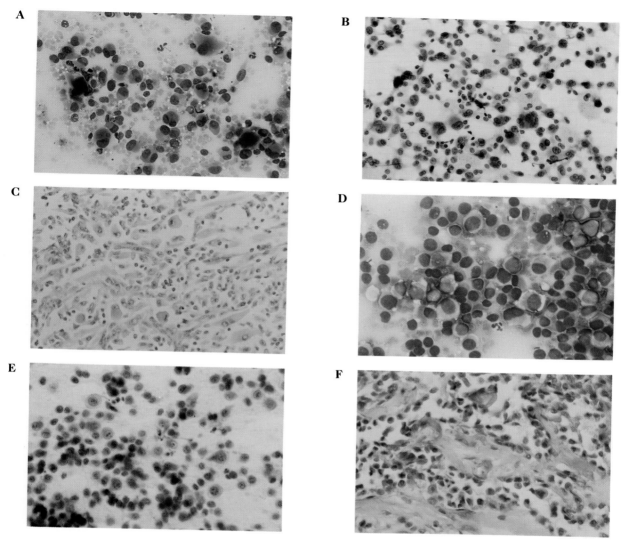

Color Plate XXXV. **A–C.** A 34-year-old woman noticed during pregnancy a rapidly growing tumor of the right hip. The mass stopped growing after childbirth. Preoperative FNA showed abundant sarcoma cells. Note dense blue "myogenic" cytoplasm of round rhabdomyoblasts in MGG stain **(A)** and perinuclear, slight eosinophilic tinge in P stain **(B)**. In paraffin sections this rhabdomyosarcoma showed obvious signs of muscular differentiation **(C)**. **D–F.** A 25-year-old man presented with a poorly circumscribed tumor with a diameter of 6 cm in the medial malleolar region of the right foot. FNA showed abundant small round tumor cells with somewhat irregular nuclei **(A)**. In P stains the nuclei contained enlarged nucleoli and showed prominent chromocenters **(E)**. A small cell malignant tumor with the above mentioned cytologic characteristics in a peripherally located soft tissue tumor in an adult was consistent with alveolar rhabdomyosarcoma. This diagnosis was confirmed histopathologically **(F)**.

COLOR PLATE XXXVI.

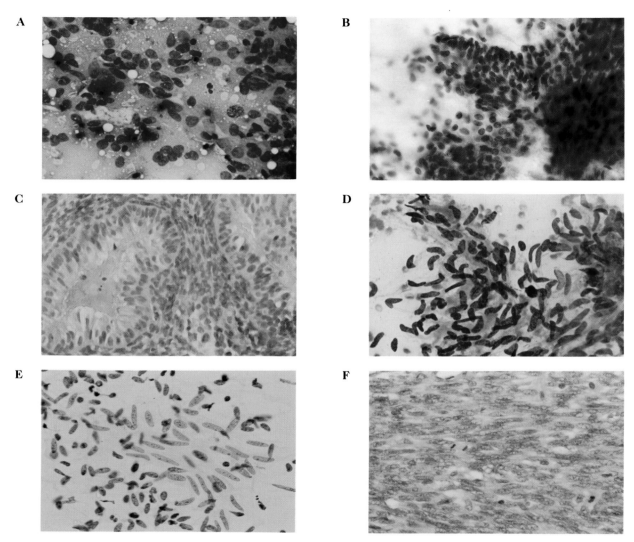

Color Plate XXXVI. **A–C.** A 46-year-old woman presented at the surgical orthopedic clinic with a fist-sized soft tissue mass in the upper lateral part of the left thigh. Fine needle aspiration revealed monomorphous spindle cells, with oval nuclei **(A)**, in which clusters of parallelly oriented cells could be discerned, best visible in the upper central part of the picture **(B)**. Histopathologically, the diagnosis of synovial sarcoma of biphasic type was confirmed **(C)**. **D–F.** A 28-year-old woman sought medical attention for a left hypochondric mass which was diagnosed initially as splenomegaly. CBC and bone marrow were normal. FNA surprisingly disclosed a monomorphous spindle cell soft tissue sarcoma **(D)**. The nuclei were slender and had a finely dispersed chromatin pattern **(E)**. Mast cells were easily visible in the smears and sections **(F)**. Obviously a monophasic synovial sarcoma was diagnosed cytologically. The surgical specimen showed a monophasic synovial sarcoma of the abdominal wall.

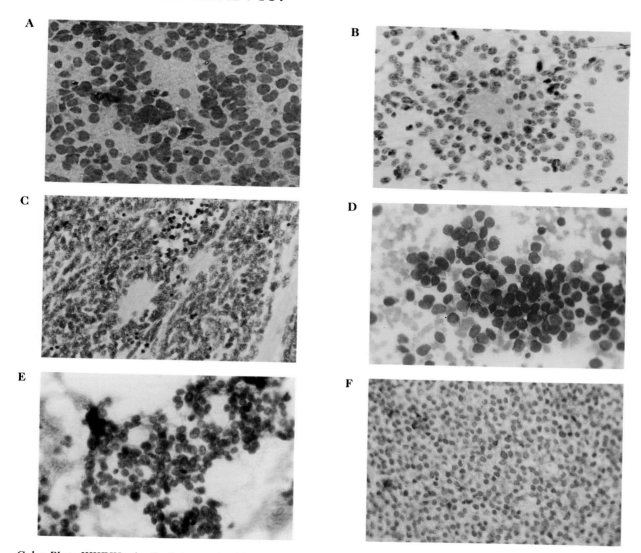

Color Plate XXXVII. **A–C.** A 4-month-old baby girl who at the age of 2 months was operated for neuroblastoma presented with clinical and radiological evidence of a local recurrence. Aspiration biopsy cytology was performed under fluoroscopic control of a dorsal intrathoric mass. FNA showed abundant material from a small cell tumor displaying a coarse chromatin pattern and pseudorosettes, in which the cells are concentrically arranged around granular substance **(A–B).** The cytologic diagnosis of neuroblastoma was confirmed histopathologically **(C). D–F.** A 23-year-old male complaining of pain and signs of urinary obstruction was found to have a large tumor in the small pelvis engaging the prostate. FNA showed haphazardly arranged small malignant mesenchymal tumor cells with monomorphous nuclei and finely dispersed chromatin **D** and **E.** Cytologically Ewing's sarcoma was diagnosed and confirmed histopathologically **(F).**

COLOR PLATE XXXVIII.

A

B

C

D

E

F

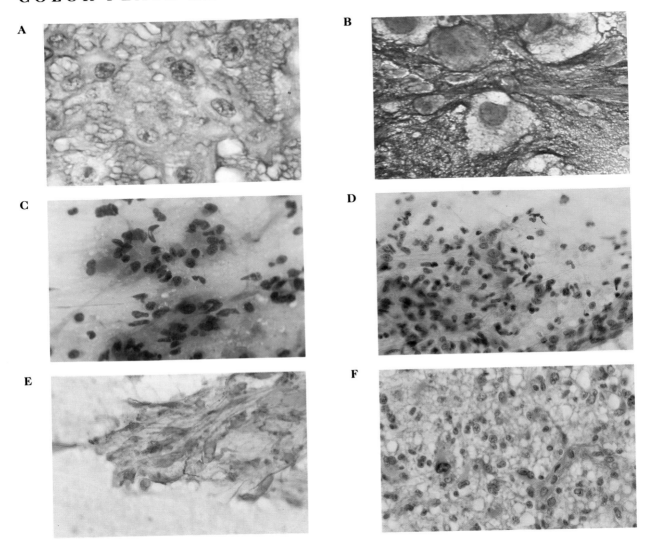

Color Plate XXXVIII. **A–B.** FNA of an osteolytic lesion in the region of the sella turcica of a 78-year-old woman showed large round to polygonal cells with bubbly cytoplasm embedded in mucinous substance **(A)**. The cell picture is typical for chordoma and the diagnosis was confirmed histopathologically **(B)**. **C–F.** Material from a glioblastoma in a 60-year-old male **(C–D)**. A polymorphic cell population is seen in MGG and P stained slides. The tumor cells contain GFAP as demonstrated immunocytochemically **(E)**. Histologically the diagnosis of glioblastoma grade IV was confirmed **(F)**.

COLOR PLATE XXXIX.

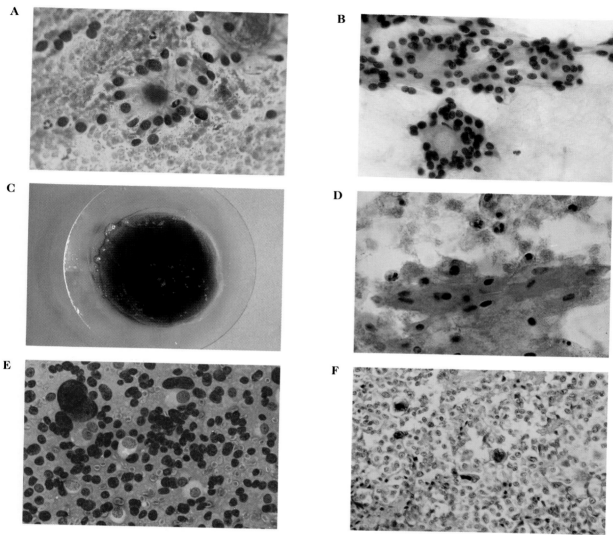

Color Plate XXXIX. **A–B.** Material from a cauda equina tumor in a 17-year-old male. Ependyma rosettes with myxoid material (red in MGG stain) are seen. The diagnosis was a myxopapillary ependymoma grade I. **C–D.** Fine needle aspiration of an intracellular lesion yielded several milliliters of an oily brown fluid **(C)**. Smears showed macrophages and a few clusters of squamous cells with signs of keratinization **(D)** demonstrating a cystic craniopharyngioma. **E–F.** FNA of a large destructive lesion of the sella turcica and the base of the skull in a 45-year-old man showed numerous cells mostly stripped of their cytoplasm with round monomorphous nuclei. Polyploid giant-sized nuclei were common. The cell picture is consistent with pituitary adenoma **(E)** as was shown in paraffin sections **(F)**.

COLOR PLATE XL.

A

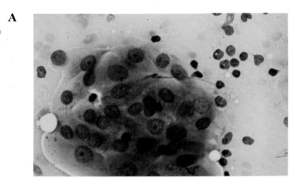

B

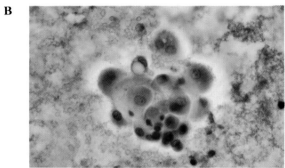

Color Plate XL. **A.** A cluster of malignant cells with prominent nucleoli and abundant cytoplasm. (R stain) **B.** Strong thyroglobulin-positive cytoplasmic stains identify the tumor cells to be of thyroid origin. (Thyroglobulin immunoperoxidase).

CHAPTER 10

Pelvic Fine Needle Aspiration Cytology in Gynecology

MEHRDAD NADJI

BERND–UWE SEVIN

It is now more than 17 years that fine needle aspiration (FNA) cytology has been in use for the diagnosis and management of gynecologic neoplasms.[11] Although it was originally restricted to the diagnosis of ovarian tumors,[1,2,6,10–12] aspiration cytology is now applied to other gynecologic conditions of the female pelvis as well.[3,4,7–9,14,17,20,21] Nevertheless, in recent years, the use of aspiration cytology in gynecology has not paralleled the growth of application of this technique to the diagnosis of pathologic conditions of other organs. There are several reasons for the apparent underutilization of aspiration cytology in the diagnosis of gynecologic disorders.

First, most clinicians prefer to investigate a pelvic mass, by exploratory laparotomy. If one selects this approach, there is little reason for performing a pre-operative needle aspiration. This in part may be because of fear of possible spillage of tumor cells into the peritoneal cavity following perforation of a cystic tumor by needle aspiration. There is, however, no firm evidence that FNA causes dissemination of a potentially benign or localized disease of the female genital system.

Second, because of the structural complexity of the female pelvis, aspiration of this region is technically difficult and requires special skills and experience. This becomes even more difficult in patients who have received irradiation to the pelvis and then undergo needle aspiration for detection of residual or recurrent disease. We therefore feel that the technique should be performed by gynecologists because they have the experience of evaluating the female pelvis. Unless the person who is sampling the lesion is also interpreting the cytologic specimen, a close cooperation between the clinician and the pathologist is mandatory to increase the diagnostic value of aspiration cytology in gynecologic disorders.

Finally, the female pelvis is a region with many organs, including portions of the urinary tract, the lower gastrointestinal tract, and the female reproductive system. These structures are embedded in connective tissue with an extensive blood vessel and lymphatic network. In addition to the normal structures in the pelvis, the female genital tract is the site of an enormous variety of pathologic processes, including inflammatory changes and benign and malignant tumors. These lesions differ in the vulva, vagina, uterine cervix, corpus, fallopian tube, and ovary. The ovary alone is the site of a multitude of malignancies of epithelial, stromal, or germ cell origin, further subdivided by type and grade. Nongynecologic tumors such as lymphomas or metastatic neoplasms can also be found in the pelvis. The contents of the peritoneal cavity, with its mesothelial lining and peritoneal fluid, are within reach of the fine needle. The variability of normal structures and pathologic changes of

261

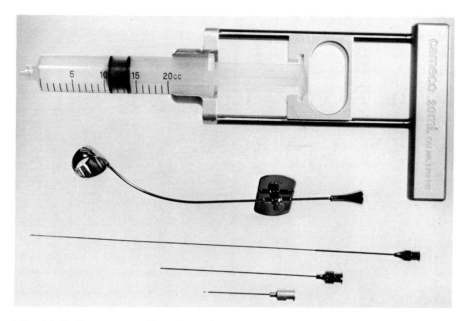

Fig. 10-1. Syringe pistol handle with 20-cc syringe, 22-gauge needles, and needle guide, designed by Franzen, for transrectal and transvaginal aspiration.

the female pelvis makes the cytologic interpretation of FNAs of this region a difficult task.[5,13,16–19] The pathologist, should not only be familiar with the cytologic appearance of normal and pathologic conditions but should also be able to recognize the morphologic changes seen in normal and neoplastic cells following irradiation and chemotherapy. Therefore, cytologic interpretation of pelvic aspiration requires extensive experience. Aspiration cytology, nevertheless, has become an important tool not only in the primary diagnosis of pelvic masses, but also in early detection of residual disease after surgery, radiation, or chemotherapy.

This chapter will discuss clinical and cytologic aspects of normal and pathologic elements within the female pelvis, including benign and malignant neoplasms as well as nonneoplastic lesions. Morphologic changes due to radiation or chemotherapy will also be described.

Technique

Pelvic organs can be aspirated by different approaches: through the vagina, through the rectum, or transabdominally. Intra-operative aspiration cytology has become more popular in recent years. We use

a syringe pistol (Cameco AB), a 20-ml syringe, and 22-gauge needles of appropriate length (Fig. 10-1). The needle guide designed by Franzen for prostate aspirations is used in most transvaginal and transrectal aspirations. Sampling the parametrium in previously irradiated patients requires that multiple passages be distributed in a fanlike fashion to optimize tissue sampling. Transrectal aspiration usually causes discomfort to the patient, especially after radiation therapy, and should preferably be done under general anesthesia. Other aspirations are usually well tolerated with local or no anesthesia. Multiple transrectal aspirations can cause bacterial seeding from the colonic flora and should be performed under prophylactic antibiotic coverage, starting before the procedure.

The aspiration itself has been described by Zajicek as a back-and-forth movement under continuous suction.[22] We feel that we get less blood contamination when we create suction while advancing the needle but not while withdrawing it. This appears to be particularly true when the pelvis is sampled at random, without evidence of clearly defined masses, and the needle passes through the richly vascularized structures of the pelvis. It is important to remember that *en route* to the target mass or organ, the needle may pass through other structures and pick up cells from these tissues (Fig. 10-2). Anterior to the uterus and parametrium is the urinary bladder, which is filled with fluid, lined with transitional epithelium, and supported by smooth muscle layers. It is advisable for the patient to empty the bladder before aspiration, thus avoiding significant fluid contamination. Fluid may become a problem in patients with increased intraperitoneal secretions. While persistent fluid aspiration, it may be advisable to prepare the aspirated fluid using conventional cytopreparatory techniques (centrifugation, filtration, or gravity sedimentation). Occasionally with transrectal aspiration, contamination with fecal material can occur. A cleansing enema before the procedure improves palpation of the target structure and the quality of the cytologic smear.

We prefer the Papanicolaou (P) staining method for pelvic FNA. In general, pathologists are most familiar with this stain. Rapid fixation of the cell sample in 95% alcohol provides cellular preservation superior to air-dried specimens stained by one of the Romanowsky (R) techniques. Air-dried specimens can also be used for special stains. A rapid version of P takes less than 2 minutes, allowing a preliminary report in 10 minutes.

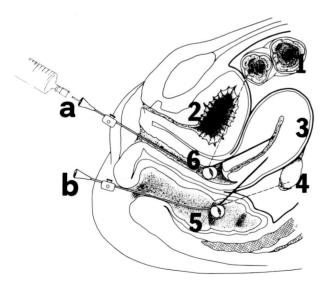

Fig. 10-2. Schematic longitudinal hemisection of the female pelvis to demonstrate the various structures that can be sampled with transvaginal (*a*) and transrectal (*b*) fine needle aspiration. (*1*) Small intestine, (*2*) urinary bladder, (*3*) uterus, (*4*) ovary, (*5*) rectum, (*6*) vagina.

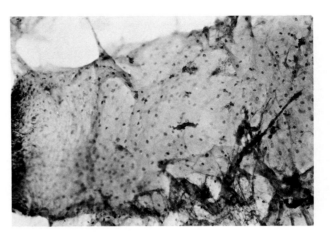

Fig. 10-3. Sheet of normal squamous epithelium. Transvaginal aspirate. There is a normal maturation from the basal layer (*left*) to superficial cells (*right*). (P stain, ×160)

NORMAL CONSTITUENTS OF FEMALE GENITAL TRACT ASPIRATIONS

Epithelial Elements

The most common normal epithelial components of female pelvic aspirations include squamous epithelium, colonic mucosa, urothelium, mesothelium, and occasionally small intestinal mucosa.

Squamous Epithelium

Small numbers of normal squamous epithelial cells are seen in samples from transvaginal or transcutaneous aspirates. Although isolated cells are occasionally encountered, squamous elements usually occur in two-dimensional groupings that include basal, intermediate, and superficial cells and represent microbiopsies of the normal maturating squamous epithelium (Fig. 10-3). The presence of numerous isolated anucleated and superficial squamous cells in an aspirate from the female pelvis should alert the examiner to the possibility of a benign cystic teratoma.

Colonic Mucosa

Colonic epithelial cells are almost always present on cytologic material from transrectal aspirations. These cells are usually arranged in small or large two-dimensional groups. Isolated cells are uncommon. Sheets of normal colonic epithelium are composed of

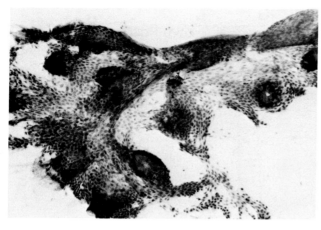

Fig. 10-4. Sheet of normal colonic epithelium. These are uniform columnar cells with abundant cytoplasm and oval nuclei, a common finding in transrectal needle aspirates. Note the concentration of cells around the openings of the colonic glands. (P stain, ×125)

uniform columnar cells with eosinophilic cytoplasm and oval to elongated nuclei, exhibiting evenly distributed chromatinic material. Tubular structures representing colonic glands are commonly found extending from the surface of cellular aggregates (Fig. 10-4). Variable amounts of mucinous material are always present in close association with colonic cells. This material forms an amorphous film on the slide,

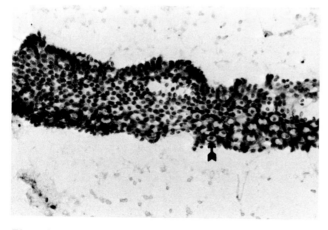

Fig. 10-5. Sheet of normal small intestinal epithelium. Transabdominal aspirate. Individual goblet cells (*arrow*) are distinguishable from absorptive cells by their larger size, clear cytoplasm, and centrally located nuclei. (P stain, ×160)

with frequent peripheral streaking that is quite distinctive from proteinaceous material with its granular texture and focal clumpings. With P, mucin stains anywhere from strongly eosinophilic to pale basophilic, depending on the thickness of the mucinous material. Fecal material is not commonly found in transrectal aspiration.

Small Intestinal Mucosa

Epithelial cells of small intestinal origin are an uncommon finding in FNAs of the female pelvis. They occur in two-dimensional sheets of uniform cells with honeycomb arrangements. Individual goblet cells are identified within the sheets of absorptive cells by their larger cell size and clear cytoplasm, which appears as a halo around a centrally located nucleus (Fig. 10-5). There is no free mucin associated with cells of small intestinal mucosa.

Transitional Epithelium

Normal urothelium may incidentally be aspirated from the bladder or ureters. These cells appear in small- to medium-sized sheets, usually composed of at least two distinct but intimately related cell populations. Predominantly, groups of smaller round to oval cells are mixed with few larger polygonal cells. The scanty cytoplasm of the smaller cell type contrasts with the abundant cytoplasm of the large cell type, which is homogeneous, dense, and eosinophilic. The nuclei of both cell types are uniform and round to oval with finely dispersed chromatin. Binucleation is common in larger polygonal cells. The small and large cells represent the parabasal and superficial (umbrella) urethelial cells, respectively (Fig. 10-6). The cytologic picture of epithelium can present diagnostic difficulties to the inexperienced eye, since it may easily be misinterpreted as cells derived from squamous cell carcinoma. Note, however, that normal transitional epithelium appears as monolayer sheets of cells with distinctive cellular borders, features not associated with squamous cell carcinomas.

Mesothelium

Only occasionally are mesothelial cells aspirated from the pelvis. They are more often seen in patients with chronic pelvic irritation or slow-growing pelvic masses. Mesothelium always occurs in two-dimensional sheets of large polyhedral cells with distinct borders and abundant, faintly eosinophilic cyto-

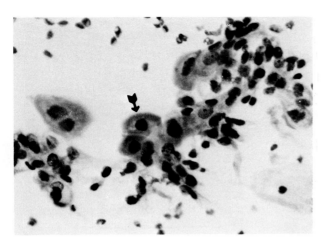

Fig. 10-6. Normal transitional epithelium. Transvaginal fine needle aspirate. Two populations of cells, larger and smaller, are seen. Although the nuclei of both types are comparable in size and shape, the amount and texture of cytoplasm are different. Larger cells, which represent superficial "umbrella" cells of urothelium, have abundant dense and eosinophilic cytoplasm *(arrow)*. (P stain, ×400)

plasm. Oval nuclei are centrally located and contain finely granular chromatin (Fig. 10-7). Binucleation, multinucleation, and prominent nucleoli are common in reactive mesothelial cells. In long-standing irritation, squamous metaplasia of mesothelium may occur. These cells differ from normal mature squamous cells by their smaller size and thickened ectoplasm (Fig. 10-8).

Other Epithelial Elements

With pelvic aspiration, epithelial elements from ovarian follicles, corpus luteum, or endosalpinx may be obtained on rare occasions.

Mesenchymal Elements

Normal mesenchymal elements are frequently encountered with FNAs and most commonly include fibroadipose tissue, collagen, smooth and skeletal muscle, and blood vessels.

Fibroadipose Tissue

Loose connective tissue is composed of spindle or stellate nuclei with abundant, faintly cyanophilic, fibrillar cytoplasm usually mixed with large, empty-appearing adipose cells (Fig. 10-9).

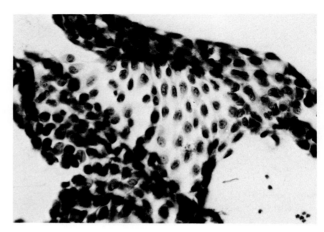

Fig. 10-7. Mesothelial cells. Transvaginal aspirate. This partially folded sheet of mesothelial cells shows abundant cytoplasm and uniform nuclei. Small nucleoli are evident. (P stain, ×400)

Collagen

We have used the term *collagen* to describe fragments of dense connective tissue frequently found in pelvic aspirations, especially after radiation therapy. These elements are usually irregular blocks of orangeophilic tissue containing few pyknotic nuclei (Fig. 10-10).

Smooth and Skeletal Muscle

Smooth muscle cells are commonly found; they occur in tight three-dimensional aggregates. The cells are elongated and contain abundant eosinophilic cyto-

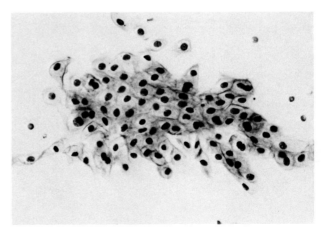

Fig. 10-8. Squamous metaplasia of mesothelium. These cells differ from normal squamous cells by their smaller size and prominent thickened ectoplasm. (P stain, ×312)

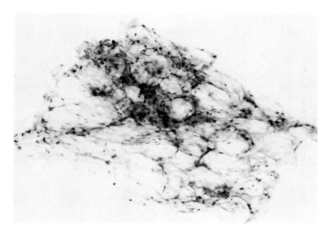

Fig. 10-9. Loose connective fibroadipose tissue mixed with empty-appearing adipose cells. Transrectal aspirate. (P stain, ×125)

Fig. 10-11. Normal smooth muscle. Transrectal aspirate. There is fibrillar cytoplasm and parallel rows of fusiform nuclei. (P stain, ×312)

plasm. Myofibrils are detectable upon fine focusing of the microscope (Fig. 10-11). Skeletal muscle is an uncommon finding; the cells form well-preserved rectangular blocks of orangeophilic fibrillar material with few oval nuclei in the periphery of each fiber (Fig. 10-12). Cross-striations are easily detectable.

Blood Vessels

Small segments of capillaries with a double row of endothelial cells are occasionally seen. They may contain erythrocytes. Endothelial cells show oval nu-

clei with finely granular chromatin and indistinct, pale, eosinophilic cytoplasm.

NONNEOPLASTIC LESIONS OF GYNECOLOGIC FNAs

Nonneoplastic lesions include inflammatory processes, endometriosis, and nonneoplastic epithelial elements from paratubal or para-ovarian inclusions or cysts.

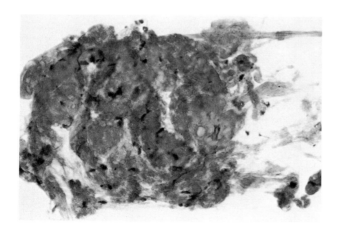

Fig. 10-10. Dense connective tissue. Transrectal aspirate. There are blocks of amorphous, deeply orangeophilic material with occasional pyknotic nuclei. (P stain, ×312)

Fig. 10-12. Normal skeletal muscle, an uncommon finding in gynecologic aspirates. Transabdominal aspirate. There are rectangular blocks of orangeophilic fibrillar muscle with peripheral localization of nuclei. (P stain, ×160)

Inflammation

The cytologic picture of aspirations from inflammatory changes is variable and depends on the type and severity of the process. Acute inflammatory contains numerous neutrophils, nuclear debris, and fibrinous material, whereas chronic inflammation is characterized by the presence of lymphocytes, plasma cells, and histiocytes. The true nature of the inflammatory process, however, should be determined by clinical and microbiologic studies. Part of the FNA sample can be used for bacterial cultures and antibiotic sensitivities.

Endometriosis

Unequivocal diagnosis of endometriosis can only be made when endometrial cells are found outside the endometrial cavity. Endometrial cells are small with scanty cytoplasm and frequently occur in loosely bound sheets. The nuclei are round to oval with uniform, finely or coarsely granular chromatin and occasional chromocenters. Cellular borders are usually indistinct (Fig. 10-13). Endometrial stromal cells may also be present and appear as isolated, denuded, oval nuclei. Fresh or hemolysed blood as well as hemosiderin-laden macrophages are frequently encountered in fine needle aspirates from endometriosis. Endometrial cells can be differentiated from normal colonic cells, which are larger and have abundant eosinophilic cytoplasm and distinct cell borders. Epi-

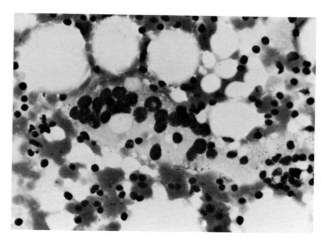

Fig. 10-14. Lymphangiographic changes. In this transabdominal aspirate from a pelvic lymph node, in addition to lymphocytes and fat globules of the background, a lipid-containing multinucleated histiocyte with empty vacuoles is seen. (Wright–Giemsa [WG] stain, ×400)

thelial cells aspirated from mesonephric or paramesonephric nests and inclusions show the general benign morphology of normal endosalpingeal epithelium. Since these structures are small and few, the chance of one's being aspirated is extremely low (see the following case report).

■ *Case 10-1*

A 31-year-old white woman was examined because of chronic pelvic pain. A small submucosal mass was palpated in the rectovaginal septum. A transrectal FNA showed glandular epithelium with endometrial-type stroma (Fig. 10-13). With the diagnosis of endometriosis, the patient was treated with progestin for 6 weeks to facilitate the surgical removal of endometriosis. Exploratory laparotomy with resection of the nodule between the vagina and rectum confirmed the diagnosis of pelvic endometriosis.

Lymphangiographic Changes in Lymph Node Aspirates

In gynecology, lymphangiography alone or in conjunction with FNA cytology is used to detect metastatic lymph node involvement.[7] Peri-aortic and pelvic lymph nodes as well as groin nodes are sampled. The needle is guided into the lymph node by fluoros-

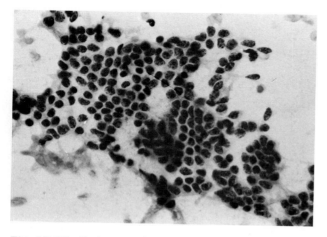

Fig. 10-13. Endometrial cells, loosely bound, aspirated from a nodule in the rectovaginal septum. Note round nuclei, coarse chromatin, and indistinct cytoplasm. (P stain, ×400)

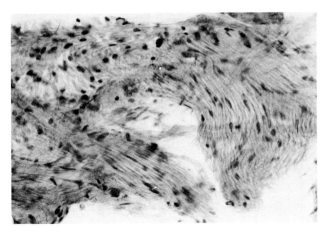

Fig. 10-15. Uterine leiomyoma. This fragment of smooth muscle was aspirated from a uterine leiomyoma transvaginally. Although there is a suggestion of whorling, the morphologic appearance of these cells is not different from those aspirated from normal smooth muscle elements of the pelvis. (P stain, ×312)

copy or direct palpation. The lipid-based contrast medium produces a pronounced foreign body reaction that facilitates cellular recognition. In addition to normal and reactive lymphocytes, a large number of fat-containing multinucleated histiocytes and free fat globules are found on the slide (Fig. 10-14).

CELLULAR MANIFESTATIONS OF BENIGN NEOPLASMS

Benign gynecologic neoplasms can be divided into neoplasms of mesenchymal and epithelial origin.

Benign Mesenchymal Neoplasms

Uterine Leiomyomas

Uterine leiomyomas are by far the most common type of benign neoplasms. Since these are solid firm tumors, cellular material aspirated from them is extremely limited. The aspiration cytology of a leiomyoma is identical with the cytomorphology of normal smooth muscle, as described previously (Fig. 10-15). The cellularity varies among patients and even in the same tumor. Since smooth muscle ele-

ments may normally occur in pelvic aspirations, a meaningful interpretation can only be made in conjunction with precise clinical information. Presence of smooth muscle on an aspirate from a firm pelvic mass in which advancement of the thin needle is difficult suggests the presence of a uterine leiomyoma.

Ovarian Thecomas

Ovarian thecomas are more cellular than myomas. Thecoma cells are spindled and, in larger aggregates, may form interlacing bundles of nuclei (Fig. 10-16). Cytoplasm is pale cyanophilic and often indistinct. Multiple bare nuclei are common findings in thecomas as well as in other benign or malignant mesenchymal neoplasms.

Benign Epithelial Neoplasms

Almost all benign epithelial neoplasms are of ovarian origin. Most of these tumors form cystic ovarian masses of variable size and consistency. Since patients with persistent cystic ovarian masses require exploratory surgery for diagnostic or therapeutic reasons anyway, we see no indication for FNA before surgery. Our cytologic material from benign epithelial neoplasms is therefore limited to that incidentally aspirated.

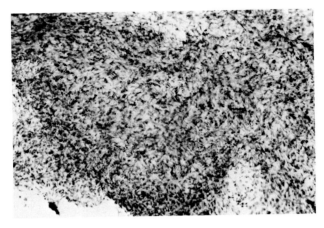

Fig. 10-16. Ovarian thecoma. This is a three-dimensional grouping of fusiform nuclei. The interlacing arrangement of cells mimics normal ovarian stroma. Isolated denuded nuclei are also found on the slide. (P stain, ×125)

Adult Cystic Teratoma (Dermoid Cyst)

Cystic teratomas and epidermal cysts usually originate in the ovary, but also may originate in the retroperitoneal space of the pelvis, presenting as a fixed mass in the presacral region of young females. This neoplasm is of multiple germ layer origin but is included in this section because it always contains cystic structures lined with epithelia. Consequently, in aspirations from dermoid cysts, epithelial elements such as squamous cells, columnar cells, and occasionally metaplastic squamous cells predominate. Sebaceous material in the form of amorphous amphophilic substance is frequently encountered (Fig. 10-17). We have not found hair shafts, a frequent component in teratomas of the ovary, in needle aspirations. This might be because of the small caliber of the needle (22-gauge). Cells derived from the mesodermal elements of cystic teratomas are infrequently seen in FNA. In samples from chronically ruptured dermoid cysts, mononuclear or multinucleated histiocytes may be found (see the following case report).

■ *Case 10-2*

A 30-year-old white woman had an exploratory laparotomy and right salpingo-oophorectomy for increasing pelvic pain 4 months before admission. On rectovaginal examination a fixed, firm

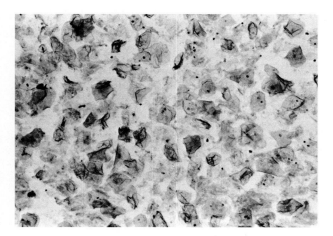

Fig. 10-17. Adult cystic teratoma (dermoid cyst). Transrectal aspirate. Anucleated squames and mature squamous cells are seen on a background of amorphous sebaceous material. (P stain, ×160)

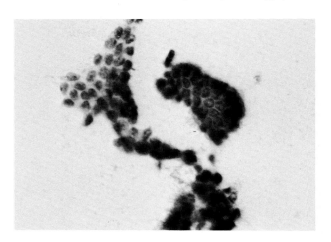

Fig. 10-18. Serous cystadenoma. Transvaginal aspirate. A small sheet of uniform cells and a papillary grouping are present. Note the small cell size, high nuclear : cytoplasmic ratio, and uniform chromatin pattern. (P stain, ×500)

---

mass was palpated on the right side and appeared to be retroperitoneal. On transrectal FNA anucleated squamous cells were found in abundance, with amorphous sebaceous background material (see Fig. 10-17). With the cytologic diagnosis of a presacral dermoid cyst, the patient underwent exploratory surgery; and, with careful dissection, the mass was removed without rupture, confirming the diagnosis of a mature cystic teratoma.

Serous Cystadenoma

Cells aspirated from serous cystadenomas are generally limited. They usually are arranged in sheets, with occasional papillary formations (Fig. 10-18). The cells are small, with oval nuclei and scanty eosinophilic cytoplasm. The background of the slide contains abundant serous fluid appearing as granular eosinophilic material.

Mucinous Cystadenoma

In aspirates from mucinous cystadenomas, isolated cells and aggregates of columnar mucin-producing cells are found. Depending on the angle viewed, the cells may appear as tall and columnar with a side-by-side arrangement and basally located nuclei or with a honeycomb arrangement, that is, having round cellular borders and central nuclei (Fig. 10-19). The cy-

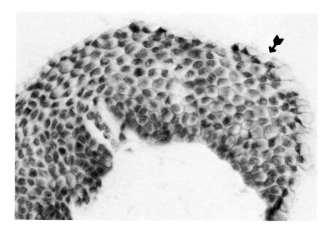

Fig. 10-19. Ovarian mucinous cystadenoma, two-dimensional sheet. When viewed from above the cells appear in honeycomb arrangement, whereas at the border of this group the columnar configuration of cells is evident (arrow). (P stain, ×400)

toplasm is clear and contains abundant mucin with mucicarmine stain. Abundant free mucin is characteristically present in aspirates from mucinous cystoadenomas, indistinguishable from mucinous material previously described in association with normal colonic mucosa.

Brenner Tumor

Benign Brenner tumor is characterized by two cell types — epithelial cells similar to transitional cell epithelium and mesenchymal cells. Epithelial cells are round and occur in small sheets with moderate amounts of eosinophilic cytoplasm and uniform round nuclei. The chromatin is finely granular, and micronucleoli are often present. Nuclear membrane folding, similar to the histologic appearance of this tumor, is only occasionally seen. Most of the mesenchymal components of Brenner tumor appear as isolated bare nuclei similar to those described in ovarian thecoma. The background of Brenner tumors contains eosinophilic granular material.

MALIGNANT NEOPLASMS OF FEMALE GENITAL TRACT

With the variety of normal organs and tissues in the female pelvis, the diversity of possible neoplasms is not surprising. Since many of these neoplasms from different parts of the female genital tract present with similar histopathology, it is essential that the pathologist be aware of the exact site of aspiration for a more specific diagnosis. In this section, the cytology of those malignant tumors common to several sites of origin will be discussed first. Specific tumors, particularly those of the ovary, will be presented separately.

MALIGNANT EPITHELIAL NEOPLASMS

Squamous Cell Carcinoma

Squamous cell carcinoma may arise from the vulva, vagina, or cervix. Cytomorphology of keratinizing and nonkeratinizing varieties of squamous cell carcinomas differ significantly. In the keratinizing group, keratin in the form of concentric pearls or, more frequently, amorphous orangeophilic material is commonly observed (Fig. 10-20). Abnormal isolated squamous cells, similar to those described in exfoliate cytology of the cervix, as well as mature squamous cells, characteristic of keratinizing squamous cell carcinoma, are present. These cells have orangeophilic or eosinophilic cytoplasm with slightly irregular nuclei. On several occasions we have observed calcific material and foreign body–type giant cells associated with this variety of squamous cell carcinoma (see the case report below).

■ *Case 10-3*

A 23-year-old white woman had a radical hysterectomy and bilateral pelvic lymphadenectomy for Stage IB keratinizing squamous cell carcinoma of the cervix. Four months later she returned with the complaint of right pelvic pain. On examination, a mass was palpated deep in the right pelvis. She had a negative Papanicolaou's (Pap) smear on vaginal scrape. A transvaginal FNA, sampling the right parametrium to the pelvic sidewall, produced cellular evidence of keratinizing squamous cell carcinoma (see Fig. 10-20A). Without further exploration, the patient was treated with radiation therapy and is free of disease 3 years later.

In the nonkeratinizing variety, the cells are arranged in large syncytial groups. Isolated cells are also always present. The cells are rather uniform in

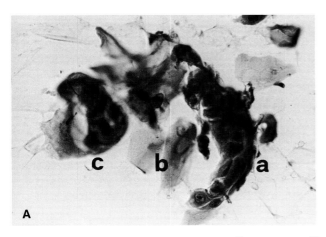

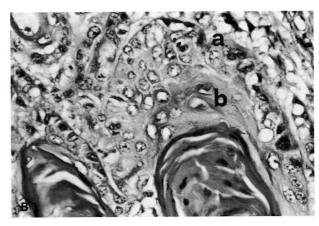

Fig. 10-20. A. Keratinizing squamous cell carcinoma. Transvaginal aspirate of parametrium showing small aggregate of cell representing *(a)* basal layer of tumor, *(b)* maturating malignant squamous cells, and *(c)* malignant pearl (keratin). (P stain, ×500) B. Histologic appearance of primary tumor in cervix with regions *a*, *b*, and *c* corresponding to areas described in A. (Hematoxylin-eosin [H & E] stain, ×500)

size and shape and contain oval nuclei and dense basophilic cytoplasm. The nuclear:cytoplasmic ratio is high, and cytoplasmic processes commonly radiate from the periphery of larger groups (Fig. 10-21). The chromatin is coarse and irregularly granular. Nucleoli are usually not prominent in nonkeratinizing squamous cell carcinoma. Poorly differentiated adenocarcinoma can be differentiated on the basis of finely granular or vacuolated cytoplasm, eccentric nuclei, prominent nucleoli, and smooth border of cellular aggregates. In both types of squamous carcinoma, the slide background usually contains degenerated and necrotic tumor cells, cellular detritus, and lysed blood.

Small Cell Carcinoma

Cell samples from small cell carcinomas of the cervix are characteristically fairly cellular. Both isolated forms and syncytial groupings are evident. The cells are small, with round to oval nuclei and scanty cyanophilic cytoplasm. The nuclei are hyperchromatic and exhibit irregularly dispersed coarsely granular chromatin. Crushing artifacts of nuclei, similar to those in histologic sections, are frequently observed.

Adenocarcinoma

Possible sites of origin for adenocarcinoma in the female pelvis are numerous. In a study of 21 adeno-

carcinomas from the female genital tract, only 9 could be subclassified according to their specific anatomic site of origin. The remaining 12 were all cytologically poorly differentiated and at times showed secondary changes due to irradiation or chemotherapy.[17] The cytomorphology varied among patients; however,

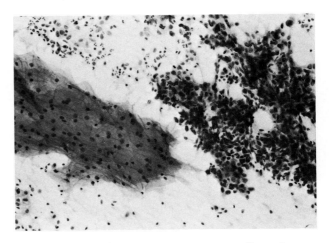

Fig. 10-21. Nonkeratinizing squamous cell carcinoma. Transvaginal aspirate. Compare sheet of normal squamous epithelial cells *(left)* to syncytia of carcinoma cells *(right)*. Malignant cells are smaller and have a high nuclear:cytoplasmic ratio and indistinct cytoplasmic borders. Isolated abnormal cells are also found on the slide. (P stain, ×312)

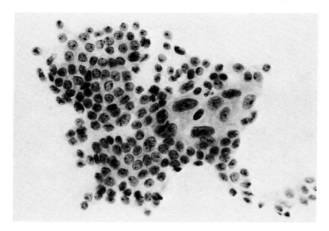

Fig. 10-22. Well-differentiated cervical adenocarcinoma of the endocervix with honeycomb arrangement of cells. Transvaginal aspirate. High nuclear:cytoplasmic ratio, prominent nucleoli, and focal atypia of the nuclei differentiate this neoplasm from normal endocervical cells. (P stain, ×500)

certain characteristic arrangements of adenocarcinoma were always present (papillary, tubular, or acinar groupings).

Endocervical Adenocarcinoma

Well-differentiated adenocarcinoma of the cervix characteristically shows a side-by-side or honeycomb arrangement of cells (Fig. 10-22). Cells contain granular, pale eosinophilic cytoplasm and round nuclei with prominent nucleoli. A high nuclear:cytoplasmic ratio and nuclear atypia differentiate this tumor from normal endocervical cells. The slide background may contain degenerated blood but is usually clean.

Endometrial Adenocarcinoma

Cytomorphology of endometrial adenocarcinoma is identical to endometrioid carcinoma of the ovary. In well-differentiated cases, the cells have a columnar configuration with eccentric nuclei and frequently form acinar patterns (Figs. 10-23 and 10-24). Abundant granular eosinophilic cytoplasm, resulting in a low nuclear:cytoplasmic ratio, finely granular chromatin, and nucleoli, differentiate endometrioid adenocarcinoma from serous cystadenocarcinoma (see the following case report).

■ *Case 10-4*

A 50-year-old white woman had a simple hysterectomy 10 years before admission for a stage I grade 1 adenocarcinoma of the endometrium. She developed a small pelvic mass above the apex of the vagina. FNA cytology revealed cellular evidence of adenocarcinoma (see Fig. 10-23). She underwent exploratory surgery for possible surgical removal of that mass, which confirmed the FNA diagnosis of recurrent adenocarcinoma of the endometrium. The tumor was unresectable, and the patient was successfully treated with radiation therapy.

Serous Adenocarcinoma

Certain general rules apply to the interpretation of fine needle aspirates from ovarian cystadenocarcinomas.[5,16] First, it is extremely difficult to separate low malignant potential tumors from well-differentiated cystadenocarcinomas by cytology alone. Second, the presence of benign epithelial cells does not exclude the possibility of malignant components in other areas of cytologic sample. Similarly, two or more different cell types may be seen in mixed carcinomas of

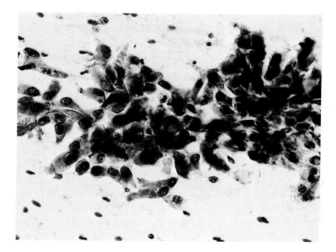

Fig. 10-23. Well-differentiated endometrial adenocarcinoma is represented here by a population of columnar cells with eccentrically located nuclei and abundant, finely granular cytoplasm. Transrectal aspirate. Nuclei are uniformly oval and show finely granular chromatin and micronucleoli. (P stain, ×400)

the ovary; usually, one cell type predominates. Finally, only better differentiated ovarian cystadenocarcinomas lend themselves to subclassification. High-grade carcinomas are difficult to subtype unless they are associated with better differentiated components.

Cells of serous adenocarcinoma occur in papillary groups with frequent branching (Fig. 10-25). Psammoma bodies may be present but are not pathognomonic of serous adenocarcinomas, since they may be found in other malignant or even benign processes. The cells are cuboidal or oval and contain scanty cyanophilic cytoplasm. The nuclei, which are arranged in the periphery of cell groups, are hyperchromatic and show irregularly distributed coarsely granular chromatin. Nucleoli are not prominent, probably because of the density of chromatin material (see the case report below).

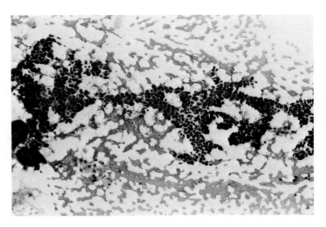

Fig. 10-25. Serous adenocarcinoma of the ovary. Transvaginal aspirate. The cells of serous adenocarcinoma are relatively small and have a high nuclear:cytoplasmic ratio and hyperchromatic nuclei. They are commonly arranged in papillary groups with frequent branchings. (P stain, ×400)

■ *Case 10-5*

A 66-year-old white woman was treated for stage I serous adenocarcinoma of the ovary with total abdominal hysterectomy followed by whole abdominal radiation. Twenty months later a mediastinal mass was seen on routine chest x-ray film. A transcutaneous FNA through the third right intercostal space, next to the sternum, revealed cellular evidence of adenocarcinoma. The patient was treated with combination chemotherapy and showed only moderate improvement. However, the diagnosis of metastatic cancer was made without a more invasive diagnostic procedure, avoiding unnecessary cost and delay of definitive therapy.

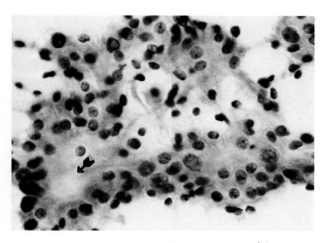

Fig. 10-24. Endometrioid adenocarcinoma of the ovary. Transabdominal aspirate. Although cell borders are not as distinct as in Fig. 10-23, the eccentricity of nuclei, abundant granular cytoplasm, and occasional acinar arrangement *(arrow)* suggest endometrial or endometrioid adenocarcinoma. (P stain, ×400)

Mucinous Adenocarcinoma

Mucinous adenocarcinoma of the ovary contains large cells in papillary groups and cell balls. Many cells exhibit vacuolated cytoplasm (Fig. 10-26). Depending on the degree of differentiation, the number and size of mucinous vacuoles vary among patients. Their content can be stained positively with mucicarmine stain. The mucin background, if present, is scanty in contrast with mucinous cystadenomas.

Clear Cell Carcinoma

Clear cell carcinoma is of Müllerian origin and may arise from the vagina, cervix, endometrium, or ovary. The cytologic picture, however, is the same regardless of the site of origin. FNA specimens from clear cell carcinomas characteristically contain large cells with prominent pleomorphism, but not all cellular components show clear cytoplasm. The cells of this

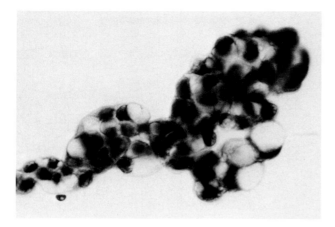

Fig. 10-26. Mucinous adenocarcinoma of the ovary. Transabdominal aspirate from a pelvic mass. This is a papillary grouping of malignant cells, with the small and large cytoplasmic vacuoles characteristic of mucinous adenocarcinomas. (P stain, ×500)

tumor exhibit the lowest nuclear:cytoplasmic ratio of all adenocarcinomas. Peripheral localization of nuclei and their protrusion above the margin of cellular groups give them the "hobnail" appearance similar to that observed in histologic sections. The nuclei are pleomorphic and frequently show foldings of the nuclear membrane (Fig. 10-27). Binucleation is common (see the following case report).

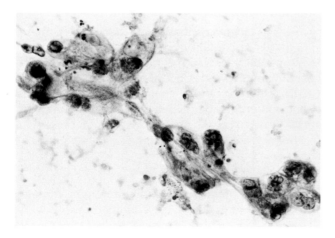

Fig. 10-27. Clear cell carcinoma from an inguinal lymph node of a patient with a history of clear cell carcinoma of the vagina. Transcutaneous aspirate. Note large cell size, clear cytoplasm, low nuclear:cytoplasmic ratio, pleomorphism, and binucleation. (P stain, ×500)

■ *Case 10-6*

A 47-year-old white woman had undergone radical surgery for a clear cell adenocarcinoma of the rectovaginal septum 1 year earlier. The appearance of a 3-cm × 2-cm mass in her left groin prompted the suspicion of metastatic disease. FNA cytology confirmed this suspicion (see Fig. 10-27), as did subsequent excisional biopsy. The patient was treated with combination triple chemotherapy and showed a good clinical response. Serial FNAs during chemotherapy (see Fig. 10-39) confirmed the clinical impression of tumor regression from cytotoxic chemotherapy.

Mixed Adenosquamous Carcinoma

Mixed adenosquamous carcinoma may be of cervical, endometrial, or ovarian origin. The cytologic picture is a combination of an adenocarcinoma of variable differentiation and a keratinizing or nonkeratinizing squamous carcinoma. The number of cells representing each component varies considerably among patients. The malignant glandular and squamous elements may be found in continuation with each other or, more commonly, separate on different areas of the slide.

Malignant Mesenchymal Neoplasms

Malignant mesenchymal tumors include leiomyosarcoma, endometrial stromal sarcoma, and malignant mixed mesodermal tumor.

Leiomyosarcomas

Unlike exfoliate cytology of sarcomas, in which most abnormal cells occur singly, FNA specimens from leiomyosarcomas commonly show three-dimensional groupings of malignant mesenchymal cells. The number of cells and their arrangement, however, depend on the type and degree of differentiation of a given tumor. In well-differentiated leiomyosarcomas of the uterus, the number of aspirated cells is often limited. Abnormal cells frequently occur in aggregates with spindle-shaped nuclei arranged in bundles (Fig. 10-28). The cytoplasm is indistinct. Few isolated cells are also present on the slide. In poorly differentiated leiomyosarcomas, in addition to malignant

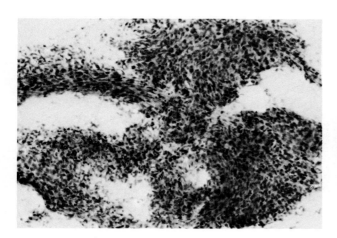

Fig. 10-28. Leiomyosarcoma showing three-dimensional grouping of spindle cells with nuclei arranged in fascicles. Transvaginal aspirate of a pelvic mass. This was a well-differentiated leiomyosarcoma of the uterus. Aspiration of a large number of cells from a well-differentiated leiomyosarcoma, however, is uncommon. (P stain, ×160)

spindle cells, abnormal bizarre configurations and malignant giant cells may be found (Fig. 10-29). These abnormal cells usually occur singly.

Endometrial Stromal Sarcomas

Endometrial stromal sarcoma is an uncommon tumor of endometrial origin. In a study of more than 600 FNAs from the female genital tract, we have seen 3 such neoplasm. In endometrial stromal sarcoma, abnormal cells occur mainly in isolated round to oval forms. The cytoplasm is pale cyanophilic with indistinct cellular borders. Nuclei are round, oval, or spindle shaped with irregularly distributed finely granular chromatin. Nucleoli are not prominent. Round intranuclear inclusions, which appear to be invagination of cytoplasm, are visible (Fig. 10-30). The background of slides contains granular eosinophilic material and blood.

Malignant Mixed Mesodermal Tumors

Histologically, malignant mixed mesodermal tumors of the uterus or ovary are composed of malignant epithelial as well as malignant mesenchymal elements. The epithelial component commonly presents as a poorly differentiated adenocarcinoma, whereas mesenchymal elements may be undifferentiated or

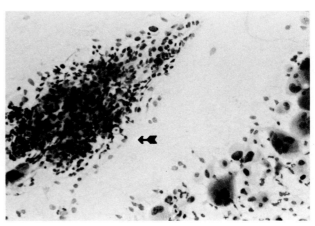

Fig. 10-29. Poorly differentiated leiomyosarcoma. Transabdominal aspirate. Note group of malignant spindle cells on the left *(arrow)* and multinucleated tumor cells on the right. The cytomorphology of a given tumor may vary depending on the area sampled. (P stain, ×200)

differentiated toward homologous (leiomyosarcoma, endometrial stromal sarcoma) or heterologous (rhabdomyosarcoma, chondrosarcoma) elements. In FNAs from malignant mixed mesodermal tumors, variable proportions of epithelial and mesenchymal elements may be found in continuity or separate from each other (Fig. 10-31). Not uncommonly, the local pelvic extension of mixed mesodermal tumors is composed of malignant epithelial elements only, whereas dis-

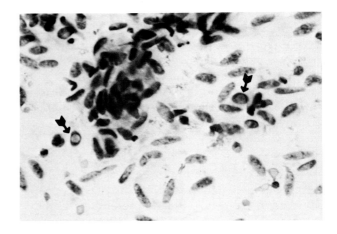

Fig. 10-30. Endometrial stromal sarcoma. Transabdominal aspirate. Loosely arranged spindle cells with faintly staining fibrillar cytoplasm and indistinct cellular borders are evident. Occasional intranuclear inclusions are also evident *(arrow)*. (P stain, ×500)

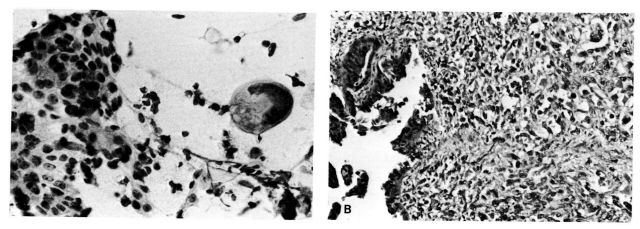

Fig. 10-31. **A.** Malignant mixed mesodermal tumor. Transvaginal aspirate. Note syncytial arrangement of poorly differentiated adenocarcinoma cells *(left)* and a single large isolated sarcoma cell *(right)*. Abnormal nucleus and giant nucleolus are evident. (P stain, ×400) **B.** Histologic section from a fungating tumor of the uterus in the same patient. Poorly differentiated adenocarcinoma *(left)* is mixed with poorly differentiated sarcoma *(right)*. (H & E stain, ×200)

tant metastasis may be of both cell types or the mesenchymal component only (see the following case report).

■ *Case 10-7*

An 83-year-old black woman had a total abdominal hysterectomy and bilateral salpingo-oophorectomy for a malignant mixed mesodermal tumor of the uterus. The tumor was confined to the uterus and completely removed. Nine months later, during a routine follow-up examination, a mass was palpated above the vaginal apex in mid-pelvis. A transvaginal aspirate clearly showed the recurrence of the primary tumor (see Fig. 10-31). The patient was treated with chemotherapy and showed a good clinical response.

Malignant Lymphoma

Extranodal malignant lymphomas may occasionally be found in the pelvis, usually presenting as an expanding pelvic mass or, at times, retroperitoneally. Recognition is important, since they may enter into the differential diagnosis of more common inflammatory lesions. Lymphomas in general exhibit a uniform population of isolated noncohesive cells with indistinct cell borders (Fig. 10-32). The size and mor-

phology of nuclei, however, largely depends on the cytologic type of lymphoma, that is, lymphocytic, histiocytic, or undifferentiated (see Chap. 14). Inflammatory infiltrates, on the other hand, are composed of a polymorphic cellular make-up, with lymphocytes, histiocytes, and polymorphonuclear leukocytes as their main constituents (see the following case report).

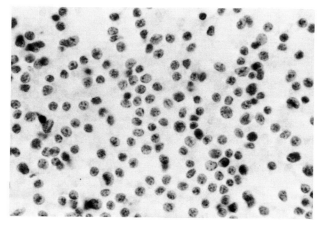

Fig. 10-32. Malignant lymphoma. Transabdominal aspirate from pelvic mass. This is a uniform population of isolated cells with indistinct cytoplasm. The large cell size (when compared to a normal lymphocyte) is consistent with histiocytic lymphoma. (P stain, ×500)

■ *Case 10-8*

A 40-year-old white woman was admitted to the neurosurgery department with increasing right hip pain. A neurologic evaluation showed neurologic deficits originating at L–5 and S–1. Physical examination revealed an 11-cm × 8-cm firm irregular mass filling the right pelvis from the sacrum to the pelvic ramus and pelvic sidewall. The mass appeared fixed, possibly retroperitoneal. An FNA was done transabdominally, and the diagnosis of a histiocytic lymphoma was made within 20 minutes (see Fig. 10-32). A metastatic work-up confirmed the existence of a large pelvic–abdominal mass. The patient was treated without further delay with combination chemotherapy. The lower abdominal mass disappeared within 48 hours. The patient currently has no gross evidence of disease.

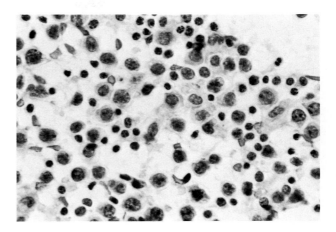

Fig. 10-33. Dysgerminoma of the ovary. Transvaginal aspirate. Loosely bound epithelial cells with large nuclei and fairly abundant cytoplasm are evident. Normal lymphocytes are initimately associated with tumor cells. (P stain, ×500)

Special Tumors of the Ovary

The cytomorphology of germ cell and sex cord tumors of the ovary are discussed below.

Germ Cell Tumors

The most common germ cell tumor of the ovary, the benign cystic teratoma, has already been discussed.

Dysgerminoma of the Ovary

Dysgerminoma, a malignant neoplasm of germ cell origin, is analogous to the seminoma of the testis. FNA dysgerminoma reveals small syncytial arrangements of loosely cohesive cells with a high nuclear:cytoplasmatic ratio and indistinct cell borders (Fig. 10-33). Many isolated abnormal cells are usually present. The morphology of isolated cells from a dysgerminoma may be similar to the cells derived from histiocytic lymphoma. It is therefore important to examine the entire slide carefully, since the presence of cohesive cells and syncytial grouping speaks for the diagnosis of dysgerminoma. Mature lymphocytes usually accompany dysgerminoma cells, but their absence does not rule out the diagnosis. Since infarction and necrosis frequently occur in dysgerminomas, it is not uncommon to find a hemorrhagic slide background with granular eosinophilic material and nuclear debris.

Malignant Mixed Germ Cell Tumors

Malignant mixed germ cell tumors of the ovary are a group of highly malignant tumors of germ cell origin with differentiation along several cell lines. The most common histologic pattern is a combination of embryonal carcinoma, malignant yolk sac elements (endodermal sinus tumor), and malignant trophoblastic elements (choriocarcinoma). Needle aspirates from these tumors generally contain abundant hemorrhagic and necrotic background material. The aspirates are fairly cellular and exhibit a pleomorphic population of highly abnormal cells with an increased nuclear:cytoplasmic ratio and many bizarre tumor giant cells. Cells derived from embryonal carcinoma or yolk sac elements usually present patterns like those seen in poorly differentiated adenocarcinomas (Fig. 10-34), whereas malignant trophoblasts occur singly and have multiple irregular nuclei with thick-textured amphophilic cytoplasm. We have been able to demonstrate human chorionic gonadotropin (HCG) and α-fetoprotein in the cytoplasm of malignant trophoblastic and yolk sac elements with immunoperoxidase technique on needle aspirate specimens.[17] A definitive diagnosis of malignant mixed germ cell tumor of the ovary based on cytology alone is extremely difficult. The association of elevated serum levels of HCG and α-fetoprotein in a young

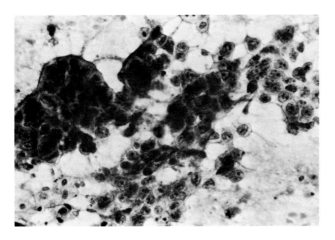

Fig. 10-34. Embryonal carcinoma of the ovary. Transthoracic aspirate from a patient with metastatic embryonal carcinoma of the ovary to the lung. The appearance of this tumor is that of a poorly differentiated adenocarcinoma with a high nuclear : cytoplasmic ratio. Multiple macronucleoli are characteristic of this tumor. Note the necrotic background. (P stain, ×500)

patient with cytologically poorly differentiated carcinoma is highly suggestive of a malignant mixed germ cell tumor.

Sex Cord Tumors

Sex cord tumors are usually functioning ovarian tumors (hormone producing) that are capable of differentiating in either a female (granulosa type) or a

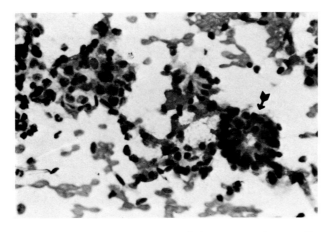

Fig. 10-35. Sex cord tumor of the ovary. These cells were aspirated from a Sertoli's cell tumor of the ovary (arrhenoblastoma). Clinically, the tumor was producing male hormones. Note uniform oval nuclei and prominent tubular arrangement *(arrow).* (P stain, ×400)

male (Sertoli's cell type) direction. Granulosa or Sertoli's cell tumors are usually of low malignant potential with late peritoneal recurrences and rare distant metastasis. With FNA cytology we were unable to differentiate granulosa from Sertoli's cell tumor. Cytologic specimens are usually fairly cellular. The cells are oval, cuboidal, or low columnar and arranged in cords, trabeculae, or acini (Fig. 10-35). The nuclei are uniform and oval and show finely granular chromatin pattern with small nucleoli. Although the cells are small and uniform, differentiation of these tumors from well-differentiated adenocarcinoma of the ovary in general, and ovarian carcinoids in particular, may prove to be exceedingly difficult with cytology alone.

EFFECT OF RADIATION AND CHEMOTHERAPY ON GYNECOLOGIC FNA CYTOLOGY

Needle aspiration cytology is well established as a simple and reliable method for the detection of persistent or recurrent disease after radiation or chemotherapy for gynecologic malignancies.[21] Several aspects of FNA cytology, however, should be discussed in relation to the history of previous irradiation or chemotherapy. Ionizing radiation and chemotherapeutic agents induce secondary changes in the female pelvis that may interfere with the sampling technique as well as with the cytologic interpretation.

Effects on Sampling Technique

Radiation therapy to the pelvis may produce extensive fibrosis of normal soft-tissue components. With palpation of pelvic structures after radiation therapy, it may at times be impossible to differentiate fibrosis from tumor recurrence. Advancing the needle through these firm structures is difficult, since the barrel of the needle easily fills with obstructing sclerotic tissue. Multiple introductions of an empty needle are necessary to ensure negative suction when aspirating these pelvic structures. Furthermore, dense fibrotic tissue causes the needle to deviate from its desired path, decreasing its sampling accuracy. Chemotherapy, especially when effective in causing cell death, may produce necrosis with liquefaction in the central part of the tumor, while the surrounding host tissue may produce a firm fibrotic capsule. Aspiration of such a tumor mass may produce mainly necrotic debris with few viable cells.

Effects on Cytologic Interpretation

Cytomorphologic changes in normal and abnormal cells secondary to irradiation and chemotherapy in exfoliative cytology have received considerable attention. In FNAs of the pelvis, these changes are more significant because of the wide variety of normal tissue components as well as benign and malignant lesions. Cytologic changes seen in normal epithelial tissue include increasing cytoplasmic and nuclear area, vacuolization and amphophilia of cytoplasm, phagocytosis, and multinucleation (Fig. 10-36). Although these changes may occasionally produce diagnostic difficulty, with good clinical history and experience they should not be confused with malignant changes.[21] Reactive changes of mesothelium and urothelium in particular may at times look very disturbing to the examiner.

Cytomorphologic changes in malignant cells from radiation or chemotherapy are of lesser importance (Figs. 10-37 and 10-38). The established histologic diagnosis before the therapy may aid the classification of aspirations if done to detect persistent or recurrent disease, and a precise histologic subclassification is not as essential as it is in primary diagnosis.

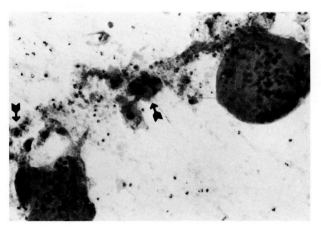

Fig. 10-37. Radiation effect on squamous cell carcinoma. Transrectal aspirate. Foreign body reaction to degenerated keratin may be seen in irradiated keratinizing squamous cell carcinomas. Besides multinucleated histiocytes, few abnormal squamous cells are present *(arrow)*. (P stain, ×200)

We have used serial FNA cytology to monitor the effect of chemotherapy on gynecologic malignancies. These samples are used for flow cytometric studies as well as for conventional light microscopy. In cytologic monitoring of the response to chemotherapy, the following parameters should be evaluated: the relative quantity of abnormal cells, the ratio of viable to necrotic tumor cells, the nature of the slide back-

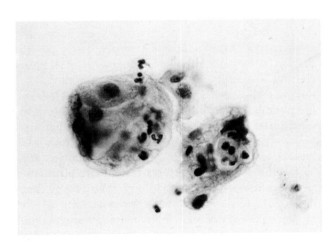

Fig. 10-36. Radiation effect on normal squamous epithelial cells. Transvaginal aspirate. Note enlarged cytoplasmic volume, multinucleation, and phagocytosis. These changes should not be mistaken for malignancy. (P stain, ×500)

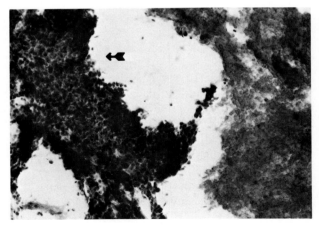

Fig. 10-38. Radiation effect on squamous cell carcinoma. Transvaginal aspirate 6 months after therapy. In this photograph a contrast between viable *(arrow)* and necrotic, nonkeratinizing, squamous cell carcinoma can be seen. (P stain, ×200)

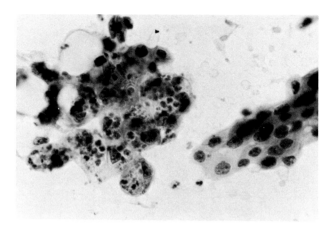

Fig. 10-39. Chemotherapy effect on metastatic clear cell carcinoma of the vagina. Transcutaneous aspirate of lymph node. Tumor cells are enlarged. On the left, pleomorphism, vacuolization, and phagocytosis are evident. Compare these cells to the nonaffected group on the right. (P stain, ×400)

ground, and morphologic changes in individual malignant cells. Extreme cellular and nuclear pleomorphism, cytoplasmic vacuolization, phagocytosis of nuclear debris and polymorphonuclear leukocytes by neoplastic cells, and, finally, cell death are features of effective chemotherapy (Fig. 10-39). Similarly, with increasing cytotoxic effect the total number of neoplastic cells decreases in serial aspirations from the same tumor.

CLINICAL APPLICATION

Clinical relevance of pelvic FNA must be demonstrated in order to make this technique not only acceptable to the clinician but, at times, the procedure of choice. Many gynecologic tumors are accessible to inspection, palpation, and direct biopsy, making a histologic diagnosis easy. Frequently, however, the tumor is located within the pelvis or abdomen and thus is accessible only by exploratory surgery. Patients with benign pelvic pathology (e.g., endometriosis) or with malignancies that are primarily treated with radiation therapy or chemotherapy would benefit tremendously from FNA because the risk, time delay, and cost of surgery can be avoided and the patient treated immediately. This is usually not the case for ovarian malignancies, because the mainstay of therapy is a combination of cytoreductive surgery and chemotherapy. In these cases, especially when the tumor mass is cystic and mobile, most clinicians prefer to investigate it by exploratory laparotomy, therefore bypassing needle aspiration. On the other hand, literature from the Scandinavian countries proposes needle aspiration as the first diagnostic step for cystic ovarian tumors.[1,2,6] In our opinion the decision whether to aspirate a cystic ovarian mass should be made by the gynecologist treating the patient. This decision is aided by clinical and ultrasonographic findings, which provide information on size, location, consistency, and mobility of a pelvic mass. Solid or semisolid tumor masses in the pelvis can be aspirated transvaginally or transrectally without fear of spillage.

In patients previously treated with radiation or chemotherapy, diagnostic exploratory surgery is particularly disadvantageous, since it is associated with significant morbidity. FNA cytology offers an accurate and relatively atraumatic alternative to surgery biopsy in many instances. This is particularly true when tumor nodules are palpable on examination. To appreciate the wide range of applicability of FNA cytology, Table 10-1 summarizes indications for FNA in our first 200 patients in relation to the origin of primary or recurrent tumor. In the earlier part of our series, the primary diagnosis of untreated patients predominated. Later, the evaluation of patients already treated for gynecologic malignancies gained more attention. Particularly, patients treated with radiation therapy to the pelvis turned out to benefit most from this procedure. Currently, every postradiation patient is routinely followed with serial FNAs starting 3 months after conclusion of therapy. Following a chest x-ray and an intravenous pyelogram, the patient is examined under anesthesia and multiple transrectal FNAs are done. If the first evaluation is negative, repeat FNAs are performed at 6-month intervals. Only patients with negative exfoliative cytology of the vagina and cervix are evaluated this way, since positive exfoliative cytology, followed by diagnostic biopsy, renders FNA unnecessary. If the biopsy is not diagnostic, FNA cytology is used and frequently has proved superior in revealing the existence of tumor growth. Of 120 patients with negative exfoliative cytology, 78 were negative by aspiration cytology while 42 had positive FNAs (Table 10-2). Thirteen patients initially had negative aspirations, but were later found to have tumor (clinical false-

Table 10-1. Indications for Fine Needle Aspiration Diagnosis of Treated and Untreated Patients

CLINICAL DIAGNOSIS	FINE NEEDLE ASPIRATIONS				
	Total	Untreated	Treated with		
			Surgery	Radiotherapy	Surgery/Chemotherapy
Cancer of the cervix	91	4	10	73	4
Cancer of the vulva/vagina	10	3	4	2	1
Cancer of the endometrium	29	5	5	18	1
Cancer of the ovary	33	11	10		12
Other: R/O malignancy	37	34	3		
Total	200	57	32	93	18

negative). Of these, 9 patients had recurrent malignancies outside the pelvis (*e.g.*, in abdomen, and in para-aortic and supraclavicular lymph nodes). Two of the four patients with pelvic recurrence had positive cytology on repeat needle aspirations (see the following case report).

■ *Case 10-9*

A 67-year-old woman was treated with radiation therapy for Stage IIB squamous cell cancer of the cervix. In the year after therapy, she had two negative pelvic FNAs. The second one showed a few atypical squamous cells. A third FNA 3 months later revealed the presence of necrotic cancer. On exploration, early recurrent squamous cell cancer was found, confined to the uterus. A simple hysterectomy was performed, and the patient is free of disease 18 months later.

Cytologic diagnosis of recurrent pelvic malignancies can be difficult because cytologic changes due to radiation therapy may mimic malignancy. Reactive changes of mesothelium and transitional epithelium of the bladder can resemble cells derived from squamous cell carcinoma. Also, radiated mature squamous epithelium of the vagina can present diagnostic difficulties. To avoid passing the needle through vaginal epithelium, we prefer the transrectal approach, which also allows better sampling of the entire small pelvis, especially in patients with extensive radiation fibrosis or a stenotic vagina. Obstruction of the needle with dense fibrous tissue may also be a technical problem leading to false-negative results. Emptying needle barrel between repeated passages may prevent this problem.

During and after chemotherapy, FNAs have been very helpful in planning the patient's treatment, and we are currently evaluating the use of flow cytometry for monitoring chemotherapy response.

Historically, the final diagnosis of malignancies is based on surgical biopsy material that defines precise histologic subclassification. For the clinician as well as the pathologist to accept FNA cytology as an equally good basis for diagnosis and eventually treatment, the high accuracy of this method must be proved. Not only must the differentiation between benign and malignant conditions be secured, but specific subclassification of tumors with precise histologic terminology should be attempted. With a modified P stain we have found a 94.5% accuracy of FNA when benign and malignant aspirates were compared with their histologic counterparts. Aspirations of malignant tumors showed a specificity of 97.9% when compared with histology. The overall reliability in differentiating benign and malignant conditions was 96.4%. The most common cause for false-

Table 10-2. Fine Needle Aspiration Following Radiation Therapy

FNA RESULT	FOLLOW-UP

Positive: 42	Positive: 42
	  Surgical: 21
	  Clinical: 21

Negative: 78	Negative: 65
	  Surgical: 10
	  One FNA: 32
	  Two or more FNAs: 23
	Positive: 13
	  Extrapelvic: 9
	  Intrapelvic: 4

negative FNAs of the pelvis in our patients was a sampling problem; that is, the aspirating needle did not reach the target lesion.[20]

In our experience with more than 600 gynecologic aspirations, the most important factors contributing to the reliability and accuracy of FNA cytology are the proficiency and enthusiasm of the clinician performing the technique and the experience of the pathologist evaluating the results. As clinicians become more experienced with this technique and pathologists gain confidence in correct evaluation of samples, FNA cytology of gynecologic lesions may prove to be one of the most valuable and acceptable tools in diagnostic gynecology.

REFERENCES

1. Ångström T: Fine needle aspiration biopsy in diagnosis and classification of ovarian tumors. In DeWatteville H (ed): Diagnosis and Treatment of Ovarian Neoplastic Alterations, pp 67–76. Basel, Excerpta Medica, 1975
2. Ångström T, Kjellgren O, Bergman F: The cytologic diagnosis of ovarian tumors by means of aspiration biopsy. Acta Cytol 16:336, 1972
3. Belinson JL, Lynn LM, Papillo JL et al: Fine needle aspiration cytology in the management of gynecologic cancer. Am J Obstet Gynecol 139:148, 1981
4. Flint A, Terhart K, Murad T et al: Confirmation of metastases by fine needle aspiration biopsy in patients with gynecologic malignancies. Gynecol Oncol 14:382, 1982
5. Ganjei P, Nadji M: Aspiration cytology of ovarian neoplasms. A review. Acta Cytol 28:329, 1984
6. Geier G, Kraus H, Schuhmann R: Die Punktionszytologie in der Diagnostik von Ovarialtumoren. Geburtshilfe Frauenheilkd 35:48, 1975
7. Göthlin JH: Post-lymphangiographic percutaneous fine needle biopsy of lymph nodes guided by fluoroscopy. Radiology 120:205, 1976
8. Helkamp BF, Sevin BU, Greening SE et al: Fine needle aspiration cytology in gynecologic malignancies. Gynecol Oncol 11:89, 1981
9. Jensen HK, Gram NC, Francis D: Finnalspunctur som diagnostik hjaelpemiddel ved gynaekologiske tumoren. Ugeskr Laeg 136:586, 1974
10. Kjellgren O, Ångström T: Transvaginal and transrectal aspiration biopsy in diagnosis and classification of ovarian tumors. In Wied GL (ed): Monographs in Clinical Cytology, vol 7, pp 80–103. Basel, S Karger, 1979
11. Kjellgren O, Ångström T, Bergman F et al: Fine needle aspiration biopsy in diagnosis and classification of ovarian carcinoma. Cancer 28:967, 1971
12. Kovačič J, Rainer S, Levičnik A et al: Cytology of benign ovarian lesions in connection with laparoscopy. In Wied GL (ed): Monographs in Clinical Cytology, vol 7, pp 57–79, Basel, S Karger, 1979
13. Mintz M, Dupré–Froment J, DeBrux J: Ponctions de 94 kysts para-uterins sous célioscopic et étude cytologique des liquides. Gynaecologia 163:61, 1976
14. Moriarty AT, Glant MD, Stehman FB: The role of fine needle aspiration cytology in the management of gynecologic malignancies. Acta Cytol 30:59, 1986
15. Nadji M: The potential value of immunoperoxidase techniques in diagnostic cytology. Acta Cytol 24:442, 1980
16. Nadji M: Aspiration cytology in the diagnosis and assessment of ovarian neoplasms. In Roth LM, Czernobilsky B (eds): Tumors and Tumor-Like Conditions of the Ovary, pp 153–173. New York, Churchill Livingston, 1985
17. Nadji M, Greening SE, Sevin BU et al: Fine needle aspiration cytology in gynecologic oncology. II. Morphologic aspects. Acta Cytol 23:380, 1979
18. Ramzy I, Delaney M: Fine needle aspiration of ovarian masses: I. Correlative cytologic and histologic study of celomic epithelial neoplasms. Acta Cytol 23:97, 1979
19. Ramzy I, Delaney M, Rose P: Fine needle aspiration of ovarian masses: II. Correlative cytologic and histologic study of non-neoplastic cysts and noncelomic epithelial neoplasms. Acta Cytol 23:185, 1979
20. Sevin BU, Greening SE, Nadji M et al: Fine needle aspiration cytology in gynecologic oncology. I. Clinical aspects. Acta Cytol 23:277, 1979
21. Sevin BU, Nadji M, Greening SE et al: Fine needle aspiration cytology in gynecologic-oncology. Early detection of occult persistent or recurrent cancer after radiation therapy. Gynecol Oncol 9:351, 1980
22. Zajicek J: Aspiration biopsy cytology. I. Cytology of supradiaphragmatic organs. In Wied GL (ed): Monographs in Clinical Cytology. Basel, S Karger, 1974

CHAPTER 11

Aspiration Biopsy Cytology of the Prostate Gland

JOSEPH A. LINSK

SIXTEN FRANZEN

The adult prostate gland is a firm, solid, indented ovoid mass. It is composed of a variety of tubular glands with infoldings and dilatations imbedded in a fibromuscular stroma. These structural characteristics result in vulnerability to acute and chronic infection and to the formation of concretions that produce palpable clinical findings. Glandular and stromal hyperplasia, both diffuse and nodular, occur with regularity under hormonal influence.[23] The gland may also exhibit normal irregularities of shape that simulate pathologic changes. Both these benign pathologic conditions, as well as normal alterations, produce palpatory findings that become targets for fine needle aspiration (FNA).

Carcinoma produces solitary nodules, local induration, and nodular and diffuse enlargements, all resulting in abnormal palpable clinical findings that are targets for aspiration.

The schematic anatomic division of the prostate into posterior, lateral median, and anterior lobes[38] is based on prior anatomic studies.[37] The posterior (dorsal) lobe presses against the rectum and is therefore easily palpable and accessible to needle biopsy. It is the site of most prostate cancers and because it remains *in situ* after suprapubic prostatectomy it continues to be a cancer site and a target for aspiration.

Diagnosis and Biopsy

Transrectal digital palpation of the posterior "lobe" and formulation of a clinical impression are the initial diagnostic manuevers.[11,15] The gland will be deemed normal, pathologic with benign changes, suspicious for malignancy, or clinically malignant.

In several studies in which palpatory findings, including stony induration, were reported as being suspicious or malignant, the diagnosis was correct in about 50% of cases.[11,26,40] Emmet has summarized about 2800 cases in which pre-operative clinical evaluation indicated benign disease. From 4.6% to 29.4% were found to be carcinoma.[11] In a more recent study, digital rectal examination was found to have a higher degree of accuracy.[22] By carefully grouping palpatory findings, Esposti also increased the accuracy of digital examination.[15] He divided his cases as follows:

Group I   Clinically benign
Group II   Cancer, slightly probable
Group III   Cancer, probable
Group IV   Clinically cancer

Cytomorphologically, he determined the percentages of cancer to be 62.4% and 89.0% in Groups III and IV, respectively. In Group I, only 1 of 175 tumors was cytologically cancer.[15]

Methods of obtaining tissue or cells for morphologic diagnosis have included open perineal biopsy, transperineal and transrectal routes, and transurethral resection.[2,10-12,25,29,35,40] Biopsy of metastatic sites, particularly bone marrow, combined with acid phosphatase studies may also yield positive results.[7,35,49] Exfoliative cytology expressed by massage of a clinically suspicious gland may provide diagnostic cells, but this procedure has been considered to be of limited value.[30]

Aspiration biopsy of the prostate was first reported by Ferguson in 1930. He used a sterile Record syringe with a 15-cm 18-gauge needle and transperineal approach. He obtained a demonstrable tissue sample in 70% of patients. The material was evaluated cytologically. This method did not receive general acceptance.[20] It was not until 1960 that a flexible transrectal fine needle method was introduced by Franzen and colleagues.[21] Since then, many reports of its use have appeared and thousands of biopsies have been done.[1,4,12,14,15,21,24,31,45]

The ease and safety of the transrectal fine needle method (see below), combined with the difficulties in clinically predicting carcinoma in many hands (see above), suggest that biopsy should be performed without hesitancy on minimal clinical evidence. In fact, performance of a routine biopsy on men over 50 has been suggested because of the high prevalence of carcinoma (14%) that escapes clinical recognition.[49] Although the yield of cancer will increase with better patient selection, a more general acceptance of the method will lead to increasing biopsies and greater total numbers of carcinomas, even though the percentage of malignant cases may fall. Nonmalignant pathologic findings, including varieties of prostatitis, provide additional useful clinical data. Finally, clinically normal glands may be punctured in search of a primary source for adenocarcinomatous metastases of unknown origin, although the yield is low.[41] At present, the diagnosis of carcinoma by aspiration biopsy cytology (ABC) has been accepted as definitive despite the absence of histologic confirmation. Many studies correlating cytologic diagnosis with biopsy histology, autopsy histology, and clinical course have confirmed the accuracy of the method.[1,10,12,13,15,21,24,31,35]

Over 22,000 aspirations with detailed follow-up study have been carried out at Karolinska Hospital since the original 100 patients were reported on in 1967.[21] Over 700 transrectal aspiration biopsies are carried out at this institution each year.

Instruments and Procedure

The procedure presented here in detail is based on the instruments devised by Franzen. The essential apparatus is a long, fine needle (20 cm), composed of a 22-gauge Luer lock needle to which a 22-gauge needle extension has been attached. It is flexible and strong but can be easily plugged. Patency must always be checked. In addition, there is a curved needle guide with a flared proximal end allowing easy entrance of the needle. Its length is equal to the length of the needle extension. A ringlike holder is attached to the distal end of the guide for insertion of the index finger, and a mobile flat plate is attached to its midportion to steady it against the palm (Fig. 11-1).

The syringe can be the one-handed type originally designed by Franzen or any one-handed holder into which a 10-ml or 20-ml disposable plastic syringe can be attached (see Fig. 10-1). Figure 11-2 illustrates the apparatus *in situ*. It is desirable but not essential that the patient empty the rectum. The procedure is carried out as follows and is depicted in Figure 11-3:

1. Wear a plastic glove on the guide hand.
2. Insert the index finger into the ring and rest the flat plate of the guide against the palm.
3. Roll a finger cot over the guide to hold it to the finger and protect the guide from being plugged with feces.
4. Use adequate lubricant.
5. Insert the finger and guide into rectum. Gentleness is essential.
6. Palpate the nodule and then extend the palpating finger beyond the target, since the needle will emerge 0.3 cm to 1 cm below the fingertip (Fig. 11-4).
7. Take the syringe, with needle attached, and insert the needle into the guide.
8. Thrust the needle through the layer of rubber into the target, judging the depth of thrust from experience and from previous observation of the extension of the needle point beyond the end of the guide (Fig. 11-2).
9. Apply full suction and, at the same time, move the needle tip through the target several times. Rotating the needle may help dislodge cells.
10. Gently release suction to equalize pressure and pull out the needle and guide.
11. Quickly detach the needle, draw in air, re-attach, and press the material onto the previously prepared slides.

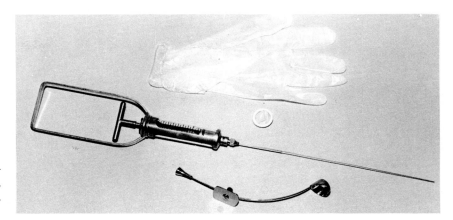

Fig. 11-1. Apparatus for transrectal biopsy includes needle guide, one-hand syringe and long needle, and plastic glove and finger cot.

12. Gently press out smears as previously described and air-dry them (Romanowsky [R]) or fix them with 95% alcohol (Papanicolaou [P]).

Positioning the Patient

The lithotomy position with the patient's legs in stirrups has been used most often. Puncture is then directed to the anterior rectal wall (Fig. 11-5). Figure 11-6 depicts puncture of the prostate schematically.

Alternate positions are knee–chest at the bedside or standing and bending forward as for a routine rectal examination in the office or clinic. The latter has the advantage of rapidity and convenience, since the patient does not have to entirely disrobe and mount a table. The prostate is more prominent in the flexed position, but control is better maintained in the lithotomy position.

Clinical Presentation and Palpation

Careful palpation around the lateral edges and proximal poles of the lobes (proximal closest to the anus) must be carried out to delineate small circumscribed nodules. Large hard masses or diffuse induration are easily palpated. The presence of a furrow without localized nodules is evidence for a benign lesion.

The large, rounded, rubbery gland of benign prostate hypertrophy (BPH) without localized induration is fairly characteristic. It indicates glandular hyperplasia in the anterior and lateral lobes that is pressing back against the posterior lobe. Where the hyperplasia is mainly stromal, there may be diffuse or nodular induration that becomes a target for puncture.

Acute inflammation produces tenderness and should dictate caution. It is not a suitable target for aspiration because of the hazards (see below) and

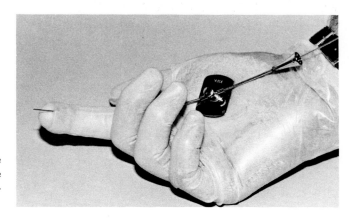

Fig. 11-2. Apparatus for transrectal biopsy. Note the ring at the end of the needle guide. It goes over the plastic glove and is covered by the finger cot. The needle emerges through the finger cot.

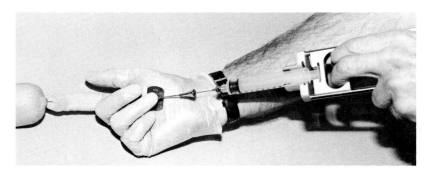

Fig. 11-3. Apparatus for transrectal biopsy. Note finger palpating nodule and needle emerging from guide to pierce nodule.

because the diagnosis is evident clinically. Where suspicious induration is also palpated, however, the acute inflammation should be treated before aspiration.

Chronic inflammation may produce localized or diffuse induration that becomes a target for puncture. Granulomatous inflammation may also produce localized induration. This disorder was clinically confused with cancer in 69% of a series of cases on initial examination.[43]

In clinical practice the cytologist will not normally see the patient until he has been evaluated by the clinician, often a urologist. The clinician has usually decided, based on clinical findings, particularly palpation, that a biopsy is indicated.

Where slides are submitted for reading, clinical data supplied by the clinician are helpful (see Chap. 1). More important, however, is the assurance that the aspiration and the smears were technically satisfactory. Fixation artefacts, particularly important in the small volume aspirates from prostate, are discussed in Chapter 1. Technical problems can be improved by cooperation between cytologist and clinician.

Clinical selection will determine the yield of carcinoma. At the Karolinska Hospital, approximately 30% of patients are diagnosed as having carcinoma.[15] In contrast, Linsk, who receives all his referrals from three urology sources found that well over 60% of the patients had cancer. It is evident that where the method is freely used and the volume of aspirations is high, a larger number of carcinomas will be found although the percentage will be smaller.

THE PROSTATIC SMEAR— GENERAL CONSIDERATIONS

Transrectal FNA of the prostate is clearly an adjuvant clinical method designed to directly diagnose or rule out cancer to allow the urologist to plan therapy. After early reluctance, many urologists have now become dependent on the method. Urologists are requesting some cytopathologists to demonstrate the technique. Initial smears produced by beginning aspirators are extraordinarily poor in our experience. With noteworthy exceptions, the best smears are produced by cytopathologists who recognize the importance of adequate yield, proper smears, rapid fixation, and repeated aspirations after immediate reading.

The prostatic smear, even when prepared from adequate material, contains many surprises as well as disappointments. Classical diagnostic smears are the exception. The smear may be dominated by amorphous debris. Cells are often traumatized, artifactually enlarged, and may simulate cancer cells. Cells exhibit varying degrees of degeneration and lose definition. There may be osmotic distortion when urine

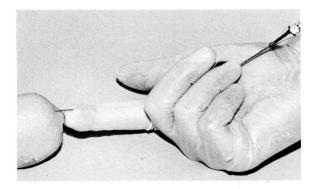

Fig. 11-4. Apparatus for transrectal biopsy. Note finger on nodule and needle emerging above and missing the target.

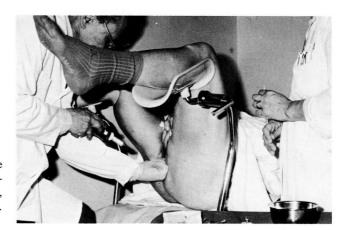

Fig. 11-5. Transrectal aspiration of the prostate. The patient's legs are in firm stirrups, and he holds his genitalia. The operator has excellent control. (From Linsk JA, Franzen S: Fine Needle Aspiration for the Clinician, p. 223. Philadelphia, JB Lippincott, 1986)

or cystic fluid is extracted with cells. Bloody aspirates will contain distorted cells trapped in fibrin clots. Cellularity may be extremely sparse in the face of clinically unequivocal cancer. Although sparse cellularity is generally considered nondiagnostic, a few convincing cancer cells coupled with the physical findings and clinical data favoring cancer may be quite enough for a cancer diagnosis (Fig. 11-7) (see False-Positive below).

A benign smear in the presence of clinical cancer (palpation) must be repeated. Smears with atypical cells extracted from clinically benign glands

may be considered benign with recommendation for follow-up.

The cytopathologist and cytotechnologist must be prepared to review poor smears, discuss findings with the clinician, and insist on repeat preparations when necessary. Every effort should be made to obtain both fixed and air-dried smears for P and R staining.

With all the above pitfalls and difficulties in mind, the procedure will nevertheless yield abundant material with easily identified cancer in the majority of tumor cases.

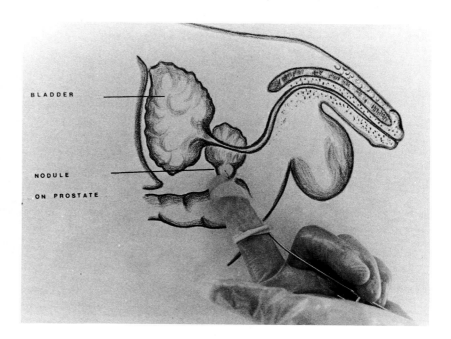

Fig. 11-6. Schematic drawing of aspiration. Coronal section.

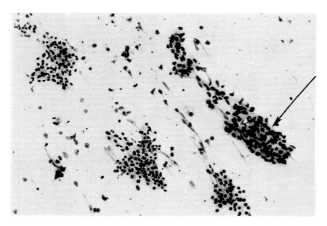

Fig. 11-7. Prostate smear with groups of benign epithelium and rare malignant group *(arrow)* sufficient for diagnosis of carcinoma. (R stain, ×190)

BENIGN LESIONS

Acute Inflammation

In acute inflammation the yield consists of viscid, sometimes voluminous fluid. The smears contain many granulocytes. There may be clusters of degenerated epithelial cells (Fig. 11-8). Histiocytes with phagocytic activity and a background consisting of cellular debris may be evident (Fig. 11-8). Granulocytes may be poorly preserved.

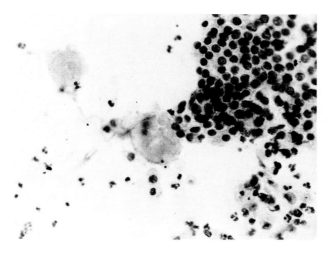

Fig. 11-8. Prostatitis, acute. Large clusters of benign epithelium, somewhat faded out, are seen within a background of fluid and polymorphonuclear leukocytes, histiocytes, and amorphous material. (P stain, ×190)

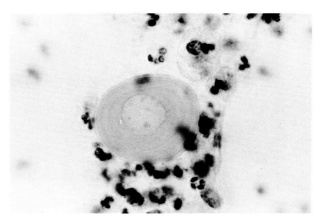

Fig. 11-9. Prostatitis, chronic. Note corpora amylacia surrounded by histiocytic and polymorphonuclear leukocytes. (P stain, ×600)

Chronic Inflammation

The material from chronic inflammation consists of a drop or droplets of bloody or cloudy fluid. Smears contain groups of benign epithelial cells, sometimes faded and degenerated. Laminated bodies (corpora amylacia) and a few inflammatory cells may be seen (Fig. 11-9). The background may stain pink (with R) with granular debris. A recent study has demon-

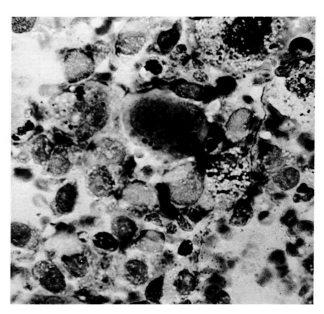

Fig. 11-10. Prostatis, granulomatous. There are numerous histiocytes with vacuoles and phagocytic material. The smear is crowded and "dirty." (R stain, ×300)

strated that the specific diagnosis of chronic prostatitis requires the presence of epithelial atypias in addition to inflammatory findings.[33]

Granulomatous Prostatitis

Numerous acute and chronic inflammatory cells are found in granulomatous prostatitis, including neutrophils, eosinophils, and histiocytes, some with phagocytosed material and multiple vacuoles (Fig. 11-10). Normal prostate epithelium may be absent, or appear as scattered islands of benign, sometimes atypical epithelium scattered through the inflammatory exudate.

Six cases of granulomatous prostatitis induced by Bacille-Calmette-Guerin (BCG) immunotherapy of bladder cancer occurred 5 to 17 months after treatment.[42] Four were diagnosed by FNA; in that group, metaplastic cells containing dyskaryotic nuclei of unequal size and shape, irregular borders, and clumped uneven chromatin were found, suggestive of malignancy.

Epithelioid cells are present singly or in small or large groups (Fig. 11-11). They are, in general, elongated and have a fairly dense, royal blue cytoplasm

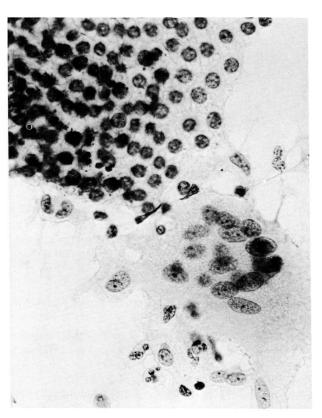

Fig. 11-12. Prostatitis, granulomatous. A multinucleated giant cell is present adjacent to a cluster of benign epithelium. The nuclei in the giant cell are elongated and epithelioid. (P stain, ×300)

with R stain. Haphazard groups of cells with elongated, curved semilunar nuclei are characteristic of granulomatous prostatitis. Some of the epithelioid cells blend with the histiocytes in a series of transformations. There are scattered multinucleated giant cells, some with 20 to 30 nuclei (Figs. 11-11 and 11-12). Nuclei are elongated and similar to those of epithelioid cells. There is a distinct impression of a destructive process (see Fig. 11-10). Multinucleated giant cells occur also in malignant smears (Figs. 11-13 and 11-14) and must be evaluated in light of the surrounding cellular pattern.

Benign Prostatic Hypertrophy

Benign prostatic hypertrophy will yield abundant cells of the glandular variety, in contrast with stromal hyperplasia in which cells are scarce (Fig. 11-15).

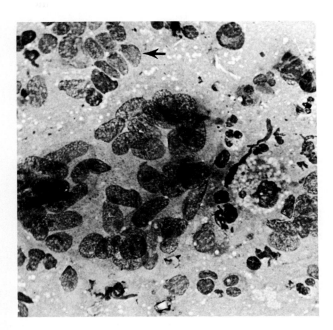

Fig. 11-11. Prostatitis, granulomatous. A group of epithelial cells (arrow) is adjacent to a multinucleated giant cell. Note the similarity of nuclei. (R stain, ×300)

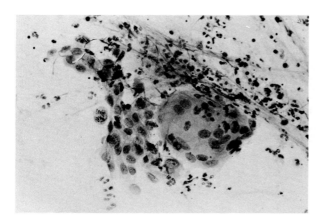

Fig. 11-13. Moderately differentiated carcinoma, prostate, together with acute inflammatory and granulomatous changes. (R stain, ×800)

Stromal hyperplasia may produce a firm nodular gland simulating cancer.

On high power, there is a distinctively orderly arrangement of round to slightly ovoid well-defined nuclei, giving a honeycomb appearance to benign epithelium (Fig. 11-16 and Color Plate XXII). With minimal overlapping, cells appear in irregularly shaped sheets (Fig. 11-17) of varying size. Cytoplasmic borders are ordinarily well defined with P and less well perceived with R stain. Where benign cells are layered clusters, monolayers of the same cells are often nearby (Fig. 11-18).

Cytoplasm is gray and fluffy with R and yellow-tan with P stain. Granules, black, purple, and

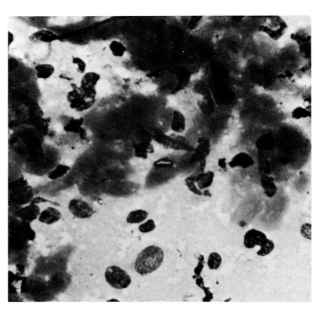

Fig. 11-15. Benign cell yield, stromal hyperplasia. This is an uncommon finding, since stromal cells are not easily dislodged. There are randomly distributed and distorted cells with elongated nuclei, poorly defined cytoplasm, and collagen masses. (R stain, ×330)

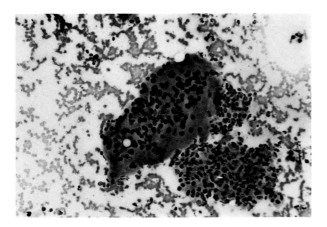

Fig. 11-14. Giant cell with innumerable nuclei adjacent to moderately differentiated carcinoma. (R stain, ×600)

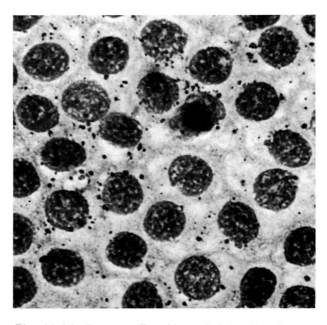

Fig. 11-16. Benign cell yield, epithelial cells. This is a monolayer of uniform cells with round central nuclei and visible cell borders. Large red to black granules overlay the cytoplasm and nuclei. (R stain, ×1000)

magenta and varying in size, may be present with R (see Fig. 11-16). This is an important differentiating point since granules are rarely found in carcinoma cells. They are not unknown, however, since Söderström has noted them in metastatic sites in well-differentiated carcinomas.[41] Cytoplasm may be markedly vacuolated, giving the appearance of a well-differentiated renal tumor (see Fig. 8-67).

Individual cells, often stripped or degenerated, may be present in the background and should not be confused with poorly differentiated cancer cells (see below). The nuclear chromatin pattern is fine and ill defined. Nucleoli are rarely or faintly seen. Sheets of benign cells may be accompanied by prostatic debris, probably inspissated secretions within glandular structures.

Abundant clear fluid may indicate aspiration of a cyst or penetration deep enough to aspirate urine or even fluid from an indwelling catheter balloon.

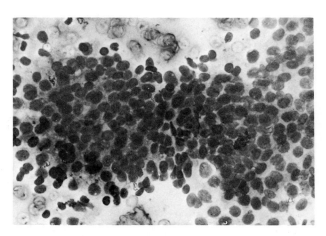

Fig. 11-18. Benign cell yield. Epithelial cells. This is an aggregated cluster of benign cells with dispersed cells at the periphery. This smear is from the same specimen as Figure 11-17. (R stain, ×300)

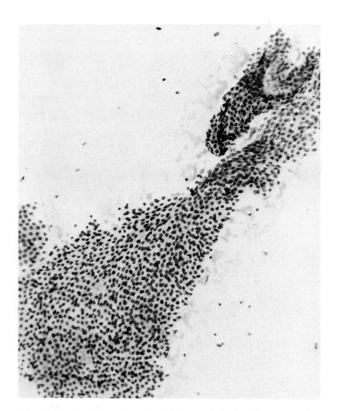

Fig. 11-17. Benign cell yield, epithelial cells. Irregular sheets of benign cells with some overlapping are seen in low power. (P stain, ×90)

Liquid expelled onto the slide lies in a pool that should be decanted with smearing of any visible particles. Superficial puncture due to tangential entrance bulging gland may yield mucus and rectal epithelium (see below).

MALIGNANT LESIONS

Prostate Carcinoma

The aspirate from prostate carcinoma will contain many cells if the needle is well placed. There may be blood. In the absence of blood, the smear presents a dense pattern of tumor cells (Fig. 11-19). A mixture of benign epithelium with scattered groups of cancer cells is not uncommon because of penetration by the needle into varying portions of the gland. Cancer is usually quite evident, and scanning is usually not necessary. In experienced hands, isolated tumor groups in a predominantly benign smear may be enough to define the diagnosis as malignant (Figs. 11-7 and 11-20). Obvious cancer cells may blend directly with a sheet of benign epithelium (see Fig. 11-19) or stand in direct contrast and juxtaposition with groups or sheets of benign cells. Familiarity with benign smears (see above) aids in recognition of the contrasting tumor cells. With direct puncture of a carcinomatous nodule or densely infiltrated gland, a thick pudding of cellular material is expelled on the

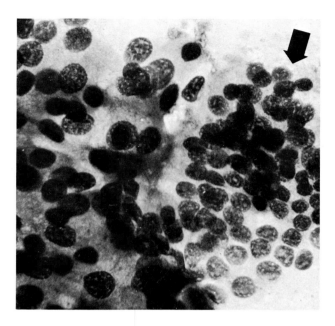

Fig. 11-19. Prostatic carcinoma, dense tumor pattern. This is a cluster of tumor cells with crowding and overlapping present throughout the smear. A contrasting benign cluster is indicated *(arrow)*. Nuclei are oval with granular chromatin. Nucleoli are not prominent with the stain. Cytoplasmic borders are indistinct. (R stain, ×400)

slide; the smear consists of solid, overlapping cancer cells in a distinctly disorderly arrangement. The cell groups tend to form interconnected strands and islands (Fig. 11-21 and Color Plate XXII). The surrounding space is peppered with many individual, isolated, stripped polymorphic nuclei with prominent nucleoli with P that are less evident with R stain. When there is marked stromal or desmoplastic reaction, aspiration may yield too few cells for diagnosis in an obvious clinical carcinoma.

Cancer cells may be grossly crowded together; and, under high power, careful examination at the edge of a dense cluster will reveal the cytologic changes of cancer (Fig. 11-22). Stripped nuclei must be considered cautiously, however, and only in the context of the whole smear (Fig. 11-23). Distorted benign nuclei may simulate cancer cells.

The diagnosis of carcinoma emerges with recognition of cytologic variations from normal epithelium. Where contrasting groups of cells are seen (Figs.

11-19 and 11-20), this is simplified. Nuclear atypia with a distinct irregular chromatin pattern; the appearance, prominence, and irregularity of nucleoli; an increasing nuclear:cytoplasmic ratio; and cell configuration and distribution will produce a diagnostic pattern. Mitoses and molding of cells may be useful criteria.[1,12] They have not been considered essential in our diagnostic studies, however. All cytologic findings must be collated in rendering a diagnosis of carcinoma. Adenomatous structures are considered below.

Prostatitis may be reflected in a "dirty" background consisting of cell debris, fluid, and bacteria upon which both benign and malignant groups may be seen (Fig. 11-24). In contrast, well-defined irregular clusters and strung-out islands of tumor cells, together with dissociated and sometimes stripped tumor nuclei, may occur on a perfectly clean background. A few red blood cells (RBCs) may be present.

The possibility of a small cell carcinoma of the prostate must be kept in mind.[5] It is an uncommon tumor and we do not have one in our archive, but

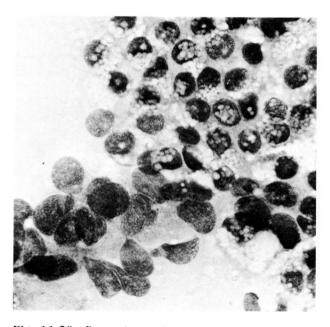

Fig. 11-20. Prostatic carcinoma, tumor group in predominantly benign smear. The cellular content was largely sheets of vacuolated benign cells. The tumor group was in sharp contrast and was diagnostic. Note large, irregular, finely granulated nuclei with cell overlapping. (R stain, ×330)

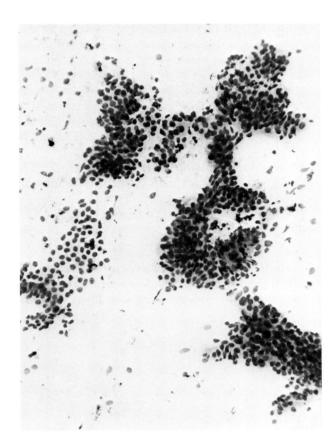

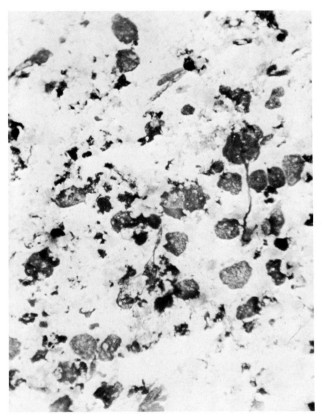

Fig. 11-21. Prostatic carcinoma, interconnected strands. This low-power view is representative of the entire smear. Such material is obtained in direct puncture of a solid carcinomatous nodule. (R stain, ×90)

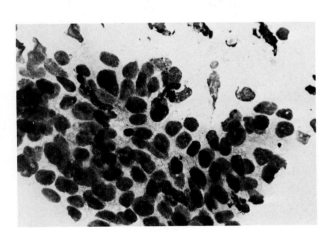

Fig. 11-22. Prostate carcinoma. Large clusters of densely crowded carcinoma cells are considered moderately differentiated. Careful examination of cells at the edge of the group may reveal cytologic changes of malignancy obscured in the main mass. (R stain, ×600)

Fig. 11-23. Prostatic carcinoma, artefactual smear. These are scattered, stripped, suspicious but nondiagnostic cells in a background of inflammation and debris. (P stain, ×400)

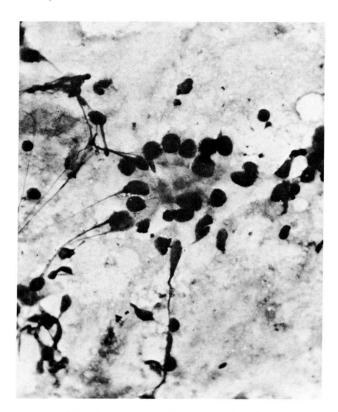

Fig. 11-24. Prostate carcinoma, well differentiated. This is an incomplete acinus on "dirty" background. Isolated smaller cells of benign origin are present for comparison. (P stain, ×300)

identification by FNA should be routinely accomplished. A discussion of small cell tumors appears in Chapter 8.

Malignancy Grading

Grading of prostatic carcinoma on histologic section correlates best with survival when the cytologic changes in section form the basis for the grading.[27] This parallels cytologic grading (see below). An attempt at histologic grading based on growth pattern was unsuccessful because of variation within the section.[44]

Cytologic grading of malignancy has great clinical significance, guiding the clinician in the choice of therapy.[15,16] Patterns have been established clearly by Esposti, although not all observers agree.[12,13]

Grade I: Well-Differentiated Cancer. The distinction of well-differentiated cancer cells from benign cells is demonstrated in Plate XXII A. Identification of grade I cancer may be simplified by the presence of numerous adenomatous structures on the smear (Fig. 11-24). Nuclei are enlarged and may be mildly pleomorphic. They form a complete or incomplete circle with the cytoplasm enclosed (Color Plate XXII B).

Nuclear chromatin is irregular, and nucleoli of varying size and shape may be seen. There may be scattered isolated cells (Fig. 11-25). Atypia in the absence of the adenomatous changes is considered insufficient alteration for the specific diagnosis of well-differentiated carcinoma.

Grade II: Moderately Differentiated Cancer. In grade II carcinoma there is some retention of the microadenomatous structure but with greater cellular

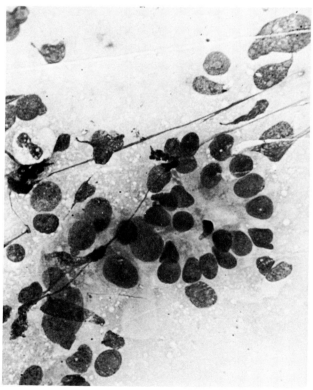

Fig. 11-25. Prostate carcinoma, well differentiated. Higher power view of isolated alveolar structure with cytoplasm occupying center of acinus. Individual cells are mildly atypical. (R stain, ×600)

atypia, larger nucleoli, and, in general, a greater deviation from benign morphology (Figs. 11-20 and 11-26). Cells, however, do appear in clusters and masses, with marked overlapping of nuclei containing one or more nucleoli (Fig. 11-27). Figure 11-19 provides further examples of moderately differentiated carcinoma.

When significant blood is obtained, the large tumor groups may appear as islands in a sea of blood (Fig. 11-28). Cytoplasm is poorly defined and usually contains no granules; however, scattered purple granules may be seen rarely in these tumor cells. On the gross stained smear, the large masses of cells may appear as though peppered on the glass and are seen as multiple groups on low power.

Cell groups may be tightly agglutinated, containing strung-out nuclei evident at the periphery.

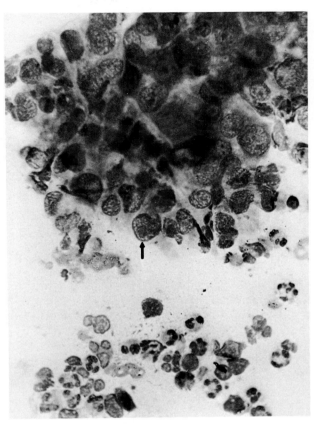

Fig. 11-27. Prostate carcinoma, moderately differentiated. This is a large dense cluster of obvious carcinoma cells. There is marked cellular variability. The chromatin is granular with evident enlarged and irregular nucleoli (*arrow*). There is a background of acute prostatitis. (R stain, ×600)

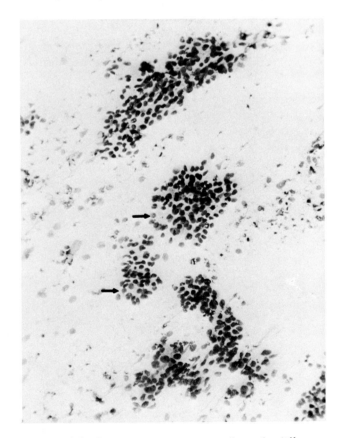

Fig. 11-26. Prostate carcinoma, moderately differentiated. This is a low-power view in which multiple groups are seen, some with residuals of the alveolar structure (*arrow*). (R stain, ×150)

There may be contrasting groups of tumor cells and small regular benign cells (Fig. 11-20). Isolated cells are often poorly preserved, and diagnosis should not be based on changes in these cells. (See Color Plate XXII C–E.)

Grade III: Poorly Differentiated Cancer (**Color Plate XXIII**). Cells are enlarged, irregular in size and shape, and dissociated in grade III cancer. Often only nuclei can be delineated. Cells are sometimes fragile and may be smeared. The overall picture is reminiscent of a large type of oat cell carcinoma or poorly differentiated histiocytic lymphoma (Figs. 11-29 to 11-31). Nucleoli are large, irregular, and

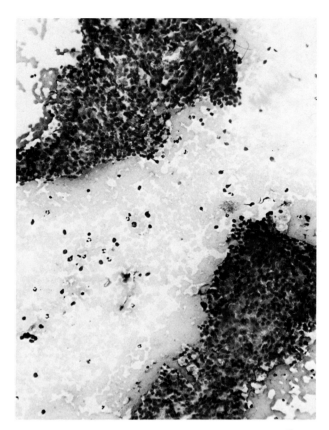

Fig. 11-28. Prostate carcinoma, moderately differentiated. Large tumor groups are sharply outlined in a background of red blood cells. (R stain, ×90)

multiple. They are pale or blue with R stain. Cells also appear in masses with overlapping and blend with the moderately differentiated variety. Individual cell cytology must be assessed. Cytoplasm, when seen, is irregular, blue-gray, and murky. Necrosis is often present. There may be retention of adenomatous structures, but these are poorly formed and composed of markedly polymorphous cells (see Fig. 11-29).

Cytologic grading of tumors is not an empty exercise. It has therapeutic and prognostic value. In a study by Faul and associates[19] 5-year follow-up of 496 patients demonstrated the prognostic validity of a grading system. One third of well-differentiated, 54.1% of moderately differentiated, 60.4% of low-differentiated, and 94.1% of anaplastic carcinomas died of the cancer. The figures are shown to be in

close agreement with those of Esposti[16] and show that prognosis is largely determined by the degree of cytologic malignancy.

DNA studies that define cytological grading by ploidy will more specifically separate tumors into prognostic and ultimately into therapeutic groups (see Chapter 17). While Faul and associates[19] questioned the use of androgen suppression in poorly differentiated tumors, Esposti indicates that 10% of the poorly differentiated cases might respond, making treatment worth trying (personal communication).

Table 11-1 presents a short guide for microscopy with the cytologic findings in benign and malignant disorders.

False-Positive Results

False-positive cytologic diagnoses (cases not confirmed by histologic examination when it is sought) have been reported.[4,10,31,34] Since a cytologic diagnosis of cancer does not require histologic confirma-

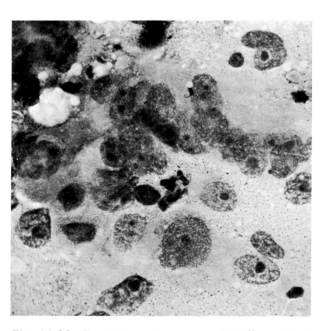

Fig. 11-29. Prostate carcinoma, poorly differentiated. Nuclei are irregular in size and shape. They have a delicate chromatin network with large prominent nucleoli. Cytoplasm is indistinct. The cells are fragile and easily smeared. Partial retention of an alveolar structure is evident. (P stain, ×400)

tion (see above), the determination of a false-positive result will depend on long-term clinical follow-up study with no sign of cancer and a careful autopsy. Of 210 patients diagnosed as positive for prostate carcinoma cytologically, 205 were confirmed histologically.[15] Of the five not confirmed (false-positive result), tissue was available from transurethral resection in four and from perineal needle biopsy in one. Because of the posterior location and focal distribution of most prostate carcinomas, the tumor could easily have been missed with such surgical biopsies.[38]

A false-positive cytologic diagnosis is however a critical error, which may lead to radical therapy including radical surgery and primary radiation. Because a cytologic diagnosis of "suspicious for cancer"

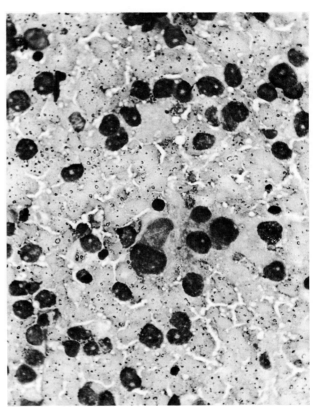

Fig. 11-31. Prostate carcinoma, poorly differentiated. A "dirty" background is more evident with this stain. There are dispersed polymorphic cells. (R stain, ×330)

is of limited use to the urologist, the cytopathologist is under increased pressure to categorize smears as cancer, benign, or unsatisfactory. This will tend to increase the number of false positives. In a series of 158 patients with histologic glandular hyperplasia, 12 were cytologically diagnosed cancer and 26 suspicious.[31] While such findings may occur during early experience with the method, even in experienced hands the possibility must not be discounted. The following caveats are recommended:

1. Familiarity with and use of both Giemsa and P staining are clearly indicated. There are numerous instances of failure to diagnose with one stain while the other is diagnostic.
2. With rare exceptions do not diagnose cancer on

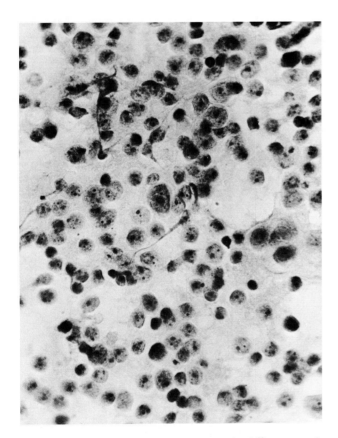

Fig. 11-30. Prostate carcinoma, poorly differentiated. Cells are dispersed with poorly defined cytoplasm suggestive of histiocytic lymphoma. Variability of size and nucleolar prominence are evident. (P stain, ×330)

Table 11-1. Aspiration Biopsy Cytology of Prostate Gland: A Short Guide for Microscopy*

DIAGNOSTIC GUIDELINES

	Benign Features	Malignant Features
Pattern	Monolayered Sheets	Multilayered Clusters, solid, microglandular
Cytoplasm	Preserved cell cohesiveness Cell borders well defined Red cytoplasmic granules	Cell dispersal Cell borders ill defined Vacuolization of cytoplasm (clear cells)
Nuclei	Well preserved, round or oval Normochromatic Nucleoli rare	Fragile, enlarged, polymorphic Hyperchromatic Nucleoli often prominent

MALIGNANCY GRADING OF PROSTATE CARCINOMA ON ASPIRATES

		Well	Moderately	Poorly
Pattern	Microglandular pattern	+++	++	+
Cytoplasm	Cell dispersal	+	++	+++
Nuclei	Nuclear enlargement	+	++	+++
	Polymorphy	+	++	+++
	Nucleoli	+	++	+++

ABC PROSTATITIS

Acute	Granulocytes
Chronic	Histiocytes, giant cells, lymphocytes, plasma cells Reactive epithelial changes (Look for granulomatous prostatitis)

ABC PERIPROSTATIC TISSUE

Rectum and anus	Columnar or squamous cells Mucus, bacteria, feces
Seminal vesicle	Conspicuous large cells with hyperchromatic nuclei and dark cytoplasmic granules Spermatozoa

* Contributed by Dr. T. Löwhagen

sparse samples. *Repeat samples must be requested.*

3. Communication with the clinician is important. In practical terms, cytologic aspiration to confirm prostatic carcinoma in a patient with bone metastases and an elevated acid phosphatase requires less rigor than the establishment of a primary diagnosis of prostate cancer by cytologic smears.

False-Negative Results

False-negative results occur in most reported cytologic series as well as in core histologic biopsy series.[1,15,29,31,34,40] They vary from 10% to 30%. In experienced hands, they are primarily a result of nonrepresentative sampling, particularly with small lesions.[15] They may also result from underestimation of atypical cellular findings.[8]

The sampling problem is reduced with increasing experience and by the ease of repeat aspiration at the same clinical encounter or on sequential follow-up examination, depending on the level of suspicion. "Missed" or underestimated carcinomas producing false-negative results are reduced by recognition of discrete atypia (Fig. 11-32), which may occur in the midst of an otherwise normal smear. Such changes should be considered suspicious for carcinoma. Repeat aspiration of clinically suspicious glands with negative initial reports will yield carcinoma in over 10% of patients in experienced hands.[14]

Since the operator carrying out the aspiration will examine and formulate his own diagnostic impression, he has the option of examining his smears after a rapid stain method and of re-aspirating. The final report, where negative for cancer, should suggest to the referring physician that further aspirations

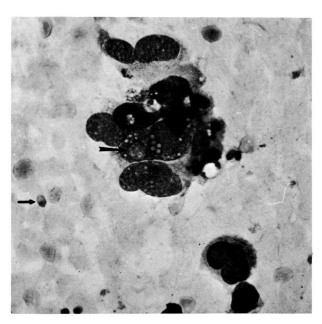

Fig. 11-33. Atypical prostate smear patterns, seminal vesicle. Cells are quite bizarre and not characteristic of poorly differentiated carcinoma. A large nucleolus (*arrow*) as well as isolated spermatozoa are noted. (R stain, ×600)

can be carried out if clinical suspicion of cancer persists. The results of core biopsy parallel those of FNA biopsy, and it is unusual to request core biopsy because of negative FNA.[1,12,31] Concomitant core biopsy and FNA has been suggested, but this appears unwarranted because of the ease and convenience of the fine needle and the resulting minimal additional diagnostic yield that simultaneous use would provide.[12]

Morphologic patterns that can be confused with carcinoma occur in the following disorders and situations.

Atypical Hyperplasia. Atypical hyperplasia has been characterized as a pitfall in the diagnosis of cancer, and indeed such a benign proliferation may be mistaken for poorly differentiated cancer and result in a false-positive diagnosis.[8,31] Criteria that identify carcinoma and must be sought include coarse, uneven chromatin; large irregular nucleoli; and irregular hyperchromatic nuclei with P stain (see Figs. 11-30 and 11-32).

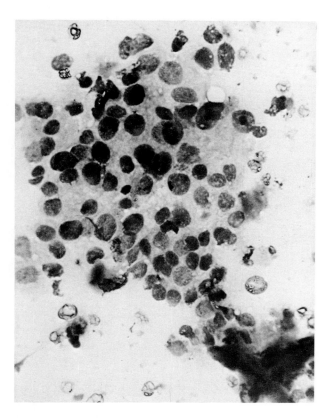

Fig. 11-32. Atypical prostatic smear, discrete atypias. Cluster of cells suggests evidence of cancer. Such groups must be considered in the context of the entire smear. Follow-up smears are indicated, combined with clinical data. (R stain, ×250)

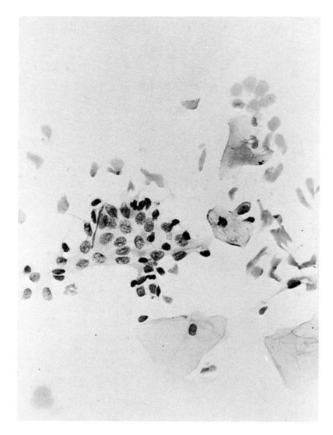

Fig. 11-34. Atypical prostate smear patterns, hormonal effect. There is a cluster of epithelial cells not recognizable as carcinoma. Squamoid cells are evident in juxtaposition. (P stain, ×190)

Granulomatous Prostatitis. Atypical epithelium suspicious for carcinoma has been reported in granulomatous prostatitis.

Seminal Vesicle Puncture. Atypical, sometimes bizarre cells are seen in seminal vesicle punctures (see below) and must be distinguished from cancer (Fig. 11-33).

Cytochemistry

Immunocytochemical staining for prostate-specific antigen (PSA) and prostate-specific acid phosphatase (PSAP) will identify prostatic cells both benign and malignant, regardless of tumor differentiation.[28]

While it may be useful to distinguish seminal vesicle, transitional, and rectal epithelium, it is of no use in separating benign from malignant prostatic epithelium in prostatic aspirations, since both will stain positively. Its major use is in identifying prostate cancer in metastatic sites[47] although a careful digital rectal examination followed by transrectal aspiration biopsy will disclose most primary prostatic cancers. Failure to carefully examine the prostate is the major cause of failure to identify metastatic tumors of that origin.

Hormonal Effect

Administration of female hormones may produce cellular alteration (Fig. 11-34 and Color Plate XXIV) in hormonally sensitive tumors (well-differentiated and moderately differentiated types).[45,48] Large glycogen or squamoid cells are often attached to a still intact tumor group (Fig. 11-35). Nuclei are small and

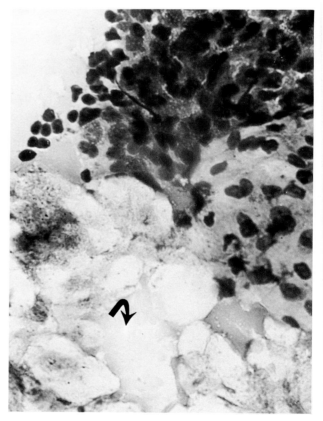

Fig. 11-35. Atypical prostate smear patterns, hormonal effect. This smear illustrates the evolution of a dense cluster of cancer cells to nucleated squamoid cells and finally to anucleated structures (*curved arrow*). (R stain, ×300)

compact, and the cytoplasm is clear with linear structures and occasional granules. Response may take several months and may be complete or incomplete on follow-up puncture. Failure to convert to glycogen cells is a sign that the tumor is not hormonally responsive.

Nonprostatic Epithelium

Transitional Cell Carcinoma

Malignant transitional epithelium is infrequently harvested by transrectal aspiration. These tumors arise from the prosthetic urethra and prostatic ducts and proliferate in the central portion of the prostate gland. Deep penetration of the needle will reach such tumors. Adenocarcinoma may occur simultaneously, and there may be a mixture of transitional and adeno-

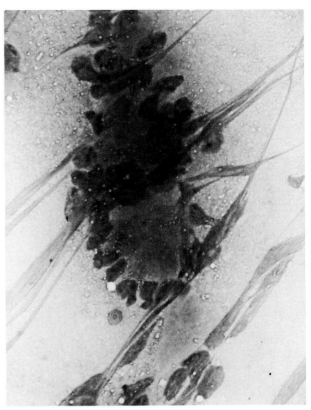

Fig. 11-37. Rectal mucosa, benign. Benign rectal mucosa intermixed with mucin may be extracted by a superficial aspiration. (R stain, ×132)

carcinoma cells on the smear. Transitional cells are recognized by the eccentric nucleus and well-defined cytoplasm, which may assume a pointed geometric pattern (Fig. 11-36).

Seminal Vesicle Epithelium

Inadvertent puncture of the seminal vesicle is an unlikely event although 50 examples were identified in 3300 prostatic aspirations.[32] If it does occur, it may yield large atypical cells that must be identified and distinguished from poorly differentiated prostatic cancer (see Color Plate XXIV). Transrectal prostatic aspirations may yield a variety of cytologic elements from the seminal vesicles. Spermatozoa and cells with large bizarre nuclei, prominent nucleoli, nuclear infolding and multinucleation, and cytoplasmic pigment granules should identify the seminal vesicle as the source of the material (see Fig. 11-33).

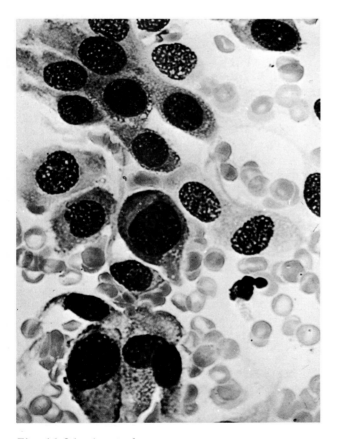

Fig. 11-36. Atypical prostatic smear patterns, transitional cells. This is a moderately differentiated transitional cell carcinoma. Nuclei tend to be eccentric, and cytoplasm may be elongated, producing a geometric pattern; other examples are seen in Chap. 9. (R stain, ×600)

Rectal Epithelium

In early experience with this method, rectal epithelium intermixed with mucin may be extracted (Fig. 11-37). If the prostate is quite enlarged and rounded, it may be difficult to direct a needle into the substance of the gland. The needle may be inserted superficially to extract rectal epithelium. Examples of colonic epithelium and debris are seen in Figure 11-37.

Nonepithelial Tumors

On rectal examination, an 8-cm mass in the region of the prostate gland was palpated in a 38-year-old man in a case presented by Cookingham and Kumar.[6] FNA revealed leiomyosarcoma. A schwannoma was diagnosed by FNA of a mass in the region of the prostate. This case was reported in *Fine Needle Aspiration for the Clinician*.[36]

Invasion of the prostate by bladder, rectal, and gastric carcinoma are cited by Willis.[46] The cytopathologist must be alert to all the above possibilities. Physical examination and clinical correlation are invaluable.

Hazards

Patients with soft, tender glands suggesting acute prostatitis should not be routinely punctured, since bacteremia and even septicemia may result.[18] Septicemia may also occur after routine puncture of patients without prostatitis.[9] Infection, however, both in our experience and in reported series is rare.[10,15,17,24] This contrasts with core biopsy in which infection is a distinct hazard and various programs of antibiotic prophylaxis and bowel preparation have been advocated.[3,39]

Postpuncture bleeding from the penis is also rare. When it does occur, it may signal the presence of large veins in the gland, which is useful information for the urologic surgeon. Such bleeding is always self-limited.

CLINICAL APPLICATION

Transrectal core biopsy failed to gain general acceptance for years because of fear of infection or fistula and, presumably, because of clinician resistance to change.[11] With the fine needle technique, fistula is nonexistent and infection is not a clinically significant event. This technique provides diagnostic material in inflammatory and in neoplastic diseases of the prostate with a high degree of specificity. The procedure is essentially atraumatic, and repeat punctures are readily accepted by patients. It is particularly useful as an outpatient screening procedure.

REFERENCES

1. Alfthan O, Klintrup HE, Koivuniemi A et al: Cytological aspiration biopsy and Vim–Silverman Biopsy in the diagnosis of prostatic carcinoma. Ann Chir Gynaecol 59:226, 1970
2. Andersson L, Jönsson G, Brunk U: Puncture biopsy of the prostate in the diagnosis of prostatic cancer. Scand J Urol Nephol 1:227, 1967
3. Babaian RJ, Lowry WL, Finan BF: Intraluminal antibiotic regimen for patients undergoing transrectal needle biopsy of prostate. Urology 20:253, 1982
4. Bishop D, Oliver JA: A study of transrectal aspiration biopsies of the prostate with particular regard to prognostic evaluation. J Urol 117:313, 1977
5. Bleichner JC, Chun B, Klappenbach RD: Pure small-cell carcinoma of the prostate with fatal liver metastasis. Arch Pathol Med 110:1041, 1986
6. Cookingham CL, Kumar NB: Diagnosis of prostatic leiomyosarcoma with FNA cytology. Acta Cytol 29:170, 1985
7. Crisp J: Random bone marrow biopsy in the staging of carcinoma of the prostate. Br J Urol 48:265, 1976
8. DeGaetani CF, Trentini GP: Atypical hyperplasia of the prostate: A pitfall in the cytologic diagnosis of carcinoma. Acta Cytol 22:483, 1978
9. Edson RS, Van Scoy RE, Leary FJ: Gram-negative bacteremia after transrectal needle biopsy of the prostate. Mayo Clin Proc 55:489, 1980
10. Ekman H, Hedberg K, Persson PS: Cytological versus histological examination of needle biopsy specimens in the diagnosis of prostatic cancer. Br J Urol 39:544, 1967
11. Emmet JL, Barber KW Jr, Jackman RJ: Transrectal biopsy to detect prostatic carcinoma. A review and report of 203 cases. J Urol 87:460, 1962
12. Epstein NA: Prostatic biopsy—A morphologic correlation of aspiration cytology with needle biopsy histology. Cancer 38:2078, 1976
13. Epstein NA: Prostatic carcinoma: Correlation of histologic features of prognostic value with cytomorphology. Cancer 38:2071, 1976

14. Esposti PL: Cytologic diagnosis of prostatic tumors with the aid of transrectal aspiration biopsy. A critical review of 1110 cases and a report of morphological and cytochemical studies. Acta Cytol 10:182, 1966
15. Esposti PL: Aspiration biopsy cytology in the diagnosis and management of prostatic carcinoma. Stockholm, Stähl and Accidens Tryck, 1974
16. Esposti PL: Cytologic malignancy grading of prostatic carcinoma by transrectal aspiration biopsy. Scand J Urol Nephrol 5:199, 1971
17. Esposti PL, Franzen S, Zajicek J: The aspiration biopsy smear. In Koss LG (ed): Diagnostic Cytology and its Histopathologic Bases, 2nd ed. Philadelphia, JB Lippincott, 1968
18. Esposti PL, Elman A, Norlen H: Complications of transrectal aspiration biopsy of the prostate. Scand J Urol Nephrol 9:208, 1975
19. Faul P, Schmidt E: Prognostic significance of cytological differentiation grading in estrogen-treated carcinoma diagnosed by fine-needle aspiration biopsy. Int Urol & Nephrol 12:347, 1980
20. Ferguson RS: Prostatic neoplasms: Their diagnosis by needle puncture and aspiration. Am J Surg 9:507, 1930
21. Franzen S, Giertz G, Zajicek J: Cytological diagnosis of prostatic tumours by transrectal aspiration biopsy: A preliminary report. Br J Urol 32:193, 1960
22. Guinan P, Bush I, Ray V, Vieth R, Rao R, Bhutti R: The accuracy of the rectal examination in the diagnosis of prostatic carcinoma. N Engl J Med 303:499, 1980
23. Ham AW, Cormack DH: Histology, 8th ed. JB Lippincott, Philadelphia, 1979
24. Hosking DH, Paraskevas M, Hellsten PR et al: The cytological diagnosis of prostate cancer by transrectal fine needle aspiration. J Urol 129:998, 1983
25. Hudson PB, Finkle AL, Jost HM et al: Prostatic cancer X: Comparison of open and punch biopsy techniques. Arch Surg 70:508, 1955
26. Jewett HJ: Significance of the palpable prostatic nodule. JAMA 160:838, 1956
27. Kahler JE: Carcinoma of the prostate gland: A pathologic study. J Urol 41:557, 1939
28. Katz RL, Raval P, Brooks TE et al: Role of immunocytochemistry in diagnosis of prostate neoplasia by fine needle aspiration biopsy. Diag Cytopathol 1:28, 1985
29. Kaufman JJ, Rosenthal M, Goodwin WE: Methods of diagnosis of carcinoma of the prostate: A comparison of clinical impression, prostatic smear, needle biopsy, open perineal biopsy and transurethral biopsy. J Urol 72:450, 1954
30. Kaufman JJ, Schultz JI: Needle biopsy of the prostate: A re-evaluation. J Urol 87:164, 1962
31. Kline TS, Kohler FP, Kelsey DM: Aspiration biopsy cytology (ABC): its use in diagnosis of lesions of the prostate gland. Arch Pathol Lab Med 106:132, 1982
32. Koivuniemi A, Tyrkkö J: Seminal vesicle epithelium in fine-needle aspiration biopsies of the prostate as a pitfall in the cytologic diagnosis of carcinoma. Acta Cytol 20:116, 1976
33. Leisteschneider W, Nagel R: The cytologic differentiation of prostatitis. Pathol Res Pract 165:429, 1979
34. Lin BPC, Davies WEL, Harmata PA: Prostatic aspiration cytology. Pathology 11:607, 1979
35. Linsk JA, Axilrod HD, Solyn R et al: Transrectal cytologic aspiration in the diagnosis of prostatic carcinoma. J Urol 108:455, 1972
36. Linsk JA, Franzen S: Fine Needle Aspiration for the Clinician. Philadelphia, JB Lippincott, 1986
37. Lowsley OS: The development of human prostate gland with reference to the development of other structures of the neck of the urinary bladder. Am J Anat 13:299, 1912
38. Mostofi FK, Price EB: Tumours of the Male Genital System. Washington, D.C., Armed Forces Institute of Pathology, 1973
39. Sharke KR, Sadlowski RW, Finney RP et al: Urinary tract infection after transrectal needle biopsy of the prostate. J Urol 127:225, 1982
40. Sika JV, Lindquist HD: Relationship of needle biopsy diagnosis of prostate to clinical signs of prostatic cancer. An evaluation of 300 cases. J Urol 69:737, 1963
41. Söderström N: Fine needle aspiration biopsy. Stockholm, Almqvist & Wiksell, 1966
42. Stilmant M, Siroky MB, Johnson KB: Fine needle of granulomatous prostatitis induced by BCG immunotherapy of bladder cancer. Acta Cytol 29:961, 1985
43. Taylor EW, Wheeler RF, Correa RJ Jr et al: Granulomatous prostatitis: Confusion clinically with carcinoma of the prostate. J Urol 117:316, 1977
44. Vickery AL, Kerr WS: Carcinoma of the prostate treated by radical prostatectomy. Cancer 16:1598, 1963
45. Williams JP, Still BM, Pugh RCB: The diagnosis of prostatic cancer. Cytological and biochemical studies using the Franzen biopsy needle. Br J Urol 39:549, 1967
46. Willis R: The Spread of Tumours in the Human Body. London, Butterworth & Co, 1973

47. Yam LT, Winkler CF, Janckila AJ et al: Prostatic cancer presenting a metastatic adenocarcinoma of undetermined origin—Immunodiagnosis by prostatic acid phosphatase. Cancer 51:283, 1983

48. Zajicek J: Aspiration biopsy cytology. II. Cytology of infradiaphragmatic organs. In Wied GL (ed): Monographs on Clinical Cytology. Basel, S Karger, 1979

49. Zinke H, Campbell JT, Utz DC, Farrow GM, Anderson MJ: Confidence in the negative transrectal needle biopsy. Surg Gynecol Obstet 136:78, 1973

CHAPTER 12

Aspiration Biopsy of the Testis

JOSEPH A. LINSK

SIXTEN FRANZEN

Lesions of the scrotal contents have traditionally presented unique problems to the clinician in that heavy reliance has most often been placed on clinical evaluation in determining the need for surgical exploration, ending as a rule in ablation of the testis.

We believe that presurgical fine needle aspiration (FNA) offers practical advantages. Gross palpable lesions can immediately be identified as malignant or nonmalignant. This guides the clinician in obtaining pre-operative cell markers (human chorionic gonadotropin and alpha-fetoprotein) and the surgeon in selecting his operative procedure. It is also important to identify the group of early lesions that may lack sufficient clinical criteria to suggest surgical intervention, and eliminating delay is a useful advantage.

With respect to biopsy, suspicious testicular lesions are considered untouchable before definitive ablative surgery.[12] No data have been offered to support this view. On the contrary, the series by Miller and Seljelid failed to demonstrate any statistical disadvantage to surgical biopsy before ablation.[11] In a review of 100 cases of FNA carried out on malignant tumors, no evidence of spread could be attributed to the method.[23] Franzen has carried out FNA of the testis for 25 years and has not observed any local seeding of tumors. The general experience in the cytology laboratory at Karolinska Hospital supports these findings as well as the diagnostic usefulness of the method. From 1976 to 1978 for example, 152 palpable lesions were presented for FNA. Of this group 86% were determined to be benign, and only 14% were tumor, reducing considerably the surgical requirements.

Experience with testicular cytology will aid identification of metastatic masses as well as primary midline germinal tumors. The cytology of inflammatory and infertility disorders also contributes useful clinical data.

The following list summarizes the advantages of FNA of the testis and epididymis:

1. Definitive diagnosis of a neoplasm confirming the need for immediate surgery, or the option to defer surgery (adenomatoid tumor)
2. Subclassification of neoplasms
3. Cytologic skills in identifying metastatic lesions and primary midline germinal tumors
4. Definitive diagnosis of an inflammatory lesion
5. Determination of spermatogenesis occurring in the presence of azoospermia (evidence of blocked excretory ducts, which may be correctible)

The development of testicular ultrasonography (see Chapter 2), now allows highly accurate diagnosis of intratesticular masses. For most urologists, this information is sufficient to mandate radical orchiectomy. FNA may then be reserved for all lesions not specifically clarified by this imaging method or where clinical data mandates pre-operative specific diagnosis (see Lymphoma below). This includes early subtle indurations (discussed below under Neoplasms).

The pathology of testicular and epididymal disease is varied and includes inflammatory and neoplastic lesions. FNA does not attempt to define cytologically all the subclassifications that emerge from pathologic studies. There are numerous classifications of tumors.[4,5,12,13,18] The classification that will be addressed in this chapter is presented in the list below. It is based on the clinical cytologic experience at Karolinska Hospital, and descriptions are rendered only of those lesions that have appeared in the files with sufficient frequency for thorough study. Although cytologic smears of malignant tumors are usually diagnostic of one tumor type, the presence of a second or third germinal component is not excluded.

Classification of Lesions Diagnosed by FNA
Cystic masses
    Hydrocele
    Spermatocele
    Hematocele
Inflammatory masses
    Acute epididimo-orchitis and orchitis
    Chronic epididimo-orchitis and orchitis
    Granulomatous orchitis
    Tuberculosis
Neoplasms of the testis
    Germinal
        Seminoma (usual)
        Seminoma (spermatocytic)
        Anaplastic
        Embryonal
        Teratocarcinoma
        Choriocarcinoma*
    Nongerminal
        Leydig cell tumor
        Lymphoma
Neoplasms of the epididymis
    Adenomatoid

Procedure

Puncture of the testis is carried out with the usual 22-gauge needle. We have not found local anesthesia to be necessary. Puncture must be preceded by meticulous palpation and localization of the lesion, which might be quite ill defined. Fixation of the testis and highlighting of the target are essential. Several areas of the target area should be entered, with aspiration carried out during oscillatory movement of the needle in the same channel during each puncture. This will provide a higher cell yield.

In fertility studies, occasionally entire seminiferous tubules are extracted and may protrude from the scrotal skin. They can be removed without ill effect and may be smeared. Very careful smearing is indicated because of the fragility of the cells.

CLINICAL DISORDERS

Fluid-Filled Sacs

Fluid in a spermatocele or hydrocele can be suspected clinically and can often be confirmed by transillumination. Spermatocele may present as a small tight sac simulating a tumor. Aspiration consists of cloudly fluid containing many spermatozoa.

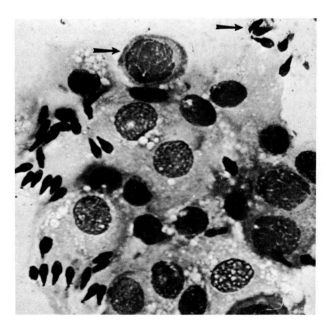

Fig. 12-1. Testis, developmental cytology. Spermatogonia *(arrow)* of varying size with developmental progression to spermatozoa *(arrow)* are seen. Nuclear chromatin is granuloreticular with perinuclear clear zone. Nucleoli are not readily seen with the stain. Such rounded, immature cells can be confused with primitive lymphoma cells. (Romanowsky [R] stain, ×800)

---

* Samples were not available.

Hydrocele yields clear fluid that can be discarded. Cloudy fluid will more likely suggest chronic inflammation than tumor and should be examined bacteriologically. Intrascrotal fluid may obscure a testicular tumor, and careful palpation must follow its removal.

Hematocele occurs secondary to trauma. The possibility of scrotal hernias in all clinical evaluations must be considered carefully before puncture. Puncture of a hernia is not attended by any serious consequences (see Chap. 9).

Infertility

Azoospermia may be due to testicular failure or blockage of the excretory ductal systems. Puncture of the testis will yield the full developmental cytology[14]

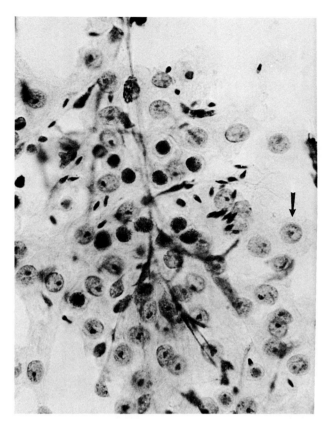

Fig. 12-3. Testis, Sertoli cells. This smear is dominated by Sertoli cells, which are pale and oblong with round, bland nuclei, each containing a single nucleolus. A few spermatogonia (dark cells) and spermatozoa are evident. (P stain, ×400)

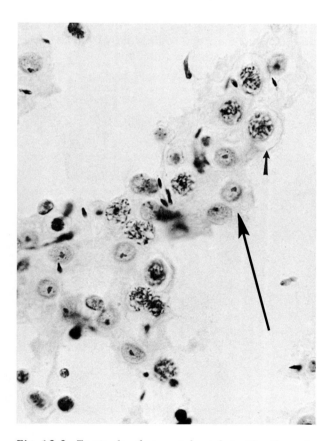

Fig. 12-2. Testis, developmental cytology. Needle aspirate. This is a mixed smear of spermatogonia *(small arrow)* and Sertoli cells *(long arrow)*. (Papanicolaou [P] stain, ×400)

from spermatogonia to spermatozoa in patients with obstruction (Fig. 12-1). In atrophy, congenital or acquired, as well as in childhood, Sertoli cells will dominate the picture (Figs. 12-2 and 12-3). Sertoli cells are uniform, with round to ellipsoidal vesicular nuclei and usually one large central nucleolus (see Figs. 12-2 and 12-3). Cytoplasm, where intact, is abundant, often vacuolated, and pale blue with Romanowsky (R) stain (Fig. 12-4). Cells are fragile, with frequent naked nuclei intermingled with spermatogonia, which dominate the smear in the adult (see Fig. 12-1). Spermatogonia are round cells with central nuclei and enlarged nucleoli. The nuclei have a finely granular or reticular texture that simulates primitive lymphoid cells. They are spherical or ovoid. Cyto-

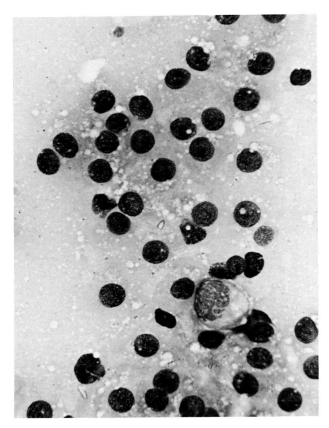

Fig. 12-4. Testis, Sertoli cells. Cytoplasm is pale blue and vacuolated with poorly defined cell boundaries. Nuclei are uniform and round with ill-defined nucleoli. A spermatogonium is evident. (R stain, ×800)

granulation (R stain). Rod shaped crystals of Reinke, staining red with P and pale gray with R stain, are occasionally seen (see Fig. 12-5). (See Color Plate XXVI J.)

Inflammation

Acute orchitis and acute epididimo-orchitis

Acute orchitis secondary to systemic disease such as mumps rarely becomes a target for aspiration, since the diagnosis is not in doubt. Acute epididimo-orchitis is ordinarily secondary to regional infection involving the urethra, bladder, and seminal vesicles. Where acute inflammation characterized by an enlarged, often exquisitely tender mass accompanied by cellulitis of the scrotum is present, puncture should not be done. This is the only contraindication of FNA of the scrotal contents. Such inflammation may mask a tumor, however, and a residual palpable lesion present after the inflammation subsides should be aspirated.

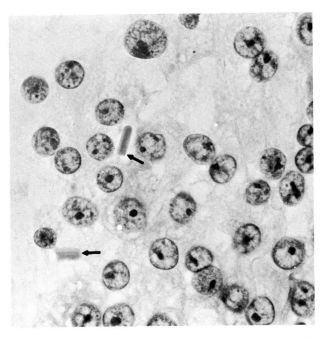

Fig. 12-5. Leydig cell tumor, testis. There are dispersed well-defined nuclei, each bearing a single nucleolus. Cytoplasm is pale or invisible. Crystals of Reinke *(arrows)* are red in this stain and colorless. (P stain, ×800)

plasm is homogeneous and nonvacuolated with well-defined borders and a perinuclear clear zone. Differential diagnosis with stem cell lymphoma must be considered, but the progression of the spermatogonia to smaller cells and finally spermatozoa will settle the issue (see Fig. 12-4).

Leydig cells, infrequently seen in routine smears, are relatively increased in atrophic smears. They are large polygonal cells with central spherical or oval nuclei. They may be binucleated. The nuclear chromatin has a fine reticular pattern and usually an inconspicuous nucleolus, which may be more evident with Papanicolaou (P) stain (Fig. 12-5). Cytoplasm is abundant and eosinophilic with gray-blue

Where aspiration of acutely inflamed scrotal contents has been carried out, however, the material consists of granulocytes, phagocytes, and fibrin without testicular cells. It is a nonspecific inflammatory exudate.

Chronic and granulomatous epididimo-orchitis

The epididymis may present as a cord descending from its head at the cephalad pole of the testis posteriorly and vertically. This cord can be palpated and moved with reference to the testis. It is slightly tender. In the presence of a palpable normal testis, this is strong clinical evidence of a benign tumor. Where the epididymis is a firm, fixed vertical mass not separable

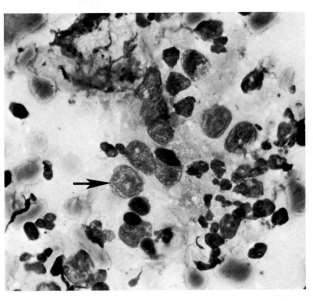

Fig. 12-7. Testis, chronic granulomatous orchitis. This smear was extracted from a firm, enlarged testis. There is evidence of a destructive lesion with cell debris and scattered polymorphonuclear and mononuclear inflammatory cells. Enlarged immature nuclei *(arrow)* are residuals of Sertoli cells and can be confused with tumor cells. Remnant of a giant cell is seen, but a typical epithelioid component is not evident in this portion of the smear. (R stain, ×800)

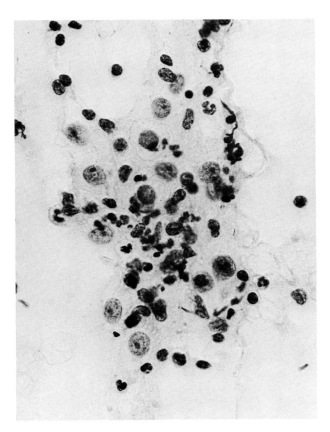

Fig. 12-6. Testis, chronic epididymo-orchitis. Sertoli cells with round to ellipsoid bland nuclei and single nucleoli are evident. Germinal epithelium is absent (atrophy). There are numerous polymorphonuclear and mononuclear inflammatory cells. (P stain, ×300)

from the testis, it may be difficult to distinguish from a tumor. The inflammatory process involves the testis by contiguity. The tissue changes in this pathologic alteration may show purely chronic inflammatory changes or varying degrees of granulomatous reaction that may clinically simulate a tumor.

FNA is thus indicated and yields a variety of cytologic findings (Figs. 12-6 and 12-7). Multinucleated giant cells, epithelioid cells, and mononuclear chronic inflammatory cells may all be present. There is some necrosis, with islands of epithelial (Sertoli's) cells. The testicular and epididymal structures have been infiltrated and disrupted. Inflammatory cells may be present in the absence of granulomatous change, producing otherwise the same picture. Tuberculosis of the epididymis is a rare entity and will yield caseous granulomatous reaction. The testis is not involved with tuberculosis in the absence of epididymal involvement. Granulomatous and lymph-

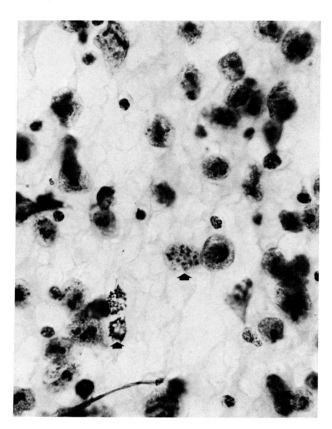

Fig. 12-8. Testis, seminoma. Note scattered primitive cells with numerous mitoses *(arrows)*. Fragile cells are distorted by smearing. Chromatin is punctuated with prominate nucleoli. Scattered lymphocytes and a pale, foamy background are not as prominent with this stain. (P stain, ×300)

oid stroma occur in seminomas, and such smears must be scrutinized carefully (Figs. 12-8 and 12-9).

Histologically, the main differential point to be considered in suspected granulomatous orchitis is malignant lymphoma or Hodgkin's disease.[12] This is less likely to occur with cytologic preparations in which extraction of a monomorphic immature lymphoid population will be more specific for lymphoma, whereas giant, epitheloid, and inflammatory cells will be extracted from an inflammatory lesion.

Neoplasms

The most important clinical aspects of testicular tumors are the insignificant early signs and symp-

toms, which, with the availability of FNA, should never be ignored. Slight changes in testicular consistency, often focal, may be the earliest change, followed by gradual enlargement with or without pain. These focal changes may consist of soft to firm to hard subcapsular swellings. Soft swellings, moreover, may have a texture no different from a normal testis and may be cancerous. This may include symmetric swelling of the testis that retains its normal shape. With further progression, there may be nodular enlargements due to subcapsular spreading of the tumors. Finally, gross nodules and diffuse swelling, with or without a heavy sensation or pain, may occur.

Because of the nonspecific nature of the swelling, germinal tumors cannot be distinguished clinically from lymphomas and Leydig's cell and adenomatoid tumors.

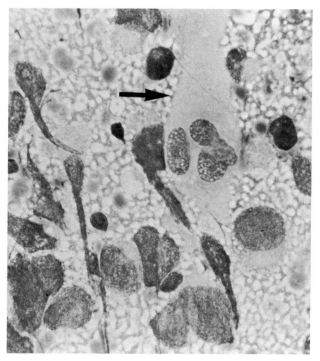

Fig. 12-9. Granulomatous stroma, seminoma of testis. Note lacy background. Seminoma cells vary in size and shape. A multinucleated giant cell *(arrow)* is evident, but epithelioid cells are not seen in this area. (R stain, ×1100)

CYTOLOGY OF TUMORS

Seminoma

The most striking aspect of the smear from a seminoma is the characteristic foamy, lacelike background staining pink, gray, and purple with R (Color Plate XXV A; Figs. 12-10 and 12-11). This background is pale and poorly visible with P stain (Fig. 12-8). This picture is present in about 70% of smears and is an identifying marker in metastatic lesions.[10] This background material is periodic acid–Schiff (PAS) positive in contrast with that in non-Hodgkin's lymphoma, which is PAS negative. A lacy pinkish gray background was observed in smears of an interstitial cell tumor stained with P.[6]

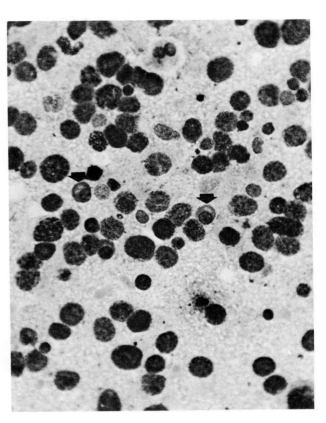

Fig. 12-11. Testis, seminoma. These are the dispersed tumor cells seen in Fig. 12-10. The scattered plasma cells (*arrows*) may suggest inflammation. (R stain, ×300)

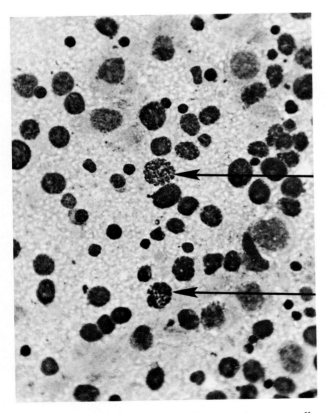

Fig. 12-10. Testis, seminoma. Dispersed tumor cells with finely granular chromatin and poorly defined cytoplasm are present. Two mitotic figures are evident (*arrows*). The characteristic foamy background is present, as are scattered stromal lymphocytes. The background can simulate lymphoglandular bodies. (R stain, ×400)

Seminoma cells are large, round to oval, dispersed fragile cells with round nuclei and fine granuloreticular chromatin with P stain. Nucleoli, seen best with P stain, are large or small (Fig. 12-12), single or multiple structures. Cytoplasm is pale, well defined with R stain, large, and irregular. There may be cytoplasmic fragments between cells that simulate the lymphoglandular bodies (see Fig. 12-10) of lymphoid smears, adding to the difficulty of differentiating seminoma and poorly differentiated lymphoma.

Although cells are generally dispersed, cell aggregates in some smears may raise the question of an embryonal tumor (Fig. 12-13).

Mature lymphocytes are a frequent component of the intercellular stroma of the usual seminoma and are readily observed in cytologic smears (see Fig. 12-8). They may be confused with small seminoma

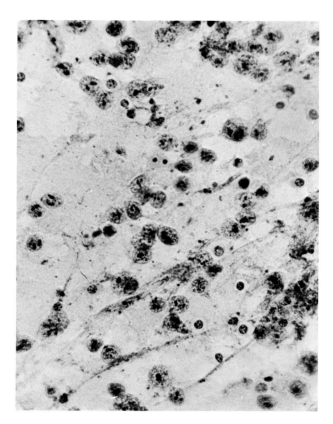

Fig. 12-12. Testis, seminoma. The dispersed tumor cells are fragile, with delicate chromatin network and disrupted nuclear strands. Nucleoli are prominent. (P stain, ×300)

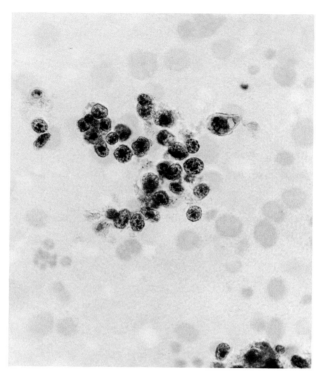

Fig. 12-13. Testis, seminoma. This is a loose cluster of tumor cells that must be distinguished from embryonal carcinoma (see Fig. 12-16). (P stain, ×300)

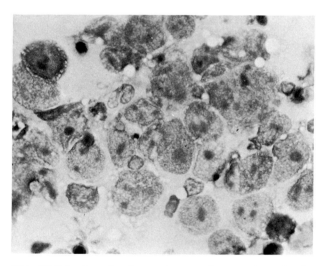

Fig. 12-14. Anaplastic seminoma of testis. A markedly polymorphous smear composed of primitive cells with large nucleoli. Cytoplasm is scant and not foamy. The typical lacy background of seminoma is not evident here, and an unequivocal diagnosis of seminoma must be cautiously made. (R stain, ×800)

cells. Multinucleated giant cells and epithelioid cells may also be seen and may help to identify the tumor as a seminoma (Fig. 12-9). The granulomatous stroma may be scanty in the smear. Embryonal cancers and teratocarcinomas do not share these stromal changes.

In anaplastic seminomas, cells are polymorphous with large nuclei and nucleoli and with mitoses, which do not immediately suggest seminoma in the absence of the usual background (Fig. 12-14). The presence of this background, however, will distinguish the anaplastic seminoma from embryonal carcinoma.

In spermatocytic seminoma, which carries an even better prognosis than the usual type, there is anisocytosis instead of the uniform cell size and absence of a stromal component (Fig. 12-15).[17] Well-defined, regular nuclear borders and single nucleoli with P suggest a lesser malignant potential. Binucleated cells are seen.

Embryonal Tumors

Embryonal tumors yield cytologic smears with greater pleomorphism than seminomas. Cells are usually in tight clusters with large overlapping cells suggesting carcinoma (Fig. 12-16 and Plate XXV B). The characteristic background of seminoma is missing and is replaced by fine granules in a gray matrix (R stain). There is marked variation in individual cell size and shape and a high nuclear:cytoplasmic ratio. Nuclei simulate those of immature lymphoid cells. Chromatin is finely granular and appears fibrillar where it is compressed or strung out (R stain). Nucleoli are large, pleomorphic, and often multiple (Fig. 12-16). The cells often consist primarily of nuclei with minimal cytoplasm. Where scanty cytoplasm is evident, it is gray, thin, and vacuolated (R stain). Single cells may contain more abundant cytoplasm. With P stain, there is marked hyperchromasia with

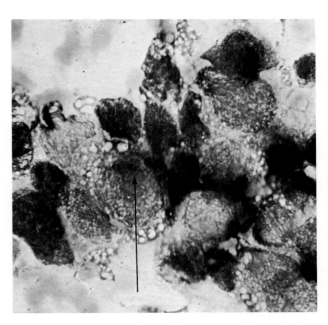

Fig. 12-16. Testis, embryonal carcinoma. These are immature large tumor cells from an overlapping cluster. The nuclear:cytoplasmic ratio is large. Macronucleoli are evident with this stain (*arrow*). Cytoplasm is foamy. (R stain, ×1200)

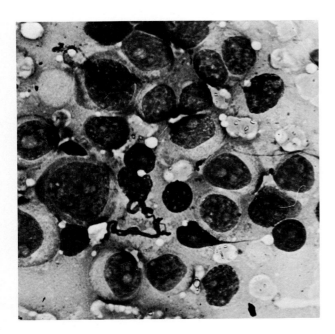

Fig. 12-15. Testis, spermacytic seminoma. The smear is richly cellular with neoplastic germ cells that vary in size. Chromatin is granuloreticular. There are numerous nucleoli, and the cytoplasm is well defined. The background is not foamy. (R stain, ×1200)

clumping of chromatin and cytoplasm appears more compact (Fig. 12-17).

In a less-differentiated variety, cells are larger. Nuclear chromatin is dense, with one large central nucleolus in each cell. Grouping is less evident.

Teratocarcinoma

Well-differentiated teratoma is clinically a hard mass that yields a very scanty aspirate for which consistent cytologic criteria have not yet been established. In teratocarcinoma, the necrosis that may be present in all malignant testicular tumors is more likely to appear. All necrotic smears without identifying cellular content require repeat puncture, usually from a more peripheral area of the tumor mass.

Cellular smears of teratocarcinoma are composed of sheets of dissociated undifferentiated cells. Nuclei are irregularly round with a well-defined chromatin pattern, numerous mitoses, and small nucleoli with P stain. Cytoplasm is scant or invisible. There are tangled stromal masses in juxtaposition (Plate XXV C). Clearly, these connective tissue

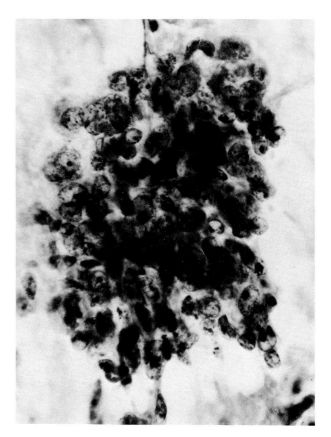

Fig. 12-17. Testis, embryonal carcinoma. This is a large, haphazard cluster of immature pleomorphic tumor cells. The chromatin pattern is coarse, and nucleoli are prominent. Cytoplasm is scant. (P stain, ×300)

meshes are not part of the supporting stroma of the tumor but are rather a dyssynchronously matured part of the teratoid structures. They consist of connective tissue fibers and contain numerous narrow and plump spindled nuclei. Other mesodermal elements, particularly chondroid cells, may be present. Epithelial structures seem well in histopathologic sections are not easily demonstrated in smears. However, unstructured epithelial groups may be seen. They may form cystic structures that yield fluid. Such a finding must always be viewed with caution.

In the absence of mesodermal elements, differentiation from embryonal carcinoma may be a problem. However, the cells do not form tight groups, and absent cytoplasm is a helpful point. Embryonal elements may, of course, be part of a teratoid tumor. Willis states there may also be confusion with seminoma.[20]

Lymphoma

Malignant lymphoma is the most common testicular tumor above the age of 60, although it is not confined to that age group.[8,19] Histologically, it is almost invariably poorly differentiated, lymphocytic or histiocytic, and may be associated with multicentric involvement of skin, central nervous system, and Waldeyer's ring.[8] This is potentially a nonsurgical lesion requiring treatment primarily with chemotherapy with or without radiation. Traditionally, diagnosis has been surgical following orchiectomy, but the cytologic identification of lymphoma of the testis is readily accomplished. Cytology of lymphoma is described in Chapter 14. Differentiation from seminoma may be a problem in the younger age group because the germinal cells in the midst of a lymphoid stroma may be mistaken for lymphoid cells (see Figs. 12-11 and 12-12). The frothy tigroid background seen in R stain (Fig. 12-10) is an important differentiating point.

Case 12-1 illustrates a problem in differential diagnosis of lymphoma and germinal tumors.

■ *Case 12-1*
A youth aged 19 years with palpable glands in the left supraclavicular fossa and right inguinal region was referred for aspiration. The smear consisted of polymorphous, mostly dissociated malignant cells with a clean background. It resembled poorly differentiated non-Hodgkin's lymphoma, particularly in the absence of an obvious primary tumor. The testes were negative. The patient died 3 months later. At autopsy, a 3-mm tumor was found in the left testis and was interpreted as probable anaplastic embryonal carcinoma.

Patients with germ cell tumors in clinically normal testes have been reported.[3]

Extrascrotal Germinal Masses

A mass palpable in the inguinal region or identified on computerized tomography in the pelvis may be an undescended testicle. Failure to palpate the scrotal sac and identify an absent testis is not unusual in routine clinical practice. Finding Sertoli cells (see Fig. 12-2) on aspiration may be a useful surprise finding. Ten of 158 cases of metastatic germ cell tumors were diagnosed by aspiration.[9]

Occult seminoma was identified by finding PAS positive tumor cells with granulomatous reaction in a left supraclavicular node on histology.[15]

Identification of germ cells cytologically is aided by a study of testicular cytology.

Acute Leukemia

The testis is a known sanctuary for leukemic blast cells.[1] Preliminary studies have shown that routine follow-up testicular biopsies in children with acute leukemia in remission can be successfully performed with the fine needle. At Karolinska Hospital, the children are anesthetized for bone marrow, lumbar puncture, and testicular puncture carried out at the same time. The normal testicular aspirate in childhood consists of sheets of Sertoli's cells, which can be readily distinguished from the primitive blast cells of acute leukemia.

Rupp and associates reported FNA findings in a 28-year-old man with leukemic relapse in a testicle. Cells consistent with leukemic infiltration were identified using P stain.[16] The addition of R stain is useful in identifying hematopoietic cells and lymphoglandular bodies.

Leydig Cell Tumors

Leydig's cell tumors cannot be clinically distinguished from other benign or malignant scrotal masses. They are composed of Leydig cells (described above), which contain crystals of Reinke in 40% of patients (see Fig. 12-5).[12] Intranuclear inclusions characteristically present in thyroid and adrenal tumors were noted in an interstitial cell tumor.[6] In the American Testicular Tumor registry, their incidence was almost 3%; and of these, 10% were malignant. In the malignant group, the cytologic changes are characterized by polymorphism, nuclear hyperchromatism, and increased number of mitoses. This is a rare tumor, and examples were not available for review.[2]

Metastasis to the Testis

While metastases to the testis are rare,[21] the cytopathologist must be prepared for any potential lesion. Hepatocellular carcinoma metastatic to the testis was recently reported.[22] These findings stress the usefulness of pre-operative cytologic diagnosis.

Adenomatoid Tumor

The smears from an adenomatoid tumor consist of masses of epithelioid elements without stroma, arranged in solid cords and tight glandular or tubular structures (Fig. 12-18). Extraction of epithelial elements from connective tissue stroma can be visual-

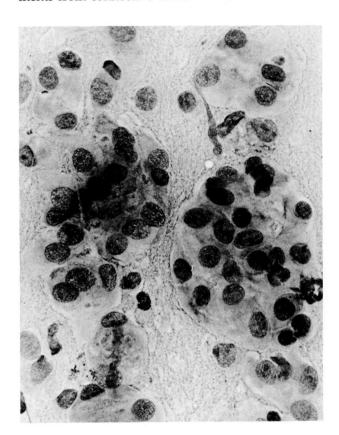

Fig. 12-18. Epididymis, adenomatoid tumor. Tumor cells are in epithelioid groups. They have been stripped away from the stromal tubules seen in histologic section (Fig. 12-19). Cells are uniform, with ovoid granular nuclei and single nucleoli. Cytoplasm is abundant and finely granular. The background is characteristically stippled with red granules. (R stain, ×600)

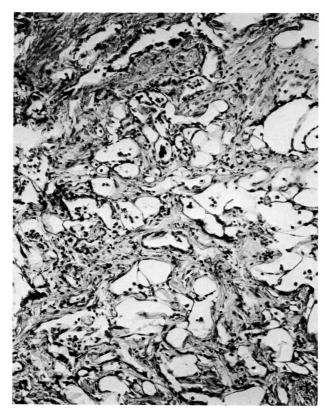

Fig. 12-19. Epididymis, adenomatoid tumor. Tissue section. Dense connective tissue stroma has formed numerous irregular tubules lined by epithelial tumor cells. (H & E, ×190)

ized from the tissue sections in Figure 12-19. There are numerous binucleated cells in the smear. Eccentric vesicular nuclei with small nucleoli and fine chromatin pattern (with P) without atypia are present. Cytoplasm is finely granulated with R. There are no mitoses. An electron microscopy study of these tumors demonstrated mesothelial origin in two and angiomatous origin in one.[7]

The nuclear : cytoplasmic ratio is low, and there is a superficial resemblance to both plasma and mesothelial cells. The background is characteristically composed of fine red granules.

CLINICAL APPLICATION

Puncture of the testis with a fine needle for diagnostic smears is a useful technique in the diagnosis of palpable disease of the scrotal contents and the scrotum.

Extensive experience has demonstrated that there is minimal discomfort, no proven hazard, and a source of data to help decide on operative intervention. Diagnostic difficulties during early experience with the method cannot be minimized. Doubtful smears should be followed by repeat aspiration or by surgical biopsy (orchiectomy), depending on the clinical context.

Recognition of gonadal neoplastic cytologic patterns in metastatic as well as extragonadal primary sites is an extremely useful diagnostic skill.

REFERENCES

1. Askin FB, Land VJ, Sullivan MD et al: Occult testicular leukemia: Testicular biopsy at three years continuous complete remission of childhood leukemia. Cancer 47:470, 1981
2. Azer PC, Braunstein MD: Malignant Leydig cell tumor. Cancer 47:1251, 1981
3. Burt ME, Javadpour N: Germ-cell tumors in patients with apparently normal testis. Cancer 47:1911, 1981
4. Collins DH, Pugh RB: The Pathology of Testicular Tumors. Edinburgh, E & S Livingston, 1964
5. Coulson WF: Surgical Pathology. Philadelphia, JB Lippincott, 1978
6. Crucioli V, Fulciniti F: Fine needle aspiration of the interstitial cell tumor of the testis — Letters to the editor. Acta Cytol 31:199, 1987
7. Davy CL, Chik-kwun T: Are all adenomatoid tumors mesothelial in origin? Lab Invest 42:110, 1980
8. Doll DC, Weiss RB: Malignant lymphoma of the testis. Am J Med 81:515, 1986
9. Kapila K, Hajdu SI, Whitmore WF et al: Cytologic diagnosis of metastatic germ-cell tumors. Acta Cytol 27:245, 1983
10. Lopes–Cardoza P: Atlas of Clinical Cytology. Targa b.v.'s-Hertogenbosch, 1976
11. Miller A, Seljelid R: Histopathologic classification and natural history of malignant testis tumors in Norway — 1959–1963. Cancer 28:1054, 1971
12. Mostofi FK, Price EB: Tumors of the Male Genital System. Washington, D.C., Armed Forces Institute of Pathology 1973
13. Mostofi FK, Sobin LH: Histological Typing of Testis Tumours. Geneva, World Health Organization 1977

14. Persson PS, Åhrén C, Obrant KO: Aspiration biopsy smears of testis in azoospermia. Scand J Urol Nephrol 5:22, 1971
15. Richter HJ, Leder L–D: Lymph node metastases with PAS-positive tumor cells and massive epithelioid granulomatous reaction as diagnostic clue to occult seminoma. Cancer 44:245, 1979
16. Rupp M, Hafiz MA, Hoover L et al: Fine needle aspiration in the evaluation of testicular leukemic infiltration. Acta Cytol 31:57, 1987
17. Talerman A: Spermatocytic seminoma. Cancer 45:2169, 1980
18. Teilum G: Special Tumor of Ovary and Testis. Copenhagen, Ejnar Munksgaards Forlag, 1971
19. Turner RT, Colby TV, Mackintosh FR: Testicular lymphomas. A clinicopathologic study of 35 cases. Cancer 48:2095, 1981
20. Willis R: Pathology of Tumours. Scarborough, Ont, Butterworth and Co, 1960
21. Willis RA: The Spread of Tumours in the Human Body. London, Butterworth & Co, 1973
22. Young RH, Van Patten T, Scully RE: Hepatocellular carcinoma metastatic to the testis. Am J Clin Pathol 87:117, 1987
23. Zajicek J: Aspiration biopsy cytology. II. Cytology of supradiaphragmatic organs. In Wied GL (ed): Monographs in Clinical Cytology. Basel, S Karger, 1979

CHAPTER 13

Melanomas and Skin Nodules

JOSEPH A. LINSK

SIXTEN FRANZEN

MELANOMAS

Melanoma is a malignant, usually pigmented, and highly unpredictable tumor occurring most often in a cutaneous region. It is detected visually or by palpation. Most surface or subsurface melanomas occur in three locations. Primary melanoma arises almost without exception from the dermal-epidermal junction.[31] Dermal or subcutaneous deposits are almost invariably metastatic. Likewise, palpable lymph nodes shown to be occupied by melanoma are of metastatic origin, even in the absence of a discernible primary tumor (see Case 13-2).[8]

Primary Melanoma

Pigmented nevi are very common, and malignant melanomas are infrequent by comparison.[27] They are not rare lesions, however, and have been increasing at a rate exceeded only by bronchogenic carcinoma.[5] These statistics suggest that the volume of these tumors will be rapidly rising and heightens the importance of prompt recognition and biopsy.

With the exception of rare lesions developing in pigmented foci in the eye or in the mucosa of the respiratory, genitourinary (GU), or gastrointestinal (GI) tract, primary melanomas are surface tumors. Clinical suspicion of cutaneous melanoma is aroused by the presence of irregular configuration and notching of the lesion and brown and black pigmentation mixed with red, white, and blue.[33] These findings aid selection, but the index of suspicion must be quite high in considering all pigmented lesions. On the one hand it is neither feasible nor practical to biopsy every pigmented lesion by surgical excision. On the other hand any suspicious lesion can be biopsied by excision or incision without altering patient survival.[16,34a] In light of these data, fine needle aspiration (FNA) of primary cutaneous melanoma or any suspicious pigmented lesion, while not a presently recommended procedure, is perfectly feasible.

Three clinical types of melanoma have been described:

1. Lentigo maligna, which is flat
2. Superficial spreading type, which is raised and palpable in some portions
3. Nodular, which is by definition raised and palpable

Any palpable skin elevation, including pigmented melanoma, can be aspirated, although specific indications for aspiration of known, suspected, or potential surface melanomas include the following:

1. Nonpigmented skin nodules not clinically identifiable as specific lesions.
2. Large nodular or fungating masses clinically identifiable as melanoma. Definitive surgery can be undertaken after cytologic identification.
3. Screening of minimally suspicious pigmented lesions in a patient with known melanoma. Multiple primary tumors occurred in 5.3% of 712 patients in one study.[24]
4. Aspiration of the nodular portion of a large facial freckle for a quick, nonscarring diagnosis.
5. Pigmented lesions of the vulva, penis, scrotum, anus, scalp, nares, and ears, which are all potential sites for aspiration. These punctures are carried out with a 25-gauge needle with minimal disturbance of the tissue.

6. FNA of a lesion about which a question is raised but which the clinician has declared benign. Obtaining cells for examination may be less hazardous than observation.

7. Aspiration of known recurrent melanoma under treatment in which evaluation by FNA is carried out to evaluate the effects of treatment and preserve the measurements of the mass.[8]

Practically speaking, aspirating malignant melanoma cells with rapid staining and reporting will encourage the patient to secure definitive therapy. Puncture of superficial lesions was not recommended by Fontanieré and colleagues; however, since incisional biopsy is quite satisfactory,[16,34a] fine needle puncture poses an even smaller hazard of cell dissemination, if any.[7]

In the differential diagnosis of pigmented lesions, one group appears to be of nonnevic origin. It includes basal cell carcinoma, squamous carcinoma, neurofibromas, and metastatic nodules.[31] Aspiration will exclude melanoma.

Metastatic Melanoma

In Skin Nodules

Skin and subcutaneous tissue are the most common sites of metastatic nodules due to melanoma.[2] Typical dermal nodules are often multiple and pigmented and may occur as a sprinkling of pigmented lesions between a resected primary site and a regional lymph node site. They are spread through the dermal lymphatic network.[36] They may be flat or slightly elevated. With a 25-gauge needle and delicate technique, the smallest of these can be aspirated for diagnosis. This is a useful preliminary maneuver when intralesional treatment with bacille Calmette Guérin (BCG) or dinitrochlorobenzene is being considered.[3] With rapid stain, the injection can be done immediately after the aspiration diagnosis.

Subcutaneous nodules of hematogenous origin usually elevate the skin. The skin can often be moved over the nodule and is discolored bluish purple to gray-black. It may simulate a vascular lesion. Aspiration is simple and efficient.

In Lymph Nodes

Clinical accuracy in determining that palpable nodes are metastatic increases with experience but is never definitive.[11] Palpable lymph nodes are easy targets

that yield ample cytologic material for study with a high degree of accuracy, and therapeutic decisions should be based on morphology and not on clinical acumen.[4,13]

The classification of primary melanomas according to levels of micro-invasion is useful in evaluating potential lymph node metastases. Such metastases occurred in 62% of lesions greater than or equal to 4 mm and in no lesions less than 0.76 mm in thickness.[1] Careful palpation and searching aspiration of subtle adenopathy are indicated in lesions with deeper levels of penetration.

Palpable lymph nodes without an immediately apparent primary tumor may suggest or be diagnostic of melanoma. No primary tumor was found in 8 of 74 patients diagnosed as having melanoma by aspiration biopsy cytology (ABC).[8] Such findings are illustrated in the following case reports.

■ *Case 13-1*

A 79-year-old man presented with a fairly rapidly enlarging mass in the right parotid region. Aspiration yielded black liquid initially interpreted as hemosiderin in a broken-down hematoma. Aspiration of residual induration yielded intensely pigmented tumor cells (Fig. 13-1). On more detailed history, it was deter-

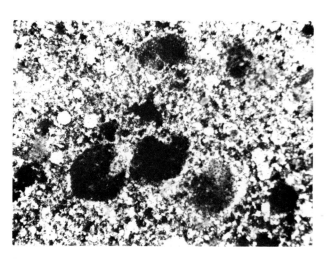

Fig. 13-1. Melanoma with intense pigmentation. Pigment production is massive, obscuring tumor cells, which might be missed if not considered in the evaluation. (Romanowsky [R] stain, ×90)

mined that a small, pigmented, level II melanoma had been excised from his scalp 6 months earlier.

■ *Case 13-2*

A 42-year-old man presented with an enlarged right axillary node. Aspiration yielded moderately pleomorphic clustered and dispersed nonpigmented cells (Fig. 13-2). On excision, the node was heavily invaded by tumor and interpreted as amelanotic melanoma (Fig. 13-3). Follow-up axillary dissection was negative for further tumor. Several complete examinations failed to discover a primary tumor up to 6 years after excision.

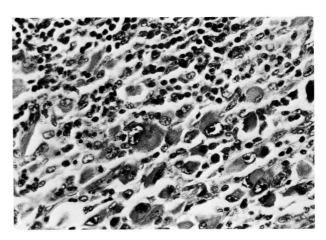

Fig. 13-3. Melanoma, lymph node metastasis. Tissue section. Tumor cells with eccentric nuclei and prominent nucleoli have infiltrated a lymph node. Same melanoma seen in Fig. 13-2. (Hematoxylin and eosin [H & E], stain, ×190)

Differential diagnosis of palpable nodes includes the monolayer variety of breast carcinoma (see Fig. 6-34), which may be extracted from axillary nodes in the absence of a palpable primary tumor. Intranuclear inclusions are present in papillary thyroid gland carcinoma, which may metastasize from an occult thyroid gland primary tumor to lateral cervical nodes. Such cytologic preparations may be confused with melanoma cells, which also may contain intranuclear inclusions (see Fig. 13-9).[7,14,15,37]

Cytology

Pigment

The presence of melanin staining black with Romanowsky (R) is a highly suspicious finding (see Fig. 13-6). Pigment *per se* is not diagnostic, since it is present in a variety of benign cells.[31] The problem of

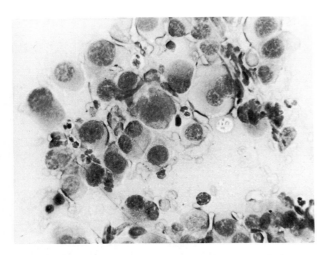

Fig. 13-2. Melanoma, lymph node metastasis. Polymorphous cells have formed a monolayer. Nuclear as well as cytoplasm outlines are sharply defined, and nuclei are eccentric. These cells have been extracted from a lymph node stroma. (R stain, ×600)

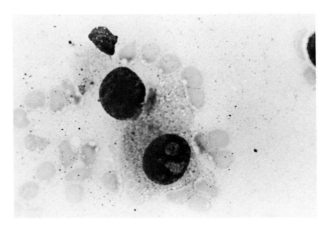

Fig. 13-4. Melanoma, dermal lesion. There are two dense nuclei, one containing irregular macronucleoli. The pigment simulates salt and pepper deposits. Cytoplasm is poorly defined, partially because of the extraction process from a dermal lesion through a 25-gauge needle. (R stain, ×600)

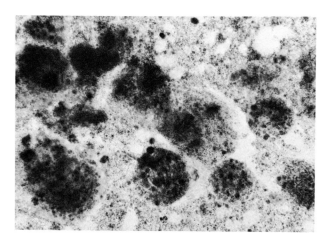

Fig. 13-5. Melanoma, subcutaneous nodule. This is a densely pigmented smear. Tumor cells can be obscured and were variably identifiable in other areas of the smear. (R stain, ×400)

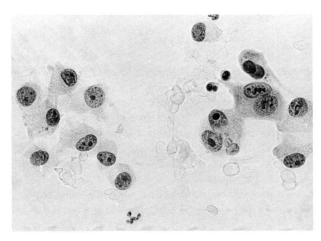

Fig. 13-7. Melanoma, subcutaneous nodule. There is faintly visible sandy granulation. Multiple nucleation and nucleoli are evident. There are neural and dendritic morphologic alterations. (P stain, ×400)

differentiation of tumor cells from reactive macrophages was noted by Perry and colleagues.[28] However, pigment in undifferentiated cancer cells is almost invariably a sign of melanoma. In apparently amelanotic undifferentiated tumor specimens, a careful search must be made for fine pigment, which may be present in a few cells (Figs. 13-7 and 13-9). Pigment usually varies from a salt-and-pepper texture to gross black masses (Figs. 13-4 and 13-5). The pigment may lie over the nucleus and cytoplasm or lie free of the cells (Fig. 13-6). Melanoma pigment overshadowed tumor cells in 6 of 23 cases described by

Gupta and associates.[12] Pigment may discolor the unstained smear brown-black, which strongly suggests the diagnosis.

Note that pigment may be present in the primary lesion and not in the metastases and, likewise, may be present in metastases and not in the primary tumor.[2] Melanin may be patchy in its distribution and missed by the needle, adding another variable to the data derived from aspiration.

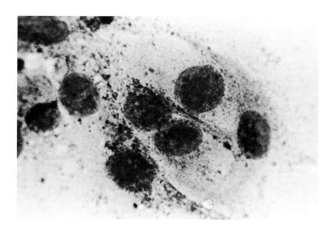

Fig. 13-6. Melanoma, subcutaneous nodule. Pigment overrides nucleus and cytoplasm and lies free of cells. Macronucleoli are evident. (R stain, ×600)

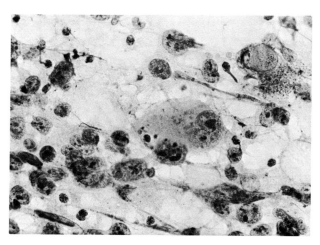

Fig. 13-8. Melanoma, subcutaneous nodule. Pigment is evident in dense masses. There are marked polymorphous and irregular macronucleoli. The cells have invaded fat, evident in the background. (P stain, ×400)

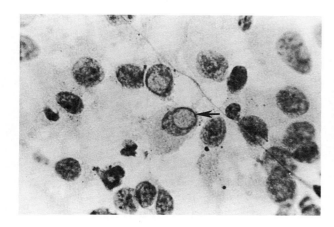

Fig. 13-9. Melanoma, lymph node metastasis. Sharply defined intranuclear inclusions are readily seen *(arrow)*. (Compare this smear to Figs. 4-39 and 8-30.) (R stain, ×600)

With Papanicolaou (P) stain, pigment may be more difficult to see and tends to stain greenish brown; however, it may stain heavily brownish black (Figs. 13-7 and 13-8). In the series of 151 melanoma aspirates reported by Perry and colleagues,[28] melanin was undetected in 60%. In contrast, amelanotic melanomas constitute a minority of histologic diagnoses. The P stain is not satisfactory for consistent identification of pigment. Both R and P stains are complementary and in melanoma diagnosis, critical.

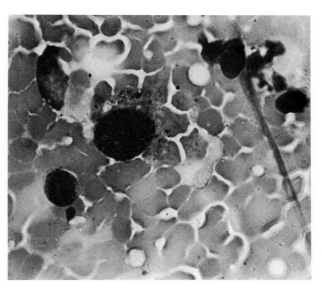

Fig. 13-11. Melanoma, dermal lesion. Significant blood has been extracted, within which are scattered somewhat traumatized tumor cells. (R stain, ×500)

Amelanotic melanoma may require electron microscopy (EM) for definitive diagnosis (see Chap. 18). However, the presence of intranuclear inclusions, fairly well-defined dispersed cells with a low nuclear:cytoplasmic ratio, vacuolated cytoplasm, and a halo effect are clues to the identity of melanoma (Figs. 13-9 and 13-10).

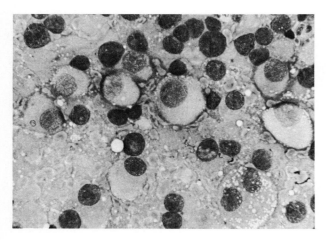

Fig. 13-10. Melanoma, subcutaneous nodule. Well-defined tumor cells with vacuolization are evident. The cytoplasm produces a fairly characteristic halo effect. (R stain, ×400)

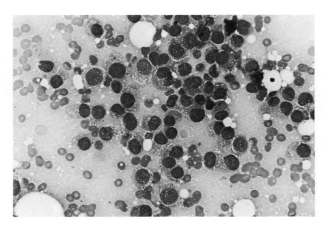

Fig. 13-12. Melanoma, subcutaneous nodule. This is a portion of a monolayer spread of melanoma cells. At this power, the smear can be confused with myeloma (see Fig. 13-25) and the monolayer breast carcinoma pattern (see Fig. 6-34A). (R stain, ×300)

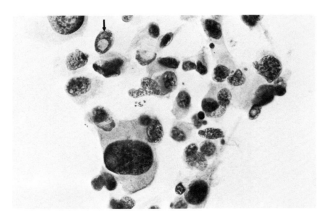

Fig. 13-13. Melanoma, subcutaneous nodule. There is marked variation in size and shape of nuclei as well as cytoplasmic outline. Cytoplasm forms connecting bonds. One macroinclusion helps identify this as melanoma in the absence of obvious pigment. Note nucleolus sharing nucleus with inclusion (*arrow;* see also Color Plate XVII). (P stain, ×300)

Cellularity and Cell Distribution

Cellular smears are generally obtained without difficulty, even from small dermal lesions. Occasionally, scattered singly but diagnostic cells mixed with blood are obtained (Fig. 13-11). Aspirated cells are usually dissociated and dispersed, consistent with the reduced cohesiveness of malignant melanoma. When quite cellular, they form sheets (Fig. 13-12), not un-like lymphoma or the monolayer type of breast carcinoma. Clusters are infrequently seen.

In large masses with necrosis a mixture of necrotic debris and identifiable pigmented cells may be seen. The necrosis may be liquefied and black (see Case 13-1).

Individual Cell Cytology

Melanoma cells have been defined as epithelial-like[20] and epithelioid.[35] The latter term is more often associated with benign juvenile melanoma (Spitz), a lesion composed of epithelioid or spindle cells or a mixture of the two. Sixty percent of such lesions were amelanotic.[35] Familiarity with such lesions is important because a number are malignant and because primary pigmented or nonpigmented lesions may become targets for FNA.

Individual cell cytology may vary from spherical to polyhedral to fusiform to pleomorphic types (Figs. 13-13 to 13-15). Most commonly, however, they are round or ovoid. Dendritic types have been clearly demonstrated in the report by Friedman and associates[8] and are seen in Figure 13-16.

Nuclei

Nuclei are usually oval or round, well defined, and eccentric. They may have a flattened surface (Fig. 13-17). Nuclear chromatin is often finely granular or

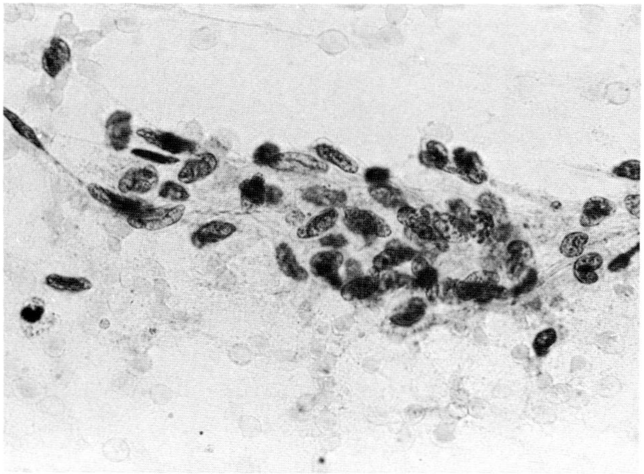

Fig. 13-14. Melanoma, subcutaneous nodule. This is a cluster of fusiform cells, suggestive but not diagnostic of melanoma in the absence of discernible pigment. (P stain, ×190)

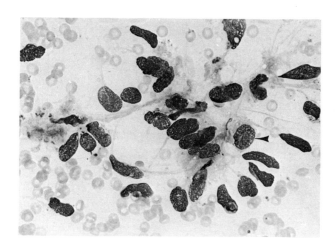

Fig. 13-15. Melanoma, dermal lesion. Note clusters of irregularly fusiform cells with reticulogranular nuclear chromatin. Nucleoli *(arrow)* is stained blue. Some pigment is evident. (R stain, ×250)

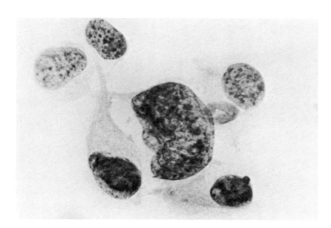

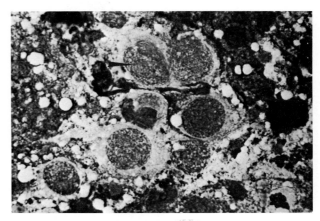

Fig. 13-16. Melanoma, lymph node metastasis. This is a markedly pleomorphic smear with irregular nuclei and coarse chromatin pattern. Dendritic cytoplasmic processes are evident. (P stain, ×600)

Fig. 13-18. Melanoma, lymph node metastasis. This is a group of atypical melanoma cells with round, fairly central, reticulogranular nuclei. In the absence of pigment, such cells may be indistinguishable from undifferentiated carcinoma or poorly differentiated lymphoma. Enlarged nucleoli *(arrow)*, however, are characteristic. (R stain, ×600)

smooth with R but may be dense, hyperchromatic, and without a well-defined chromatin pattern with P stain (Figs. 13-7, 13-8, 13-18, and 13-19). Nuclear pleomorphism may be striking with P stain. Binucleation, multinucleation, and bizarre nuclei are present particularly, in giant cell forms; but they may also be seen in otherwise relatively uniform smears (Figs. 13-20 and 13-21). Both normal and abnormal mitoses may be found.

Nucleoli are predominantly seen with R stain and are usually large, blue, and multiple (Fig. 13-22).

A macronucleolus and a nuclear inclusion are seen in Figure 13-13. The nuclear vacuoles or inclusions are of cytoplasmic origin, as has been demonstrated for papillary carcinoma of the thyroid gland (see Chap. 4). Their frequency is inversely related to the amount of pigment.[37] Hepatocytes with nuclear inclusions and bile could suggest melanoma cells (Fig. 8-30). Bile stains black with R and may be present in hepatoma cells, creating a problem in differential diag-

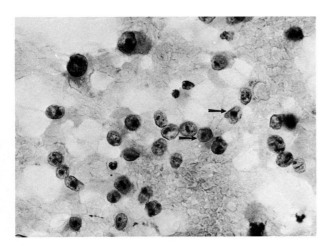

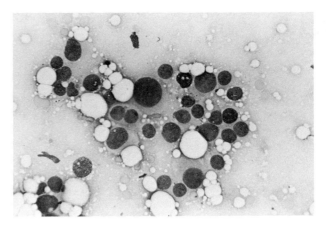

Fig. 13-17. Melanoma, subcutaneous nodule. Well-defined nuclei exhibit flattened surfaces *(arrow)* characteristic of melanoma. (P stain, ×300)

Fig. 13-19. Melanoma, subcutaneous nodule. This is a group of melanoma cells intermixed with fat. Nuclear chromatin is smooth and homogeneous. Cytoplasm is pale, and pigment is sparse. (R stain, ×300)

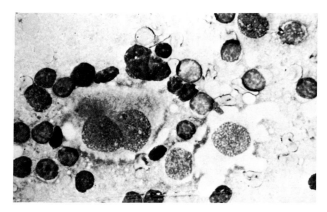

Fig. 13-20. Melanoma, lymph node metastasis. This is a binucleated cell with granular cytoplasm and sparse pigment surrounded by lymphocytes. (R stain, ×330)

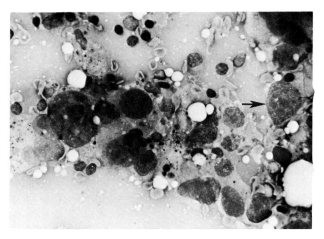

Fig. 13-22. Melanoma, subcutaneous nodule. There are numerous macronucleoli, some multiple *(arrow)*, staining blue with this stain. Marked variation in size and shape of nuclei is present. Pigment is focally distributed. (R stain, ×300)

nosis with metastatic melanoma to the liver (Figs. 13-23 and 13-24). Hepatoma cells, however, have other characteristics (see Chap. 8).

Cytoplasm

Cytoplasm is abundant, and the nuclear:cytoplasmic ratio may be low (see Fig. 13-7). With R stain, cytoplasm is blue-gray and often uniform with well-defined cell boundaries. It may be vacuolated (see Fig. 13-10). Cytologic vacuolation was recently stressed by Gupta and colleagues.[12] Nonpigmented, disso-

ciated tumor cells with a reduced nuclear:cytoplasmic ratio and vacuolated, well-defined cytoplasm should suggest melanoma. With P stain, cytoplasm is pale green and varies between being fairly solid to thin and wispy (see Fig. 13-8). With P stain it is often too sparse to be useful in diagnosis,[28] and the addition of R-stained smears is important in laboratories that rely on P for all of their diagnostic reporting.

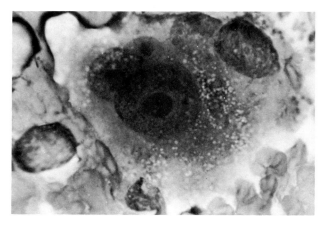

Fig. 13-21. Melanoma, subcutaneous nodule. This is a bizarre irregular cell with finely vacuolated cytoplasm. There are multiple nucleoli within multiple nuclei. Pigment is not prominent. (R stain, ×480)

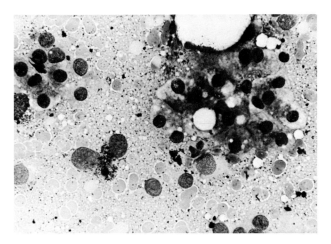

Fig. 13-23. Melanoma in liver. Primitive nuclei with nucleoli distinguish these cells from hepatocytes. Stippled pigment spills over from the cytoplasm. (R stain, ×300)

Differential Diagnosis

Spindle-cell forms, rare in exfoliative smears, are not uncommon in aspirates.[14] They may present a problem in differential diagnosis in smears composed of nonpigmented cells from metastatic sites with an unknown primary tumor. Other epithelial tumors with spindle-cell subtypes are renal cell carcinoma (see Fig. 8-79), medullary carcinoma of the thyroid gland (see Fig. 4-45A), and oat cell carcinoma (see Chap. 7).

Poorly differentiated histiocytic lymphoma may be difficult to distinguish from amelanotic melanoma in tissue section. This is less of a problem cytologically. The nuclei do not have the lymphoid appearance that may be simulated by some nasopharyngeal (see Chap. 3) and bronchogenic (see Chap. 7) carcinomas. The presence or absence of lymphoglandular bodies (see Chap. 14) is an important diagnostic clue in the identification of lymphomas.

Multiple myeloma may also be considered in differential diagnosis with amelanotic melanoma (Fig. 13-25).

SKIN NODULES AND INFILTRATIONS

FNA of dermal and subcutaneous skin nodules is a simple and efficient diagnostic tool. Nodules are distributed singly, in crops, generally disseminated, and *en cuirass*. Four morphologic patterns, illustrated in Figure 13-26, become targets for superficial puncture as follows:

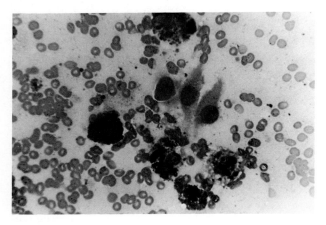

Fig. 13-24. Melanoma in liver. This is more typically melanoma with spindle cells. (R stain, ×300)

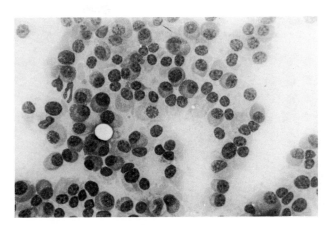

Fig. 13-25. Myeloma, bone lesion, lytic bone area. This spread of atypical plasma cells can simulate melanoma (see Fig. 13-12). (R stain, ×250)

1. *Dermal nodules* have an intact surface and move with the skin.
2. *Subcutaneous nodules* have an intact cutaneous surface that can be moved over the nodules.
3. *Smooth diffuse infiltration*, which elevates the intact skin and may impart a purplish color, is characteristic of lymphoma.
4. *Ulcerated skin with a nodular base or margin* occurs in primary and metastatic lesions.

Clinical targets for FNA appear as a result of the following mechanisms:

1. Direct permeation of the surface by an underlying tumor.
2. Spread of motile tumor cells through a dermal lymphatic, forming collections and then nodules.[36] Examples are chest wall recurrences after mastectomy.
3. Hematogenous spread resulting in subcutaneous nodules.
4. Implantation of tumor nodules in scars after tumor surgery.
5. Implantation nodules after thoracentesis and paracentesis of malignant effusions. Implantation nodules after FNA of solid tumors are extremely rare.[32]

The extraction of few or possibly distorted epithelial cells may not provide satisfactory criteria for the specific diagnosis of malignancy (Fig. 13-27). Their presence, however, is sufficient indication for

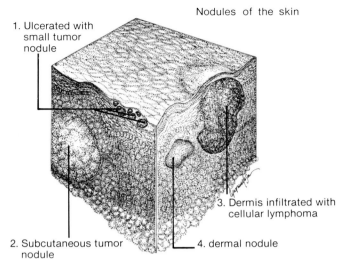

Nodules of the skin

1. Ulcerated with small tumor nodule

2. Subcutaneous tumor nodule

3. Dermis infiltrated with cellular lymphoma

4. dermal nodule

Fig. 13-26. Three-dimensional section of skin illustrating four lesions that may become targets for fine needle aspiration.

1. Ulcerated skin with a nodular base is aspirated tangentially with a 25-gauge needle.

2. Subcutaneous nodules lie within the fat layer and elevate the full thickness of skin, which moves above the nodule.

3. Infiltration of the dermal–epidermal region with neoplasm, particularly lymphoma. The skin surface is often pink or violaceous and smooth.

4. Dermal nodules palpated as elevations with intact skin, move with the skin. Breast is a frequent primary.

surgical biopsy. Such cells are aspirated from dermal lymphatics, as illustrated in Figure 13-28. Dermal deposit of metastatic breast carcinoma is illustrated in Figure 13-29.

The aspiration is innocuous and essentially painless and should be undertaken to clarify any unexplained skin elevation or nodulation that is not clearly a simple cutaneous lesion. The physician may be reluctant to incise the skin of the nares pinna, nipple, vagina, or anus for lesions with a low index of suspicion. FNA may extract malignant cells with minimal trauma at these sites or may yield inflammatory exudates that will clarify the lesion and guide therapy.

BENIGN LESIONS

Inflammatory Nodules

Skin or subcutaneous nodules appearing in patients with known primary or metastatic malignancies are suspicious for cutaneous metastases. Such nodules are, however, generic "tumors."[18]

Inflammatory smears were aspirated from 42 patients. The majority had a history of malignancy and the target lesion was a presumed residual or recurrent tumor.[30] The cytologic diagnoses included abscess, a variety of cysts, actinomycosis, and granulomatous reactions. A number of specific bacterio-

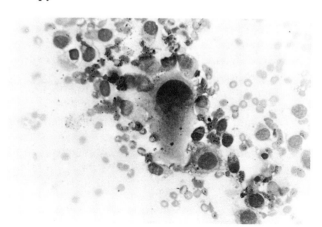

Fig. 13-27. Skin infiltrate, carcinoma cells. This is a faded, distorted, obviously abnormal cell cluster extracted from a carcinomatous infiltrate. (R stain, ×300)

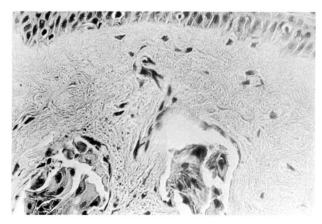

Fig. 13-28. Skin infiltrate, carcinoma cells. Tissue section. Groups of tumor cells are present in dermal lymphatics. (H & E stain, ×190)

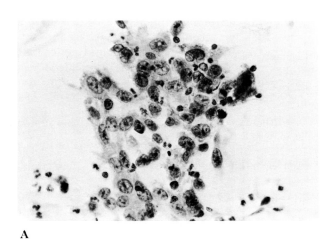

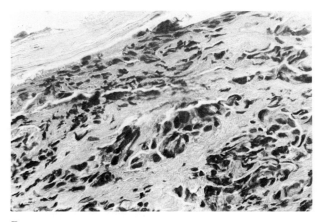

A   B

Fig. 13-29. A. Nodular skin infiltrate, metastatic breast carcinoma. This is an abundant cell yield of carcinoma cells, consistent with duct carcinoma, extracted with a 25-gauge needle (see Fig. 13-46). Tumor cells with nuclear variation and nucleoli in haphazard arrangement are present. (P stain, ×300) **B.** Nodular skin infiltrate, metastatic breast carcinoma. Tissue section. Tumor cells are infiltrating skin. Surface epithelium is partially eroded. (H & E stain, ×290)

logic diagnoses including *M. tuberculosis, Staphylococcus aureus,* and Nocardia were found.

Unexpected suppuration in afebrile patients may occur in multiple body sites. Large cervical and inguinal masses yielding sterile suppurative exudates were reported in *Fine Needle Aspiration for the Clinicians.*[18]

Nodules adjacent to incisions suspicious for recurrent cancer may yield little cellular content or histiocytes, fibroblasts, few inflammatory cells, and occasional multinucleated giant cells. Such lesions may be classified as suture granulomas.

In converse of the above, metastatic skin nodules may appear clinically to be acute inflammatory reactions in skin.

FNA will yield carcinoma cells intermixed with polymorphonuclear cells (see Fig. 3-5).

Calcifying Epithelioma of Malherbe (Pilomatrixoma)

Pilomatrixoma is a benign adnexal (hair matrix) tumor of the skin. It appears as a slow-growing dermal or subcutaneous mass under 3 cm, usually located in the head or neck. A rare malignant variety has been reported.[19]

The cytologic smear contains dense masses of basal cells with poor cell definition; a carpeting of more readily defined basal cells with poorly visible cytoplasm; and, occasionally, partially necrotic masses of cells. The nuclei are bland with absent or very small nucleoli. Keratinization is not evident, and the cells do not form the tight plugs seen in basal cell carcinoma. With R stain there may be some pink-staining acid mucopolysaccharide arising from other adnexal structures (Fig. 13-30).

Endometriotic Nodules

Endometriosis results in deposition of endometriotic nodules at various sites within the pelvic cavity as well as in the vagina and rectovaginal septum. Subcutaneous deposits are rare. Two cases were reported by Griffin and Betsill.[10] Smears consisted of syncytial clusters of epithelial cells with a background of neutrophils and histiocytes, some containing hemosiderin.

The epithelial cells were easily characterized by the authors as of endometrial origin. The clinical context is essential and would allow identification of the smear in Figure 13-31 extracted from a cutaneous nodule in the perineum.

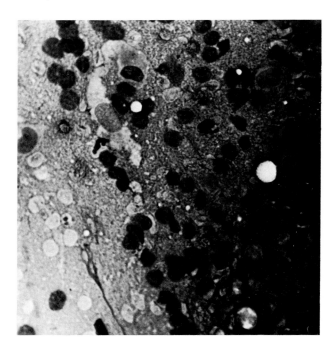

Fig. 13-30. Primary skin lesion, calcifying epithelioma of Malherbe. This lesion yields a dense agglomerate of basal cells intermixed with and emerging from a thick pink stroma. (R stain, ×300)

Ganglion Cysts

Ganglion cysts are generally easily identified clinically. Aspiration yields a pale syrupy, sometimes gelatinous, fluid and may require a larger bore needle.

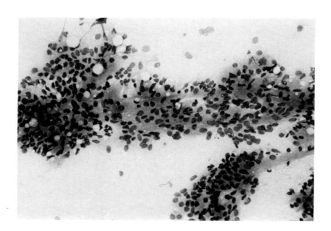

Fig. 13-31. Endometriotic nodule. Loose irregular cluster of round to oval endometrial cells forming a syncytium. (R stain, ×300)

The puncture technique remains the same as with a fine needle. Cells resembling histiocytes are present. An ultrastructural study of Oertel and colleagues demonstrated the fibroblastic and/or histiocytic character of the cells.[26]

Nodular Fasciitis

Nodular fasciitis, a primary subcutaneous lesion, is considered in Chapter 15.

Sebaceous and Dermoid Cysts

Sebaceous and dermoid cysts are considered in Chapter 3.

PRIMARY LESIONS

A miscellaneous series of primary benign and malignant lesions have now become suitable targets for identification by FNA. Primary lesions may present as nodules or infiltrations. Aspiration may be a preliminary maneuver to demonstrate the need for an appropriate surgical biopsy, or it may yield a definitive diagnosis in preparation for radiation or complete surgical excision. It may also confirm a benign lesion not requiring further therapy. A lipoma (see Chap. 15) is such a lesion (Fig. 13-32). New observations continue to be reported.

MALIGNANT LESIONS

Basal Cell Carcinoma

Lesions diagnosed clinically as basal cell carcinoma may be squamous carcinomas, mixed basal and squamous carcinoma, or a variety of benign tumors or inflammatory lesions. Malberger and associates[21] reported 107 cases diagnosed clinically as basal cell carcinoma. Sixty-one of these were ultimately diagnosed both cytologically and histologically as basal cell carcinomas. In 21, no tumor was revealed by smear or surgical biopsy. In this latter group, it was noteworthy that epithelial elements were scarce, while material was ample in the basal cell and squamous groups. With careful FNA, a neoplastic lesion is easily identified and a benign lesion will be spared surgery.

Cytologically, there are cohesive plugs of small cells (Fig. 13-33). Cells appear cuboidal in tight formation, with oval nuclei and large nuclear : cytoplasmic ratio evident in higher power. Plugs of cells may be sharply demarcated or splayed out and intermixed with pink-staining stroma (R) (Fig. 13-34). Peripheral palisading may be evident in some fragments. There is minimal cytoplasm, and nucleoli are largely absent. Smears that appear grossly sparse may contain adequate diagnostic material on microscopic examination.

Squamous Carcinoma

Primary squamous carcinoma may be aspirated by an oblique puncture. The purpose is to distinguish basal cell carcinoma, squamous carcinoma, and nonmalignant keratoses.[21] Cytologic patterns are described in Chapter 3. Squamous differentiation is illustrated in the following case report.

■ *Case 13-3*

A 69-year-old man with advanced arteriosclerosis had a crusted infiltrated skin lesion on his left shoulder. Aspiration yielded keratinizing squamous carcinoma (Fig. 13-35). Hard, palpable axillary nodes yielded nonkeratinizing carcinoma on aspiration (Fig. 13-36).

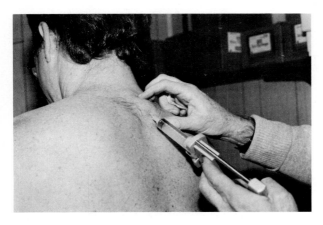

Fig. 13-32. Round delimited mobile rubbery subcutaneous mass clinically consistent with lipoma yields only fat (see Chap. 15).

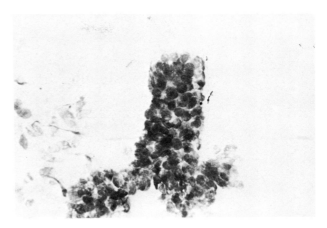

Fig. 13-33. Primary skin lesion, basal cell carcinoma. This is a characteristic tight plug of small cuboidal cells. (R stain, ×280)

These parallel findings illustrate the process of metastasis of a less-differentiated portion of a tumor to regional nodes. The importance of aspiration of possible malignant primary lesions of the scalp is discussed in Chapter 3.

While squamous and basal cell carcinomas are usually subjected to surgical biopsy, recurrent lesions, particularly basal cell carcinomas, may be aspirated at the advancing edge of the tumor to document the presence of malignancy without further surgical biopsy. In the context of multiple squamous

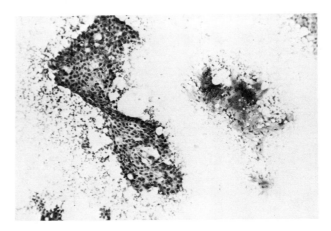

Fig. 13-34. Primary skin lesion, basal cell carcinoma. This is a low-power view of a basal cell cluster with an adjoining stromal mass staining pink with this stain. This cellular arrangement is similar to that of fibroadenoma. (R stain, ×90)

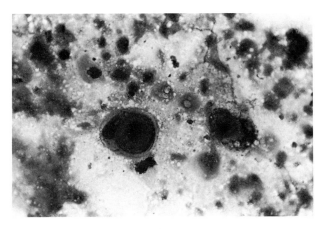

Fig. 13-35. Primary skin lesion, squamous carcinoma. Well-differentiated tumor cells are present in a background of debris. Rounded keratin masses displacing nuclei are evident. (R stain, ×300)

carcinomas, FNA should suffice for diagnosis after surgical biopsy has demonstrated squamous carcinoma in one of the lesions.

Merkel Cell Carcinoma

Merkel cell carcinoma is a rare tumor the cytology of which has been described in several recent case reports.[22,29,34] A clinical report of reddish-purple painless nodules is of importance in the current climate of

immune deficiency disease associated with Kaposi's sarcoma, which can be cytologically distinguished (see below).

Cytologically, there are dispersed small cells with some loosely cohesive clusters. Chromatin is finely dispersed, and nucleoli are small or absent. Cytoplasm, absent or poorly perceptible with P stain, may be evident in R, demonstrating eccentric position of nuclei. Molding, if any, is minimal in contrast with metastatic oat cell carcinoma.

Differential diagnosis with lymphoma, melanoma, neuroblastoma, oat cell carcinoma, and carcinoid is stressed. Ultrastructural study may be necessary for definitive diagnosis. The reader is referred to the original articles for further details.

Cutaneous Lymphoma

Primary cutaneous lymphoma may yield characteristic or suggestive smears (Figs. 13-37 to 13-39). Aggressive tumors with multifocal infiltrated, elevated lesions presenting as smooth, pink-purplish tumors are usually composed of intermediate or less mature lymphoma cells. Aspiration is particularly useful in identifying fresh lesions for radiotherapy or chemotherapy after prior surgical biopsy of a lesion and establishment of cell and tissue types of lymphoma. Extraction may be difficult, with resulting cellular trauma (see Figs. 13-38 and 13-39). With well-differ-

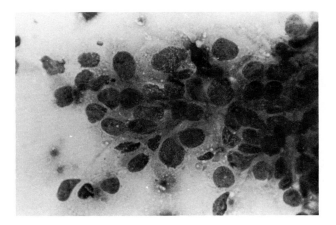

Fig. 13-36. Metastatic squamous carcinoma, lymph node. This is the same carcinoma seen in Fig. 13-35, illustrating metastasis of less-differentiated tumor cells to regional nodes. Keratinization is absent. (R stain, ×300)

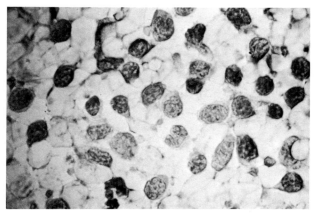

Fig. 13-37. Primary skin lesion, cutaneous lymphoma. This smear of intermediate malignant lymphoid cells extracted with subcutaneous fat. Well-differentiated lymphocytes would be indistinguishable from a benign infiltrate. (P stain, ×600)

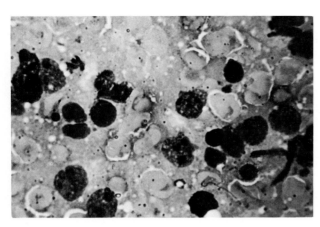

Fig. 13-38. Primary skin lesion, cutaneous lymphoma. This smear was extracted from a deeper lymphomatous swelling of the forearm in a patient with multiple cutaneous lesions. There is some distortion of the cells and a large admixture of red blood cells. (R stain, ×600)

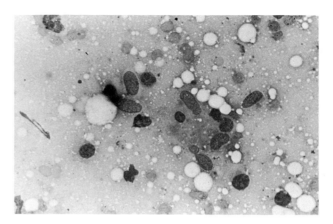

Fig. 13-40. Primary skin lesion, mycosis fungoides. These are scattered, oval, mononuclear cells on a background of tissue juice, intermixed with red blood cells and lymphocytes. (R stain, ×190)

entiated lymphocytic smears, a malignant process cannot be readily distinguished from a benign infiltration of lymphocytic cells. Clinical data are important. The cytology of lymphoma is considered in Chapter 14.

The diagnosis of mycosis fungoides can be suggested by the presence of oval mononuclear cells with granular nuclei on a background of tissue juice and lymphocytes. This cell yield from a lesion clinically similar to lymphoma cutis can raise the suspi-

cion of myocosis fungoides (Fig. 13-40). Lymphoid cells with convoluted nuclei present in T-cell lymphomas may also be extracted from cutaneous nodular lesions of mycosis fungoides.

Kaposi's Sarcoma

Kaposi's sarcoma (KS) may present as a fairly characteristic elevated purple papule. It is difficult to distinguish clinically from lymphoma and Merkel cell

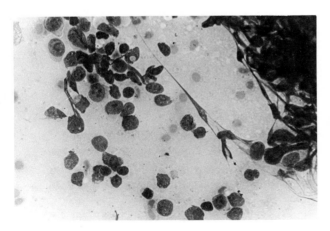

Fig. 13-39. Primary skin lesion, cutaneous lymphoma. Cells were extracted with difficulty. A tangled lymphoid mass is seen, as are pleomorphic cells with nucleoli. (R stain, ×190)

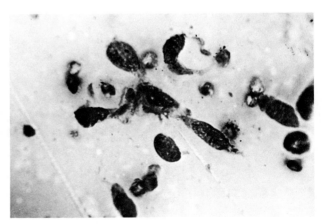

Fig. 13-41. Primary skin lesion, Kaposi's sarcoma. These cells were extracted from a red papule on the toe. Note delicate endothelial cells, some with fine pigment. Diagnosis was confirmed with surgical biopsy. (R stain, ×600)

tumor (see above). The basic cellular structure has been determined to be of endothelial origin.[25] FNA yields delicate spindle cells (Fig. 13-41), easily distinguishing this lesion from a lymphoid infiltrate (Figs. 13-37 to 13-39). Similar cells aspirated from a lymph node involved with KS were described in a recent case report.[9] With the rapid spread of the acquired immune deficiency syndrome (AIDS) and the frequency of lymphadenopathy in homosexual men, awareness of these cells in primary lesions and enlarged lymph nodes will become more important.

Metastatic Lesions

Surface lesions are easily aspirated with a 25-gauge short (1-cm) needle, and diagnostic cells are obtained in a high percentage of patients. In the context of a known primary tumor, this procedure renders surgical biopsy and histopathologic examination unnecessary.

Permeation of lymphatics and tissue spaces by tumor cells producing edema and thickening of the skin without nodularity is characteristic of inflammatory breast carcinoma (see Chap. 6). Less well known is the presence of breast carcinoma cells in the lymphatics of the edematous postmastectomy arm. Cells can be extracted by FNA, altering the therapeutic plan. Particularly in aspiration of extremities, striated

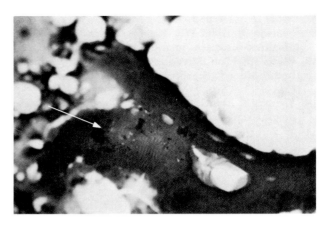

Fig. 13-43. Striated muscle. Striations are evident *(arrow)* in this high-power view. (R stain, ×1000)

muscle may occasionally be extracted with tumor cells (see Fig. 9-1). Cytoplasm of striated muscle is royal blue with R and when fragmented may suggest keratin and a possible squamous lesion (Fig. 13-42). With focusing, striations can be seen (Fig. 13-43).

Where the diagnosis of malignancy occurs in the absence of a known primary tumor, the cytologic pattern may guide the search for such a tumor. Iden-

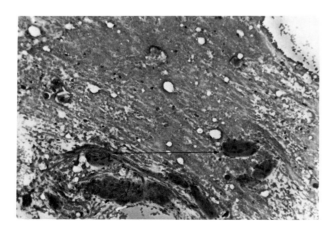

Fig. 13-42. Striated muscle. Suspected tumor deposits may yield fragmented muscle that stains royal blue with this stain and may be mistaken for squamous material. Muscle clusters are trapped in blood clot *(arrow)*. (R stain, ×190)

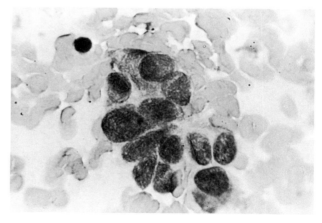

Fig. 13-44. Metastatic skin lesion, adenoid cystic carcinoma. This is a vaguely alveolar cluster of poorly differentiated carcinoma cells extracted from a facial skin nodule in a patient with prior resection of adenoid cystic carcinoma of the parotid gland. This is not an identifiable pattern in the absence of clinical context. (R stain, ×600)

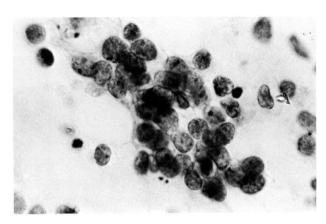

Fig. 13-45. Metastatic skin lesion, undifferentiated carcinoma, left temporal nodule. Abundant cell masses were extracted but failed to give a clue to the primary source. (R stain, ×400)

tifiable cell patterns include melanoma, renal cell carcinoma, thyroid gland carcinoma, breast carcinoma (some types; see Chap. 6), small cell (oat cell) carcinoma, and squamous carcinoma. A large number of cell yields; however, will be identified only as malignant cells of adenocarcinoma and not otherwise specified (Fig. 13-44). Poorly differentiated carcinoma with an unknown primary lesion aspirated from a subcutaneous mass is illustrated in Figure 13-45.

Typical local recurrence of breast carcinoma is illustrated in Figure 13-46. Such lesions are easily aspirated, eliminating the need for confirmation of an

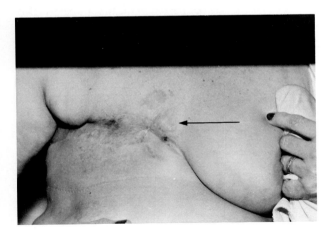

Fig. 13-46. Elevated skin lesions due to dermal infiltration of recurrent breast carcinoma. Lesions are easily aspirated by tangential insertion of a 25-gauge needle.

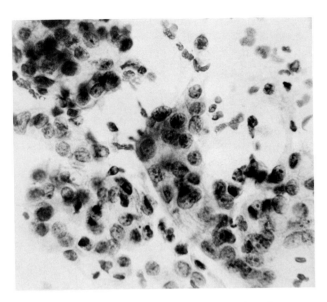

Fig. 13-47. Metastatic breast carcinoma, locally recurrent. Irregular clusters of polymorphous tumor cells. Chromatin pattern and nucleoli are prominent. (P stain, ×600)

obvious clinical state by surgical biopsy. Abundant cell yield is often obtained (Fig. 13-47) (see Fig. 13-29).

REFERENCES

1. Balch CM, Murad TM, Soong S–J et al: Tumor thickness as a guide to surgical management of clinical stage I melanoma patients. Cancer 43:883, 1979
2. Balch CM, Soong S–J, Murad TM et al: A multifactorial analysis of melanoma IV. Prognostic factors in 200 melanoma patients with distant metastases (stage III). J Clin Oncol 1:126, 1983
3. Cohen MH, Jessup JM, Felix EL, et al: Intralesional treatment of recurrent metastatic cutaneous malignant melanoma. Cancer 41:2456, 1978
4. DeVita VT Jr, Fisher RI: Natural history of malignant melanoma as related to therapy. Cancer Treat Rep 60:153, 1976
5. Elwood JM, Lee JAH: Recent data on the epidemiology of malignant melanoma. Semin Oncol 2:149, 1975
6. Evans RW: The Histological Appearances of Tumours. Edinburgh, E & S Livingston, 1968
7. Fontonieré B, Noel P, Mayer M et al: The place of cytodiagnosis in black tumors of the skin: Its

value and limits. In Proceedings of the 9th International Pigment Cell Conference, Houston, 1975. Basel, S Karger, 1975

8. Friedman M, Forgione H, Shanbhag V: Needle aspiration of metastatic melanoma. Acta Cytol 24:7, 1980

9. Gagliano EF: Fine needle aspiration of Kaposi's sarcoma in lymph nodes—a case report. Acta Cytol 31:25, 1987

10. Griffin JB, Betsill WL: Subcutaneous endometriosis diagnosed by FNA cytology. Acta Cytol 29:584, 1985

11. Gumport SL, Harris MN: Results of regional lymph node dissection for melanoma. Ann Surg 179:105, 1974

12. Gupta SK, Rajwanshi LK, Das DK: FNAC smear patterns of malignant melanoma. Acta Cytol 29:983, 1985

13. Hafström L, Hugander A, Jönsson P et al: Fine-needle aspiration cytodiagnosis of recurrent malignant melanoma. J Surg Oncol 15:229, 1980

14. Hajdu SI, Hadju EO: Cytopathology of Sarcomas. Philadelphia, WB Saunders, 1976

15. Hajdu SI, Savino A: Cytologic diagnosis of malignant melanoma. Acta Cytol 17:320, 1973

16. Knutson CO, Hori JM, Spratt JS Jr: Melanoma. In Current Problems in Surgery, pp 1–55. Chicago, Year Book Medical Publishers, 1971

17. Krementz ET, Cerise EJ, Foster DS, et al: Metastases of undetermined source. Curr Probl Cancer 4(5), 1979

18. Linsk JA, Franzen S: Fine Needle Aspiration for the Clinician. Philadelphia, JB Lippincott, 1986

19. Lopansri S, Mihm MC Jr: Pilomatrix carcinoma or calcifying epitheliocarcinoma of Malherbe. Cancer 45:2368, 1980

20. Lund HZ, Kraus JM: Melanotic tumors of the skin. Atlas of Tumor Pathol Sec Fasc 3. AFIP, Washington DC, 1962

21. Malberger E, Tillinger R, Lichtig C: Diagnosis of basal-like carcinoma with aspiration cytology. Acta Cytol 28:301, 1984

22. Mellblom L, Åkerman M, Carlen B: Aspiration cytology of neuroendocrine (Merkel-cell) carcinoma of the skin: Report of a case. Acta Cytol 28:297, 1984

23. Moore GE, Woods LK, Quinn LA, et al: Characterization of cell lines from four undifferentiated human malignancies. Cancer 45:2311, 1980

24. Mosely HS, Giuliano AE, Storm FK, et al: Multiple primary melanoma. Cancer 43:939, 1979

25. Nadji M, Morales PR, Ziegeles–Weissman J et al: Kaposi's sarcoma. Arch Pathol Lab Med 105:274, 1981

26. Oertel YC, Beckman ME, Engler WF: Cytologic diagnosis and ultrastructure of fine-needle aspirates of ganglion cysts. Arch Pathol Lab Med 110:938, 1986

27. Pack GT, Scharnagel IM: The prognosis for malignant melanoma in the pregnant woman. Cancer 4:324, 1951

28. Perry MD, Gore M, Seigler HF et al: FNAB of Metastatic Melanoma. Acta Cytol 30:385, 1986

29. Pettinato G, DeChiaro A, Insubato L et al: Neuroendocrine (Merkel-cell) carcinoma of the skin. Acta Cytol 28:283, 1984

30. Pontifex AH, Roberts FJ: FNA biopsy cytology in the diagnosis of inflammatory lesions. Acta Cytol 29:979, 1985

31. Shaffer B: Identification of malignant potentialities in melanocytic (pigmented) nevus. JAMA 161:1222, 1956

32. Sinner WN: Pulmonary neoplasms diagnosed with transthoracic needle biopsy. Cancer 43:1533, 1979

33. Sober AJ, Fitzpatrick TB, Mihm MC et al: Early recognition of cutaneous melanoma. JAMA 242:2795, 1979

34. Szpak CA, Bossen EH, Linder J et al: Cytomorphology of primary small-cell (Merkel-cell) carcinoma of the skin in fine needle aspirates. Acta Cytol 28:290, 1984

34a. Urist MM, Balch CM, Milton GW: Surgical management of the primary melanoma. In Balch CM, Milton GW (eds): Cutaneous Melanoma. Philadelphia, JB Lippincott, 1985.

35. Weidon D, Little JH: Spindle and epithelioid nevi in children and adults. A review of 211 cases of Spitz nevus. Cancer 40:217, 1977

36. Willis RA: The Spread of Tumours in the Human Body, 3rd ed. Scarborough, Ont., Butterworth & Co, 1973

37. Yamada T, Itou U, Watanabe Y, et al: Cytologic diagnosis of malignant melanoma. Acta Cytol 16:70, 1972

CHAPTER 14

Diseases of the Lymph Node and Spleen

JOSEPH A. LINSK

SIXTEN FRANZEN

With a section on Immunochemistry by Edneia M. Tani and
Lambert Skoog

Lymph nodes were early targets for aspiration yielding diagnostic smears in a variety of inflammatory and neoplastic disorders.[13,22,23,24,40,43] The geometric progress and acceptance of fine needle aspiration (FNA) since these early reports and indeed since the first edition of this book have been particularly striking in the diagnosis of lymph node disease. Numerous papers have appeared reporting percentages of malignant versus benign aspirates, lymphomatous versus nonlymphomatous smears, specialized staining, and morphologic observations.[9,32,46–50,53] It has become evident that understanding lymph node cytology, in addition, aids in the diagnosis of the considerable number of extranodal lymphomas (discussed in Chaps. 4, 5, 7, and 12) including splenic lymphoma (see below).

The diagnosis of lymphomas by cytologic technique remains, however, one of the major controversial areas in aspiration biopsy cytology (ABC).[25] Critical comment ranges from the categorical statement that it must never be attempted to the realistic observation that it is difficult.

In recent reports, a correct diagnosis confirmed histologically was reported in 42 of 49,[49] 170 of 180,[48] and 53 of 59[53] cases. Lopes–Cardozo reported on 1023 primary lymphoma cases he diagnosed initially on cytologic smears.[37] Based on the author's hands-on experience with primary lymphoma, there is no question that the diagnosis can be made with ease by

FNA. In fact, as pointed out by Lopes–Cardozo,[37] cytologic smears may reveal the diagnosis before histology is "positive" in some cases and provide essential evaluation of individual cell cytology not evident in histologic sections.

The caveats remain as follows: experience, clinical correlation, and a ready willingness to defer to surgical biopsy as needed. Difficulties are reduced immediately by dividing cases into *de novo* and known lymphomas. Aspiration of an inguinal and femoral node in a patient with Hodgkin's disease established in a cervical or axillary region will often yield readily identified Hodgkin's smears. The hallmark is the Sternberg–Reed (S–R) cell (Fig. 14-1) and in the fact its presence in the appropriate clinical setting makes the diagnosis of Hodgkin's one of the least difficult.

THE NODAL MASS

Cervical and other regional nodes must be viewed as generic tumors[36] until proved to be of lymph nodal origin by examination of cytologic or histologic preparations under the microscope. The cytologist must be prepared at all times to recognize glandular (submandibular gland), cystic (branchial cyst), soft tissue (lipoma, carotid body tumor, hibernoma[28]), and metastatic carcinoma.

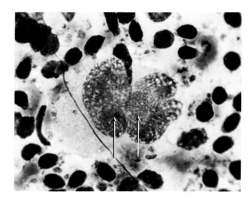

Fig. 14-1. Aspirate of lymph node shows Sternberg–Reed cell with giant nucleoli *(arrows)*. (R stain, ×1000) (Linsk JA, Franzen S: Fine Needle Aspiration for the Clinician, p. 54. Philadelphia, JB Lippincott, 1986)

USES OF LYMPH NODE CYTOLOGY

Useful lymphoid aspirates are yielded in the following clinical contexts:

1. Primary tumors (lymphoma) of lymphoid tissue arising as a monoclonal proliferation from many of the cellular components of the lymph node.

2. Regional and remote nodes or extranodal tumors simultaneously present or sequentially appearing. Smears from such nodes allow nonsurgical staging and can establish the progression of the disease to a less-differentiated state.

3. Lymphadenopathy as a reflection of a regional or systemic disorder (infectious mononucleosis, cat scratch disease, sarcoidosis, tuberculosis, acute inflammations, primary cancer, leprosy, toxoplasmosis, trypanosomiasis, anticonvulsant therapy, vaccinations, and acquired immune deficiency syndrome [AIDS]).[6,11,16,22,30,42]

4. Extraction of lymphocytes for study of immunologic markers and the effects of functional and physiological stimulation (see pp. 358–360).[54,64,65]

5. Intra-operative FNA of lymph nodes as an adjuvant or substitution for frozen section.[61]

The Clinical Context

ABC has its optimal use in the hands of the interested clinician with cytologic training and the pathologist with clinical training, who are in the best position to integrate morphologic findings and clinical impression. Knowledge of the clinical context or direct examination of the patient and target areas is an invaluable aid to cytologic diagnosis. The following questions should be answered:

Is the target a regional drainage site for infection or tumor?

Is there a known lymphoma in which multicentric or progressive disease is likely to have occurred?

Is the node in an atypical site, such as the posterior cervical triangle, suggesting *de novo* disease?

Is there a systemic disease known to be accompanied by lymphadenopathy?

Is there a dermatitis?

Did the patient receive a vaccination recently?

Is the patient receiving anticonvulsants?

Are carotid body tumor and submandibular salivary glands in cervical areas a possibility?

Is there venereal disease of external genitalia or a small hernia (when dealing with an inguinal or femoral node).

A positive answer to any of these questions will help the cytopathologist evaluate the smears.

The cytology report should clearly state that the lesion is an inflammatory disorder (*e.g.,* granulomatous, pyogenic), that the lesion appears to be hyperplastic and surgical biopsy can be deferred pending results of medical treatment, or that the lesion is a lymphoma that requires surgical biopsy for classification. *All inconclusive cytologic reports carry the potential for false-negative results, and further diagnostic measures, including surgical biopsy, must be used as deemed appropriate.*

At the point of excision, the interested clinician must synchronize his efforts with the surgeon's and the laboratory technicians' to obtain cytologic information by aspirating the excised node. The specimen should be kept in normal saline solution when secured and hand-delivered. For the purpose of aspiration, place the node on a clean piece of gauze, hold it down with the thumb and index finger, and carry out aspiration as usual. If done quickly, this step will not interfere with the preservation of tissue in formalin, which follows immediately. Table 14-1[33] summarizes basic cytologic and clinical correlations.

Time is a major ingredient in lymphoma diagnosis. When morphologic and clinical findings leave

Table 14-1. Basic Cytologic and Clinical Correlations

PHYSICAL FINDINGS	FINDINGS ON ASPIRATION	LABORATORY INVESTIGATION	EXAMPLES
Unilateral or symmetric, freely movable regional LN; no hepatosplenomegaly	Specific or reactive lymphadenitis; polymorphic smears; epithelioid cells; histiocytic reaction	Atypical lymphocytes in blood smear; specific serologic tests	Viral "nonspecific" lymphadenitis; mononucleosis; toxoplasmosis; AIDS-related complex.
Unilateral fluctuant LN; absence of generalized lymphadenopathy, no hepatosplenomegaly	Lymphadenitis with abscess formation; abundant PMNs; caseation (necrosis); giant cells; epithelioid cells	Gram's stain and acid-fast smear and culture of aspirate	Pyogenic infection; "sterile" abscess; tuberculosis; cat scratch disease
Regional or multicentric LN enlargement; fever, weight loss, hepatosplenomegaly, or a combination	Malignant lymphomas (HL or NHL); monomorphic smears; pleomorphism in the presence of Sternberg–Reed cells	Anemia, lymphopenia, or lymphocytosis; mediastinal adenopathy on x-ray film. Bone marrow surgical biopsy.	Malignant lymphomas
Multiple regional LN enlargement; hepatosplenomegaly	Prolymphocytes and plasmacytoid smears	Specific serum protein analyses (IgM or other monoclonal protein spike)	Waldenström's macroglobulinemia; heavy chain disease; gastrointestinal lymphoma

Key: *LN*, lymph node; *PMN*, polymorphonuclear leukocyte; *HL*, Hodgkin's lymphoma; *NHL*, non-Hodgkin's lymphoma.

doubt, it is best to wait a suitable period (up to many weeks in some cases) and repeat the aspiration. Immediate therapy is often not critical. Waiting is preferable to surgical biopsy in many instances. The ease of FNA lends itself to this approach. Also as stressed by Lopes-Cardozo[37] excision of a node leaves no marker to gauge progression or regression of the pathologic process.

Method and Procedure

The technique of FNA described in Chapter 1 is fully applicable to lymph nodes and related masses. An important variation in the technique is the use of the 25-gauge or larger needle to puncture small nodes in atypical sites such as the posterior cervical triangle. Such nodes may be the initial presentation of non-Hodgkin's lymphoma. Greater control is exercised with small mobile targets if the needle is inserted manually and the syringe subsequently attached. Fine needle sampling without aspiration has been stressed by Zajdela.[64] Smears should be prepared with more than the usual care and instructions should be addressed to referring clinicians carrying out the procedure. It has been established that FNA leaves no significant effect on benign lymph node histology.[1]

Thin aspirates consisting of sanguineous tissue fluid may contain a surprising number of cells. Such smears may be all that is attainable, particularly in children. Although not the most satisfactory mate-

rial, a report can be issued stating that the smear is lymphoid, is mixed, contains Sternberg–Reed cells, is monomorphic, and so on. It is important to note that non-Hodgkin's lymphoma in children is almost invariably poorly differentiated[31,63]; therefore the presence of mature lymphocytes should largely exclude that diagnosis. Convoluted cells where identified (see below) will suggest lymphoblastic lymphoma of childhood.

A well-placed needle should produce a gelled droplet of lymphoid material. Smears for Romanowsky (R) stain can be made, and sufficient time will be allowed to produce fixed smears for Papanicolaou (P) before the material dries. Smears are initially inspected under low magnification, and a good specimen will reveal heavily stained areas that spread more evenly toward the periphery. Because of variability in available material, the cytopathologist may be faced with thick smears with cellular distortion, dense cell aggregates, and traumatized cells. Clues to the diagnosis may sometimes be evident on examination of the periphery of the dense areas (Fig. 14-2A and B).

Strands of nuclear chromatin are often present, stretching between delicate traumatized cells. They are more commonly seen with immature cells; however, their presence very much depends on care in smearing, and they may be seen with all lymphoid smears. Such smears are of limited diagnostic value as a rule (Fig. 14-2C). However there may be intact groups of cells that are diagnostic.

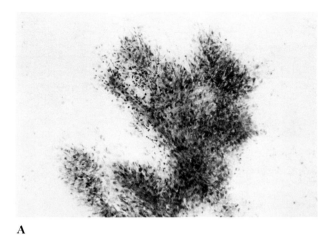

A

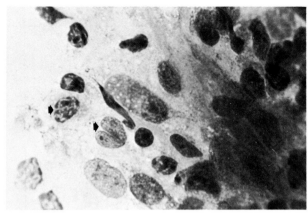

B

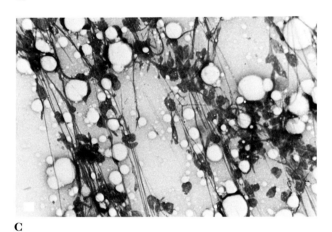

C

Fig. 14-2. A. Dense lymphoid aggregate with feathered edge. **B.** Lymphadenitis. A high power view of the edge of a dense smear may be necessary when an evenly spread smear is not available. Clefted (cleaved) cells are evident *(arrows).* **C.** Traumatized smear with nuclear strands. (R stain; A: ×90; B: ×1000; C: ×330)

CYTOLOGIC COMPONENTS OF NORMAL LYMPH NODES AND BENIGN LYMPHADENOPATHIES

The normal lymph node is not a target for FNA. Its components may be studied in lymph nodes harvested at autopsy. Benign lymph nodes draining surgical specimens may display varying degrees of reaction including sinus histiocytosis and are therefore not normal.

The progressive morphologic changes during lymphocytic proliferation are sampled by the aspirating needle, which will produce smears with individual components in varying concentrations. The mixed population of lymphadenopathies can yield the following cells:

Mature lymphocyte
Prolymphocyte[20]
Small cleaved cell
Large cleaved cell
Immunoblast (centriblast, transformed cell, follicular center cell)
Histiocyte
Plasma cell
Mast cell

Monoclonal proliferation of any of these components will yield a monomorphic smear characteristic of the full spectrum of reticulo-endothelial malignant disorders with the exception of Hodgkin's disease.

Additional reactive cells include the following:

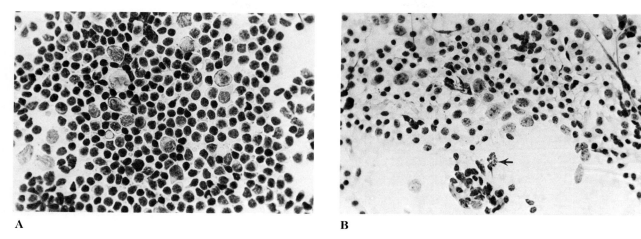

A   B

Fig. 14-3. A. Lymphadenitis, reactive. The gradation from small to intermediate to transformed lymphocytes is evident in this smear. (R stain, ×300) **B.** Lymphadenitis, reactive. This smear illustrates the characteristic pleomorphic aspect of a reactive lymph node which allows presumptive identification. Mitosis is identified *(arrow)*. (P stain, ×190)

Eosinophils
Neutrophils
Epithelioid cells (characteristic of granulomatous reactions) (see Fig. 14-8)
Phagocytes (tingible body cells)
Giant cells

Benign giant cells are reactive types— Langerhans', foreign body, Warthin–Finkeldey (measles) which must be distinguished from the giant Sternberg–Reed (S–R) cell. The reactive giant cell is multinucleated in contrast with the S–R cell, which is binucleated or has one large nucleus (Fig. 14-1).

Fatty replacement of lymph nodes can mimic relapse of lymphoma.[56]

Benign Cytology

A pleomorphic picture is considered essential to the cytology of benign lymphadenitis (Plate XXVII). Mature, small cells, cleaved or noncleaved; intermediate prolymphocytes; and large cleaved and noncleaved immature lymphoid cells (follicular center cells, transformed cells, centriblasts) are all represented (Fig. 14-3). The highly reactive lymph node with follicular hyperplasia will yield increased transformed (follicular center) cells intermixed with the full spectrum of lymphoid cells (Fig. 14-4). Although lymph-

oid cells make up the major component, phagocytic cells with engulfed debris and variable numbers of neutrophils, plasma cells, and eosinophils are commonly seen. Eosinophils make differentiation between Hodgkin's lymphoma and benign enlargements in atypical smears difficult, but non-Hodgkin's lymphoma is usually free of these cells.[51] Tingible body cells, while rare in non-Hodgkin's lymphoma, do not exclude the diagnosis (Fig. 14-5).

The classic disorder that yields a granulomatous cytologic smear is sarcoid. Acid-fast, fungal, syphilitic, and various bacterial infections may produce granulomas with and without necrosis, caseation, and acute inflammation. Be careful not to overlook the possibility of similar and confusing reactive changes as a part of the cytologic picture of Hodgkin's disease. Large histiocytes, sometimes multinucleated, may simulate S–R cells but lack macronucleoli (Fig. 14-6). Such atypical smears, particularly with eosinophils, mandate surgical excision. Contrast with true S–R cell (Fig. 14-7).

Multinucleated giant cells are a known part of caseating and noncaseating granulomas. The hallmark of the granulomatous smear is the presence of clusters of epithelioid cells (Fig. 14-8A) scattered through the lymphoid smear. These are usually elongated, often semilunar cells with royal blue cytoplasm with R stain that are contrasted with the sur-

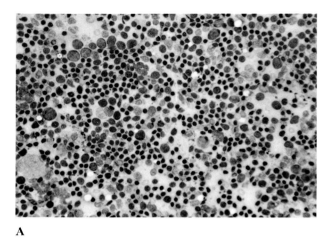

A

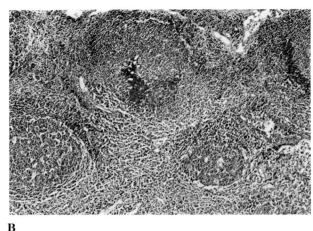

B

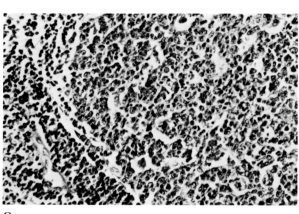

C

Fig. 14-4. **A.** Reactive lymphadenitis. This is a mixed smear with numerous transformed cells intermediate and small lymphocytes. Histiocytes and tingible body cells were present. Lymph node enlargement persisted and surgical biopsy was carried out. (R stain, ×300) **B.** This is a low-powered view of the follicular changes in the reactive lymph node from which the aspirate (**A**) was taken. (H & E stain, ×90.) **C.** This is a high-powered view of the follicle center showing the variety of cells which can be extracted. (H & E stain, ×280)

rounding lymphocytes. With P stain, fine granular nuclear chromatin is evident. Multinucleated giant cells are occasionally seen in conjunction with the epithelioid groups. They may be seen as isolated cells in an inflammatory exudate on a background of polymorphonuclear leukocytes (see Fig. 14-8B). In caseating granulomas, the background may be granular necrosis.

In aspiration of a fluctuant node, several additional smears are prepared for bacterial staining (Gram's and acid-fast stains). The smear may be dominated by an intensely neutrophilic population, almost to the exclusion of lymphoid elements (see Fig. 14-9).

Lymphoid Enlargements

The determination that a lymph node enlargement is benign is based on clinical evaluation; cytologic aspi-

ration; and, finally, surgical excision with histopathologic diagnosis.[55]

Enlarged regional nodes draining an obvious focus of infection should pose no diagnostic problem. Lymphadenopathy appearing *de novo* or remaining after the primary inflammatory focus has subsided is clearly a diagnostic problem. In children, it appears reasonable to observe such nodes without surgical intervention while carrying out medical therapy. A protracted period of observation even after treatment may be in order, since such nodes in children may regress slowly. However, all persistent nodes must be viewed with suspicion. Remember that malignant lymphoma may be preceded by a long period of reactive lymphadenopathy, and patients with a history of several lymph node biopsies before the definitive histologic diagnosis is made are not rare.

Lymphoma must be suspected in nodes appearing in atypical sites. A single posterior triangle

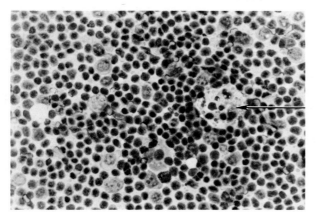

A

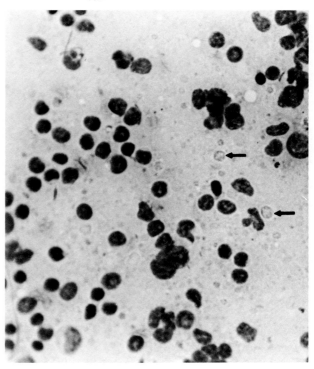

B

Fig. 14-5. **A.** Tingible body cell *(arrow)* contrasted with non—phagocytic-transformed cells. (R stain, ×800) **B.** Lymphoglandular bodies *(arrows)*, mixed lymphocytic—histiocytic non-Hodgkin's lymphoma. Lymphoglandular bodies are free fragments of lymphocyte cytoplasm shed within a preserved cell membrane. They are abundant in organized lymphatic tissue, including lymphomas, and offer a fine identification criterion for such tissues. (MGG stain, ×1000)

node in an adult with no scalp infection, with no other regional nodes, and no symptoms can easily be the sentinel node of Hodgkin's or non-Hodgkin's lymphoma. Clinical evaluation is followed by ABC. A mixed cytologic smear containing histiocytes, occasional granulocytes, and eosinophils without classical S–R cells (Fig. 14-6) is nevertheless suspicious for Hodgkin's disease. Compare with Fig. 14-7. A similar isolated node with a uniform cell population is highly suspicious for non-Hodgkin's lymphoma. The location and isolation of the node are important diagnostic data.

Cytologic Interpretation — Practical Problems

Problems of histologic interpretation of benign and malignant lymphoid lesions are well documented.[8,19,30,59,60] Tissue fixation, thinness of sections, and variation in staining all produce variables that will interfere with uniform and consistent interpretations. These variables yield to careful and directed histologic technique.

The cytopathologist must be warned about the intractable variation in the quality of smears he will receive. It is extremely difficult to obtain optimum smears from the cross section of clinicians. Expectation that smears received in clinical practice will present with clearly defined monomorphic and mixed patterns with neatly dispersed cells allowing subclassification will rarely be fulfilled. The quality of smears is improved when they are prepared by the cytology staff, but even here the nature of the target node and the puncture technique will introduce variables. The cytologist therefore must be prepared to improve his skills by selecting the best part of the smear[33] rejecting patently unsatisfactory areas or smears, and most importantly insisting on clinical correlation. Theoretically, repeat aspirations can be requested, but practically are not always available because of patient noncompliance or remote location.

Cytologic clues joined with clinical correlation may not yield a definitive cytologic diagnosis, but these data together with judicious suggestion may readily lead the clinician to the final patient diagnosis.

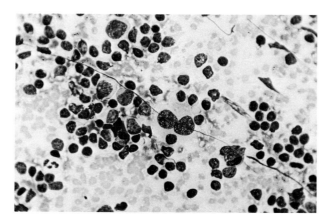

Fig. 14-6. Lymphadenitis, reactive. This smear contains numerous enlarged histiocytes. One has formed a double nucleated cell simulating a Sternberg–Reed cell but without giant nucleoli and with voluminous cytoplasm. Contrast with Fig. 14-7. Surgical biopsy is indicated, depending on the clinical context. (R stain, ×300)

Clinical context should divide the cytopathology into two groups—nonspecific reactive lymphadenopathy and specific lymphoid reaction.

Nonspecific Reactive Lymphadenopathy

In practical terms, the designation "reactive lymphoid smear" is the most common cytologic diagnosis made. The nodal reaction results from exposure to a variety of antigens and allergens including bacteria, viruses, and environmental and autoreactive substances. The cytologic yield is composed in varying concentrations of the cells listed above (pp. 340–341) and illustrated in Figures 14-2B, 14-3, 14-4A, and 14-6. The concentration of individual cellular components will depend on the extent of change in the node (time since nodal stimulation) and the fortuitous location of the needle at the time of puncture (sampling variation). Identification of the etiologic agent in the reactive lymphoid hyperplasias can be made in less than 10% of cases.[30]

Specific Lymphoid Reaction

The histology of 27 specific types of lymphadermitis and lymphadenopathy has been assembled by Ioachim.[30]

No systematic cytologic study has been made of these subgroups, although there are individual case reports and small groups of cases.[6,16] It is critical that

clinical data be coupled with the cytologic smears. A smear rich in immunoblasts, for example, should be confidently diagnosed if a history of vaccination followed by regional lymphadenopathy is provided.

Bacteria such as gonococcus, tuberculosis, and leprosy can be detected on smear or culture. The easy identification of mycobacterium intercellularae in smears from AIDS-related lymphadenopathy has been a useful recent observation (see below).

Summary of Cytopathologist's Diagnostic Plan for Benign Lymphadenopathies

1. Obtain full clinical data from referring physician or perform focused history and physical examination at time of aspiration.
2. Separate smears into metastatic carcinoma, lymphoma, or benign lymphoid proliferation by adhering to cytologic criteria (see Chap. 3).
3. Separate benign smears into the nonspecific reactive category or specific lymphadenitis based on cytologic and clinical criteria (for example, recent cat scratch), if possible.
4. Review the variety of lymphadenopathy etiologies and offer diagnostic guidance to the referring physician or to the patient.

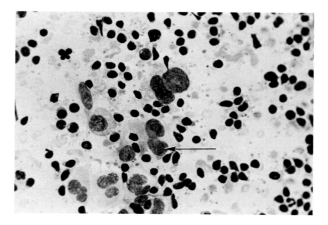

Fig. 14-7. Hodgkin's lymphoma, lymphocyte rich. In this portion of the smear, there is a cluster of Sternberg–Reed cells surrounded by small non-neoplastic lymphocytes. Lymphocytes are indistinguishable from those of well differentiated NHL. Some of the Sternberg–Reed cells show classical macronucleoli *(arrow)*. Others may be the large Sternberg–Reed-like cell without macronucleoli cited by Kardos.[32] (R stain, ×800)

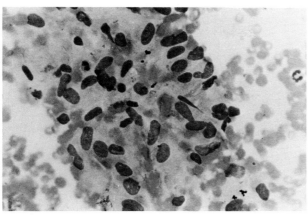

A

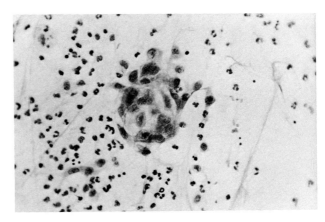

B

Fig. 14-8. A. Granulomatous aggregate. This is a loose cluster of elongated semilunar epithelioid nuclei in a syncytium of blue cytoplasm, part of a chronic granulomatous reaction. Nuclear chromatin is homogeneous. (R stain, ×120) **B.** Lymphadenitis, granulomatous, acute. An epithelioid cluster, possibly forming a giant cell, is seen on a background of granulocytes. Nuclear chromatin is fine dusted. (P stain, ×100)

MALIGNANT LYMPHOMA

Histopathologic Considerations

Histopathology of malignant lymphoma has been the basis for therapeutic decision making although the large number of classifications has blunted the preciseness of this process.[2,3,12,18,21,35,39,41,44,51] Fol-

lowing the cytologic diagnosis, nodal excision is recommended to determine nodular versus diffuse patterns in non-Hodgkin's lymphoma and subtype (lymphocyte rich—nodular sclerosing—mixed cellularity and lymphocyte poor) in Hodgkin's disease. Therapy varies considerably depending on histologic subtype and may even be omitted in some nodular lymphomas. There has been erosion in this mandate

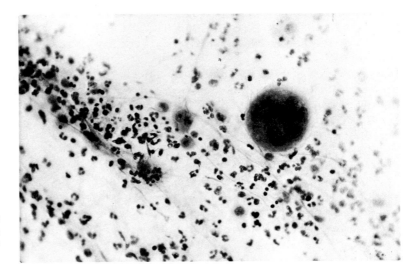

Fig. 14-9. Lymphadenitis, acute. Smear is heavily infiltrated with granulocytes. There is one massive multinucleated giant cell. (P stain, ×150)

 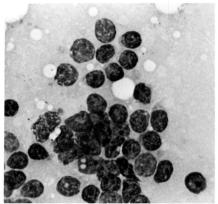

A   B

Fig. 14-10. Non-Hodgkin's lymphoma, lymphocytic, well differentiated. A. Low-power view illustrates a uniform carpet of cells. (R stain, ×33) B. Higher power view of cells morphologically not distinguished from mature lymphoid cells. Of importance is the minimal variation in cell size. Chromatin is clumped, and nucleoli are absent. The mottled gray background is composed of packed lymphoglandular bodies. (R stain, ×180)

to obtain histopathology, particularly in some extranodal lymphomas (see below). Laparotomy in an elderly patient with x-ray proven lymphadenopathy and cytologically proven non-Hodgkin's lymphoma may not be in the patient's best interest. The full panoply of clinical, including therapeutic, considerations should be reviewed as a team effort.

Reliance on histopathologic diagnosis as well as subclassification of lymphomas must be tempered. "In 37–53% of biopsied lymph nodes, a definitive diagnosis is not reached."[30] An initial diagnosis of Hodgkin's disease in 600 cases was not confirmed by Symmers in 40% of the cases;[59] and of 226 cases diagnosed reticulum cell sarcoma, 27% were not confirmed.[60]

A detailed evaluation of four of the major histologic classifications of lymphoma has been recently reported.[35] There is considerable overlap and a variety of advantages and disadvantages noted for each classification. The paper concluded that pathologists should continue to use that classification with which they feel most comfortable.

The aspiration (or imprint) smear[14,37] provides cytologic definition, which cannot be recognized in histologic sections. Final diagnostic and therapeutic decision should rest on cytologic smears, histopathology, and clinical data.

Cytologic Considerations

After cytologic separation of benign and malignant smears, focusing on specific etiologies in benign lymphadenitis (see above), morphologic (and theoretically therapeutic) subtyping in malignant lymphomas can take place.

The details of a cytologic classification vary with the interest and experience of the cytopathologist and a number have been published.[34,37,47]

Non-Hodgkin's Lymphoma

Division into small, intermediate, and large cell types is the first determination and is the most practical. Cleaved cell pattern means merely that some of the cells in the smear, small or large, will be clefted (see below). Prognostic and therapeutic considerations based on such detailed cell types continue to undergo modification.

Small cell lymphoid patterns are extracted from well-differentiated, diffuse, non-Hodgkin's lymphoma; chronic lymphocytic leukemia; and lymphocyte-rich Hodgkin's disease (Figs. 14-10 and 14-11).

Diffuse plasmacytoid lymphoma associated with Waldenstrom's macroglobulinemia is histologically classified as one of the small cell lymphomas.[51]

Plasmacytoid lymphocytes, however, appear somewhat larger than the lymphocytes of small cell lymphocytic lymphoma (Fig. 14-12 and Plate XXVIII).

Lymphoplasmacytic cytohistology is frequent in primary lymphoma of the gastrointestinal tract. Aspiration of plasmacytoid cells from an abdominal mass offers a clue to point of origin of the target (see Table 14-3).

Intermediate lymphoid smears are not representative of a specific histologic category and are probably a larger variety of relatively well-differentiated non-Hodgkin's lymphoma. Many smears cannot be categorized as other than intermediate (Fig. 14-13A and B), based on cell size and atypicality of the nucleus.

Diffuse large cell lymphoma in which curative chemotherapy has been undertaken[45] is a heterogeneous group both morphologically and by marker studies.[8] Consensus regarding immunoblastic versus nonimmunoblastic subtypes was reached in only 142 of 327 cases (43%) by a group of expert hematopathologists.[8] The diffuse histiocytic category, while frequently cited, has been reduced to less than 5% of non-Hodgkin's lymphomas in the working formulation (Fig. 14-14).[33,44] The cytopathologist must be

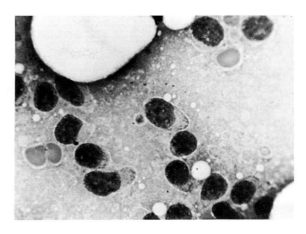

Fig. 14-12. This smear was extracted from an enlarged axillary node in a patient with Waldenström's macroglobulinemia. Plasmacytoid cells are evident. (R stain, ×330)

practical and not overreach into subcategories that to date have not been established histologically or found consistently useful therapeutically.

Non-Hodgkin's lymphoma (large cell) poorly differentiated (Fig. 14-15) is the common large cell lymphoma. Follicular center cell, centriblastic, and large cleaved cell are terms applied (Plate XXIX). As noted, few cleaved cells may be evident cytologically.

The term *lymphocytic-histiocytic* (Fig. 14-14) is loosely applied to lymphoid tumors composed of generally an equal distribution of larger and smaller cells. Histologically, there may be a mixed diffuse pattern with large follicular center cells and small cleaved cells. More commonly, mixed pattern occurs in follicular lymphomas. Cytologic identification of mixed small cleaved and large cell lymphoma is accomplished only with difficulty. More important is the recognition that sporadic and often frequent large cells are intermixed with small cells in benign reactive lymphadenopathies (see Figs. 14-3A and 14-4A) and should not be confused with lymphoma. On the other hand, in monomorphic patterns characteristic of lymphoma, cell size variation is common (see Figs. 14-13B and 14-15B), and attempts at distinguishing a diffuse from a mixed pattern may not be warranted. Microscopic focusing initially with low power will usually provide an accurate overview of monomorphic distribution or suspicion of it.

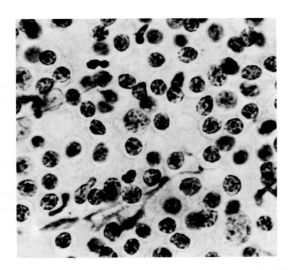

Fig. 14-11. Hodgkin's disease, lymphocyte rich. Dispersed lymphoid cells with distinct punctuate chromatin but no well-defined nucleoli. There is mild variation in cell size and the shadow of cytoplasmic boundaries is evident. Mitoses are absent. Sternberg–Reed cells are present in other fields. (P stain, ×250)

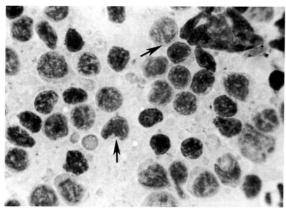
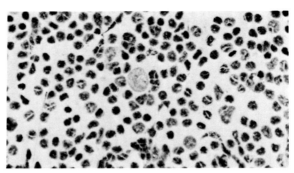

A   B

Fig. 14-13. A. Non-Hodgkin's lymphoma, intermediate differentiation cells are some-what larger than small mature lymphocytes. Nuclear chromatin is looser. Nucleoli are inconspicuous in this stain. Clefted and cleaved cells are indicated (arrows) and lym-phoglandular bodies are apparent in the background. (R stain, ×400) B. Non-Hodg-kin's lymphoma, intermediate differentiation. This is a carpet of somewhat irregularly shaped, occasionally clefted intermediate lymphoid cells, punctuated by an occasional enlarged histiocytic cell. Such a smear must be distinguished from lymphadenitis (compare to 14-3A) on the one hand and from Hodgkin's disease (compare to 14-7) on the other hand. The entire smear was more consistent with a monomorphic pattern. (R stain, ×600)

Diffuse and nodular lymphomas cannot be cy-tologically distinguished with any assurance. Histo-pathology is necessary for this separation.

Progression of well-differentiated lymphocytic lymphoma to intermediate or poorly differentiated types on subsequent aspiration of regional or multi-centric nodes has been clinically useful in dictating change of therapy. Awareness in general of the cyto-logic classification of a primary nodal target helps refine the diagnosis of smears from enlarged remote nodes.

Individual Cell Cytology

R stains are particularly suitable for identification of lymphoid and plasmacytoid cells. Distinction of poorly differentiated lymphoma from poorly differ-entiated carcinoma may be more easily accomplished with this stain. The hallmark of lymphoid smears is the presence of lymphoglandular bodies described by Söderström.[57] They are round, pale, basophilic struc-tures that have either smooth borders or projections (Fig. 14-16). The presence of lymphoglandular

bodies in large numbers is highly characteristic of lymphoid tissue.

Poorly differentiated carcinomas primarily bronchogenic may display cytoplasmic fragments that can be distinguished from lymphoglandular bodies with experience. Other criteria of carcinoma (clumping and molding) are usually present. In acute leukemia, both myeloblastic and lymphoblastic, the blast cells shed cytoplasmic fragments, but the clini-cal picture is decisive.[7]

Because of the fragility of lymphoid cells and the problems of smearing, there is considerable varia-tion of cell morphology within each subclassification.

Nuclei are small and round or oval in well-dif-ferentiated cells (see Figs. 14-7, 14-10B, and 14-11). As differentiation lessens, nuclei increase in size and shapes vary (see Fig. 14-13). In our experience, con-voluted or cerebriform nuclei occur infrequently in routine smears obtained in the general hospital set-ting. T-cell lymphoma with the presence of at least some convoluted nuclei occurred in only 27 of 278 cases studied cytologically by Das and associates.[9] Over 50% presented with mediastinal compression (Fig. 14-17). Where the number of convoluted nuclei

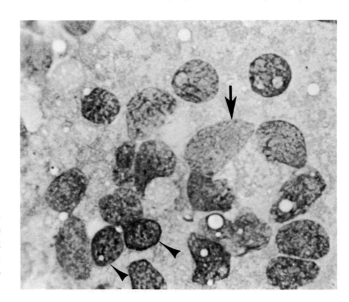

Fig. 14-14. Non-Hodgkin's lymphoma (large cell), histiocytic type. Clusters of enlarged, elongated, plump and notched nuclei with finely granular chromatin *(arrow)*. Smaller varieties with similar chromatin pattern *(arrowheads)* suggests mixed lymphocytic–histiocytic subtype. (R stain, ×900)

in the smear is small, cytochemical analysis with acid phosphatase, periodic acid-Schiff, and alpha naphthyl acetate esterase should be applied to confirm the T-cell origin (see below). Convoluted nuclei are more prominent in P but may also be identified in R stain. The nucleus is well defined in R but the nuclear membrane is not sharply etched as it is in P stain (see Figs. 14-13A and 14-15B). Mitotic figures are occasionally seen in benign reactive smears (see Fig. 14-3B) but they are rare in well-differentiated lymphoma smears, and are uncommon generally.

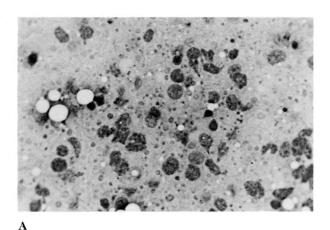

A

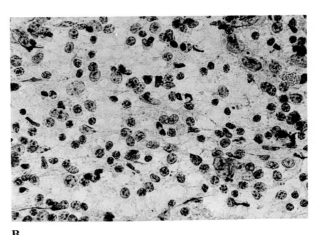

B

Fig. 14-15. **A.** Non-Hodgkin's lymphoma (large cell), poorly differentiated type. Dispersed, fragile, moderately pleomorphic nuclei with clumped, blurred chromatin are seen on a diffuse bed of lymphoglandular bodies and opaque cytoplasmic sap. A clear background is not characteristic. (R stain, ×800) **B.** Dispersed cells with well-defined nuclear membranes are again seen on a (mottled) grey background, with ill-defined lymphoglandular bodies. Nucleoli are present in some cells. Mitoses are not evident. (P stain, ×900)

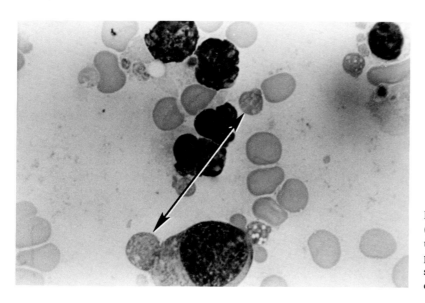

Fig. 14-16. Non-Hodgkin's lymphoma (large cell), poorly differentiated. Immature lymphoid cell is seen shedding lymphoglandular body *(arrow)*. Others are seen intermixed with smaller lymphoid cells. (R stain, ×1000)

Chromatin is fairly dense and clumped in well-differentiated cells (see Figs. 14-10B and 14-11) and loosens up in the intermediate cells (see Fig. 14-13A). It appears granular and finely divided in both benign transformed (centriblastic) cells and in the cells of poorly differentiated lymphoid (see Fig. 14-15) as well as immunoblastic and histiocytic tumors (see Figs. 14-14 and 14-18A, and Plate XXXI).

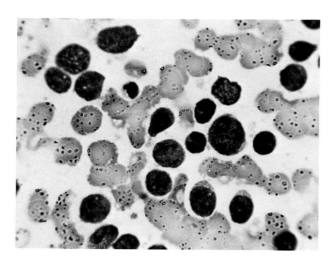

Fig. 14-17. Non-Hodgkin's lymphoma, lymphoblastic type. Smear was aspirated from a mediastinal mass in a 14-year-old girl with lymphoblastic leukosarcoma. The "characteristic" convoluted cells were rare. Note small rims of cytoplasm. (R stain, ×800)

Nucleoli are not usually evident in benign lymphocytes (see Fig. 14-5B) and in well-differentiated lymphocytic lymphoma cells (see Figs. 14-10B and 14-11). Small nucleoli may be seen with P stain. Nucleoli are more prominent with decreasing differentiation (see Fig. 14-18A). In R stain they are royal blue, single, and vary considerably in size.

Cytoplasm is often absent, invisible, or may be present as a small partial rim (see Fig. 14-18A and Plate XXX). It is often stripped completely from the poorly differentiated cells (see Figs. 14-14 and 14-15A and B). In plasmacytic cells, the nucleus is eccentric and cytoplasm is more prominent (see Fig. 14-12).

Poor cytoplasmic staining with P stain is also noted in imprint cytology.[14] Lymphoglandular bodies generally not evident with P may be sparsely evident in poorly differentiated or immunoblastic smears (see Fig. 14-15B). Differentiated histiocytosis, which runs a malignant course, is illustrated in a smear (Fig. 14-19) extracted from a lymph node from a child with Letterer–Siwe disease. Malignant histiocytic cells have fine granular chromatin, elongated or indented nuclei, pale blue nucleoli, and poorly defined cytoplasm (see Fig. 14-14).

Hodgkin's Lymphoma

A number of observers have addressed the utility of FNA in the diagnosis of Hodgkin's disease.[10,17,32,34]

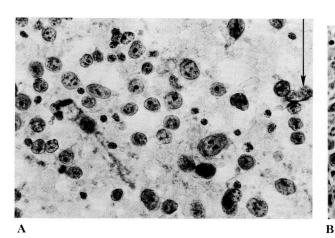
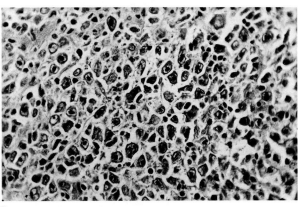

A   B

Fig. 14-18. A. Non-Hodgkin's lymphoma (large cell), immunoblastic type. Cells are large with round to oval nuclei and coarsely granular and punctate chromatin. Nucleoli are large and may be massive. Cytoplasm is absent or appears "pasted on" in some cells. Pale lymphoglandular bodies are poorly perceived in the background. Mitoses may be numerous. One cleaved cell is seen (*arrow*). (P stain, ×1000) **B.** Tissue section highlighting pleomorphism and large nucleoli. (H & E stain, ×400)

Although primary diagnosis is always followed by surgical biopsy, staging can be carried out by FNA of regional palpable nodes (*e.g.*, inguinal adenopathy in a patient with biopsy-confirmed Hodgkin's disease above the diaphragm).

The cytology of Hodgkin's lymphoma contrasts sharply, for the most part, with that of non-Hodgkin's lymphoma. Lymphocyte-rich Hodgkin's lymphoma may simulate non-Hodgkin's lymphomas in histopathology. Monomorphic smears should therefore be screened carefully for Sternberg–Reed cells. They may be more evident in smears than in section, and cytology should complement pathology (see Fig. 14-7). The usual Hodgkin's lymphoma smear, however, is pleomorphic rather than monomorphic and must be differentiated from lymphadenitis (see Figs. 14-3B and 14-8). Hodgkin's lymphoma and lymphadenitis have neutrophils, eosinophils, phagocytes, and atypical histiocytes in common. The Sternberg–Reed cell is the distinguishing feature. This cell is large, with bilobed (Fig. 14-1), sometimes mirror image (Fig. 14-20A), and monolobed nuclei (Fig. 14-20B). Nuclear chromatin is finely granular (R) or hyperchromatic and clumped (P). What distinguishes the cell from atypical histiocytes is the presence of macronucleoli. This distinction is not always

clear cut. A low-power view may reveal multiple clear-cut Sternberg–Reed cells in a background of lymphocytes (Fig. 14-20C).

A careful study of Hodgkin's cytology by Kardos, and associates[32] has established the impor-

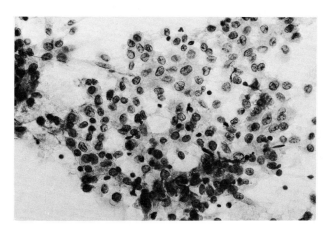

Fig. 14-19. Histiocytosis, differentiated. Letterer–Siwe disease. This is a large aggregate of fairly uniform histiocytic cells with folded nuclei and dusty chromatin. Nucleoli are small. Cells are not cytologically malignant. (P stain, ×330)

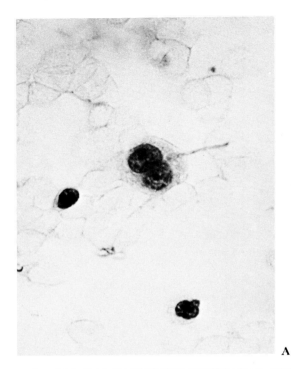

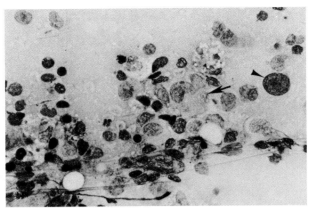

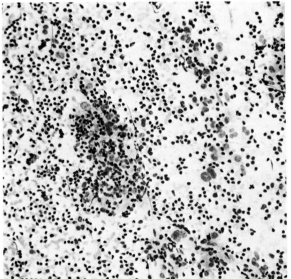

Fig. 14-20. Hodgkin's disease. **A.** An isolated Sternberg–Reed cell with mirror-image bilobed nucleus. Nucleoli are prominent. (P stain, ×330) **B.** In this smear, the pleomorphism, phagocytes, and atypical histiocytes *(arrow)* precluded a diagnosis of Hodgkin's disease initially. Large, rounded nuclei with blue macronucleoli *(arrowhead)* were highly suspicious. Hodgkin's disease was confirmed on surgical biopsy. (R stain, ×600) **C.** Panoramic view of numerous Sternberg–Reed cells scattered on a background of lymphocytes. (R stain, ×132)

tance of the polyploid cell, present in all cases reviewed by them. This is a large Sternberg–Reed-like cell without macronucleoli. In addition, the presence of metachromatic material, appearing only in R-stained material, was found to be a useful indicator of Hodgkin's disease.

Lymphadenopathy and Malignant Lymphoma in AIDS

In the AIDS-related complex or pre-AIDS syndrome, the patient may have generalized adenopathy. Immediate diagnostic clues are provided by careful history designed to rule out other systemic disorders such as infectious mononucleosis and secondary syphilis and to establish positive risk factors (homosexuality, drug usage, hemophilia, Haitians, background, etc).

Histologically, there is florid follicular hyperplasia. The cytologic picture is typical of other reactive hyperplastic nodes with a mixed lymphoid pattern including immunoblasts, tingible body phagocytes, plasma cells, neutrophils, and benign giant cells.[5,15] A specific cytologic diagnosis of AIDS cannot be made; but with appropriate clinical data, a diagnosis of "consistent with AIDS" can be offered.

With progression of immune deficiency and multiple infections, lymph nodes may become "burned out" of lymphoid elements.[52] This will account for sparse lymphoid yield on aspiration. We have observed enlarged histiocytes, some with pseudo-Gaucher appearance in both lymphoid and splenic aspirates. The bubbly cytoplasm is packed with acid-fast bacilli on special stain. Mycobacterium avium intracellulare is the dominant organism. The possibility of other organisms appearing in lymph node aspiration smears must be continually entertained. A review of FNA cytology in lymphadenopathy of AIDS has recently been published.[5]

In a study of 30 cases of malignant lymphoma in AIDS, 90% were extranodal and 67% were exclusively extranodal. All non-Hodgkin's lymphomas were of the diffuse type with the greatest number (49%) large follicular center cell types.[11] The cytologic yield from these lymphomas should parallel the findings in lymphoma patients without AIDS.

Kaposi's Sarcoma

Careful aspiration of large easily targeted nodes in a patient with disseminated Kaposi's yielded blood but no lymphoid content on several passes. This finding should exclude lymphoma and suggests the increased vascularity of a florid Kaposi's lesion.[27] Fifteen cases of Kaposi's sarcoma were studied by Hales and associates by FNA. Five samples were extracted from lymph nodes. Loosely cohesive clusters of spindle cells (see Chap. 13) as well as intact tissue fragments composed of fragile spindle cells were extracted.[26]

Table 14-2. Clinical Presentation of Extranodal Sites of Lymphoma

LOCATION OF LESION	METHOD ASSISTING CYTOLOGIC DIAGNOSIS
Brain	Scan
Salivary gland mass	Palpation only
Thyroid mass	Palpation, scan, or both
Pulmonary solitary lesion (lung or pleura) or interstitial pattern	Radiography
Hepatomegaly	Palpation, scan, or both
Gastrointestinal (GI) lesion	Palpation, radiography, or both
Renal mass	Palpation, radiography, ultrasonography
Breast	Inspection and palpation
Skin lesion	Inspection and palpation

Nodal Versus Extranodal Disease

Most lymphomatous lesions appear as palpable nodal enlargements in a regionally localized or multicentric pattern. Aspiration should be carried out from the clinically dominant and accessible site. If lymphoid smears are obtained, the main problem becomes that of distinguishing a benign from a neoplastic process. In contrast, a lymphoid smear obtained from a palpable or image-identified extranodal mass may be diagnostic of a primary lymphoma, extension of a known lymphoma, or a benign or pseudolymphomatous proliferation. A clinical presentation of extranodal sites of lymphoma is summarized in Table 14-2.

Extranodal lymphoid aspirates from the salivary gland, thyroid gland, lung, skin, breast, brain, Waldeyer's ring, spleen and testis are considered in Table 14-3. Although specific nonmalignant diagnoses may be made cytologically on the basis of criteria discussed in other chapters of this text, malignant lymphoma should readily form part of the differential diagnosis. Since lymphoid cytology is structurally part of the normal picture in many of the organs mentioned in Table 14-3, special attention is required before a primary extranodal lymphoma can be diagnosed. The histology of extranodal lymphomas parallels that of primary lymph node lymphoma. This holds true for the cytology.[29] It is also important to be aware that apparent benign lesions can transform or progress to malignant lymphomas (*e.g.,* Hashimoto's thyroiditis—see Chap. 4).

Table 14-3. Extranodal Lymphoid Aspirates

SITE OF ASPIRATION	SPECIFIC LESION
Salivary gland (Chap. 5)	Sjögren's disease; papillary cystadenoma lymphomatosum (Warthin's tumor); lymphoma
Thyroid gland (Chap. 4)	Hashimoto's thyroiditis; lymphoma
Lung (Chap. 7)	Pseudolymphoma; lymphoma
Skin (Chap. 13)	Atypical lymphoid infiltrates; lymphoma
Breast (Chap. 6)	Intramammary lymph node; lymphoma
Testis (Chap. 12)	Seminoma; lymphoma
Brain (Chap. 17)	Lymphoma
Waldeyer's ring (Chap. 3)	Benign lymphoid hyperplasia (tonsil); lymphoma
Spleen (Chap. 14)	Hypersplenism; lymphoma
Abdomen (Chap. 8)	Lymphoma (liver, kidney, lymph nodes, gastrointestinal tract [plasmacytoid lymphoma])

COORDINATION OF CYTOLOGIC DATA TO FORMULATE A REPORT

The following example of a solitary lymph node in an otherwise asymptomatic patient highlights the cytologic report and indicates the difference between a benign and a malignant aspirate.

Benign Aspirate

The smear presents a pleomorphic mix of lymphocytes, many small and mature, and a fair number of intermediate types with a rim of basophilic cytoplasm. Transformed cells (germinoblasts), follicular center cells, and "lympho-blasts" will be present on an average 1:1 oil-immersion field. Tingible body cells (phagocytic histiocytes), nonphagocytic vacuolated histiocytes, neutrophils, plasma cells, and occasional giant cells of various types together with cellular debris may be seen. Cells are generally intact in well-done smears. The nuclear:cytoplasmic ratio is normal. Lymphoglandular bodies are present. This smear suggests a *reactive lymphadenitis.*

Malignant Aspirate

Contrasting with the benign smear is a monomorphic smear in which the cell component may be a small, intermediate, or large cell (lymphocytic, interme-diate, poorly differentiated lymphocytic, immunoblastic) or a mixed population of small and large cells, usually in comparable numbers. Both small and large cleaved cells may be seen. Mast cells (may be) present under low-magnification scanning. The nuclear:cytoplasmic ratio is increased, and there is a tendency for cells to "smear," with strands and cytoplasmic fragments (lymphoglandular bodies) often larger than those seen in benign smears. Sternberg–Reed cells are the hallmark of Hodgkin's disease, and plasma cells and eosinophils may be present. Such smears suggest *lymphoma.*

SPLEEN

The anatomy and different compartments of this organ can be found in standard textbooks.

Cytology of the "Normal" Spleen

The cytology and differential cytologic diagnosis of splenic "aspirates" has been studied in great depth by Söderström[58] and the following is from his contribution to the first edition.

The spleen, which is the main lymphatic organ of the abdomen could hardly be expected to offer identification problems were it not for the occurrence of abdominal lymphomas that are not engaging the spleen. Specific features proving the splenic origin of a specimen are related to the typical subdivision of the smears into white and red pulp. The cytology of white pulp areas is nearly identical with that of lymph nodes, but it is often evident in the smears that the lymphatic cells are attached in a grape-like way to small blood vessels (Fig. 14-21). The red pulp areas are rather poor in cells but contain the double-tailed endothelium cells of splenic sinusoids (Figs. 14-21 and 14-22) and the large platelet aggregates that are specific signals of the splenic origin of the specimen.

Remember that the presence of a few "foreign" usually epithelial cells in otherwise normal spleen aspirates does not necessarily mean metastasis or contamination by double puncture. Liver epithelium cells have been seen in spleen aspirates from patients with aggressive hepatitis. In one aspirate, tubular epithelium cells were present in the spleen after renal transplantation and nephrectomy. Thyroid follicular cells appear in spleen aspirates for no apparent reason. Osteoblasts may accompany myeloid metaplasia

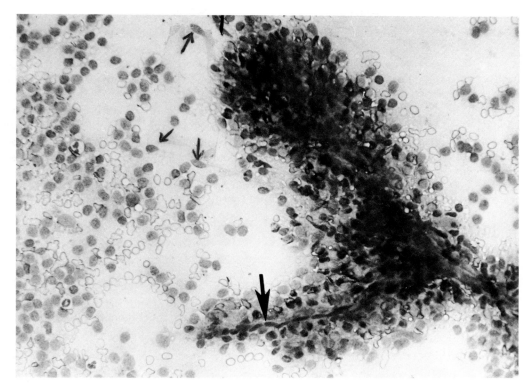

Fig. 14-21. Spleen. The white pulp is represented by lymphocyte masses clinging to small arteries *(arrow)*. (MGG stain, ×330)

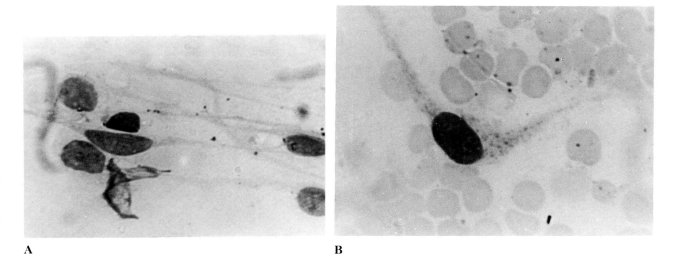

A   B

Fig. 14-22. Spleen. Delicate, elongated, sinusoid endothelial cells, some with cytoplasmic granules, are part of the normal spleen cytology. This is from the same smear as Fig. 14-24. (MGG stain, ×1300)

of the spleen in patients with myelosclerosis and osteosclerotic cancer metastasis and will appear in splenic aspirates. Finally, cancer cells are now and then seen in spleen aspirates and do not mean metastasis, although cancer metastasis to the spleen are rather common. Such metastases are seen on careful sectioning of spleens at autopsy.[4] Clinical metastasis to the spleen is discussed below.

The explanation for the presence of such foreign cells may be that the spleen is a net placed in the circulation to pick up and process foreign particles including circulating tissue cells. If indications were wider for the aspiration of normal-sized spleens, findings like those mentioned above might prove rather common.

The spleen may present as a palpable left upper quadrant mass easily simulated by other masses in this region (see Chap. 8), particularly renal tumors. It thus becomes a suitable target for FNA needle aspiration. Disease that has not enlarged the spleen below the costal margin requires intercostal puncture. FNA of the spleen provides distinctive cytologic patterns suggestive or diagnostic of reticuloendothelial (leukemias, lymphomas, myelofibrosis), storage, (Gaucher's, Niemann–Pick) and granulomatous (sarcoid) diseases. Metastatic cells and, in addition, patterns that suggest or, by exclusion, confirm a diagnosis of hypersplenism may be aspirated.

Method and Procedure

The aspiration technique varies from puncture of a palpable abdominal mass (see Chap. 8) to direct passage of the fine needle (0.7-mm outer diameter) 4 cm to 5 cm into the splenic substance in one continuous movement without the oscillation usually performed in FNA (see Fig. 1-2).[58] The latter method reduces the potential for tearing the splenic capsule. Söderström has seen no noteworthy morbidity. We have used the latter method in significantly fewer patients and have observed one splenic "leak" that required transfusions but not splenectomy.

To proceed, instruct the patient to perform respiratory apnea after several deep breaths. Use inspiratory apnea for subcostal puncture and expiratory apnea for intercostal puncture. Infiltrate the skin with local anesthetic. Briskly insert a fine needle attached to a one-hand syringe through the abdominal or thoracic wall into the spleen, simultaneously applying a full 10 ml of suction to the syringe and releasing suction before full withdrawal. Needle length can vary from 3 to 10 cm depending on the thickness of the body wall. Shorter needles, less flexible, are more easily controlled. Thoracic puncture is best carried out in the ninth intercostal space between the mid and anterior axillary lines. Carry out abdominal puncture in the mid-portion of the palpable target to avoid passing the needle through the thinner peripheral part of the spleen into surrounding organs.

Clinical and Cytologic Considerations

The splenic parenchyma increases in bulk and produces splenomegaly as a result of various disorders.

Myeloid Metaplasia

Myeloid metaplasia (extramedullary hematopoiesis) accompanies myelofibrosis of the bone marrow as well as chronic myelogenous leukemia and is part of the primary process. Megakaryocytes as well as myeloid and erythroid precursors may be seen (Fig. 14-23).

Lipidoses and Tumor Cell Infiltrations

Lipids are stored in the reticuloendothelial cells, distending them and producing massive enlargement. Gaucher's disease is the classic example of lipidoses and tumor cell infiltrations (Fig. 14-24).

Metastatic cells have been found in a large number of postmortem spleens, but clinical enlargement is rare.[4] In addition to the sources of metastases cited by Willis[62] (*e.g.*, breast, melanoma, and chorioe-

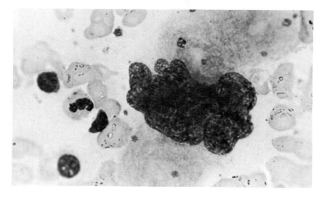

Fig. 14-23. Myeloid metaplasia, spleen. Massive megakaryocytes as well as myeloid cells are evident. (R stain, ×660)

pithelioma), we have seen enlargement from bronchogenic carcinoma and one patient with massive splenomegaly due to metastatic bladder carcinoma (Fig. 14-25).

Lymphoma and Leukemia

Splenic puncture is usually not undertaken in patients with leukemia, since the diagnosis is determined by clinical, peripheral blood, and bone marrow findings. Aspirates from non-Hodgkin's lymphoma are densely cellular and may be monomorphic or of mixed cellularity. Such smears cannot be distinguished from lymph node smears with similar pathologic changes (see above). A *left upper quadrant mass must be definitively identified as spleen to diagnose splenic lymphoma, since enlarged nodes may be present as left upper quadrant masses.* The cytologic yield in Hodgkin's lymphoma is generally nondiagnostic. Granulomatous reaction and atypical histiocytes should raise suspicions. Classical Sternberg–Reed cells are diagnostic.

Hyperplasia of the Reticuloendothelial Structures (RES)

Hyperplasia of the RES may occur in response to systemic stimuli (infection, inflammation, collagen disorders, and congestion with blood as in acute splenic vein blockage or congestive heart failure). The cytologic yield will be composed of the normal cellular elements noted above and may be scanty. In the absence of metaplastic cells (myeloid metaplasia), foreign cells (Gaucher's disease) metastatic cells, and

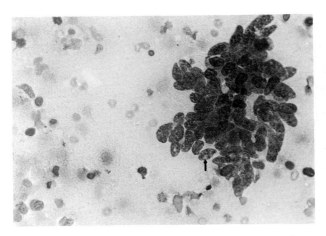

Fig. 14-25. Bladder carcinoma, metastatic, spleen. Tight cluster of plump, elongated tumor cells with nucleoli *(arrow)* and scant cytoplasm. The usual abundant cytoplasm (Chapter 9) has been shed by the extraction. Scattered splenic cells are evident. (R stain, ×132)

monomorphic cellular lymphoid patterns (lymphoma), a diagnosis of reactive splenomegaly can be earnestly entertained. The presence of phagocytic histiocytes containing red blood cells (RBCs) and hemosiderin (hemolysis), platelets (thrombocytopenia), and histiocytes with lipid vacuoles (thrombocytopenia treated with steroids) will confirm the diagnosis in appropriate clinical contexts.

Cell Types

Splenic cell types extracted by FNA have, for the most part, been described in this and other chapters. Plasmacellular reactions have been cited by Söderström.[58] They are seen in systemic lupus erythematosis (SLE), rheumatoid arthritis (Felty's syndrome), hemolytic anemias, and chronic aggressive hepatitis and in some patients with myeloma. Rare primary vascular tumors of the spleen may yield sarcomatous smears, described in Chapter 15. Cytologic findings are listed in Table 14-4.

Hazards and Contraindications

Patients with suspected polycythemia vera and megakaryocytic myelosis with thrombocytosis should probably not be punctured, since splenic tear with consequent splenectomy may produce refractory thrombocytosis. Patients with infectious mononucle-

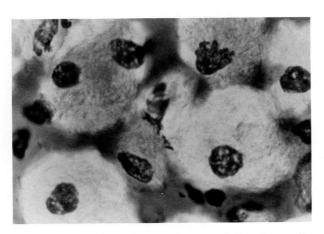

Fig. 14-24. Spleen. Dense clusters of Gaucher cells. Cells are large and lipid-laden with crinkled cytoplasm and small nuclei. (R stain, ×2000)

Table 14-4. Cytologic Findings in Spleen Puncture

DIAGNOSIS	CYTOLOGY
Myelofibrosis with myeloid metaplasia	Megakaryocytes; clusters of erythroid cells; myeloid cells (see Fig. 14-23)
Non-Hodgkin's lymphoma	
Well differentiated	Sheets and clusters of small lymphocytes; mast cells; cell fragility and lymphoglandular bodies
Poorly differentiated	Sheets and clusters of intermediate and large lymphocytes; cell fragility and lymphoglandular bodies
Histiocytic lymphoma	Sheets and clusters of primitive lymphoid cells; cell fragility with cytoplasmic fragments
Hodgkin's disease	Granulomatous reaction; atypical histiocytes; Sternberg–Reed cells; scanty yield
Gaucher's disease	Gaucher's cells (see Fig. 14-24)
Niemann–Pick disease	Niemann–Pick cells
Sarcoidosis	Granulomatous reaction (see Fig. 14-8)
Carcinoma, melanoma, and sarcoma	Appropriate malignant cell types (Fig. 14-25)
Congested spleen, hypersplenism, and related splenomegaly	Scanty yield; lymphocytes; plasma cells (see text); histiocytes with and without phagocytic inclusions (see text)

osis and probably related viral splenomegaly (cytomegalovirus [CMV]) should not be punctured because of reported ease of splenic rupture. Coagulation disorders and thrombocytopenias that are not correctable are relative contraindications, as are uncooperative and psychotic patients. We have not had experience in puncture of parasitized spleens due to malaria, leishmaniasis, and trypanosomiasis and thus cannot comment on the hazards associated. The chief hazard in spleen puncture is hemorrhage, and vital signs should be observed for hospitalized patients and appropriate instructions given to ambulatory patients.

Clinical Application

The spleen frequently participates in generalized systemic disorders in which the diagnosis is determined by a series of noninvasive methods. For example, typical chronic myelogenous leukemia with splenomegaly rarely requires spleen puncture for confirmation.

Discrete splenomegaly, however, is a suitable target, and aspiration smears may readily define the pathology. Such punctures should always be preceded, however, by blood and bone marrow studies, which may establish the diagnosis.

Splenic cytology, which seeks to separate pathologic disorders into broad categories, is not difficult to learn. For example, the distinction between lymphocytic lymphoma or chronic lymphocytic leukemia and Gaucher's disease should be straightforward. Similarly, the presence of bone marrow cells (megakaryocytes and erythroid and myeloid cells) is rather easily detected and confirms myeloid metaplasia.

The identification of isolated splenic lymphoma or the establishment of splenic involvement in a patient with known lymphoma is one of the most valuable uses of the spleen puncture technique. Perhaps the most important criterion in distinguishing lymphoma from benign hyperplastic spleen is cellularity. Hypercellularity does not, however, exclude a benign smear. The term *lymphocyte plethora*, which has been proposed by Söderström to describe lymphoma smears, is quite appropriate.[58] As with lymph nodes, monomorphic patterns signal neoplastic proliferation.

Careful splenic puncture with a fine needle appears to be free of significant morbidity. It is probably less hazardous than core biopsy of the liver and may yield a diagnosis in patients with hepatosplenomegaly.

ASPIRATION CYTOLOGY AND IMMUNOCYTOCHEMISTRY OF NON-HODGKIN'S LYMPHOMAS
by Edneia Tani and Lambert Skoog

The use of immunologic markers has had a major impact on the understanding and classification of lymphoproliferative diseases. With the aid of monoclonal antibodies it has been possible to phenotype (B-, T-, null- and true histiocytic cells) lymphoreticular neoplasms, which in conjunction with morphology has resulted in new classification schemes for non-Hodgkin's lymphomas.[34a,39,39a] The importance of this new approach to phenotyping non-Hodgkin's lymphomas lies to a large extent in the fact that it gives improved prognostic information, an ultimate requirement for designing more differentiated therapeutic regimens. In addition, the use of markers to identify monoclonality in the lymphoid population

leads to improved diagnostic accuracy in differentiating reactive lesions from some type of lymphomas, in particular those of follicular center cell type.[34a] Such marker studies have so far been performed on excised lymph nodes, and there are few reports on the application of monoclonal antibodies to cells obtained by FNA.[31a,34b,60a] At the Division of Clinical Cytology, Karolinska Hospital, marker studies have been included in the morphological work-up on FNA material from patients with lymphoproliferative disorders. Below methodological aspects are discussed as well as some guidelines in the interpretation of the results.

Methodology

For an immunologic marker study approximately 10^6 cells are sufficient. This number of cells can usually be obtained by one aspiration using a 0.6-mm diameter needle. The cells should immediately be suspended in ice-cold buffered saline (1.5 ml, pH 7.4). This is accomplished by gentle aspiration of the saline through the needle. Trypan-blue staining is used to determine cell number and viability.[66] In cellular aspirates the cell-count is adjusted to 1 to 2×10^6 nucleated cells per ml buffer solution. If the suspension contains many necrotic cells, the FNA is repeated from another area of the lesion. At least seven cytocentrifuge preparations are made (Shandon cytospin; 3 minutes 600 rpm at 20°C within 30 minutes after the FNA). After air-drying, one cytocentrifuge smear is stained with Giemsa for cytomorphology, and the remaining preparations are stored at 4°C for a maximum of 5 days. Immediately before immunological staining the cells are fixed in −20°C acetone for 10 minutes, air-dried and rinsed in buffered saline for 5 minutes. Immunocytochemical staining is done by incubation with monoclonal antibodies (mouse anti-human) for 30 minutes followed by rinsing in buffered saline for 10 minutes. Alkaline phosphatase conjugated rabbit anti-mouse Ig anti-serum is applied for 30 minutes followed by rinsing. Alkaline phosphatase conjugated swine anti-serum to rabbit Ig is then applied for 30 minutes. After rinsing and incubation in alkaline phosphatase developing solution, the cells are stained with Harris hematoxylin. A primary cytomorphologic diagnosis is made using the Kiel classification as adapted to cytology and then the markers are selected according to this evaluation. The marker panel that is most commonly used consists of anti-kappa, anti-lambda, anti-my, anti-T, anti-B and anti-α1-antitrypsin. However, when indicated by cytomorphology antibodies to cALLa, EMA, cytokeratin and subset of T-cells are used.

Cytomorphology and Immunocytochemistry

Representative cases of FNA material from lymph nodes with reactive hyperplasia and different types of malignant lymphomas are shown in Plates XXVII to XXXI. We have chosen to present lymphomas of the non-Hodgkin's type. In our experience, marker studies on Hodgkin's lymphomas are only of limited value for the diagnosis that relies on cytologic evaluation. This can partly result from the fact that the diagnostic Hodgkin's and S−R cells are relatively few in FNA from most cases of Hodgkin's lymphoma. Moreover, these cells seem to be fragile and are poorly preserved in the cytospin preparations, which in combination with the relatively unspecific antibodies that are available will markedly limit the usefulness of immunologic analysis.

Reactive hyperplasia is characterized by a polymorphous population of lymphocytes with all cell types represented in routine smears (Plate XXVII A). Marker analysis shows a variable mixture of T- and B-cells (Plate XXVII B and C). The latter express a polyclonal light chain pattern (both kappa and lambda). (Plate XXVII D and E).

Low-grade malignant lymphomas of type CLL and immunocytoma present a monotonous population of small lymphocytes in routine smears (Plate XXVIII A). The phenotyping of such cells is usually easy as well as the establishment of monoclonality as presented in Plate XXVIII B. However, it should be pointed out that in CLL the cells usually only bear surface immunoglobulins with a low density, which might result in a weak immunological reaction. This contrasts with the cells in immunocytomas, which express both surface and cytoplasmic immunoglobulins that will result in a more intensive staining reaction (Plate XXVIII C).

Low-grade malignant lymphomas derived from follicular center cells (B-cells) show a variable mixture of small and large cells in routine smears that sometimes result in significant difficulties in cytologic diagnosis (Plate XXVIII D). With the aid of markers it is possible to identify both B- and T-cells. In fact in some follicular center cell lymphomas the nonneoplastic T-cells might be the dominating lymphoid cell. The neoplastic B-cells bear moderate amounts of surface immunoglobulins whose light chain is restricted to one type (monoclonal pattern) (Plate XXVIII B and C). A majority of the cases will also express a monoclonal cytoplasmic immunoglobulin resulting in a strong immunologic staining. If the cells only bear surface immunoglobulins, the evaluation of clonal pattern requires a careful scrutiny of the immunologic staining reaction.

High grade malignant lymphomas (Plates XXIX, XXX, and XXXI) are usually readily diagnosed by cytomorphology on material, obtained by FNA. In fact, the lymphoid cells are often easier to identify on smears than in sections. Phenotyping of the cells will in most cases give an unequivocal result. It should, however, be emphasized that the high-grade malignant lymphoma cell is fragile, and care should be taken when the cell suspension is being prepared to avoid destruction of the cells. If the Giemsa-stained cytospin preparation contains many damaged cells, it is advisable to repeat the FNA. This should also be done if the aspirate contains a lot of cell debris and necrotic material, which occasionally occurs in lymphomas. In some cases of high-grade malignant lymphoma, the neoplastic cell does not express the antigens, tested for in the usual marker panel.

In these cases phenotyping will give an equivocal result. Under such circumstances it can be of value to use more general markers for lymphoid origin to test for immunoglobulin production (B-cell origin) and to test for subset of T-cells. Epithelial markers will sometimes be needed to rule out a poorly differentiated carcinoma, although this is in our experience rarely a differential diagnostic problem.

The final diagnosis of malignant lymphomas and their subclassification has so far required surgical biopsy and histologic examination. FNA biopsy has for the most part only been accepted in the differential diagnosis between lymphoma versus cancer metastasis and lymphoma versus lymphadenitis. The successful application of immunologic markers to material, obtained by FNA, has completely changed the conditions for a wide acceptance of FNA in the final diagnosis of most non-Hodgkin's lymphomas. In our experience there is an excellent correlation between the results from FNA and surgical biopsy when the marker studies are combined with the morphologic evaluation.[60a] In fact cytospin preparations

from material, obtained by FNA, offer several distinct advantages. In the first place, cytologic analysis of the cells is superior on FNA material compared to sections. Secondly, marker studies on cytospin preparations can be performed with close morphologic control, that sometimes is difficult on frozen sections. Moreover, background staining can usually be neglected in cytospin preparations. In the future excisional biopsy for histologic examination may be required only in cases of lymphomas in which the growth pattern (nodular or diffuse) is of importance in deciding initiation of therapy. Presently it is not possible to obtain this information from FNA material. However, lymphomas of low or high grade can in a majority of the cases be accurately diagnosed and phenotyped on aspirates. With an unequivocal result the patient can be spared an excisional biopsy. The possibility of determining an accurate classification of lymphomas on aspirates is of particular importance when the tumor mass is not readily accessible to surgery or in rapidly growing lymphomas.

The use of cell markers on FNA material will increase the accuracy in differentiating a reactive lesion from a neoplastic and allow phenotyping of non-Hodgkin's lymphomas.

This technique is a useful adjunct in the workup of FNA material from patients with lymphoproliferative diseases. In this context it must be emphasized however that the diagnosis of a lymphoid neoplasia on FNA is primarily based on a cytologic evaluation, which requires extensive cytomorphologic experience. Cell markers will aid cytomorphology but cannot replace it.

REFERENCES

1. Behm FG, O'Dowd GJ, Frable WJ: Fine-needle aspiration effects on benign lymph node histology. Am Clin Pathol 82:195, 1984
2. Bennett MH, Famer–Brown G, Henry K et al: Letter to the Editor: Classification of non-Hodgkin's lymphoma. Lancet 2:405, 1974
3. Berard CW, Greene MH, Jaffe ES et al: NIH conference. A multidisciplinary approach to non-Hodgkin's lymphomas. Ann Intern Med 94:218, 1981
4. Berge T: The metastasis of carcinoma with special reference to the spleen. Acta Pathol Microbiol Scand (suppl) 188:1, 1967

5. Bottles K, McPhaul LW, Volberding P: Fine-needle aspiration biopsy of patients with the acquired immune deficiency syndrome (AIDS): Experience in an outpatient clinic. Ann Int Med 108:42, 1988

6. Christ ML, Feltes–Kennedy M: Fine needle aspiration cytology of toxoplasmic lymphadenitis. Acta Cytol 26:425, 1982

7. Conjalka M, Rajdev N, Tavassoli M: Lymphoreticular fragments, the cellular debris of acute lymphosarcoma cell leukemia. Cancer 44:2124, 1979

8. Cossman J: Immunologic markers and histopathology of diffuse large cell lymphoma: Proc Symp at Marco Island, Fla. Nov 14–17, 1985. Library Congress Catalog No 86-060809

9. Das DK, Gupta SK, Datta U et al: Malignant lymphoma of convoluted lymphocytes: Diagnosis by fine-needle aspiration cytology and cytochemistry. Diag Cytopathol 2:307, 1986

10. Desmet V, Beert J, Reybrowck G: Value of the cytological diagnosis of Hodgkin's disease in comparison with histological findings. Sangre (Barc) 9:73, 1964

11. DiCarlo EF, Amberson JB, Metroka CE et al: Malignant lymphomas and the acquired immune deficiency syndrome. Arch Pathol Lab Med 110:1012, 1986

12. Dorfman RF: Letter to the Editor. Classification of non-Hodgkin's lymphomas. Lancet 22:1295, 1974

13. Engzell U, Jakobsson PA, Sigurdson A et al: Aspiration biopsy of metastatic carcinoma in lymph nodes of the neck. A review of 1101 consecutive cases. Acta Otolaryngol 72:138, 1971

14. Feinberg MR, Bhaskar AG, Bourne P: Differential diagnosis of malignant lymphomas by imprint cytology. Acta Cytol 24:16, 1980

15. Finkle HI: Letter to the Editor. The lymph node touch preparation in the diagnosis of the acquired immune deficiency syndrome (AIDS). Acta Cytol 30:695, 1986

16. Frable MA, Frable WJ: Fine needle aspiration biopsy in the diagnosis of sarcoid of the head and neck. Acta Cytol 28:175, 1984

17. Friedman M, Kim U, Shimaoka K et al: Appraisal of aspiration cytology in management of Hodgkin's disease. Cancer 45:1653, 1980

18. Gall EA, Mallory TB: Malignant lymphoma: A clinical pathologic survey of 618 cases. Am J Pathol 18:381, 1942

19. Gall EA, Rappaport H: Seminars on diseases of lymph nodes and spleen. In McDonald JR (ed): Proceedings of the Proc Seminar American Society Clinical Pathologists, 1958

20. Galton DAG, Goldman JM, Wiltshaw E et al: Prolymphocytic leukemia. Br J Haematol 27:7, 1974

21. Gerald–Marchant R, Hamlin I, Lennert K et al: Letter to the Editor. Classification of non-Hodgkin's lymphoma. Lancet 17:406, 1974

22. Grieg EDW, Gray ACH: Note on lymphatic glands in sleeping sickness. Lancet 1:1570, 1904

23. Guthrie CG: Gland puncture as a diagnostic measure. Bull Johns Hopkins Hosp 32:266, 1921

24. Hajdu SI, Hajdu E: Cytopathology of Sarcomas. Philadelphia, WB Saunders, 1976

25. Hajdu SI, Melamed MR: Limitations of aspiration cytology in the diagnosis of primary neoplasms. Acta Cytol 28:337, 1984

26. Hales M, Bottles K, Miller T et al: Diagnosis of Kaposi's sarcoma by fine-needle aspiration biopsy. Am J Clin Pathol 88:20, 1987

27. Harris NL: Hypervascular follicular hyperplasia and Kaposi's sarcoma in patients at risk for AIDS. N Engl J Med 310:462, 1984

28. Hashimoto CH, Cobb CJ: Cytodiagnosis of hibernoma. Diag Cytopathol 3:326, 1987

29. Higgins GK: The pathologic anatomy of non-Hodgkin's lymphoma. In Molander DW (ed): Lymphoproliferative Diseases. Springfield, IL, Charles C Thomas, 1975

30. Ioachim HL: Lymph Node Biopsy. Philadelphia, JB Lippincott, 1982

31. Jenkin D: The treatment of non-Hodgkin's lymphoma in childhood: In Wiernick, P (ed): Leukemias and Lymphomas. New York, Churchill Livingston, 1985

31a. Johnson A, Cavallin–Stahl E, Åkerman M: Analysis of clonal excess in fine needle biopsies. A diagnostic tool in non-Hodgkin lymphoma (NHL) in adults. Proc Am Soc Clin Oncol 5:195, 1986

32. Kardos TF, Vinson JH, Behm FG et al: Hodgkins disease: Diagnosis by fine needle aspiration. Am J Clin Pathol 86:286, 1986

33. Khoory MS: Diseases of lymph node and spleen. In Clinical Aspiration Cytology. Philadelphia, JB Lippincott, 1983

34. Koss LG, Woyke S, Olszewski W: Aspiration Biopsy Cytology Interpretation and Histologic Bases. New York, Igaku–Shoin, 1984

34a. Lennert K: Histopathology of non-Hodgkin's lymphomas. Springer-Verlag, Berlin, Heidelberg, New York, 1981

34b. Levitt S, Cheng L, DuPuis M et al: Fine needle

aspiration diagnosis of malignant lymphoma with confirmation by immunoperoxidase staining. Acta Cytol 29, 895–902, 1985

35. Lieberman PH, Filippa DA, Struas DJ et al: Evaluation of malignant lymphomas using three classifications and the working formulation. Am J Med 81:365, 1986

36. Linsk JA, Franzen S: Fine Needle Aspiration for the Clinician. Philadelphia, JB Lippincott, 1986

37. Lopes–Cardozo P: The significance of fine needle aspiration cytology for the diagnosis and treatment of malignant lymphomas. Folia Haematological (Berlin/Leipzig) 107:602, 1980

38. Lukes RJ, Collins RD: Immunologic characterization of human malignant lymphomas. Cancer 34:1488, 1974

39. Lukes RJ, Collins RD: New approaches to the classification of the lymphomata. Br J Cancer (suppl) 31:1, 1975

39a. Lukes RJ, Collins RD. Functional classification of malignant lymphomas. In Rebuck JW, Berard CW, Abell MR (eds): The Reticuloendothelial System. Baltimore, Williams & Wilkins, (Monographs in Pathology, No 16, pp 213-242), 1975

40. Martin HE, Ellis EB: Aspiration biopsy. Surg Gynecol Obstet 59:578, 1934

41. Mathé G, Rappaport H, O'Connor GT et al: International Histological Classification of Tumours, no 14. Histological and Cytological Typing of Neoplastic Diseases of Hematopoietic and Lymphoid Tissues. Geneva, World Health Organization, 1976

42. Moore RD, Weisberger AS, Bowerfind ES: An evaluation of lymphadenopathy in systemic disease. Arch Int Med 99:751, 1957

43. Morrison M, Samwick AA, Rubinstein J et al: Lymph node aspiration. Am J Clin Pathol 22:255, 1952

44. National Cancer Institute sponsored study of classification on non-Hodgkin's lymphomas: Summary and description of working formulation for clinical usage. Cancer 49:2112, 1982

45. Nissen NI, Ersboll J: Treatment of non Hodgkin's lymphoma. In Wiernik PH (ed): Leukemias and Lymphomas. New York, Churchill–Livingston, 1985

46. O'Dowd GJ, Frable WJ, Behm FG: Fine needle aspiration cytology of benign lymph node hyperplasia—Diagnostic significance of lymphohistiocytic aggregates. Acta Cytol 29:554, 1985

47. Orell SR, Skinner JM: The typing of non-Hodgkin's lymphomas using fine needle aspiration cytology. Pathology 389, 1982

48. Pontifex AH, Klimo P: Application of aspiration biopsy cytology to lymphomas. Cancer 53:553, 1984

49. Qizilbash AH, Elavathil LJ, Chen V et al: Aspiration biopsy cytology of lymph nodes in malignant lymphoma. Diag Cytopathol 1:18, 1985

50. Ramzy I, Rone R, Schultenover SJ et al: Lymph node aspiration biopsy: Diagnostic reliability and limitation—An analysis of 350 cases. Diag Cytopathol 1:39, 1985

51. Rappaport H: Tumors of the hematopoietic system. In Atlas of Tumor Pathology, Second series, fasc. v. 8. Washington, DC, Armed Forces Institute of Pathology, 1966

52. Reichert CM, O'Leary TJ, Levens DL et al: Autopsy pathology in the acquired immune deficiency syndrome. Am J Pathol 112:257, 1983

53. Russell J, Skinner J, Orell S et al: Fine needle aspiration cytology in the management of lymphomas. Aust NZ J Med 13:365, 1983

54. Sallstrom JF, Juhlin R, Stenkvist B: Membrane properties of lymphatic cells in fine needle biopsies of lymphomas. I. Cap formation after exposure to fluorescein conjugated concanavalin A. Acta Cytol 18:392, 1974

55. Sinclair S, Beckman E, Ellman L: Biopsy of enlarged superficial lymph nodes. JAMA 228:602, 1974

56. Smith T: Fatty replacement of lymph nodes mimicking relapse. Cancer 58:2686, 1986

57. Söderström N: The free cytoplasmic fragments of lymphoglandular tissue. (Lymphoglandular Bodies) Scand J Haematol 5:138, 1968

58. Söderström N: Spleen. II. Cytology of infradiaphragmatic organs. Basel, S Karger, 1979

59. Symmers W St C SR: Survey of the eventual diagnosis in 600 cases referred for a second histological opinion after an initial biopsy diagnosis of Hodgkin's disease. J Clin Pathol 21:650, 1968

60. Symmers W St C SR: Survey of the eventual diagnosis in 226 cases referred for a second histological opinion after an initial biopsy diagnosis of reticular cell sarcoma. J Clin Pathol 21:654, 1968

60a. Tani EM, Christensson B, Porwit A et al: Immunocytochemical analysis and cytomorphological diagnosis on fine needle aspirates of lymphoproliferative diseases. Acta Cytol 32:209, 1988

61. Wilkerson JA: Intraoperative cytology of lymph nodes and lymphoid lesions. Diag Cytopathol 1:46, 1985

62. Willis RA: The spread of Tumours in the Human Body. London, Butterworth & Co, 1973

63. Wilson JF, Jenkin RDT, Anderson JR et al: Studies

in the pathology of non-Hodgkin's lymphoma of childhood I. The role of routine histopathology as a prognostic factor. A report from the Children's Cancer Study Group. Cancer 53:1695, 1984

64. Zajdela A, Zillhardt P, Voillenot N: Cytological diagnosis by fine needle sampling without aspiration. Cancer 59:1201–1205, 1987

65. Zajicek J, Engzell U, Franzen S: Aspiration biopsy of lymph nodes in diagnosis and research, pp 262–264. In Ruttiman A (ed): Progress in Lymphology. Stuttgart, Georg Thieme Verlag, 1967

66. Zajicek J, Engzell U, Plesnicar S: Studies on cell viability in aspirates from human lymph nodes. Eur J Cancer 4:23, 1968

CHAPTER 15

Aspiration Biopsy Cytology of Soft Tissue Tumors

JAN–SILVESTER WILLEMS

Mesenchymal tumors have their own cytologic presentation and can, in most instances, be separated from neoplasms derived from lymphoid tissue, integumental and visceral epithelium, neural tissue, and melanocytic cells. A large proportion of soft tissue tumors derive from and resemble differentiated cells such as fibroblasts, histiocytes, and muscle, fat, or endothelial cells, thus making it possible to type them. However, some neoplasms can cytologically be recognized only as mesenchymal, although they cannot be ascribed to a particular cell of origin. Some tumors of neuroectodermal origin growing in the soft tissues are also included in this discussion, mainly for reasons of differential diagnosis and similarities in management.

Most palpable mesenchymal tumors are diagnosed in apparently healthy patients who notice the gradual or sudden appearance of a painless or tender mass in soft tissue. However, a certain proportion of mesenchymal tumors can be unexpectedly picked up by fine needle aspiration (FNA) biopsy in the course of follow-up study for a malignancy, for instance, lymphoma, carcinoma, melanoma. In such patients the appearance of a new nodule ordinarily raises the suspicion of recurrence or metastasis of the primary tumor.

The large majority of palpable soft tissue masses are benign; genuine sarcomas are rarely encountered except in centers for diagnosis and treatment of tumors of the locomotive system. Clinically, a sarcoma may present as a large, firm mass in the soft tissues; physical findings usually suggest malignancy. Other sarcomas, however, may appear as singularly innocuous swellings. For this reason sarcomas should be included in the differential diagnosis of all palpable lesions of the head and neck, breast, thorax, abdomen, genitals, and extremities. With the exception of a few characteristic clinical patterns, as in the case of a large fleshy breast tumor (cystosarcoma phyllodes) or grapelike masses in the vagina of a child (sarcoma botryoides), FNA biopsy may be the only way to confirm a clinical suspicion of malignancy before surgery.

This chapter intends to supply a systematic description of the cytology of benign and malignant mesenchymal tumors. Such a description should facilitate pre-operative diagnosis and throw light on the potential role of FNA in the managment of patients with soft tissue tumors. For references to the histopathological structure of soft tissue tumors, see Enzinger and Weiss.[20]

Inspection, Palpation, and Findings on Aspiration

Inspection of the body surface can reveal subtle shadows and asymmetries useful in diagnosis of soft tissue tumors. Palpation, which includes running the flat of the hand over the suspected tissue surface can, for example, provide vital information for diagnosis.

Superficial lesions, that is, those located in the skin, in particular the cutis, subcutis, or superficial fascia, are easy to discover clinically. Such lesions tend to be benign. The most common benign tumor encountered in the superficial soft tissues is the lipoma, which may be either single or multiple. Palpation and aspiration findings in lipoma are useful as a baseline from which to judge more ominous soft tissue lesions. Swellings with short histories and characteristics obviously differing from lipomas in size, contour, consistency, and sometimes, accompanying skin discoloration can advantageously be biopsied by FNA.

More deep-seated lesions, for example, those involving striated muscle, can be difficult to discover when still small. Often when the patient seeks medical attention, the mass may have a diameter of several centimeters, although it only vaguely distorts the normal topographic relationship of surrounding tissues. One such example is sarcoma of the thigh, which may occasionally clinically be mistaken for thrombophlebitis. Deep-seated lesions usually turn out to be malignant. They also make obtaining adequate diagnostic cell material difficult. The inexperienced operator attempting to aspirate with a thin needle of standard length can end up with misleading nonrepresentative samples, which are more common in deeper soft tissue lesions than in more easily accessible tumors. This is partly because deepseated targets may be only indistinctly demarcated by palpation. In addition, their consistency at aspiration may not differ much from the surrounding tissues. Furthermore, a standard 3-cm needle may well be too short if not introduced perpendicular to the skin and directed toward the center of the mass.

The experienced observer can, by naked eye inspection, often discern nonrepresentative material from diagnostic samples. For instance, aspirated striated muscle has a characteristic granular appearance when smeared. Aspirates from fatty tissue or from a lipoma consist macroscopically of fat globules and some solid fragments of tissue—an important finding that should serve as a standard for judging needle aspirates from other lesions. An adequate aspirate from a nonlipomatous solid tumor usually reveals a mixture of blood and of several minute tissue particles. From myxoid tumors sticky mucoid substance is obtained. Some tumors of fibrogenic type, however, yield only a drop or two of blood containing a few cells, despite vigorous aspiration. Aspirates from other lesions can consist mainly of necrotic tissue or fluid. Cell yields in fine needle

aspirates can be checked by rapid staining of airdried smears.

If for some reason an aspiration proves to be nondiagnostic, the biopsy should be repeated, preferably with particular care being taken to determine the optimal site for aspiration. Clinical data obtained through ultrasonography soft tissue x-ray pictures, computed tomography (CT) scans, and angiography are useful for this purpose. Depending on the approximate depth and diameter of the tumor, an appropriately long needle should be selected for the aspiration. If palpation proves to be inadequate for precise localization of a retroperitoneal tumor, aspiration can be performed under radiographic or ultrasonographic control.

TUMORS AND TUMORLIKE LESIONS OF FIBROUS TISSUE

Fibromatoses

Fibromatoses are proliferations of fibroblasts that can present in various clinical settings depending on the age of the patients and the site and depth of the lesions. Some are peculiar to infants and children, such as fibrous hamartoma of infancy, infantile digital fibromatosis, and congenital generalized fibromatosis. Others occur regardless of age, but more often in adults than in children, such as superficial or fascial fibromatosis of various types (palmar, plantar), musculo-aponeurotic fibromatosis (desmoid tumor) and forms of intra-abdominal fibromatosis affecting the pelvis or mesentery.

Fibromatoses have infiltrative properties, although they never metastasize. When not radically resected, they show a marked tendency to recur and often pose problems of definitive local treatment at subsequent surgical intervention. The cell density of fibromatoses in paraffin sections may vary from poorly cellular and predominantly collagenized to highly cellular growths.[41] However, a feature common to all fibromatoses is the uniformity of the spindle-shaped fibroblasts, which are associated with interstitial fibrillary collagen (Fig. 15-1). No atypical hyperchromatic or bizarre nuclei are observed. Mitotic activity is usually rare or absent.

From various types of fibromatosis, including desmoid tumors, usually a drop or so of blood is

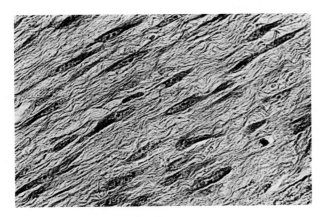

Fig. 15-1. Desmoid tumor. Uniform spindle-shaped fibroblasts separated by fibrillar collagen. (H & E stain, ×600)

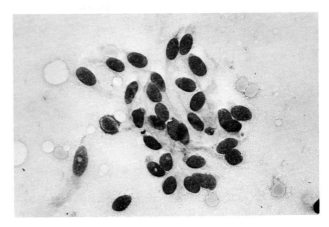

Fig. 15-2. Fibromatosis. Slender spindle-shaped fibroblasts with regular oval nuclei and delicate chromatin. (May–Grünwald–Giemsa [MGG] stain, ×600)

aspirated. From the superficial fibromatoses, in particular, the smears contain scanty material. From these fibrous lesions generally a cell picture composed of the following elements is seen in May–Grünwald–Giemsa (MGG) smears: fragments of acellular collagen, a pinkish staining background due to mesenchymal mucosubstances, damaged cells (debris), naked nuclei, and preserved fibroblasts free-lying and in clusters. The amount and size of the cell clusters are small and the number of free-lying fibroblasts varies. The clusters are thin, often with little mixture of intercellular substance, so that individual cells are well displayed. The fibroblasts are usually spindle shaped or elongated with either slender or more blunt frayed cytoplasmic ends (Fig. 15-2). Sometimes the cell body has a more rounded form. With MGG the cytoplasm stains bluish, partly with indistinct cell borders. The nuclei are oval and regular, but a few larger nuclei with visible nucleoli may be noted. The background is usually hemorrhagic and contains occasional small fragments of blood vessels. In some cases only acellular collagen fragments are found, with isolated free regular spindle cells.

Other cases of fibromatosis, especially desmoid tumor, give significantly larger tissue fragments built up of fibroblasts, lying embedded in fibrillar eosinophilic intercellular material (with MGG) smears. Between the cell clusters, abundant mucoid substance can surround numerous single fibroblasts. The latter have the same regular cell characteristics as described

earlier, but binucleated forms may occur, and there is usually less cell damage. Sometimes from areas where the desmoid tumor infiltrates atrophic muscle fibers with multiple peripheral oval nuclei and densely staining cytoplasm, mimicking neoplastic giant cells are found. The inexperienced cytopathologist may wrongly interpret these as sarcoma cells. (See Color Plate XXXII A.)

A conclusive cytologic diagnosis of fibromatosis should not be rendered on scanty material, but the general smear pattern speaks for a fibroblastic collagenous lesion. If the aspirate is cellular and contains much mucoid substance, differential diagnosis with nodular fasciitis may be problematic. Furthermore, confident differentiation between fibromatosis and low grade spindle cell sarcoma showing a minimum of nuclear and cellular atypia is not always feasible. The above-mentioned differential diagnoses cannot in every case be solved on cytologic grounds alone, but put in the context of clinical setting the cell picture can facilitate decisions about further management. (See Color Plate XXXII B and C.)

Nodular Fasciitis

Nodular fasciitis is most common in the upper extremities, trunk, and neck of young adults. Characteristically, patients present with a lump that has grown rapidly. Generally, the history is shorter than a month, but lesions of several months to a year may be encountered. Tenderness is a feature in about half

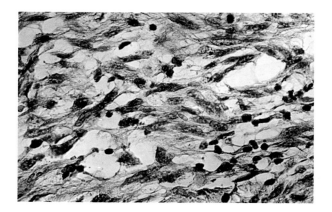

Fig. 15-3. Nodular fasciitis. Plump fibroblasts arranged in short bundles and lying in a richly mucoid matrix. Some lymphocytes in the lower right corner. (H & E stain, ×600)

of the patients. The growth occurs in direct association with superficial or deep fascia and extends in the subcutaneous fat or adjacent muscle. The lesions vary from 1 cm to 5 cm; they are self-limited and may regress even when incompletely excised.

In sections, most forms of nodular fasciitis have a loose-textured feathery appearance imparted by the abundance of myxoid matrix[17,34] surrounding the fibroblasts that are arranged in short interwoven bundles (Fig. 15-3). Mitotic figures are a constant feature. Characteristically proliferating capillaries, extravasated and a round cell inflammatory infiltrate, including erythrocytes, lymphocytes, and occasionally plasma cells, are present in varying numbers. Macrophages due to disintegration of infiltrated subcutaneous fat can be seen.

Cell yield at aspiration and smear pattern depend partly on the proliferative activity of the fasciitis and the degree of mucoid change or hyaline fibrosis of the stroma. Most aspirates from nodular fasciitis yield rich cellular material consisting of plump fibroblasts embedded in mucoid substance and appearing in large or small clusters or as free cells. Usually there is a rich amount of myxoid intercellular substance, which lies as a thick film over the smear and is visible as a bright pinkish red background, sometimes with an eosinophilic granularity due to precipitated dye (with MGG).

The myofibroblasts of nodular fasciitis are large-sized cells (Fig. 15-4 and Color Plate XXXII D), mostly with a plump asymmetric spindle shape with

well-defined cytoplasm, and are well visible with MGG. Most of these cells contain one oval, often eccentric nucleus with a visible nucleolus. Some cells are fusiform with long tapering ends, while others are nearly oval and some round. Still other myofibroblasts have broad angulated cytoplasmic bodies and either a triangular or polygonal shape, or they may show one or more angular cytoplasmic extensions resulting in a stellate appearance. Some of the latter cells may have long slender cytoplasmic processes. In some cases a sizable proportion of the stellate cells are very large with dark blue–staining cytoplasm (MGG).

The angular or stellate myofibroblast contain one or two round oval nuclei, a proportion of which may be larger than average and simulate atypia. They contain a prominent nucleolus. In Papanicolaou (P) smears, fibroblasts have an oval or round nucleus and a sharp, sometimes slightly indented nuclear membrane that stands out as a distinct black line. The chromatin is finely granular. Some mitotic figures can usually be seen.

Focally oval cells are seen containing vacuolated cytoplasm and small red granules and red hyaline globules — representing macrophages. In addition, various numbers of lymphocytes and sometimes plasma cells can be found. In some cases multinucleated monomorphic giant cells, sometimes containing fine red granules and showing cytoplasmic extensions, are seen. Small tufts of fibrillar and mucoid substance occur in the background. Also,

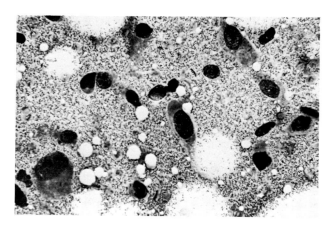

Fig. 15-4. Nodular fasciitis. Plump myofibroblasts and smaller, pointed cells against mucoid background, in this case granular because of dye precipitate. (MGG stain, ×600)

fragments having a central capillary with cells radiating away from it can be noted.

The general cell picture of nodular fasciitis is fairly characteristic in most cases, allowing recognition of this lesion in fine needle aspirates.[12] Differential diagnosis comprises desmoid tumor (see above) and low-grade sarcoma containing myxoid intercellular substance such as myxofibrosarcoma (myxoid malignant fibrous histiocytoma), which generally shows a lesser cellularity and more atypical nuclei. Sometimes rather poor cellular yields are obtained, consisting of a fibrillar pink background substance with some free-lying regular oval nuclei. Such aspirates are nondiagnostic.

Proliferative Fasciitis and Myositis

Proliferative fasciitis[10] and myositis[20] are closely related proliferations of fibrous tissue occurring in subcutaneous fat or striated muscle. They are unusual but seen in adults mainly after the age of 40 and before 70.

In paraffin sections proliferative fasciitis and myositis are characterized by the presence of a population of large basophilic ganglion cell-like mononucleated and often binucleated cells (Fig. 15-5) mixed with fibroblasts constituting this lesion. Proliferative fasciitis and myositis have in the past often been misdiagnosed as sarcoma on account of the unusual but impressive cell picture.

Fine needle aspirates from proliferative fasciitis and myositis are composed of two types of cells.[32a] Often dominating the picture are large mostly oval cells with one or two and sometimes even more nuclei lying at the periphery of the cell body. The cytoplasm is denser in the center than in the periphery, which often contains small vacuoles. The nuclei are vesicular, round to oval (Fig. 15-6), and contain prominent nucleoli, varying in size with the cell size. The second type of cells consists of spindle-shaped fibroblasts with oval nuclei, which sometimes are enlarged and hyperchromatic. Characteristically the background consists in MGG of pink-staining mucosubstances, but even fat cells or blood may be seen. In cases of proliferative myositis, fragments of striated muscle are intermingled with the ganglion cell–like[53] cells and fibroblasts. (See Color Plate XXXII E.)

In aspirates as in paraffin sections, the cell picture can be confounding and it may elicit misinter-

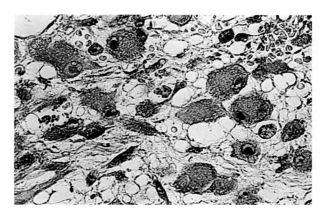

Fig. 15-5. Proliferative fasciitis. Basophilic ganglion cell-like giant cells with one or two peripheral nuclei containing large nucleoli. (H & E stain, ×600)

pretation as sarcoma, especially of the types containing pleomorphic cells such as, for example, malignant fibrous histiocytoma. However, the multinucleated giant cells, the atypical histiocytelike cells, and spindle cells in malignant fibrous histiocytoma usually have a more irregular form, nuclei, and chromatin pattern. With knowledge of the micromorphology of this clinicopathologic entity, pre-operative cytologic diagnosis is possible, avoiding surgery in these cases, which can regress completely within a matter of weeks.

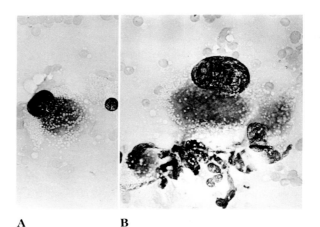

A   B

Fig. 15-6. Proliferative fasciitis. A and B. Ganglion cell-like giant cells with enlarged nuclei and abundant cytoplasm, which is dense in the center of the cell body and contains numerous small vacuoles in the periphery. (MGG stain, ×600)

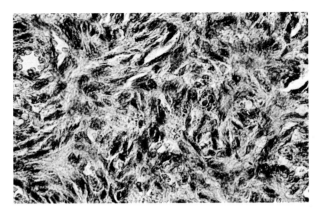

Fig. 15-7. Benign fibrous histiocytoma. Spindle cells arranged in a storiform pattern. Some cells in the center of the field contain hemosiderin particles. (H & E stain, ×600)

Fibrosarcoma

Fibrosarcoma is a malignant tumor of fibroblasts that, in addition to local aggressiveness, is capable of giving rise to distant metastases, chiefly through the bloodstream. Fibrosarcomas most commonly arise in the deeper tissues of the thigh and knee regions and the trunk. Patients usually present with a palpable, nontender, often slowly growing mass. However, tumors that involve nerves or have undergone necrosis with hemorrhage may be tender.

In paraffin sections the picture consists chiefly of interlacing, densely cellular fascicles of more or less uniform spindle cells, often forming a herringbone pattern, which is very distinct in variants showing good cellular differentiation.[17] The nuclei can vary from a rather uniform spindle shape without prominent nucleoli to hyperchromatic pleomorphic oval forms. Well-differentiated fibrosarcomas showing fibrogenesis feel hard to the aspirating needle and give poorly cellular smears. The aspirates contain some clusters of spindle cells with slightly atypical nuclei and free-lying cells of the same type. They also include some fragments of dense fibrillar intercellular substance that stain pale pink with MGG and traumatized cells. A definitive cytologic diagnosis of well-differentiated fibrosarcoma is practically impossible due to the minimal cellular and nuclear atypia. Differential diagnoses are fibromatosis (desmoid tumor) and nodular fasciitis. Mostly one has to resort to examination of the resected tumor for confident classification.

From poorly differentiated fibrosarcomas with little fibrosis, abundant material can be obtained consisting of smaller more ovoid cells with moderately pleomorphic nuclei. Although a cytologic diagnosis of sarcoma is possible, a confident differential diagnosis from other types of sarcoma such as monophasic synovial sarcoma, neurofibrosarcoma, or spindle cell varieties of malignant fibrous histiocytoma cannot be made from the smears.

FIBROHISTIOCYTIC TUMORS

Benign Fibrous Histiocytoma

Fibrohistiocytic tumors (xanthofibroma, dermatofibroma) usually present as small subepidermal nodules seldom more than 1 cm. Dermatofibromas usually persist indefinitely. They occur particularly frequently on the extremities and may be light brown to reddish because of hyperpigmentation of the skin or hemosiderin phagocytosis. Benign fibrous histiocytomas growing subcutaneously may, if located in the neighborhood of a lymph gland or a parenchymatous organ, clinically simulate a tumor of those organs. Rarely, a benign fibrous histiocytoma occurs in the deeper tissues. Histologically, the proliferation has morphologic properties of both fibroblasts and histiocytes, and these may occur in variable proportions.[17] Spindle cells mostly dominate and are arranged in a whorled or "cartwheel" pattern with interlacing and anastomosing bands or bundles (Fig. 15-7). Variable amounts of collagen form part of the lesion. In these predominantly spindle cell tumors, there often is a histiocytic cell component consisting of lipid- or hemosiderin-laden macrophages.

At aspiration the amount of cells obtained from benign fibrous histiocytomas show a great deal of variation. Abundant cell samples are obtained from histologically cellular lesions with a typical whorled growth pattern. Aspirates from lesions with a whorled pattern consist of numerous thick clusters of closely packed short fibroblastlike cells together with a large number of rather well-preserved single cells of the same type. In the clusters, as in the background, only very little intercellular substance can be discerned. The cells have a moderate amount of cytoplasm (Fig. 15-8) and usually a short stellate shape, with at least three processes in each cell. Other cells have histiocytic properties—a large amount of pale cytoplasm, parts of which have ill-defined borders. With MGG stain, the cytoplasm of a fraction of these cells show dark blue–staining pigment granules, probably consisting of hemosiderin, while a few other cells show vacuoles. The spindle cells generally have one, or sometimes two, oval nuclei, with a small but significant population of nuclei having a distinctly larger size. Chromatin is evenly distributed and finely granular. The nucleolus is small and unremarkable with MGG. Few spindle cells with elongated nuclei can be noted. Occasional giant cells with monomorphous nuclei are seen. Take care not to overestimate variation in nuclear size and overdiagnose this lesion as a low-grade sarcoma.

If a benign cutaneous fibrous histiocytoma has been present for a long time, fibrosis is conspicuous and macrophages cannot be demonstrated any longer. Cell yield is then scant at FNA, showing a few cells together with some fragments of densely woven fibrillar substance. If poor cellular yield with the above characteristics is obtained, a diagnosis of benign fibrotic lesion can be made, but no type diagnosis is permissible on the cytologic findings alone.

Aspirates from benign fibrous histiocytomas without storiform pattern but with a more prominent histiocytic differentiation built up of rounded histiocytes and histiocytic giant cells of the Touton type (Fig. 15-9) contain abundant material. The smears show spindly to rounded cells with pale cytoplasm in which hemosiderin particles and fat vacuoles are

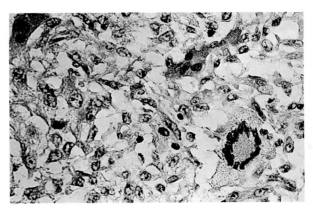

Fig. 15-9. Benign fibrous histiocytoma predominantly consisting of rounded histiocytic cells and multinucleated giant cells, some of them of the Touton type. (H & E stain, ×600)

clearly visible. Hemosiderin stains yellowish brown with P (Fig. 15-10) and blue-black with MGG (Fig. 15-11). With MGG the fat vacuoles contrast better with the gray-blue cytoplasm than with P, and they are found in the background all over the smear (Fig. 15-11). Touton giant cells are very conspicuous with both staining methods (Fig. 15-12). The nuclei are monomorphous, oval, and lack hyperchromasia. Certain low-grade forms of malignant fibrous histiocytoma resemble benign fibrous histiocytoma with predominantly histiocytic features. However, low-grade malignant fibrous histiocytoma shows nuclear

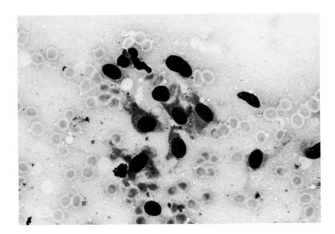

Fig. 15-8. Benign fibrous histiocytoma. Plump stellate cells with moderate amounts of homogeneous cytoplasm and regular nuclei. Histiocyte-like cells with hemosiderin granules in the cytoplasm. (MGG stain, ×600)

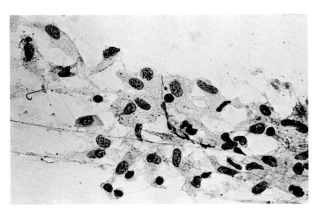

Fig. 15-10. Benign fibrous histiocytoma. Histiocytes with regular nuclei and pale cytoplasm containing hemosiderin particles and small fat vacuoles. (P stain, ×600)

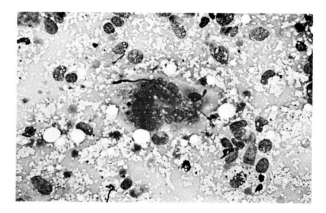

Fig. 15-11. Benign fibrous histiocytoma. Mononuclear histiocytes and Touton giant cell containing small vacuoles. Note abundance of fat vacuoles in the background. (MGG stain, ×600)

pleomorphism and mitotic figures. In doubtful cases conclusive diagnosis has to await examination of paraffin sections of the excised neoplasm.

Dermatofibrosarcoma Protuberans

Dermatofibrosarcoma protuberans generally occurs in middle-aged men, mostly on the trunk. It usually forms a protruding nodular or multinodular mass by infiltrating the entire dermis and subcutaneous fat, but it may also occur in the deeper tissues. The tumor has a tendency to recur locally after simple excision but is not known to metastasize.

Histologically, the tumor is composed of small uniform spindle cells arranged in a storiform or cart-

wheel pattern due to the arrangement of nuclei in a radial fashion (Fig. 15-13).[7,17] Giant cells are infrequent, and histiocytic cells with ingested lipids and hemosiderin are scarce or absent.

A case of dermatofibrosarcoma protuberans aspirated at Karolinska Hospital gave poorly to moderately cellular yields. The cell material constitutes small cohesive clusters of short spindle cells. The larger of these clusters show a ramifying pattern of short curved cell bundles of high density. In the background few dispersed cells are seen as stripped nuclei. The tumor cells have very little faintly staining cytoplasm enveloping the nucleus and ending as narrow cell poles. Nuclei are monomorphous, oval, and short. With P, they are pale and have an inconspicuous nucleolus. Because of their spindly shape, the tumor cells resemble fibroblastic cells more than histiocytic cells (Fig. 15-14). The differential diagnosis thus includes other monomorphous spindle cell tumors, from which unequivocal separation is not always possible. Hence, a cautiously formulated preoperative report of *tumor of monomorphous spindle cell soft tissue* should be issued, without specification of its benign or malignant nature. The decision on further management of the patient should be made with consideration of the clinical picture.

Malignant Fibrous Histiocytoma

Malignant fibrous histiocytoma most often occurs in persons aged 60 to 70 years but also in lower age groups.[16,48] The majority of tumors occurs in one of

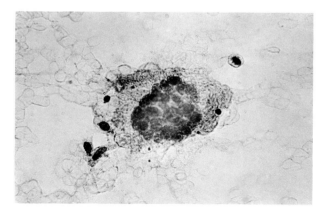

Fig. 15-12. Benign fibrous histiocytoma. Touton giant cell. (P stain, ×600)

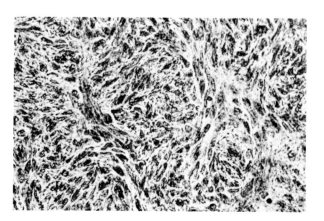

Fig. 15-13. Dermatofibrosarcoma protuberans. Very cellular spindle cell tumor displaying a storiform pattern. (H & E stain, ×600)

the extremities, the retroperitoneum, or the abdomen, with the thigh being the most common single site. The most common symptom when the tumor is located in an extremity or on the trunk is a painless enlarging mass, usually present under 6 months. Two thirds of the tumors are 5 cm or more when first examined. Malignant fibrous histiocytoma typically involves deep-lying muscle; much less commonly, it is confined to the subcutis or extends through the subcutis and involves fascia. Biologically malignant fibrous histiocytoma is a sarcoma with a high incidence of recurrence and metastasis. Metastasis is most frequently to lungs and lymph nodes, usually in a nodular fashion.

Histologic sections of malignant fibrous histiocytoma display several variants of growth, and the majority of tumors are high grade; however, there is usually one more prominent pattern in any given tumor. Thus, tumors with predominantly pleomorphic, fascicular, or storiform patterns of growth can be distinguished.[62] The pleomorphic form is by far the most common. In the pleomorphic pattern, plumper fibroblastlike cells occur in greater numbers than rounded histiocytelike cells and pleomorphic multinucleated giant cells. In the fascicular pattern, plump spindly to fibroblastlike cells and scattered giant cells occur. In the storiform pattern, plump spindle cells are arranged in short fascicles in a cartwheel pattern. However, rounded cells resembling histiocytes are also present. In all forms, both the spindle and rounded cells most often clearly display nuclear atypia.

Fine needle aspirates from malignant fibrous histiocytoma usually yield rich cell samples showing sarcoma cells partly arranged in solid clusters of variable size and partly occurring as single cells. Single tumor cells can be scarce or numerous. From tumors with extensive fibrosis, less cellular smears are obtained. Sarcoma cells from malignant fibrous histiocytoma are generally atypical spindle- or stellate-shaped fibroblastlike cells, rounded cells with features of histiocytes, and pleomorphic giant cells.[60] In the sarcomas with a dominating storiform pattern, most tumor cells resemble irregular, plump stellate fibroblasts—a few being bipolar, which are large compared to cells from benign fibrous histiocytoma. In predominantly pleomorphic malignant fibrous histiocytoma larger proportions of histiocytelike cells and cells with forms intermediate to fibroblasts and histiocytes are seen.

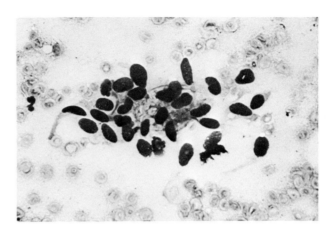

Fig. 15-14. Dermatofibrosarcoma protuberans. Thin spindle cells with scanty cytoplasm and slightly irregular nuclei occurring singly. (MGG stain, ×600)

The spindle-shaped neoplastic cells usually have one oval or round nucleus showing variable pleomorphism from case to case. With P, the nuclei are indented; with a sharply outlined, slightly thickened nuclear membrane; and in the majority of cases, a course chromatin pattern in which several chromocenters are visible. A certain number of cases have vesicular nuclei with pale chromatin. Usually nucleoli are visible. Mitotic figures can be numerous; and several, both typical and atypical, may be seen in the same high-power field. The tumor cells have faintly stainable fragile cytoplasm with indistinct cell borders (Figs. 15-15A, 15-16, and 15-17). Many cells contain numerous minute vacuoles. Similar vacuoles, cytoplasmic substance, and free nuclei are seen scattered in the background.

In malignant fibrous histiocytoma, in addition to the stellate spindly and round tumor cells, large numbers of bizarre, rounded, pleomorphic, mononuclear or multinucleated giant cells (Fig. 15-18) occur, in which the nuclei are several times the size of the more common cell population (Fig. 15-15). These nuclei are oval, ovoid, or kidney shaped, often clustering together centrally or lying in the periphery of the cell. They have a distinctly hyperchromatic coarse chromatin pattern. In some cases, cytoplasmic intranuclear inclusion bodies—*pseudonucleoli*—of variable sizes can be seen. Many of these bizarre large nuclei lie free in the background. The pleomorphic giant cells have a large amount of bluish cytoplasm with MGG stain, often containing numerous small

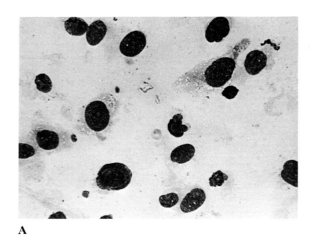
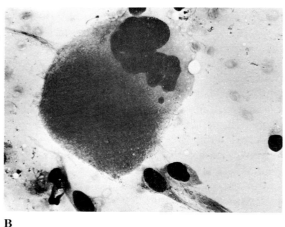

A   B

Fig. 15-15. Malignant fibrous histiocytoma. A. Dispersed atypical fibroblast-like and histiocyte-like cells with fragile, finely vacuolated cytoplasm. Note nuclear pleomorphism. B. Pleomorphic multinucleated giant cell. (MGG stain, ×600)

vacuoles (Fig. 15-19) or hemosiderin particles, red granules (Fig. 15-20) round eosinophilic hyaline bodies, and cell debris, including pyknotic nuclei.[60] Some osteoclastic giant cells with numerous monomorphous nuclei are a feature of this tumor. Between the cells, fragments of capillaries, often slender, can be seen.

In cases showing cellular pleomorphism, a diagnosis of malignancy is rather simple. The reverse is true in cases with poor cellularity or mild nuclear atypia or those in which osteoclastlike and Touton-type giant cells are seen, making a differentiation between malignant and benign fibrous histiocytoma considerably more difficult. Some areas of malignant fibrous histiocytoma may be fibrotic and yield poorly cellular samples that are insufficient for diagnosis. In the presence of a distinct firm mass that repeatedly yields nondiagnostic material, extirpation followed by histologic examination is mandatory in order not to miss a sarcoma with focal or large fibrotic areas.

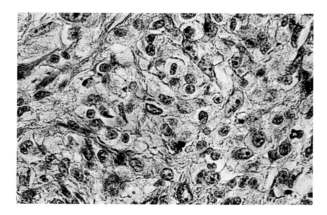

Fig. 15-16. Malignant fibrous histiocytoma. Sarcoma cells showing histiocytic characteristics with pale vacuolated cytoplasm and oval or round mildly pleomorphic nuclei. (H & E stain, ×600)

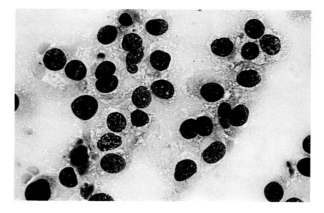

Fig. 15-17. Malignant fibrous histiocytoma. Atypical histiocyte-like cells with cytoplasm containing numerous small vacuoles. (MGG stain, ×600)

Myxoid Malignant Fibrous Histiocytoma

Myxoid malignant fibrous histiocytoma is a variety of malignant fibrous histiocytoma, in which at least half of the tumor is mixoid. This sarcoma typically arises on the extremities of adults with a peak incidence in the seventh decade. It is usually attached to fascia and grows subcutaneously or involves skeletal muscle.[16,61] This neoplasm often gives rise to repeated local recurrences, followed by metastases, but in a lesser proportion of patients than in nonmyxoid malignant fibrous histiocytoma.

The pattern in paraffin sections of this myxoid sarcoma, which is also known as myxofibrosarcoma,[5a] varies with the grade of malignancy. Low-grade sarcomas show a low cellularity and consist predominantly of fibroblastlike spindle cells with slightly pleomorphic nuclei, and they contain few giant cells. High-grade sarcomas display a higher cellularity and consist of a mixture of atypical fibroblastlike spindle cells and histiocytelike round cells showing nuclear pleomorphism. Also higher numbers of pleomorphic giant cells are seen.

Generous amounts of tumor material can be aspirated from the myxoid type of malignant fibrous histiocytoma. Smears show tumor fragments with myxoid architecture and single sarcoma cells. The clusters are found either against a hemorrhagic background or merging with mucoid, focally fibrillar background substance in which the single cells are bathed. The cellularity of the clusters is, as a rule, less than in nonmyxoid fibrous histiocytoma, and it varies

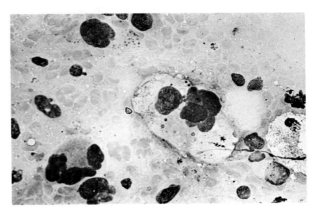

Fig. 15-19. Malignant fibrous histiocytoma. Giant cell *(center)* with histiocytic features; pale stainability due to vacuolization of the cytoplasm. (MGG stain, ×600)

from poor to moderate. In the low-grade sarcomas the tumor cells are almost exclusively fusiform, spindle shaped, or elongated stellate; plump rounded cells are uncommon. In high-grade myxoid malignant fibrous histiocytomas, plump and rounded cells are common among the fibroblastlike cells. These types of cells usually contain one oval, slightly to frankly pleomorphic nucleus depending on the grade. Their cytoplasm, which often has indistinct borders, may contain vacuoles, small rodlike structures, cellular debris, or hyaline globules.[43] A conspicuous proportion of high-grade sarcomas contain large, often rounded, multinucleated giant cells with

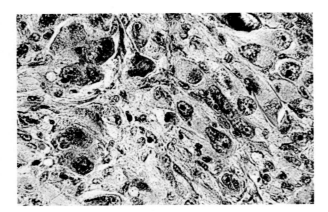

Fig. 15-18. Malignant fibrous histiocytoma. High grade sarcoma containing pleomorphic giant cells. (H & E stain, ×600)

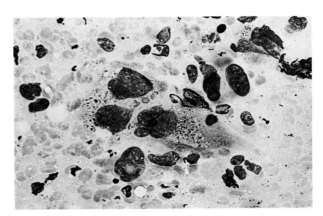

Fig. 15-20. Malignant fibrous histiocytoma. Pleomorphic histiocytic giant cells with red granules in the cytoplasm in addition to small vacuoles. (MGG stain, ×600)

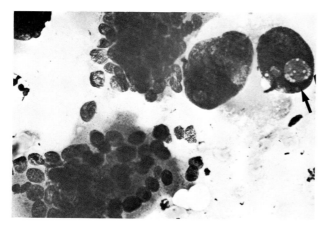

Fig. 15-21. Malignant giant cell tumor of soft parts. Binucleated pleomorphic giant cell with pseudonucleolus (*arrow*) together with osteoclast-like giant cells. (MGG stain, ×600)

bizarre pleomorphic giant nuclei. Their cytoplasm may be vacuolated. In the background, when it is hemorrhagic, single mucoid pools containing a few cells as well as free-lying capillary fragments may be found. Sometimes tumor cells can be seen abutting against a capillary wall. The myxoid type of malignant fibrous histiocytoma may be difficult to separate from nonmyxoid malignant fibrous histiocytoma, since the latter may show myxoid areas. A differential diagnosis in low-grade forms is nodular fasciitis (see above). (See Color Plate XXXIII A–F.)

Malignant Giant Cell Tumor of Soft Parts

Malignant giant cell tumor of soft parts is a variant of malignant fibrous histiocytoma in which osteoclast-like giant cells are the predominant feature.[8,26] The tumor may be superficial or deep. The superficial tumors are generally small and are found in the sub-cutis and superficial fascia, with a predilection for the legs. The deeper-lying tumors are large and involve skeletal muscle, deep fascia, and tendons. Patients with the deep-lying type tend to survive 5 years less than patients with the superficial type.[26]

In paraffin sections of malignant giant cell tumor of soft parts, the osteoclastlike giant cells are invariably associated with smaller mononuclear histiocytelike cells, atypical fibroblastlike cells, and often with pleomorphic giant cells.

Malignant giant cell tumor of soft parts yields abundant cell material consisting of solid tumor cell clusters from which single cells are rather easily detached. In addition to osteoclastlike (Fig. 15-21) cells, which are conspicuous all over the smear, bizarre pleomorphic giant cells and a large number of mononuclear cells are present in aspirates from this tumor.[5] The majority of mononuclear cells show a wide range of shapes, with stellate cells having short processes at one end of the spectrum to oval cells at the other and plump forms with curved cell borders in between. All smears show a sizable proportion of spindle cells.

Mononuclear cells have an eccentric nucleus. In some tumors nuclear pleomorphism is not prominent and nucleoli are small. In others the mononucleated cells have often giant-sized nuclei with prominent nucleoli. Mitotic figures may be frequent. The mononuclear sarcoma cells have a moderate to rich amount of palely stainable fragile cytoplasm with hazy cell borders. Many of them contain small cytoplasmic vacuoles.

Osteoclastlike cells may, besides their monomorphous nuclei, contain numerous micronuclei. Their cytoplasm is dense with distinct cell borders and reveals pointed cytoplasmic processes. They may contain small peripheral vacuoles. Multinucleated cells with bizarre hyperchromatic nuclei several times the shape of the average nuclei occur. Intranuclear cytoplasmic inclusions are numerous in some cases. In these giant cells mitoses are sometimes observed. The cytoplasmic characteristics of the pleomorphic giant cells are similar to those in malignant firbrous histiocytomas. In the background many free nuclei of all types, as well as cytoplasmic debris containing small vacuoles, can be recognized.

Cytologic diagnosis of sarcoma is easy in tumors showing nuclear pleomorphism. However, it is impossible to differentiate malignant giant cell tumor of soft parts from extraskeletal osteosarcoma when ostoid is lacking and the cell sample is small.

Inflammatory Malignant Fibrous Histiocytoma

Inflammatory malignant fibrous histiocytoma has been described as a separate entity.[35,57] Whether this tumor has a biologic behavior differing from the classic type of malignant fibrous histiocytoma is, at present, unclear. Histologically, the hallmark of the tumor is diffuse inflammatory infiltration with granulocytes, lymphocytes, and plasma cells.

When tumor material is aspirated from inflammatory fibrous histiocytoma, cell yields are abundant and consist of dense clusters and single cells. Smears have a cellular composition of inflammatory cells, mostly granulocytes and tumor cells. The latter have the same characteristics as cells in other malignant fibrous histiocytomas; that is, there are varying proportions of stellate, spindle, angular, and oval-shaped cells, mostly with one oval or round, variably pleomorphic nucleus, the oval cells forming, in some cases, the most prominent cell type. The cytoplasm stains faintly and is delicate, often containing numerous small vacuoles resembling xanthoma cells. Occasionally engulfed cell débris is visible. A sizable proportion of the oval cells may display a fine eosinophilic granularity of their cytoplasm. Multinucleated giant cells with bizarrely shaped pleomorphic nuclei occur. The background is dirty from cell debris and acute inflammatory cells. The inflammatory component may be abundant enough to obscure the neoplastic cells, causing the lesion at low-power examination to resemble granulation tissue superficially. However, nuclear pleomorphism and, especially, the bizarre giant nuclei should provide the correct diagnosis.

TUMORS OF ADIPOSE TISSUE

Lipoma

Lipomas mainly occur in the subcutaneous tissues and are encapsulated but can present anywhere and be poorly demarcated. Although usually solitary, they can be multiple and can vary enormously in size. They are benign growths made up exclusively of mature adipose tissue cells showing no evidence of cellular atypia.

In deeper sites, fatty tumors associated with capillary and venous vascular proliferations and even with proliferations of smooth muscle can be found. Occasionally, lipomas in the adrenal medulla, the presacral region, and the mediastinum may contain active bone marrow; these have been called *myelolipomas*.[17] At aspiration, lipomas yield a transparent droplet of fat containing adult fat cells with regular small oval nuclei without atypia (Fig. 15-22). With MGG, a little eosinophilic mucopolysaccharide substance is normally seen between the cells. Occasionally the aspirate is almost cell-free and consists of droplets of lipid. When a mixture of fat cells and hematopoietic cells is found in aspirates from a soft

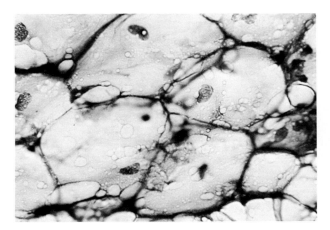

Fig. 15-22. *Lipoma. Adult fat cells with regular nuclei. (MGG stain, ×600)*

tissue mass, the diagnosis of myelolipoma can be made cytologically.

Lipomalike swellings consisting of regular fat cells cannot be cytologically separated from true lipoma.

Pleomorphic Lipoma

Pleomorphic lipoma principally occurs in men aged 50 to 80 years. The tumor shows a predilection for the back of the neck, shoulders, and back. Typically, the lesion appears as a painless circumscribed subcutaneous mass that on clinical examination resembles an ordinary lipoma. The neoplasm is characterized by an intimate mixture of variably sized mature fat cells and bizarre, pleomorphic, multinucleated giant cells. Many of the giant cells show a distinctive floretlike arrangement of their nuclei and are interlaced with strands of collagen. Multivacuolated atypical lipoblasts, which are diagnostic of liposarcoma, are lacking. Despite the pleomorphic picture of this tumor, its clinical behavior is benign.[23,55]

Pleomorphic lipoma gives poorly cellular yields consisting chiefly of adult fat cell clusters but also of a mixture of a few cell aggregates consisting of large bizarre cells with irregular hyperchromatic nuclei. Some of these are round and have a peripheral wreath of large hyperchromatic nuclei—hence the designation *floret cells*.

If the aspirate is representative for a well-demarcated tumor with the palpatory characteristics of lipoma, a cytologic diagnosis of pleomorphic lipoma

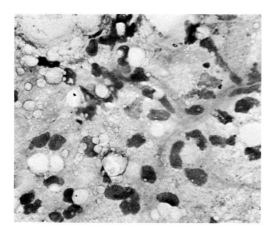

Fig. 15-23. Myxoid liposarcoma. Cluster of myxoid tissue containing multivacuolar atypical lipoblasts with nuclear excavation and monovascular lipoblasts. Note nuclear monomorphism.

can be made on the basis of the above findings. However, if similar material is obtained from deep-seated intramuscular or retroperitoneal masses, well-differentiated sclerosing liposarcoma cannot be excluded.

Liposarcoma

Liposarcoma appears most frequently as a mass in the musculature of the lower extremity, especially in the thigh and gluteal region or in the retroperitoneum. The tumor is often quite large, lobular, and bulky. It is a malignant infiltrating neoplasm, even when circumscribed, and is characterized by the presence of atypical lipoblasts.

The histologic picture of liposarcoma varies from well-differentiated lipomalike to more cellular and pleomorphic types, and biologic behavior varies with degree of differentiation.[21] Well-differentiated myxoid liposarcomas have a protracted course, often with local recurrences and a rather low percentage of metastases even after many years, with the 5-year survival rate exceeding 70%. The poorly differentiated, round cell, and pleomorphic types show frequent wide metastasis, the most common site being the lungs, pleurae, and liver. The 5-year survival rate is less than 20%.

The clue to cytologic diagnosis of liposarcoma in aspirates is the recognition of atypical multivacuolated lipoblasts. Further, depending on the amount of fat production in the tumor, the sarcoma cells are mixed with fat globules of varying sizes, which stand out more clearly against the stroma substance in air-dried MGG smears than in P or hematoxylin-eosin (H & E) preparations.

Myxoid Liposarcoma

In histologic sections, myxoid liposarcoma consists of myxoid tissue resembling embryonal fatty tissue in which atypical multivacuolated lipoblasts or signet ring–shaped lipoblasts with a single large lipid vacuole are found. The tumor cell nuclei are small and have a monomorphous appearance. Some areas of the tumor display a well-developed plexiform capillary vascular pattern.

Myxoid liposarcomas yield tumor material at aspiration, mostly displayed as solid clusters and a variable number of single tumor cells. The background of the smear can be myxoid or hemorrhagic. The clusters are moderately cellular and consist of primitive cells with small nuclei and unidentifiable cytoplasm with both MGG and P stain, so that the tumor tissue particles have the appearance of a pseudosyncytium. However, the space between the nuclei is filled by a fibrillar intercellular matrix, which stands out a bright pink-red with MGG stain. In some cases a number of free-lying oval tumor cells containing a variable number of minute optically empty vacuoles can be seen with both MGG and P stain.

The tumor nuclei are small, oval, and monomorphous, features well appreciated with P. They have a thin nuclear membrane and contain evenly distributed finely granular chromatin with visible chromocenters but an inconspicuous nucleolus. In some cases the nuclei are larger and show a slight degree of pleomorphism. In the clusters, among the more primitive cells, multivacuolated lipoblasts can best be seen with MGG stain (Fig. 15-23). These vacuoles are of variable size and can attain several times the size of the nuclei in which they make excavations. Dense ramifying networks of capillaries can be discerned traversing the cell clusters. Outside the clusters, in the background, a variable proportion of free nuclei are seen.

Recurrences and metastases usually display a picture similar to the primary tumor, even when they appear after many years. In myxoid liposarcomas that are in part composed of pleomorphic areas, pleomorphic multinucleated cells may be seen.

Differential diagnosis consists mainly of intramuscular myxoma and other monomorphous myxoid tumors, all of which show fusiform or stellate monomorphous cells with identifiable cytoplasm and abundant myxoid substance. The latter may include low-grade myxoid malignant fibrous histiocytoma and myxoid chondrosarcoma. Aspirates from intramuscular myxoma generally are characterized by very poor cellularity and are devoid of atypical lipoblasts. When clusters are obtained, they do not have fat cells incorporated in them. Moreover, they lack the prominent capillary network seen in myxoid liposarcomas. Low-grade spindle-celled myxoid malignant fibrous histiocytomas yield fibroblastlike cells with distinctly visible elongated cell bodies. Myxoid chondrosarcomas, whether arising extraskeletally or in bone, are, with MGG stain, characterized by very intense pink-to-red-staining chondroid substance in which monomorphous small round or slightly elongated chondroblasts are embedded or loosely arranged in ramifying rows, cords, or small groups. Occasional binucleated cells are seen, but no atypical lipoblasts can be identified. (See Color Plate XXXIV A.)

Round Cell Liposarcoma

Round cell liposarcoma, a high-grade sarcoma, shows closely packed round lipoblasts with more outspoken nuclear atypia as compared to myxoid liposarcoma. FNA of round cell liposarcoma yields abundant cell material. The smears show mucoid substance in which numerous tumor cells are embedded. They have round to oval more irregular nuclei than in low-grade myxoid liposarcomas, showing hyperchromasia and slight pleomorphism. The diagnosis rests on the recognition of atypical multivacuolated lipoblasts. (See Color Plate XXXIV B and C.)

Pleomorphic Liposarcoma

In sections, the picture of pleomorphic liposarcoma is that of pleomorphic sarcoma with the additional occurrence of lipoblasts or signs of fat cell production in some focal areas. In pleomorphic liposarcoma, cell yields are abundant and consist of large pleomorphic cells with abundant cytoplasm that stains pale blue-gray with MGG. The nuclei are extremely variably shaped. Gigantiform multinucleated cells with bizarre nuclei are numerous. Because of the outspoken anaplasia, liposarcoma may be suspected only in those instances where evidence of fat cell production within tumor clusters or of atypical lipoblasts can be

noted. Differential diagnosis includes, in the first place, other pleomorphic sarcomas, especially malignant fibrous histiocytoma of classic and myxoid types.

TUMORS OF MUSCLE TISSUE

Smooth Muscle

Leiomyoma

Benign smooth muscle tumors rarely occur in sites other than the uterus or gastrointestinal (GI) tract. In the deep soft tissues leiomyomas are uncommon but are occasionally observed in the broad ligament, retroperitoneum, mesentery, omentum, mediastinum, and orbit. The superficial varieties occurring in the skin and subcutaneous tissues are either totally made up of interlacing bundles of smooth muscles or have a prominent vascular component. The former are probably derived from musculi arrectores pilorum in the skin or genitals. The latter are thought to derive from the smooth muscle of blood vessels. Either variety of superficial leiomyoma may occasionally cause pain and only rarely become large. The fact that superficial leiomyomas usually are small and easily resectable is probably the reason why they are very seldom referred for FNA biopsy.

Leiomyomas usually give poor cell yields consisting of occasional solid cell clusters and single smooth muscle cells. The cell cytoplasm stains densely blue, and the nuclei are elongated and regular and show insignificant variation in chromatin content and texture.

Leiomyosarcoma

Malignant smooth muscle tumors are relatively rare neoplasms in soft tissue.[24,56,59,63] The lesions occur mostly in adults and generally affect the skin or subcutaneous tissue of the extremities, occasionally being intramuscular. The deeper lesions are found in the retroperitoneum and the walls of large vessels, mostly deep veins. Skin and subcutaneous lesions are usually single nodules varying from a few millimeters to under 5 cm, and many are either painful or tender. Deeper-lying lesions are usually more than 5 cm.

Histologically, the tumor is cellular, being composed of elongated acidophilic cells containing varying amounts of nonstriated myofibrils and a clear perinuclear space. The tumor cells tend to form sharply intersecting bundles and fascicles. Compared to benign smooth muscle tumors, well differentiated

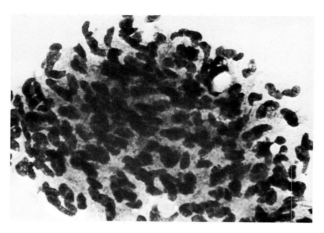

Fig. 15-24. Leiomyosarcoma. Cohesive cluster of elongated spindle cells with dense cytoplasm and thin elongated nuclei. (MGG stain, ×600)

leiomyosarcomas show a greater degree of cellularity and mitotic figures. Moderately to poorly differentiated forms show in addition cellular and nuclear pleomorphism (including tumor giant cells), and typical and atypical mitotic figures.

On repeated FNA, very poor cell yields may alternate with very rich cell samples, both from primary tumors as well as metastases. Aspirates from well-differentiated to moderately differentiated varieties of leiomyosarcoma show a smear pattern dominated by cohesive, often ramifying cell clusters built up by spindle-shaped cells (Fig. 15-24).[13,38] This feature reflects the sharply defined cell bundles of histologic sections. With MGG stain, the cytoplasm of the spindle cells stains a dense blue and stands out against a small amount of pink to red fibrillar intercellular substance in which they seem to be embedded. Their nuclei are oval or elongated and slender, often showing no pleomorphism, even with P stain. Good cell cohesivity or, conversely, limited tendency to dispersal is expressed by the small to moderate number of isolated single cells or stripped nuclei. They may, in fact, be the only finding in some aspirates. Since leiomyosarcoma shares the cohesive smear pattern and cytoplasmic tinctorial characteristics of cells from leiomyoma, for purposes of differential diagnosis emphasis should be placed on nuclear atypia. However, the degree of nuclear atypia in leiomyosarcomas varies and may be slight, and hypocellular areas occur, thus making a confident diag-

nosis of malignancy problematic. The differential diagnosis should include other spindle cell sarcomas of nonmuscle origin with rather monomorphous appearances, for instance, fibrosarcoma, some neurofibrosarcomas, synovial sarcoma, or even benign tumors such as schwannomas and desmoid tumors.

Poorly differentiated leiomyosarcoma with anaplasia shows strikingly pleomorphic nuclei and multinucleated giant cells. Such varieties of leiomyosarcoma are difficult to separate from some other pleomorphic sarcomas, such as malignant fibrous histiocytoma, liposarcoma, and rhabdomyosarcoma. However, the cytologic picture leaves no doubt of their malignancy. When the cytologic characteristics of the smear are not typical enough to suggest the specific diagnosis or even to suggest malignancy, the lesion should be extirpated for histopathologic examination, the cytologic report stating *soft tissue tumor* as the tentative pre-operative diagnosis. (See Color Plate XXXIV D–G.)

Striated Muscle

Adult Rhabdomyoma

Adult rhabdomyoma is a rare tumor mostly observed in the upper neck region, tongue, pharyngeal wall, and neighborhood of the larynx.[57] Histology is highly characteristic, with polygonal, frequently vacuolated cells and granular, strongly acidophilic cytoplasm. The tumor is firm, and cell material is difficult to obtain. Except for a drop of blood, the smears show either cohesive cell clusters or loose aggregates of fewer than ten cells dispersed in the smears. Sometimes only a few single or paired free-lying cells are obtained. Cells from adult rhabdomyoma are voluminous because of their very large amount of bluish granular cytoplasm, which is somewhat mottled by vacuoles. The tumor cells are polyhedral with blunt angles and bulging edges or irregularly oval (Fig. 15-25). In some of the cells, cross-striations can be discerned with routine stains. Single or multiple nuclei are mostly peripherally situated and have regular oval shape and size. A few cells may be bipolar or spindle shaped. The smear background is clean, and the general cytologic picture is pathognomonic.

Rhabdomyosarcoma

Histologically, there are three distinguishable types of rhabdomyosarcoma: embryonal, alveolar, and pleomorphic.[17] Embryonal rhabdomyosarcoma is the most frequent type and occurs mainly in children. The most common sites of origin are the head and neck, retroperitoneum, urogenital tract, and extremities, the last being involved in less than 5% of cases. Alveolar rhabdomyosarcoma occurs most often in the extremities of older children and young adults.[18,54] Muscles of the upper extremities, lower limbs, and trunk are almost equally affected, with a slightly lower incidence in the head and neck. Pleomorphic rhabdomyosarcoma, the rarest variety, is seen primarily in the large skeletal muscles of adults, but patients with it appear to have a better prognosis than those with the other two types.[57]

Embryonal Rhabdomyosarcoma. Embryonal rhabdomyosarcoma appears in paraffin sections as a solid tumor built up of small undifferentiated cells, among which a variable number of larger differentiated rhabdomyoblasts occur. The differentiated rhabdomyoblasts are often round with acidophilic cytoplasm, or they may show cross-striations or peripherally arranged vacuoles. Yet others may be elongated with acidophilic cytoplasm and two or more nuclei in tandem arrangement. Some areas may be myxoid with abundant mitotic figures, predominantly stellate cells, and only an occasional differentiated tumor cell.

At aspiration, cell-rich preparations can be made consisting either of highly cellular or of more sparsely populated tumor fragments; the latter have a

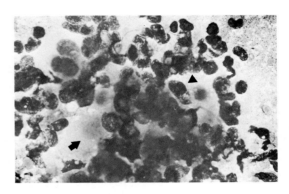

Fig. 15-26. Embryonal rhabdomyosarcoma. Atypical rhabdomyoblasts with voluminous cytoplasm showing rounded condensations *(arrows)* and nuclei at one pole, surrounded by undifferentiated small tumor cells *(arrowhead).* Note small vacuoles in the background. (MGG stain, ×600)

conspicuous component of intercellular matrix enveloping the tumor cells. In between the clusters small cell aggregates, loose sheets, and single tumor cells can be seen. The majority of the tumor cells are usually small mesenchymal cells without other cytologic signs of differentiation. They have an oval to plump spindle shape and contain an oval, often eccentric, irregular nucleus. Their cytoplasm is pale staining and often inconspicuous, usually displaying many small, optically empty vacuoles. In the background of the smears, confluent cytoplasmic substance containing large numbers of small vacuoles can be seen.

The pathognomonic cells in embryonal rhabdomyosarcoma are the round neoplastic rhabdomyoblasts, which are present in varying numbers and can be seen in the clusters or as free-lying cells. These cells are large with abundant dense cytoplasm (Fig. 15-26) and one eccentric irregular nucleus. Some are binucleated or multinucleated. Their nuclei vary in size and shape from oval to flattened or lobulated, and they are often hyperchromatic. Their cytoplasm is denser in the center than in the cell periphery, where its round margins are sharply delineated. The intense stainability of the cytoplasm of round rhabdomyoblasts can be judged with both P and MGG. It is more prominent in MGG due to the deep blue tone contrasting with the pale cytoplasm of undifferentiated cells in the smear. With P stain the round rhabdomyoblasts stain grayish green but si-

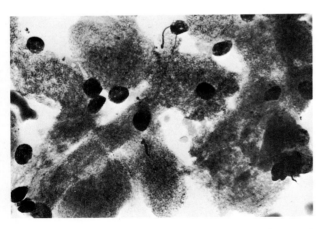

Fig. 15-25. Adult rhabdomyoma. Sharply delineated large cell bodies with dense granular cytoplasm and regular nucleus. A few elongated cells. (MGG stain, ×600)

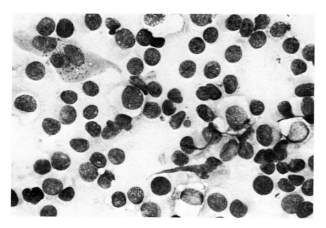

Fig. 15-27. Alveolar rhabdomyosarcoma. Circularly arranged small tumor cells with eccentric nuclei and marked hyperchromasia. Note vacuolated cytoplasm of some cells and a contaminating osteoclastic giant cell. (MGG stain, ×600)

multaneously with a more or less prominent red eosinophilic tinge. However, with these routine stains it is extremely difficult, if not impossible, to identify cross-striations. Some of the round rhabdomyoblasts may contain peripheral vacuoles.

From areas of necrosis, some very poorly cellular strands of myxoid tumor connective tissue substance, together with necrotic material, may be aspirated. (See Color Plate XXXV A–C.)

Alveolar rhabdomyosarcoma. The predominant histologic pattern of alveolar rhabdomyosarcoma consists of irregular, vaguely defined groups and nests of tumor cells with poor differentiation, marked loss of cellular cohesion, and the presence of "alveolar" spaces. Dense fibrous trabeculae surround and separate the cell nests. Tumor necrosis may be a striking feature. In a number of cases one can observe round or strap-shaped multinucleated giant cells with acidophilic cytoplasm and a marginal wreath of small nuclei.

FNA contains an abundance of small tumor cells, either in solid, large, or small clusters or as clusters in which the cells are oriented around an empty space, imitating the alveolar pattern of histologic sections. In between the clusters the smears are crowded with single cells. The dominating cell type has an oval cell body, eccentric nucleus, and scanty cytoplasm with a distinct rounded contour and stains pale blue

with MGG (Fig. 15-27). In most cells many small vacuoles are recognizable. The nuclei are round to oval. With P stain, the nuclear membrane is distinctly thickened and sometimes indented. The chromatin is granular, and a small nucleolus is seen. Some of the cells have more cytoplasm than others and are rather rectangular, their nuclei lying eccentrically at one of the shorter sides of the cell. Occasional cells have one thin pointed end, giving the cell a conical shape, with the nucleus situated at its base. A few cells are spindle shaped, with two long pointed ends. Some cells are binucleated, with nuclei eccentrically positioned in the cell body. (See Color Plate XXXV D–F.)

In some cases the characteristically large multinucleated giant cells in which the nuclei are arranged as a peripheral wreath, can also be demonstrated in aspirates. A mixture of occasional large round neoplastic rhabdomyoblasts with dense cytoplasm may occur. In smears from alveolar rhabdomyosarcoma, the background frequently consists of cytoplasmic substance and numerous small vacuoles.

In the absence of distinguishing cytologic features such as neoplastic round rhabdomyoblasts or multinucleated giant cells, a number of rhabdomyosarcomas cannot confidently be diagnosed cytologically, and other malignant small cell tumors, discussed below, should be considered in the differential diagnosis of rhabdomyosarcoma.

In *lymphocytic lymphoma* the tumor cells show less cellular cohesion and a more pronounced dispersed pattern with only rare naked nuclei. Moreover, the nuclei have a more central position in the cell, and the cytoplasm is more scanty and round in outline. The finding of so-called lymphoglandular bodies—small rounded cytoplasmic fragments in the background—is typical for lymphoma.

A number of other small cell malignant mesenchymal tumors, such as certain cases of *neuroblastoma* or *Ewing's sarcoma*, cannot always be differentiated from rhabdomyosarcoma. Under those circumstances, although the diagnosis of malignancy is obvious, final classification of the tumor must await until surgical biopsy provides more material for histopathologic examination, which includes immunocytochemical staining and electron microscopic (EM) analysis. However, it has been shown that a cytologic diagnosis of *small cell malignant tumor* in a proportion of cases can be specified before surgery or other forms of therapy by the demonstration of ultrastructural characteristics such as, for example, specific myofilaments or Z-band material in rhabdomyosarcoma or neuritic processes bearing neurosecretory granules in neuroblastoma on cell material taken by FNA biopsy.[11,40] Also the availability of immunocytochemical demonstration of desmin in rhabdomyosarcoma and positive staining for S-100 protein in neuroblastoma allow for an improvement of preoperative diagnosis.

Pleomorphic rhabdomyosarcoma. Spindle cell and pleomorphic rhabdomyosarcomas, occur mainly in adults, are histologically poorly characterized, and are probably exceedingly rare. Their diagnosis depends on identification of rhabdomyoblasts, which are not reliably demonstrable with routine P or MGG stain. Hence, this sarcoma variety cannot in practice be confidently diagnosed cytologically as myogenic and, as a rule, cannot be differentiated from other pleomorphic sarcomas in fine needle aspirates.

TUMORS OF BLOOD VESSELS

Angioma

Capillary hemangioma

Capillary hemangiomas, or "strawberry marks," first appear in infants between the third and fifth week of life and grow for a few months only, regressing spontaneously and often completely in several years. Histopathologically, capillary hemangiomas show in their growth phase solid strands and aggregates in which only a few small capillary lumina can be discerned. More mature lesions show flatter endothelial cells lining wider capillary lumina. Later, fibrosis brings about shrinkage of the lesions.

Aspirates from capillary hemangioma in neonates consist of blood and large amounts of intertwining cellular strands (Fig. 15-28) and only occasional single cells. Identification of the large cell clusters as budding and ramifying capillaries is best done with P, in which blood-filled lumina can be discerned; MGG also stains the reticulin sheaths enveloping the strands. Some clusters of endothelial cells from wider blood vessels are arranged as small monolayered sheets of about ten cells or less, which easily disperse into single cells. They have a moderate amount of delicate, light blue–staining homogeneous cytoplasm (with MGG) that in smears takes several shapes, from spindly to polygonal, showing

Fig. 15-28. Capillary hemangioma. Circular cohesive ramifying tufts of capillaries. (MGG stain, ×600)

serrated, rather poorly demarcated cell borders. Nuclei are monomorphous, small oval, regular, and normochromatic with finely dispersed chromatin and no nucleolus. Capillary hemangioma should not be cytologically mistaken for a fibroblastic proliferation, which in smears would show a much larger proportion of single, randomly spread, spindle-shaped cells with more distinctly delineated cytoplasmic borders.

Cavernous and Intramuscular Hemangioma

Cavernous hemangioma may be a large subcutaneous mass, often in association with an overlying capillary hemangioma of the skin. Histologically, a cavernous hemangioma shows large, irregular, vascular spaces located in the deeper dermis or in the subcutaneous tissue. The spaces are lined by a single layer of thin endothelial cells and have fibrous walls of varying thickness resulting from proliferation of adventitial cells.

Intramuscular hemangioma mostly occurs in young adults and consists of a painful poorly circumscribed benign vascular growth diffusely infiltrating striated muscle.[17] Either capillary or cavernous structures may predominate. Both cavernous and intramuscular hemangiomas are usually soft to the aspirating needle and predominantly yield blood. Up to several milliliters of blood are often rapidly drawn into the barrel of the syringe, and during the aspiration procedure a cavernous hemangioma may diminish considerably and fill again after aspiration. The smears are mostly disappointing and show, apart from blood cells, exceedingly poor cellularity. Occasional small sheets of usually fewer than ten cells can be found, with pale-staining delicate cytoplasm, an oval to spindly outline, and regular nuclei; they resemble the endothelial cells obtained in capillary hemangiomas.

Hemangiopericytoma

Hemangiopericytoma often presents in adults as a solitary, deep-seated lesion of varying size in the lower extremities, the pelvic fossae and retroperitoneum, head and neck, trunk, and upper extremities. Tumors occur[6,19,42] within muscle or are intimately attached to muscle and deep fascia. In histologic sections it is the general pattern that helps in recognition of the lesion, since its cellular features are less easily identifiable. A characteristic pattern is that of thin-walled dilated vascular channels together with cells with indistinct cytoplasm and oval or elongated nuclei. Differentiating benign from malignant forms may be difficult.

According to Nickels and Koivuniemi, who described FNA findings in five patients with malignant hemangiopericytoma, structures corresponding to the knoblike formations of histologic sections can be seen in cytologic preparations.[45] In addition, tumor cells show cohesiveness near capillaries and blood spaces. Their nuclei show anisokaryosis, hyperchromasia, mitosis, and prominent, often multiple nucleoli. However, hemangiopericytoma shares cytologic features with malignant fibrous histiocytoma, leiomyosarcoma, synovial sarcoma, liposarcoma, and mesenchymal chondrosarcoma in that all these tumors are histologically known to show hemangiopericytomalike areas to various extent. Furthermore, in borderline cases cytologic classification as benign or malignant can be worrisome when there is lack of frank cellular atypia.

Angiosarcoma

Angiosarcoma is a malignant tumor of vascular origin affecting a wide number of sites, namely, the skin, subcutaneous tissue, muscles, and viscera, with the female breast being one of its most common locations. When superficial, the tumor may manifest itself as a flat, bluish-tinged, diffuse mass with poorly defined margins, a nodular mass, or an ulcerated lesion. A swelling may be the only evidence of a more deeply seated tumor.

Histologically, angiosarcoma is characterized by irregular anastomosing vascular channels with a lining of one or more layers of atypical endothelial cells that often appear immature. A similar histologic picture is seen in tumors originating from lymph vessels. Lymphangiosarcoma may occur in a lymphedematous limb, particularly after mastectomy, as in Stewart–Treves syndrome.

A cytologic description of a case of well-differentiated angiosarcoma of the skin in fine needle aspirates was published by Abele and Miller.[1] The smears were characterized by scant cellularity and showed aggregates of one to ten cells with single hyperchromatic nuclei showing shallow longitudinal indentations. The cells showed erythrophagocytosis, and cytoplasmic vacuoles were evident in Wright-stained smears. In addition, enlarged cells were seen protruding from circular groups.

In a case of well-differentiated angiosarcoma of the iliac vein seen at Karolinska Hospital, similar cell findings were observed. Malignant vascular proliferations of well-differentiated type thus appear to follow the smear pattern of benign vascular proliferations, with the exception of capillary hemangioma, in that they give poor cell yields consisting of rather cohesive cell clusters. In my experience, the presence of thrombosis at the tumor site may add to the difficulty in obtaining diagnostic cell material.

In poorly differentiated angiosarcomas, richer cellular yields may be expected, with less cohesivity of neoplastic endothelial cells and more nuclear pleomorphism and cellular anaplasia.

TUMORS OF SYNOVIAL TISSUE

Synovial Sarcoma

Synovial sarcoma generally affects young adults and has a predilection for tissues of the large joints, bursae, and tendon sheaths.[17,22,44] Its most common primary site of origin is the lower extremities, in the vicinity of the knee joint, but it also arises in the upper limb, trunk, neck, and nasopharynx. When locally excised, the tumor recurs in about 50% of patients. Metastasis is predominantly to the lungs, but lymph node metastasis has been known to occur. The overall 5-year survival rate was 35% in a study of 90 patients.[44]

In typical lesions, a biphasic cellular pattern can be recognized, consisting of slitlike spaces lined by tall columnar cells, the lumen of which contains hyaluronidase-resistant mucoid substance. Between the glandlike clefts one can observe spindle cells resembling those of fibrosarcoma. Hyalinized and x-ray-demonstrable calcified areas can be seen frequently. Mast cells are conspicuous in some cases. In the monophasic form, the glandlike spaces are rare and the sarcomatous spindle cells focally appear somewhat epithelioid. Some portions of the fibrosarcomalike areas in synovial sarcoma undergo myxoid change.

On aspiration of biphasic synovial sarcoma, abundant cell material is obtained, consisting of dense clusters and numerous dispersed cells. The cell clusters are large, ramifying, and sharply delineated. They are focally twisted and folded to some extent, with tightly packed cells in them. Their periphery is often partly lined by spindle cells and partly by rows of cylindric cells with oval nuclei arranged in parallel formation, representing the cells lining clefts or glandlike structures,[33] a feature best appreciated with P (Fig. 15-29). Occasional fragments of more loosely built tissue with plenty of intercellular substance may be found. Between the solid tumor fragments, abundant free-lying cells often with stripped nuclei, are seen, with cytoplasmic fragments in the background. However, in some smears the number of single cells is small. Typically, the nuclei are monomorphous, relatively small, oval, and slightly blunt, with finely granular pale chromatin and a visible nucleolus.[28] In some cases the nuclei are somewhat elongated, and in others they are more round, but in all tumors the nuclear pattern is monomorphous. Mitotic figures are

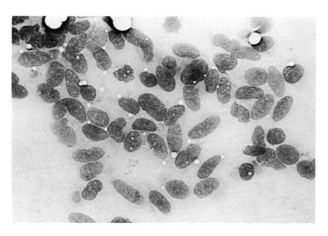

Fig. 15-30. Monophasic synovial sarcomas. Spindle cells with mildly atypical nuclei and small amount of faintly staining cytoplasm. (MGG stain, ×600)

common. The tumor cells have faintly staining gray cytoplasm; the cylindric cells are the larger of the two types and appear in air-dried smears as plump cells with rounded cell borders. Scattered mast cells can regularly be noted. Circumscribed calcified crystalline material may occasionally be aspirated. If spindle cells and a cylindric cell component are obtained at FNA, cytologic recognition of biphasic synovial sarcoma is reasonably easy. The occurrence of cylindrical cells in large cell clusters may suggest the possibility of adenocarcinoma, but adenocarcinomas as a rule lack a sizable proportion of spindle cells. (See Color Plate XXXVI A–C.)

In aspirates from monophasic synovial sarcomas, only the spindle cell type is seen, with the same nuclear and cytoplasmic characteristics as the corresponding fibrosarcomalike cells of the biphasic form (Color Plate XXXVI D, E, and F). Mast cells and calcifications in an otherwise abundantly cellular monomorphous spindle cell tumor are features supporting a sarcoma of synovial origin, namely monomorphous bipolar spindle cells with oval nuclei and a small amount of faintly staining cytoplasm (Fig. 15-30). If only a very cellular monomorphous neoplastic spindle cell population is obtained, a sarcoma of fibrosarcomatous type can still be diagnosed.

Nontypical cases, which yield poor cell material because of collagenization of certain tumor areas, entail a risk for overlooking the malignant character of the lesion, particularly because of the accompanying

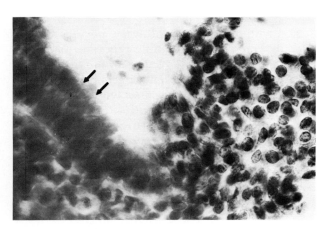

Fig. 15-29. Synovial sarcoma, biphasic type. Cylindrical cells with parallel orientation, sharply delineated from stromal cells. (MGG stain, ×600)

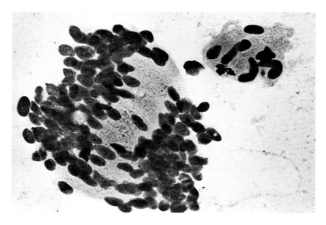

Fig. 15-31. Benign Schwannoma (neurilemoma). Verocay body and small fragment of tumor tissue consisting of homogeneous cytoplasm with indistinguishable cell borders. (MGG stain, ×600)

cellular monomorphism. However, in the presence of poor cell samples, a repeat biopsy is usually helpful for diagnosis.

TUMORS OF PERIPHERAL NERVES

Benign Schwannoma (Neurilemoma)

Schwannomas are neoplasms that arise on cranial and spinal nerve roots and also on peripheral nerves. They are generally solitary, but multiple tumors occur in von Recklinghausen's neurofibromatosis. Schwannomas grow slowly and retain their histologic benign appearance. The tumors may occur anytime in life and commonly involve major nerve trunks. They usually appear on the flexor aspects of the limbs, particularly close to the elbow, wrist, and knee. Other sites likely to be involved are the face, neck, hands, intercostal nerves, and posterior mediastinum.

The type A tissue of Antoni in histologic sections is built up of compact, interwoven bundles of long bipolar spindle cells with oval to rod-shaped nuclei. The nuclei show some variation in chromatin and contain inconspicuous nucleoli and uniformly pale cytoplasm. Focally, the cell bundles can form whorls of variable sizes and compactness. In some areas the nuclei may show a palisade arrangement, especially in those involving the spine. Type B tissue

of Antoni has a loose texture with polymorphism of the tumor cells, which have indefinite cell borders. The cells are separated by an eosinophilic matrix containing relatively scanty, evenly distributed reticulum fibers.[29]

At the moment of fine needle biopsy from neurilemomas, patients commonly experience intense pain. Moderately or very cellular aspirates are obtained. Cell yields consist of large and small tumor tissue fragments and free-lying cells. With MGG stain, the clusters stain homogeneously pink to violet because the cell boundaries are, in most cases, indiscernible and the cytoplasm between neighboring cells appears confluent, resembling a syncytium. However, in other MGG-stained cases and in most wet-fixed, P-stained smears, a fibrillar pattern can be discerned and is caused by the juxtaposition of the elongated cells. The borders of the clusters are irregularly sheared, and often long, frayed, or tapering cytoplasmic ends stick out of their periphery. The nuclei are oriented in a parallel fashion, and focally a distinct crowding or "palisading" can be discerned, especially in Antoni type A areas, which also show a higher cell density in the clusters.

Nuclei are basically oval, more or less slender, and of varying size, but some are small and round resembling those of lymphocytes. With MGG stain, they stain dark purple and have a finely woven chromatin texture. With P stain, they are homogeneously gray with a sharply accentuated, slightly thickened nuclear membrane and an inconspicuous nucleolus. In so-called ancient neurilemoma, the nuclei appear atypical, enlarged, and show anisokaryosis. The general cytologic picture is rather characteristic and confusion with other round soft tissue nodules such as fasciitis is rare. A relatively uncommon but diagnostic feature of this benign nerve sheath tumor is the occurrence of small tumor cell clusters consisting of a central core of fibrillar substance lying between two zones of palisaded nuclei (Fig. 15-31) and known as Verocay bodies.[52] Single intact cells are rarely seen. They can have extremely long slender cytoplasmic processes. Otherwise, most free-lying cells, which may constitute a moderate number, occur as naked nuclei against a clear background. A diagnostic problem is so-called ancient schwannoma, which shows a smear pattern and cytoplasmic characteristics fully consistent with schwannoma, but with the occurrence of very pleomorphic bizarre nuclei.

Neurofibroma

Solitary neurofibromas are rare. Multiple neurofibromas, as in von Recklinghausen's neurofibromatosis, may involve one to all of the following sites: the cranial and spinal nerve roots and ganglia; the major nerves of the neck, trunk, and limbs (including the sympathetic system and its ganglia); the dermis and adjacent tissues; and the visceral sympathetic plexuses.[7]

Neurofibromas are in histologic sections distinguished from schwannomas on the basis of the arrangement and form of the tumor cells and stromal features. As compared to schwannomas, neurofibromas have a looser texture and poorer cellularity. The cells are elongated, with thin tapering processes and narrow, spindle-shaped or long darkly staining nuclei that are often curved. An indefinite interstitial matrix widely separates numerous bundles of sinuous, collagen fibers.

From neurofibromas, there are small to moderate amounts of cell material obtained, consisting in smears of tumor tissue clusters and of free cells. In some cases the aspirated cell material is mixed with some edema fluid. The clusters may be small or large and consist of pinkish violet nervous tissue substance, which is less cellular than in schwannomas. In addition, pink-staining fibrous tissue fragments and damaged cells can be noted. However, in aspirates from some neurofibromas, the clusters may be quite cellular and consist of aggregates of cells with poorly preserved pale-staining cytoplasm and fragile, hyperchromatic, elongated, irregular nuclei. Among the dispersed cells, few have intact cytoplasm; most are represented by naked hyperchromatic nuclei. These cellular and nuclear characteristics may entail the risk of a false diagnosis of malignancy. Hence, in the presence of the above cell picture, a cytologic diagnosis of suspected *neurogenic tumor* can be rendered and the lesion should be further investigated histopathologically.

Neurofibrosarcoma

Neurofibrosarcomas occur sporadically but usually arise in patients with von Recklinghausen's neurofibromatosis and only rarely, if ever, develop in a benign schwannoma. In a study of 46 patients, neurofibrosarcomas were found to occur principally in the lower extremities and retroperitoneum of adults. The tumors are mostly large and may be painful. Neurofibrosarcomas are highly malignant.

The histologic group of neurofibrosarcoma shows considerable variation in growth pattern even when strict criteria, such as association with clinical neurofibromatosis, perineural origin in a sizable nerve, or evidence of Schwann's cell differentiation, are enforced.[27] Many tumors show extremely high cellularity and are built of intersecting bundles of well-arranged tumor cells that closely resemble Schwann's cells, together with wavy collagen fibers that form a compact "herringbone" pattern. Schwann's cells have cytoplasm with indistinct margins and elongated wavy or buckled, sharply outlined hyperchromatic nuclei with a "punched-out" appearance. Some tumors have alternating cellular and less cellular myxoid areas with loosely arranged tumor cells. Others show a predominantly diffuse myxoid pattern.

Cytologic experience related below stems from limited material of histologically proven neurofibrosarcoma arising in pre-existing neurofibromas. Aspirates displayed variable amounts of cell material. They showed very cellular clusters of tumor cells with badly preserved fragile cytoplasm and hyperchromatic pleomorphic nuclei. A large proportion of the tumor cells occurred as dispersed naked nuclei. When, despite nuclear pleomorphism, there is some doubt of malignancy, a tentative pre-operative cytologic diagnosis of suspected sarcoma may be made, leaving the final diagnosis to the histopathologist.

Neurofibrosarcomas occurring in neurofibroma may be missed by the aspirating needle when the malignant transformation is still localized to a limited area of the neurofibroma. Thus, in the presence of a growing tumor, cytologic diagnosis of neurofibroma should always be supplemented by histologic examination of the extirpated mass.

TUMORS OF SYMPATHETIC GANGLIA

Ganglioneuroma

Ganglioneuromas occur most frequently in children and young adults. The majority of retroperitoneal and mediastinal tumors appear to originate in ganglia of the autonomic nervous system. They occur as solitary or multiple benign neoplasms, consisting in his-

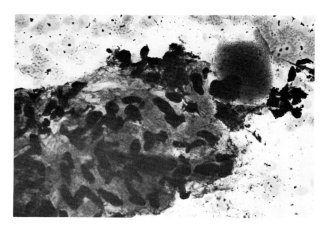

Fig. 15-32. Ganglioneuroma. Bundle of nerve tissue with elongated regular nuclei. At its end, one ganglion cell with large nucleus. (MGG stain, ×600)

tologic sections of nerve fibers and mature ganglion cells.[17]

Aspirated material from ganglioneuroma shows nervous tissue clusters staining pink-violet with MGG and containing regular oval nuclei oriented in a parallel manner with a micromorphology very similar to neurilemoma tissue. In addition, very large round cells with abundant blue cytoplasm are seen connected to and in between the clusters. The cells contain an eccentric round to oval nucleus with a prominent nucleolus and represent ganglion cells (Fig. 15-32).[50] In cases where extensive tumor necrosis and retrogressive changes have taken place, the number of diagnostic ganglion cells may be small and may escape the aspirating needle. However, even if the diagnosis of ganglioneuroma cannot be made, the features of a benign tumor of nervous tissue origin will be obvious and should be stated in the cytologic report.

Neuroblastoma

Neuroblastomas almost always occur in children under 8 years of age. They are malignant neoplasms of primitive neuroectodermal cells. The tumor commonly arises in the adrenal medulla but may also originate in the ganglia of the sympathetic nervous system and occasionally may develop in the peripheral nervous system. Tumor cells can be differentiated enough to be able to synthesize and liberate epinephrine, norepinephrine, or related substances into the bloodstream.

In sections, neuroblastomas are built up of small tumor cells containing a hyperchromatic nucleus.[29] Pyknotic nuclei are frequent. In well-differentiated neuroblastomas, cells appear to border one another and seem to be aligned circularly around a network of tangled processes. The pattern has been described as rosette or pseudorosette, and much emphasis has been given to it in the diagnosis of neuroblastoma. However, just as distinctive and more frequent are acellular, fibrillar areas resembling glial tissue.

Aspirates from neuroblastomas usually give abundant yields of small-sized tumor cells. The material has a tendency to stick to the slide as a thick layer of confluent three-dimensionl clusters from which numerous aggregated and single tumor cells detach to fill the spaces in between. In the well-differentiated varieties, the larger cell clusters show several cell-free circular areas filled with granular acidophilic staining substance around which several layers of tumor cells are arranged (Fig. 15-33). These structures, which correspond to the pseudorosettes in sections, contain fibrillar condensations best seen in wet-fixed P smears. Neuroblastomas with less organoid structures tend to give smaller cell clusters that disperse into single cells. Focally isolated rosettelike structures consisting of concentrically arranged tumor cells can be found. (See Color Plate XXXVII A–C.)

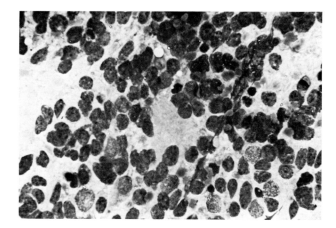

Fig. 15-33. Well-differentiated neuroblastoma. Numerous small tumor cells with irregular nuclei surrounding a focus of fibrillar nerve substance in a so-called pseudorosette formation. (MGG stain, ×600)

In general, the tumor cells are small, but they have a large nucleus that fills most of the cell body. The nuclei are irregularly round to oval and contain a visible nucleolus. Chromatin shows a coarse granularity with several chromocenters but without a thickened nuclear membrane—properties best visible with P. With MGG, chromatin stains darkly in certain cells and pale in other, often larger cells. Mitoses are common. Some tumor cells are binucleated, while others have three nuclei. Their cytoplasm is usually seen only as a rim, mostly as a short pointed cap at one pole of the cell, similar to lymphocytes. Other cells have more cytoplasm, usually accumulated at one pole. A few cells have one or two narrow tails. The cytoplasm stains moderately acidophilic and, in areas adjacent to necrosis, may contain small vacuoles. The cytoplasm may not be preserved, depending on pressure exerted while smearing. The background is finely granular, often containing necrotic cell material mixed with macrophages and crystalline calcified bodies of variable size, which stain blue with MGG.

In neuroblastoma metastasizing to bone marrow, for example, the cells aspirated may be mostly dispersed yet show a distinct tendency to cluster into small aggregates. Differential diagnosis of neuroblastoma consists of malignant small cell mesenchymal tumors, Ewings sarcoma and lymphoma (see Chap. 14). Inevitably, there will be a number of smears from undifferentiated tumors that cannot be diagnosed cytologically as neuroblastoma, but they can at least be diagnosed as malignant small cell, nonlymphomatous tumor.

Ganglioneuroblastoma

Differentiated neuroblastoma or ganglioneuroblastoma occurs predominantly in childhood.[3,29] It resembles in histologic material the primitive forms but, in addition, contains groups of or single large cells with abundant eosinophilic cytoplasm and pale, eccentric nuclei with prominent nucleoli. The large cells show some resemblance to ganglion cells, and occasionally Nissl granules can be demonstrated in them.

Aspirates from ganglioneuroblastoma show abundant cell material consisting, in representative cell samples, of immature cell forms of the same type as in neuroblastoma. In addition, large cytoplasm-rich cells with one to three large nuclei containing a prominent nucleolus and having a morphology corresponding to ganglion cells can be seen.[49] These cells can be very numerous and may be the only cell component aspirated, as in one patient examined at Karolinska Hospital. A morphologic differential diagnosis between ganglioneuroblastoma and ganglioneuroma thus cannot, in such instances, be made before histopathologic examination.

TUMORS OF DISPUTED OR UNCERTAIN HISTOGENESIS

Granular Cell Tumor

Granular cell tumor, also known as granular cell myoblastoma, is found in the dermis or subcutis and less often in striated muscle . It is also found in a host of other tissues, for instance, mucous membranes and submucous tissue, in the muscularis of the alimentary tract, larynx, bladder, uterus, vulva, and in the omentum, breast, and retroperitoneum.[2] Most lesions are small and histologically composed of large round or polygonal cells with finely granular acidophilic cytoplasm, small dense nuclei, and fibrous septa between groups of cells. There is also a malignant variety with metastatic potential, which differs from the benign form in that it shows slight pleomorphism, mitotic activity, and tendency for necrosis.

Smears from fine needle aspirates are generally rich in cells, although occasionally poorly cellular samples may be obtained.[25,39] When many cells are present, clusters as well as free-lying cells with abundant cytoplasm are seen. The cytoplasm is highly fragile; invariably, many of the cells are damaged and are visible between the clusters as free-lying nuclei. The background consists partly of a thick syncytial cytoplasmic matrix and partly of small cytoplasmic fragments and anucleated cell bodies. Well-preserved cells are large and round to oval, with smoothly curved contours and hazy cell borders (Fig. 15-34, see also Fig. 6-24A). The cytoplasm is discretely granular, especially with P. Nuclei have an eccentric position in the cytoplasm and are usually small, oval, and normochromic. Occasional cells may have a decidedly larger than average nucleus containing a conspicuous nucleolus. When nuclei are distinctly atypical against a background of necrosis in the smear, malignant granular cell tumor can be suspected.

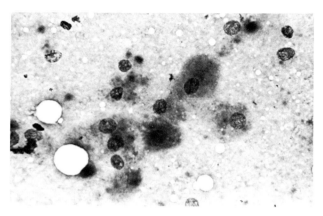

Fig. 15-34. Granular cell tumor. Tumor cells with coarsely granular cytoplasm and round-to-oval regular nuclei against a granular cytoplasmic background. (MGG stain, ×600)

In the cervical region, adult rhabdomyoma may sometimes invite differential diagnosis. However, cells of the latter are larger and better circumscribed, some being elongated with striations and square borders, and they stand out against a clean background. When the lesion is situated in the breast, large cell carcinoma, such as medullary or apocrine carcinoma, should be excluded. In the former, despite the similarity of cytoplasmic fragility and free nuclei, nuclear pleomorphism is usually more outspoken and evidence of duct cell differentiation is visible in some tumor cells. The latter has distinct and sharp cytoplasmic borders in contrast to granular cell tumor, and it may show a mixture of a malignant duct cell component.

Intramuscular Myxoma

Intramuscular myxomas are benign but often infiltrating growths occurring most commonly in the large muscles of the shoulder and thigh. Histologically, they are very poorly cellular and characterized by inconspicuous round, spindle, or stellate cells within a matrix containing abundant mucoid material. The matrix consists chiefly of hyaluronic acid, a loose meshwork of reticulin and collagen fibrils, and scanty vascular structures.[17]

At aspiration, myxomas yield a drop of glistening, viscid substance that forms a thick coat on the smears. At microscopy, a few very loosely textured cell clusters and single tumor cells appear in the abundant amount of intercellular substance, which stains a prominent pink-red with MGG. The cells of myxoma have well-preserved cytoplasm also when occurring singly. They are slender and spindle (Fig. 15-35) or stellate shaped with homogeneous cytoplasm showing an occasional small vacuole, or plump or oval shaped with some vacuoles in the vicinity of the nucleus. Cells with the appearance of macrophages may be filled with vacuoles. The nuclei are monomorphous and oval, with an evenly, finely distributed chromatin and a small nucleolus. Some cells contain two or even three nuclei.

Cytologic differential diagnosis includes myxoid lesions, with scant cellularity, especially ganglion; the latter shows few cells, and they are almost exclusively macrophagelike oval to round, often with a pointed cytoplasmic extension and containing several small vacuoles in the cytoplasm. For differentiation from low-grade myxoid malignant fibrous histiocytoma, or myxoid liposarcoma, see the specific sections.

Epithelioid Sarcoma

Epithelioid sarcoma occurs in young adults, mostly affecting the soft tissues of the hand, forearm, and pretibial region. The tumor has a nodular growth pattern along fascial structures and tendons, often showing central necrosis and ulceration of the overlying skin. The clinical course is protracted, the tumor having been noted months or years before diagnosis. It frequently recurs and tends to metastasize late.[15]

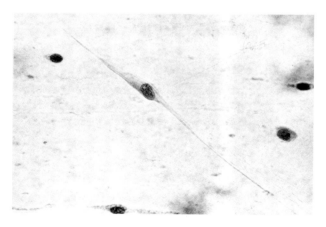

Fig. 15-35. Myxoma. A few slender spindle cells and a plump cell, all with regular nuclei, embedded in homogeneous mucoid substance. (P stain, ×600)

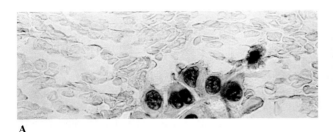
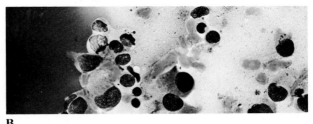

A   B

Fig. 15-36. Epithelioid sarcoma. **A.** Large polygonal tumor cells with dense cytoplasm and sharp cell borders with some resemblance to squamous cells. **B.** Hyalin and plump sarcoma cells. (MGG stain, ×600)

Histopathologically, there are irregular nodular masses of large polygonal acidophilic cells blending with spindle cells. In addition, large amounts of hyalinized collagen occur; and the histologic appearance of the tumor often resembles granuloma or carcinoma.

Aspirates from epithelioid sarcoma are usually cellular, consisting of dense clusters from which well-preserved single cells rather easily disperse during smearing. Generally, the tumor cells are large and broad with distinct cell borders having plump stellate to polygonal shape with pointed angles (Fig. 15-36A). Other cells have convex cell boundaries, sometimes ending in short cytoplasmic extensions. Spindle cells, which usually are fewer, have often a broad outline. The cytoplasm stains densely acidophilic with both MGG and P. Some of the smaller cells contain a few small vacuoles. The background is rather clean.

The majority of the cells have one mostly eccentrically positioned nucleus, but cells with two to three or even more nuclei can be observed. The nuclei show mild pleomorphism, but in some cases cells with enlarged atypical nuclei occur. Nuclear indentations and folds can be seen, together with enlarged irregular chromocenters. Prominent nucleoli are visible in the large cells. Sometimes a few multinucleated giant cells with monomorphous nuclei are present.

Fragments of fibrillar mesenchymal stromal substance in which a few tumor cells are embedded can be seen; and sometimes sharply delineated, rounded, bright eosinophilic bodies that (Fig. 15-36B) contain a few scattered tumor cells are found.

The mesenchymal nature of the cells is usually obvious cytologically, and the most important differential diagnosis conceivable is malignant fibrous histiocytoma. However squamous cell carcinoma may

sometimes be considered to be caused by the dense cytoplasm of the squamous cell. Cases with minimal nuclear atypia and in which spindle cells dominate in the smears carry a risk of being diagnosed as a benign soft-tissue lesion.

Clear Cell Sarcoma of Tendons and Aponeuroses

Clear cell sarcoma most commonly involves tendons or aponeuroses around the foot or knee of young adults and usually presents as a tumor under 5 cm. Histologically, one finds small groups, nests, or fascicles of pale spindle-shaped cells with vesicular nuclei containing large conspicuous nucleoli; multinucleated giant cells occur in more than half the cases. Because of the occurrence of melanin pigment and melanosomes in certain cases, this neoplasm has also been called malignant melanoma of soft parts. Repeated local recurrences and eventual metastatic growth after a protracted clinical course are typical for this sarcoma.[14]

From a case of clear sarcoma subjected to FNA at Karolinska Hospital, abundant cell material consisting predominantly of dispersed tumor cells was obtained. The cells were rather short, oval or spindle shaped, and had a small amount of homogeneous, pale-staining, fragile cytoplasm, which was also found as a granular and slightly vacuolated background. The nuclei were oval and showed slight pleomorphism. Characteristically, a large very prominent nucleolus was seen (Fig. 15-37). In addition, strands of acellular stromal interstitial tissue were seen, with adherent tumor cells. The cell picture most closely resembles that of spindle cell melanoma. However, most metastases of malignant melanoma

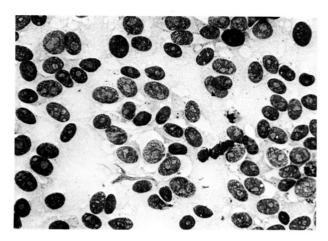

Fig. 15-37. Clear cell sarcoma. Plump spindle cells with oval and ovoid nuclei containing very large prominent nucleoli. (MGG stain, ×600)

include a component of round cells with pleomorphic nuclei. Another differential diagnosis is the fascicular variety of malignant fibrous histiocytoma, which yields more elongated atypical fibroblastlike spindle cells and bizarre giant cells. The spindle cells from monophasic synovial sarcoma are more slender and have oval nuclei without conspicuous nucleoli, and mast cells are a frequent finding.

Alveolar Soft Part Sarcoma

Alveolar soft part sarcoma is a rare tissue sarcoma with a characteristic light microscopic presentation, observed in young patients[9] between 15 and 35 years of age.[20] In adults it chiefly occurs in the thigh and in children in the head and neck region.[20] This sarcoma shows a slow growth, often without symptoms, and commonly metastasis to the lung is the first presenting manifestation, sometimes preceding the discovery of the primary tumor by many years.

In paraffin sections the tumors show a fairly constant picture consisting of solid nests or alveolar structures separated by thin fibrous septa containing thin-walled vessels. The tumor cells are large, polygonal or round, and contain vesicular nuclei. The cytoplasm is granular and eosinophilic. A pleomorphic variety exists containing cells showing large variation in nuclear size and stainability.

In aspirates from such a pleomorphic tumor in the thigh of a 24-year-old woman,[51] our findings were except for a more outspoken nuclear pleomorphism similar to those reported by others.[31,46,58] The cell picture was dominated by clusters consisting of large-celled tumor aggregates filling out the spaces in a prominent vascular network. In addition, small clusters of tumor cells and of free-lying tumor cells and naked nuclei were noted. Also capillary networks could be demonstrated. The tumor nuclei were round, hyperchromatic, and varied in size. With P they showed a coarse chromatin pattern with conspicuous chromocenters and prominent nucleoli. Also intranuclear cytoplasmic inclusions were seen. The sarcoma cells were large, round to polygonal with abundant mostly pale-staining cytoplasm containing numerous small round vacuoles (Fig. 15-38). The endothelial cells in the capillaries had spindle-shaped nuclei and pale-staining cytoplasm (Fig. 15-39), and they were enveloped by a pink-staining fine-fibrillary collagen matrix. Tumor cells from the lung metastasis, which was also aspirated, showed similar morphology.

Differential diagnosis is in the first place renal cell carcinoma, which also is built up of large polyhedral cells with granular cytoplasm and high vascularity. However in renal cell carcinoma, varieties with more cylindrical or vacuolated cells exist; and the vascular component is often not as prominent as in smears of alveolar soft part sarcoma.

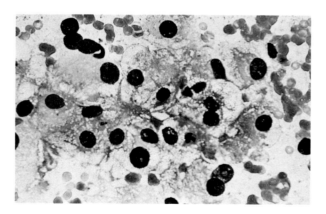

Fig. 15-38. Alveolar soft part sarcoma, large round sarcoma cells with round or oval nuclei and finely vacuolated cytoplasm showing some resemblance to carcinoma cells. (MGG stain, ×600)

NONNEOPLASTIC OR QUESTIONABLY NEOPLASTIC LESIONS

Ganglia and Bursae

Ganglia are round or oval, pseudocystic swellings that develop in the vicinity of the wrist, on the dorsum of the foot, or around the ankle or knee.[17] Their origin has been open to much discussion; possibilities include displaced synovial tissue during embryogenesis, herniation of synovia into adjacent tissues, or a process of mucoid degeneration of tendon sheaths, ligaments, or aponeuroses. The cavities are filled with clear, viscid fluid; and their walls are generally lined by highly compressed thin cells or, occasionally, by a discontinuous lining of synoviallike cells.

Bursae are cysts that have an obvious communication with the joint, although often the passage may be closed off from fibrosis. The cysts may be multilocular; they contain mucoid substance and are lined by typical synovium or dense fibrous tissue. Focal or diffuse lymphocytic infiltrates and hemosiderin-laden macrophages may be seen in their walls. Adventitious bursae develop in connective tissue or bony prominences or at sites of abnormal tissue mobility. They usually have fibrous walls enclosing mucoid substance. Occasionally, synoviallike cells are formed by a process of metaplasia of the cells lining the cystic spaces.

From ganglia and bursae, a viscous drop of glistening transparent fluid is generally obtained, which

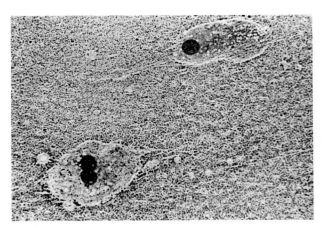

Fig. 15-40. Ganglion. A large amount of mucoid substance (granular because of precipitated dye) containing two macrophage-like cells. (MGG stain, ×600)

spreads as a thick film. The smears display an abundant amount of mucoid substance that, with MGG, appears intensely violet or pale pink. This background substance may either be homogeneous or show an eosinophilic staining granularity. With P, this mucoid substance is more translucent and stains light green. Otherwise, both P and MGG smears show extremely few cells consisting of some macrophagelike round cells and histiocytelike cells with plenty of cytoplasm and a small, regular, oval nucleus with a small nucleolus (Fig. 15-40). Some cells have two pointed extensions at opposite ends of their round cell bodies. The cells appear embedded in the slimy background, and their features may not be very prominent because of the poor staining quality of the cytoplasm with MGG. The cytologic pictures of both ganglia and bursae are similar and so consistent and characteristic that these lesions should pose no diagnostic problems.

BENIGN GIANT CELL TUMOR OF TENDON SHEATH

Benign giant cell tumor of tendon sheath develops most commonly in the knees of young adults, who experience pain and swelling of the affected joint.[17] They are also found in the fingers as lobulated, often brownish nodules of under 2 cm. In sections, villous

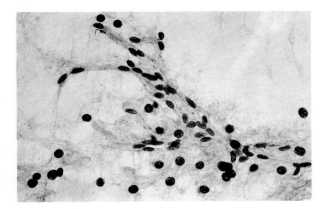

Fig. 15-39. Alveolar soft part sarcoma. Capillary vessel with some attached sarcoma cells. (P stain, ×480)

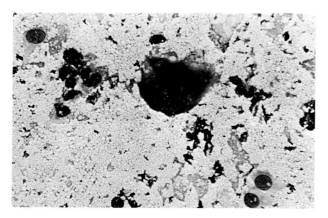

Fig. 15-41. Benign giant cell tumor of tendon sheath. A few histiocyte-like cells of moderate size with fragile vacuolated cytoplasm and regular nuclei. A multinucleated giant cell is also seen. (MGG stain, ×600)

tissue is made up of strands of vascular connective tissue with a lining of synovial cells and focal or diffuse lymphoplasmocytic infiltrates. Solid areas are composed of rounded or spindle-shaped cells with monomorphous nuclei vacuolated, lipid-bearing macrophages, and hemosiderin deposits. Multinucleated giant cells with central nuclei are sometimes prominent and resemble osteoclasts in size and appearance. The cellularity varies considerably, some nodules being fibrotic while others are densely cellular. Sometimes the pigmented tissue extends into the interstices of the joint capsule or even erodes bone. Despite these features mimicking invasion, these lesions have never been known to metastasize.

From these benign proliferations comparatively poor cellular yields of single cells and, less often moderate amounts of clusters are obtained at FNA biopsy. The smears show a component of mononucleated cells with regular oval nuclei and a moderate amount of cytoplasm (Fig. 15-41). They have different shapes, and their relative proportions vary from case to case. They may either be spindle shaped with tapering and sometimes frayed ends or plump stellate cells with more than two cytoplasmic extensions. Others have broad, squarely angulated cell bodies, sometimes containing two round basal nuclei. Some cells are round and have abundant cytoplasm. The latter may contain numerous small optical empty vacuoles, probably fat globules mostly lying in the peripheral cytoplasm, but sometimes filling the whole cell body. In such instances the background contains many anucleated cell bodies and cytoplasmic fragments. Other cells contain hemosiderin particles. Generally speaking, these cells very strongly resemble those seen in benign fibrous histiocytoma. In addition, multinucleated giant cells with dense cytoplasm showing pointed extensions are usually seen. A mixture of lymphocytes and plasma cells also occurs. In the context of the clinical setting aspirates are usually diagnostic for this benign condition. A differential diagnostic alternative is the deceptively similar but extremely rare malignant giant cell tumor of tendon sheath, which gives cell samples of much higher cellularity showing more mitotic figures than its benign counterpart.

CLINICAL APPLICATION

The field of soft tissue tumors is enormously vast and, as yet, cytologically relatively undiscovered. Fortunately, interest is not lacking, as is evident from reports in the literature as well as from current research. The present standards of technical equipment and procedure have made it possible to perform FNA biopsies not only of palpable masses but also of topographically deeply situated lesions. This improved accessibility enables the sampling of a wide diversity of soft tissue tumors.

The aim of aspiration cytology in soft tissue lesions at present is to be able to provide a reliable diagnosis of benign or malignant lesion and to assign tumors a specific histopathologic type. Retrospective assessment of accuracy of aspiration biopsy cytology (ABC) as a diagnostic method for soft tissue tumors is complicated by the fact that the principles for tumor classification have changed over the past 25 years. Åkerman and colleagues estimate that the total reliability of cytologic diagnosis in the series of 187 soft tissue lesions examined during a 5-year period is around 85%.[4] Layfield and associates report a diagnostic sensitivity and specificity of 95% for the determination of malignancy in a series of 136 primary soft tissue tumors.[36] Their false-positive rate is 2%, which is in agreement with the authors' own experience.[64] At the Karolinska Hospital, where four out of five lesions can be conclusively diagnosed, diagnostic accuracy varies between different tumor types. Factors affecting the accuracy rate negatively are deep location of the tumor, high degree of fibrosis, necrosis, and lack of nuclear and cellular pleomorphism.

In surgical material, a considerably larger volume of tissue is available for processing through a battery of histochemical reactions. Fine needle aspirates have decidedly less material at their disposal, and as of today they have usually been used for cytodiagnosis. However, the possibility exists that part of the cell sample can be subjected to complementary cytochemical and immunocytochemical reactions or processed as cell blocks when necessary.[42]

Fine needle aspirates are easily prepared for transmission EM, thus making it possible to reveal important diagnostic information on the properties of cells from soft tissue tumors.[32,37,47] Another method applicable to fine needle aspirates is the quantitative measurement of deoxyribonucleic acid (DNA) on single smeared cells or on populations of cells from suspended aspirates.[50] Such measurements can provide significant information of prognostic value about soft tissue tumors.

Finally, the alert, perceptive, and analytic clinical cytopathologist still is the most important single factor guaranteeing the maintenance of high accuracy in the cytologic diagnosis of soft tissue tumors, even if more sophisticated methods of investigation become available.

REFERENCES

1. Abele JS, Miller T: Cytology of well-differentiated and poorly differentiated hemangiosarcoma in fine needle aspirates. Accepted for publication in Acta Cytol (Baltimore), February, 1981
2. Abrikossoff A: Über myome ausgehend von der quergestreiften willkürlichen muskulatur. Virch Arch (Pathol Anat) 260:215–238, 1926
3. Adam A, Hochholzer L: Ganglioneuroblastoma of the posterior mediastinum. Cancer 47:373–381, 1981
4. Åkerman M, Idvall I, Rydholm A: Cytodiagnosis of soft tissue tumours and tumour-like conditions by means of fine needle aspiration biopsy. Arch Orthop Trauma Surg 96:61–67, 1980
5. Angervall L, Hagmar B, Kindblom L–G et al: Malignant giant cell tumor of soft tissues. A clinicopathologic, cytologic, ultrastructural, angiographic and microangiographic study. Cancer 47:736–747, 1981
5a. Angervall L, Kindblom L–G, Merck C: Myxofibrosarcoma—a study of 30 cases. Acta Path Microbiol Scand Sect A 85:127–140, 1977
6. Angervall L, Kindblom L–G, Möller NJ et al: Hemangiopericytoma: A clinicopathologic, angiographic and microangiographic study. Cancer 42:2412–2427, 1978
7. Bednár B: Storiform neurofibromas of the skin, pigmented and non-pigmented. Cancer 10:368–376, 1957
8. Christopher F: Giant cell sarcoma apparently arising from the aponeurosis of the forearm. JAMA 87:167–168, 1926
9. Christopherson WM, Foote FW Jr, Stewart FW: Alveolar soft part sarcoma. Structurally characteristic tumors of uncertain histogenesis. Cancer 5:100–111, 1952
10. Chung EB, Enzinger FML: Proliferative fascitis. Cancer 36:1450–1458, 1975
11. Collins VP, Ivarsson B: Tumor classification by electron microscopy of fine needle aspiration biopsy material. Acta Pathol Microbiol Scand (Sect A) 89:103–105, 1981
12. Dahl I, Akerman M: Nodular fasciitis. A correlative cytologic and histologic study of 13 cases. Acta Cytol (Baltimore) 25:3, 215–222, 1981
13. Dahl I, Hagmar B, Angervall L: Leiomyosarcoma of the soft tissue. A correlative cytological and histological study of 11 cases. Acta Pathol Microbiol Scand (Sect A) 89:285–291, 1981
14. Enzinger FM: Clear-cell sarcoma of tendons and aponeuroses. An analysis of 21 cases. Cancer 18:1163–1174, 1965
15. Enzinger FM: Epithelioid sarcoma. A sarcoma simulating a granuloma or a carcinoma. Cancer 26:1029–1041, 1970
16. Enzinger FM: Recent Developments in the Classification of Soft Tissue Sarcomas: Management of Primary Bone and Soft Tissue Tumors, pp 219–234. Chicago, Year Book Medical Publishers, 1977
17. Enzinger FM, Lattes R, Torloni H: Histological Typing of Soft Tissue Tumours. International Histological Classification of Tumours. No 3. Geneva, World Health Organization, 1969
18. Enzinger FM, Shiraki M: Alveolar rhabdomyosarcoma. An analysis of 110 cases. Cancer 24:18–31, 1969
19. Enzinger FM, Smith BH: Hemangiopericytoma. An analysis of 106 cases. Hum Pathol 7:61–82, 1976
20. Enzinger FM, Weiss SW: Soft Tissue Tumors, pp 840. St. Louis, CV Mosby 1983

21. Enzinger FM, Winslow DJ: Liposarcoma. A study of 103 cases. Virch Arch (Pathol Anat) 335:367–388, 1962

22. Evans HL: Synovial sarcoma. A study of 23 biphasic and 17 probable monophasic examples. Pathol Annu 15:2, 309–331, 1980

23. Evans HL, Soule EH, Winkelmann RK: Atypical lipoma, atypical intramuscular lipoma, and well-differentiated retroperitoneal liposarcoma. A reappraisal of 30 cases formerly classified as well-differentiated liposarcoma. Cancer 43:574–584, 1979

24. Fields, JP, Helwig EB: Leiomyosarcoma of the skin and subcutaneous tissue. Cancer 47:156–169, 1981

25. Franzén S, Stenkvist B: Diagnosis of granular cell myoblastoma by fine-needle aspiration biopsy. Acta Pathol Microbiol Scand 72:391–395, 1968

26. Guccion JG, Enzinger FM: Malignant giant cell tumor of soft parts. An analysis of 32 cases. Cancer 29:6, 1518–1529, 1972

27. Guccion JG, Enzinger FM: Malignant Schwannoma associated with von Recklinghausen's neurofibromatosis. Virch Arch (Pathol Anat) 383:48–57, 1979

28. Hajdu SI, Hajdu EO: Cytopathology of Sarcomas and other Nonepithelial Malignant Tumors. Philadelphia, WB Saunders, 1976

29. Harkin JC, Reed RJ: Tumours of the Peripheral Nervous System Atlas of Tumor Pathology. Second Series Fascicle 3, pp 174, Washington, DC, Armed Forces Institute of Pathology, 1969

30. Kapila K, Chopra P, Verma K: Fine-needle aspiration cytology of alveolar soft part sarcoma. A case report. Acta Cytol 29:359–361, 1985

31. Kern WH: Proliferative myositis. A pseudosarcomatous reaction to injury. Arch Pathol 69:209–216, 1960

32. Kindblom L–G: Light and electron microscopic examination of embedded fine-need aspiration biopsy specimens in the preoperative diagnosis of soft tissue and bone tumors. Cancer 51:2264–2277, 1983

32a. Kindblom L–G, Lundgren L, Willems J–S, Angervall L: Proliferative fascitis and myositis. An immunohistochemical, cytologic and electron microscopic study and a quantitative DNA analysis. Am J Surg Pathol, submitted for publication, 1988

33. Koivuniemi A, Nickels J: Synovial sarcoma diagnosed by fine-needle aspiration biopsy. A case report. Acta Cytol 22:6, 515–518, 1978

34. Konwaler BE, Keasby L, Kaplan L: Subcutaneous pseudosarcomatous fibromatosis (fasciitis): Report of 8 cases. Am J Clin Pathol 25:241–252, 1955

35. Kyriakos M, Kempson RL: Inflammatory fibrous histiocytoma. An aggressive and lethal lesion. Cancer 37:1584–1606, 1976

36. Layfield LJ, Anders K, Glasgow BJ et al: Fine-needle aspiration of primary soft tissue lesions. Arch Pathol Lab Med 110:420–424, 1986

37. Lindholm K, Nordgren H, Akerman M: Electron microscopy of fine needle aspiration biopsy from a malignant fibrous histiocytoma. Report of a case. Acta Cytol 23:399–401, 1979

38. Lopes–Cardozo P (ed): Atlas of Clinical Cytology. Distributed by JB Lippincott, Philadelphia

39. Löwhagen T, Rubio CA: The cytology of the granular cell myoblastoma of the breast: Report of a case. Acta Cytol 21:314–315, 1977

40. MacKay B, Masse SR, King OY: Diagnosis of neuroblastoma by electron microscopy of bone marrow aspirates. Pediatrics 56:1045–1049, 1975

41. Mackenzie DH: The Differential Diagnosis of Fibroblastic Disorders. Oxford and Edinburgh, Blackwell Scientific Publication, 1970

42. McMaster MJ, Soule EH, Ivins JC: Hemangiopericytoma. A clinicopathologic study and long-term follow-up of 60 patients. Cancer 36:2232–2244, 1975

43. Merck C, Hagmar B: Myxofibrosarcoma. A correlative cytologic and histologic study of 13 cases examined by fine needle aspiration cytology. Acta Cytol 24:2, 137–144, 1980

44. Moberger G, Nilssone U, Friberg S, Jr: Synovial sarcoma. Histologic features and prognosis. Acta Ortho Scand (suppl) 111:3–38, 1968

45. Nickels J, Koivuniemi A: Cytology of malignant hemangiopericytoma. Acta Cytol 23:4, 286–287, 1976

46. Nieberg RK: Fine needle aspiration cytology of alveolar soft part sarcoma. A case report. Acta Cytol 28:198–202, 1984

47. Nordgren H, Akerman M: Electron microscopy of fine needle aspiration biopsy from soft tissue tumors. Acta Cytol 26:179–188, 1982

48. O'Brien JE, Stout AP: Malignant fibrous xanthomas. Cancer 17:1445–1455, 1964

49. Palombini L, Vetrani A: Letter. Cytologic diagnosis of ganglioneuroblastoma. Acta Cytol 20:4, 286–287, 1976

50. Palombini L, Vetrani A, Vecchione R et al: Letter. The cytology of ganglioneuroma on fine-needle aspiration smear. Acta Cytol 26:259–260, 1982

51. Persson S, Willems J–S, Kindblom L–G, Angervall L: Alveolar soft part sarcoma. An immunohistochemical, cytologic and electronmicrosco-

pic study and a quantitative DNA analysis. Virchows Archiv A 412:499–513, 1988

52. Ramzy I: Benign Schwannoma: Demonstration of Verocay bodies using fine needle aspiration. Acta Cytol 21:2, 316–319, 1977

53. Reif RM: Letter to the Editor. The cytologic picture of proliferative myositis. Acta Cytol 26:376–377, 1982

54. Riopelle JL, Thériault JP: Sur une forme méconnue de sarcome des parties molles: Le rhabdomyosarcoma alvéolaire. Ann Anat Pathol (Paris) 1:88–111, 1956

55. Shmookler BM, Enzinger FM: Pleomorphic lipoma: A benign tumour simulating liposarcoma. A clinicopathologic analysis of 48 cases. Cancer 47:126–133, 1981

56. Stout AP, Hill WT: Leiomyosarcoma of the superficial soft tissues. Cancer 11:844–854, 1958

57. Stout AP, Lattes R: Tumours of the soft tissues atlas of tumor pathology, second series fascicle 1. Washington, DC, Armed Forces Institute of Pathology, 1967

58. Uehara H: Letter to the Editor. Cytology of alveolar soft part sarcoma. Acta Cytol 22:191–192, 1978

59. Varela–Duran J, Oliva H, Rosai J: Vascular leiomyosarcoma. The malignant counterpart of vascular leiomyoma. Cancer 44:1684–1691, 1979

60. Walaas L, Angervall L, Hagmar B et al: A correlative cytologic and histologic study of malignant fibrous histiocytoma: An analysis of 40 cases examined by fine-needle aspiration cytology. Diag Cytopathol 2[1]:46–53, 1986

61. Weiss SW, Enzinger FM: Myxoid variant of malignant fibrous histiocytoma. Cancer 39:1672–1685, 1977

62. Weiss SW, Enzinger FM: Malignant fibrous histiocytoma. An analysis of 200 cases. Cancer 41:2250–2266, 1978

63. Wile AG, Evans HL, Romsdahl MM: Leiomyosarcoma of soft tissue: A clinicopathologic study. Cancer 48:1022–1032, 1981

64. Willems J: Accuracy of aspiration biopsy cytology of soft-tissue tumors. Abstract. 8th International Congress of Cytology, Montreal, Canada, June 19–23, 1983

CHAPTER 16

Aspiration Biopsy Cytology of Tumors and Tumor-Suspect Lesions of Bone

JAN–SILVESTER WILLEMS

Hard tissue, such as bone, presents a natural barrier to fine needle aspiration (FNA) biopsy. However, a number of factors permit FNA biopsy of malignant bone tumors, making cytologic diagnosis and non-surgical delineation of these lesions possible. They are as follows:

Most malignant bone neoplasms cause destruction and lysis of pre-existing bone.

Most malignant bone tumors are at least in part composed of soft friable tumor tissue that can be sampled by a thin aspirating needle.

Most malignant primary bone tumors eventually perforate the cortex and invade the surrounding soft tissues, where they are accessible to biopsy.

Bone tumors, whether primary or of metastatic origin, are often complicated by pathologic fractures whereby the intra-osseous tumor site is exposed.

Even if certain destructive osteolytic lesions remain inaccessible because of their covering of cortical bone, aspiration biopsy cytology (ABC) provides an accurate and reliable pre-operative morphologic diagnosis in many instances.

Patients who may benefit from ABC of bone are those with localized skeletal pain, tenderness, warmth, or swelling, all of which may be clinical manifestations of a bone tumor. Other patients eligible for FNA biopsy are those with x-ray-documented osteolytic bone lesions, symptomatic or not, or with pathologic fractures, in whom the differential diagnosis includes primary bone tumor, metastatic malignancy, or lymphoma. Indeed, when evidence of a metastatic carcinoma or lymphoma is conclusive, diagnostic surgery may be obviated and appropriate therapy instituted on the basis of the cytologic report. Occasionally, FNA biopsy of bone will provide conclusive evidence against tumor of bone, as in cases of osteomyelitis.

The topographic site of suspected malignant bone lesions may make FNA biopsy more attractive than surgical biopsy as a first-choice method of investigation in, for example, the cervical spinal column or the base of the skull (see Chaps. 3 and 17).

Procedure

The general principles of FNA are applicable for suspected bone tumors (see Chap. 1). For superficial osteolytic lesions of the skull, cervical spinal column, ribs, and limbs, a 6-cm needle with an outer diameter of 0.7 mm should suffice. Bony lesions located deeper require a 12-cm needle. In patients with palpable bone lesions, the biopsy can be performed under palpatory control. In other patients the target area can be determined with the aid of roentgenograms. Sites in the vertebral column or the skull, which are not easily accessible, are best aspirated under fluoroscopic monitoring (see Chap. 2).[11] The chances for successful aspiration are highest when a soft-tissue component is present. However, even when the periosteum is elevated over an apparently intact cortex, aspiration will often produce diagnostic material because of extension of tumor through the interstices of bone. When aspirating a bone lesion without soft tissue involvement, start by exploring the surface of the bone with the tip of a thin needle in order to detect an area of destruction or a fracture line not shown by the x-ray film. If an opening is found in the cortex, aspiration should be performed through it with either a thin or a thick needle.

In lytic bone lesions covered by a thin lamella of bone, thin-walled needles may be passed through a perforating core biopsy needle to obtain cytologic material. Alternatively, a bone marrow aspiration needle, preferably child sized, may be used and procures plenty of material. Often this procedure provides material not only for smears but for bacteriologic cultures and paraffin blocks for histopathologic examination. Some authors have systematically used large needle biopsy for the simultaneous preparation of smears and histologic sections.[2,8]

After obtaining the cell sample, I prefer to smear and air-dry it for subsequent May–Grünwald–Giemsa (MGG) staining; if there is sufficient material, I stain supplementary wet-fixed smears with the Papanicolaou (P) method. MGG smears display the character of the intercellular substance vividly; they also show the cytoplasmic properties of the various primary bone tumors and metastatic adenocarcinomas and the general features of lymphomas.

TUMORS METASTATIC TO BONE

Among the secondary tumors causing osteolytic destruction of bone tissue, carcinomas dominate; those that most frequently metastasize to bone arise in the breast, prostate, lung, thyroid gland, kidney, and stomach. Cytologic diagnosis of metastatic adenocarcinoma, undifferentiated carcinoma, or small cell carcinoma usually always requires representative cell samples.

Stormby and Åkerman reported that in 39 of 48 patients in whom the clinical history or the roentgenologic examination was suspicious for metastatic bone lesions, cytologic confirmation of metastatic malignancy could be obtained.[15] Cytologic diagnosis was facilitated by the fact that the primary tumor was known in almost all of the patients. However, in some patients aspiration biopsy helps locate the primary site of a metastasizing tumor, as illustrated in Figure 16-1.

MALIGNANT HEMATOPOIETIC TUMORS OF BONE

Both flat and long bones may be the site of involvement by hematopoietic tumors, particularly plasma cell myeloma, which is the most common tumor affecting bone. The skull, scapula, humerus and pelvic bones not infrequently show large osteolytic lesions caused by neoplastic proliferation of plasma cells (Figs. 16-2 and 13-25) or lymphoplasmocytic cells, which clinically may be mistaken for metastatic carcinoma. Moreover, such tumors run a risk of histopathologic misinterpretation as anaplastic carcinoma if inappropriately stained. Such cases may advantageously be investigated by FNA and conclusively diagnosed pre-operatively.

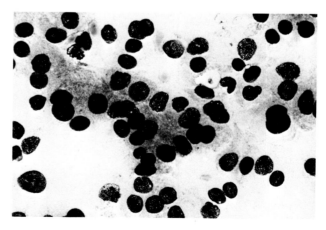

Fig. 16-1. Thyroid carcinoma, metastatic to humerus. A cohesive epithelial cluster with granular cytoplasm and monomorphous nuclei is present. (May–Grünwald–Giemsa [MGG] stain, ×600)

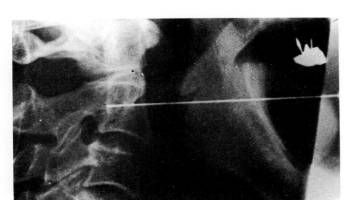

A

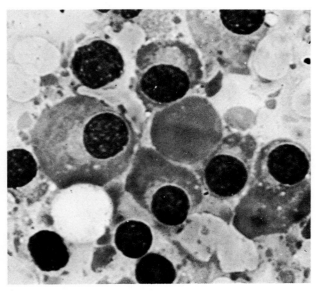

B

Fig. 16-2. A. Fine-needle aspiration through the mouth from a radiolucent lesion in a cervical vertebra. **B.** Plasma cell myeloma. Atypical plasma cells are evident. (MGG stain, ×600)

PRIMARY MALIGNANT NEOPLASMS OF BONE

Osteosarcoma

Osteosarcomas are rare malignant tumors of bone that produce material resembling osteoid, bone, and sometimes cartilage.[3,13,14] Youths aged 11 to 19 years and men are predominantly affected, the metaphy-

seal ends of long tubular bones and the jaws being the most common sites. Intermittent pain and sometimes a bulky tender swelling are common symptoms. Alkaline phosphatase is elevated in half the patients. An important subgroup of osteosarcoma consists of single or multifocal tumors developing in the older age group in Paget's disease of bone. A change in the intensity or quality of pain at a known Paget's site before roentgenographic evidence is a signal for biopsy. Such tumors may occur in the femur, humerus, and tibia and in flat bones, including the skull and pelvis.[14]

Roentgenographically, osteosarcoma may be lytic or sclerotic or a combination. Osteosarcomas commonly produce a Codman's triangle of reactive subperiosteal new bone in which neoplastic tissue may be found only extremely rarely, thus making the triangle an unsuitable site for biopsy. Histologically, osteosarcomas show a frankly sarcomatous stroma and malignant neoplastic osteoid and bone or cartilage that, on the basis of one dominant element, can be subclassified into osteoblastic, fibroblastic, or chondroblastic. The newly formed osteoid and bone display pleomorphism, nuclear hyperchromatism, and abnormal mitotic figures. Necrosis, hemorrhage, fibrosis, and calcification may be seen.

The histologic diagnosis of highly pleomorphic osteosarcoma is relatively simple, whereas the much rarer, well-differentiated cases can only with difficulty be distinguished from benign conditions. Hematogenous metastasis predominates, and pulmonary deposits are common. By the time an osteogenic sarcoma reaches medical attention, it has most often broken through the cortex of the bone, slightly or completely destroying the bony cortex. This makes it possible to sample the lesion with a fine needle.

At aspiration, very cellular yields are obtained. However, some needlings mainly give blood. Representative aspirates show abundant clusters and free-lying dispersed cells as well as some osteoid. Cell populations from osteosarcomas consist of spindle or plump stellate, angular, or rounded sarcoma cells in varying proportions. Sometimes one type dominates. These tumor cells have mostly one but occasionally two or three nuclei. In addition, pleomorphic multinucleated giant cells and many multinucleated giant osteoclastic-type cells with monomorphous nuclei are seen. The mononuclear cells vary in size but are generally large, and, irrespective of shape, have dense gray-blue cytoplasm with well-delineated cell

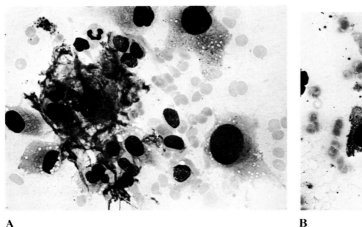
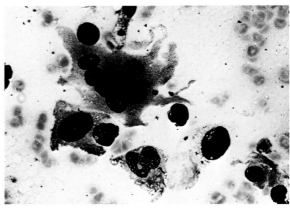

A   B

Fig. 16-3. Osteosarcoma. **A.** Clusters and dispersed plump sarcoma cells with pleomorphic nuclei. Note coarsely fibrillar osteoid lying between the tumor cells. (MGG
stain, ×600) **B.** Osteoclastic giant cell with well-defined cytoplasmic extensions and
some mononucleated pleomorphic sarcoma cells. (MGG stain, ×600)

borders (with MGG stain) (Fig. 16-3A). In some cases
rounded cells contain small, optically empty vacuoles. In others, especially those built of spindle-
shaped cells, MGG may reveal abundant, small, red-
staining granules in the central part of the cell body,
that is, around the nuclear poles.

Nuclei are irregularly oval or round and have an
eccentric position in the cell. With P stain, the nuclear
membrane often shows a shallow indentation. Chromatin is coarsely granular and hyperchromatic to
rather pale, with several irregular chromocenters.
Nuclear pleomorphism is outspoken in the majority
of cases but may occasionally be mild.

The number of multinucleated giant cells with
pleomorphic nuclei is variable, but usually a moderate amount are seen in the majority of cases. They are
few in cases with mild nuclear atypia. Osteoclastlike
giant cells with monomorphous nuclei are, as a rule,
present in most cases but may be scarce in anaplastic
tumors. Their cytoplasm is dense; and cell borders are
sharp and may show several long, sometimes tenta-
clelike cytoplasmic extensions (Fig. 16-3B). The osteoclasts have a few to a moderate number of monomorphous oval nuclei but usually not the large
number seen in giant cell tumors of bone.

Osteoid is seen in the clusters from osteoblastic
osteosarcomas between sarcoma cells but is also associated with dispersed single cells or occurs as free-
lying acellular substance. The best staining method

for revealing the osteoid substance is MGG, in which
it stands out as pinkish red, contrasting with the
blue-staining cytoplasm. It appears as a thick, mostly
hyalinized but partly fibrillar substance, ramifying
between tumor cells (see Fig. 16-3A). Stain, as a rule,
does not "leak" into the background, although the
edges may focally be blurred or serrated and somewhat granular.

In aspirates from some osteosarcomas, partly
calcified laminated bony spicules poorly stainable
with MGG, but somewhat better with P, can be seen.
In representative aspirates from osteoblastic osteosarcomas, the general cell picture, together with the
occurrence of osteoid, should provide the diagnosis.
Sometimes, however, in predominantly fibroblastic
osteosarcomas only a small quantity of osteoid can be
demonstrated. No osteoid is found in aspirates from
chondroblastic osteosarcomas. In those cases intercellular substance with the tinctorial characteristics of
chondroid is found with cells showing considerable
nuclear and cellular atypia. In cases where mononuclear cells only show mild atypia, the cytologic picture may be misinterpreted and the malignant nature
of the tumor underestimated. The general picture described above is also found in primary osteosarcoma
of soft tissue and in distant metastases. Not infrequently, metastatic deposits of treated osteosarcoma
show increased amounts of osteoid compared to the
primary tumor.

Chondrosarcoma

Chondrosarcoma is the second most common primary malignant tumor of bone occurring in adults between 30 and 60 years of age.[3,13,14] The tumor arises *de novo* in bone but can secondarily develop in pre-existing cartilaginous exostoses. The pelvic bones are the most frequent site of origin (see Figs. 9-18 and 9-19), followed by the ribs, femur, humerus, spine, scapula, tibia, fibula, sacrum, and sternum. Patients may have had some pain or swelling for months over the tumor area. Chondrosarcomas may become large. Roentgenographic examination is, in most cases of peripheral chondrosarcoma, indicative of the diagnosis. However, central chondrosarcomas can be more difficult to diagnose.

In histologic sections, the cartilaginous nature of the tumor is identifiable, but well-differentiated chondrosarcomas may sometimes be hard to differentiate from chondromas because of poor cellularity and slight atypia. Poorly differentiated varieties,[4,9] which are the least common, pose no diagnostic problems. Well-differentiated chondrosarcomas grow slowly, whereas poorly differentiated types grow rapidly and are more malignant.[4,9]

From chondrosarcomas, large amounts of viscid material obtained at FNA biopsy are smeared as a thick film showing some granularity macroscopically due to the fact that the chondroid substance cannot be flattened out completely. Staining reveals chondroid substance that partly stands out as ramifying solid fragments and partly as amorphous background substance. Chondroid displays various shades of color, reflecting differences in consistency, density, and composition. With MGG, it is bright magenta, while with P it is gray to violet. The denser areas show a faintly fibrillar structure. In addition, a pattern of groups of tumor cells emerges, lying embedded in the intercellular substance.

The cells are single and can be located in lacunae, which have a more intensely stained border than the surrounding chondroid often with a violet tinge (Fig. 16-4). If caught in lacunae in solid tissue fragments, chondrosarcoma cells are smaller than when lying free. Chondrosarcoma cells have well-demarcated, rounded cytoplasmic margins with both MGG and P. With MGG, they have a very pale blue inconspicuous cytoplasm containing several small bubble-like vacuoles, often located in the cell periphery. With

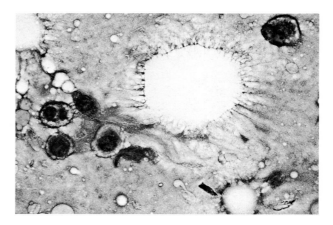

Fig. 16-4. Chondrosarcoma. Chondroid substance with some binucleated cells contained in lacunae with dark edges. (MGG stain, ×600)

P, the cytoplasm stains very faintly gray; and vacuoles are not as visible. Some chondrosarcomas have a fibroblastlike cell component in addition to the round chondrocytes.

The nuclei in well-differentiated chondrosarcoma are usually single, but binucleated cells (see Fig. 16-4) occur.[7] They are circular and oval, sometimes indented, and eccentric. P reveals evenly staining granular chromatin with some visible chromocenters. In moderately and poorly differentiated chondrosarcomas, the nuclei are generally larger and more pleomorphic; and they show more structural detail, including nucleoli. Variations in cellularity and cell and nuclear features used for grading chondrosarcomas are subtle and conclusive distinction between grades can be difficult. However in recurrent or metastatic chondrosarcomas progress towards lesser degrees of differentiation can be observed.

The cytologic picture of moderately and poorly differentiated chondrosarcoma is generally diagnostic, but cytologic diagnosis of well-differentiated chondrosarcoma has its limitations when chondroma is a differential diagnosis. As in paraffin sections binucleated cells are an important factor favoring malignancy. Chordoma, which occurs along the vertebral column (a common site for chondrosarcoma), differs from the latter in smears by forming large aggregates of trapezoid triangular or round cells lying in a homogeneous mucinous background.

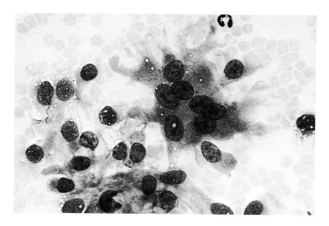

Fig. 16-5. Benign giant cell tumor of bone. Note monomorphous, partly vacuolated stromal cells and osteoclastic giant cell. (MGG stain, ×600)

Giant Cell Tumor of Bone

Giant cell tumor of bone is a neoplasm mostly occurring in adults aged 20 to 50 years with a slight predominance in women.[3,13,14] Typically, the tumor arises in the epiphyses of long bones, more than half occurring in the knee. The tumor tissue is soft and friable and produces a radiolucent lesion. Fracture in large tumors of weight-bearing bones is common. Giant cell tumor of bone is composed of giant cells of osteoclastic type and a stroma of spindle cells. The two cell types differ in proportion from case to case and in different areas of the same tumor. Features associated with increased risk of malignancy are high cellularity, prominent atypia, frequent mitoses and giant cells with few nuclei. However, tumors lacking these characteristics also may behave in a malignant fashion.

FNA biopsy usually obtains abundant cell yields mixed with blood. The smears show two populations of cells: mononuclear fibrohistiocytic cells and multinuclear osteoclastic giant cells.[11] The mononuclear cells (Fig. 16-5) generally occur in clusters or, less often, are dispersed. They have either a spindle shape or a plump, partly curved, and partly angular cell body with short, pointed cytoplasmic extensions. Their cytoplasm, which stains blue with MGG, is rather well-demarcated and contains small, optically empty vacuoles. The nuclei are oval and vary in size. They have an evenly distributed, finely granular chromatin pattern. P shows a distinct, often indented

nuclear membrane, some chromocenters, and a small nucleolus.

In every giant cell tumor of bone the population of multinucleated giant cells of osteoclastic type (Fig. 16-5) is very conspicuous, and either associated with the clusters of mononuclear cells or lying freely in the smears. They have well-demarcated dense cytoplasm with cytoplasmic extensions that sometimes have tapering ends. They contain a few to several dozen monomorphous oval nuclei with a finely granular chromatin pattern and a visible nucleolus. The nuclear membrane is sometimes indented (with P stain), and some chromocenters are visible. Lipid-laden macrophages (xanthoma cells) may be found. Osteoid is not usually seen in uncomplicated cases, nor is there any identifiable chondroid substance.

Malignant giant cell tumor of bone cannot be cytologically distinguished from its benign counterpart (described above).[16] In a case originating in the proximal tibia of a woman aged 28 years seen at Karolinska Hospital, the picture of the lung metastasis was identical to that of the local recurrence and of the primary tumor, and its malignant behavior was in no way predictable. The differential diagnosis of giant cell tumor of bone should include other bone lesions containing giant cells as for example aneurysmal bone cyst, which is composed of monomorphous fibrohistiocytic stromal cells and giant cells, which however are less numerous than in giant cell tumor of bone (see below).

Ewing's Sarcoma

Ewing's sarcoma is a primary malignant bone tumor believed to arise from undifferentiated mesenchyma of the medullary cavity.[3,13,14] The tumor occurs in childhood and adolescence but is most common in males between 10 and 30 years of age. It can involve almost any bone of the body, although the long tubular bones of the extremities and flat bones such as the ilium and ribs are the most frequent sites. Pain is an early symptom, followed by a palpable tender mass, sometimes accompanied by fever, leukocytosis, and anemia. A characteristic roentgenologic feature is the onion-skin pattern of subperiosteal bone formation, but lytic areas of bone destruction and dense regenerative foci are also seen. From gross specimens it appears that larger areas of bone are infiltrated by Ewing's sarcoma than can be anticipated from roentgenograms.

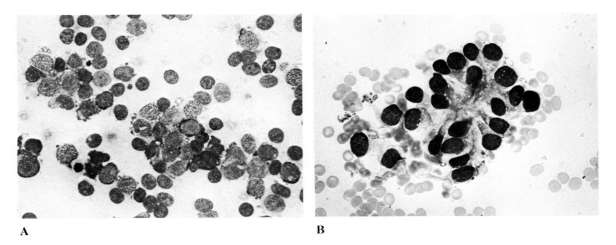

A   B

Fig. 16-6. Ewing's sarcoma. **A.** Aggregates of small tumor cells with rather monomorphous nuclei and scanty cytoplasm containing small vacuoles. (MGG stain, ×600) **B.** Rosette-like structure. Note vacuoles in the cytoplasm. (MGG stain, ×600)

Histologically, the lesion is densely cellular, consisting of compartments of rounded, uniform cells with little, thin, intercellular fibrous strands. The cells have faintly staining periodic acid–Schiff (PAS)-positive cytoplasm, indistinct cell borders, and moderately basophilic nuclei with occasional nucleoli.

Typically, cellular aspirates appear on the slides as solid clumps with a film of thin cell sheets and individual cells. The tumor cells are small and display a monomorphous cell picture, although some larger cells occur.[12] Certain cases consist almost entirely of larger immature cells of monotonous appearance. Many tumor cells appear in small, loose, almost monolayered aggregates (Fig. 16-6A). A larger number present as stripped nuclei. In some cases, many rosettelike structures (Fig. 16-6B) are scattered in the smear.

With MGG stain, the nuclei have a finely granular evenly distributed chromatin. P stain shows, in addition a few small chromocenters; one or two small nucleoli; and a thin, distinct nuclear membrane. Mitoses can be seen. The cell cytoplasm stains gray-blue with MGG. It is scanty and distributed asymmetrically around the nucleus, either with a rounded cell boundary or an irregular serrated margin. When the tumor cells form aggregates, the borders between adjacent cells are often not discernible. The cytoplasm contains small, peripherally located, optically empty vacuoles; special staining may show glycogen. The cytoplasm is fragile and can be found as a granular, partly vacuolated background substance that focally assumes a laminated appearance. The differential diagnosis of Ewing's sarcoma in bone includes lymphoma (see Chap. 14), particularly lymphoblastic (T-cell) type, Burkitt type, and lymphoblastic leukemia. Also, metastatic rhabdomyosarcoma and neuroblastoma (see Chap. 15) should be ruled out in children and adolescents. In young adults, small cell anaplastic bronchial carcinoma (oat cell type) is yet another possibility (see Chap. 7). (See Color Plate XXXVII D–F.)

Chordoma

Chordoma is an uncommon neoplasm arising from remnants of the notochord and is mostly seen in persons over 30 years of age.[3,13,14] It is frequently noted in men. The most common site is the sacrococcygeal region, followed by the spheno-occipital region of the base of the skull and the cervical vertebrae. The tumor grows slowly but progressively, and when symptoms such as pain appear, the tumor is usually large and has destroyed bone and infiltrated adjacent structures. Sacrococcygeal chordoma very often has a presacral extension that may be detected on rectal examination and, as a consequence, be biopsied by transrectal FNA (see Chap. 8).

Histologically, chordomas show areas of mucoid intercellular material containing strands and is-

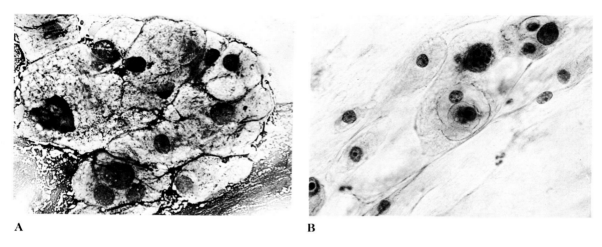

A   B

Fig. 16-7. Chordoma. **A.** Nest of large cells in a background of mucoid substance which contains small vacuoles. The largest cell has bubbly cytoplasm and a pleomorphic nucleus and represents a physaliferous cell. (MGG stain, ×600) **B.** Physaliferous cells. Note the perinucleolar halo, which is prominent in certain cells *(lower left).* (P stain, ×600)

lands of tumor cells. Some tumor cells have a particular vacuolated appearance and are thus called *physaliferous;* they are quite large, with a nucleus that may be vesicular or pleomorphic.

Aspirates from chordoma contain large amounts of gelatinous material that can be flattened out evenly on a slide by gently smearing. The abundant amount of partly fibrillar substance is intimately mixed with nests and cords of tumor cells and stains bright magenta with MGG and pink to orange with P. It forms also the background and contains many small vacuoles. Immersed in their partly fibrillar and partly vacuolated mucin nests and cords of large chordoma cells are seen. The aggregated neoplastic cells have square or curved cell bodies with rectangular, trapezoid, triangular, and sometimes round shapes, with rounded cytoplasmic corners.

With MGG, the tumor cell cytoplasm appears as sharply delineated, extremely pale gray-blue cell bodies. With P, the cells are light pink or orange. Typically, chordoma cells have abundant cytoplasm, but a proportion of the tumor cells are small and round to polygonal.[1] Some cells have cytoplasm containing multiple vacuoles of varying size (Fig. 16-7A and Color Plate XXXVIII A and B), representing physaliferous cells.[10] The majority of the cells have a single nucleus, but binucleated and multinucleated cells occur. The nuclei are eccentrically located, oval or round, and usually small. Some cells have large pleomorphic and hyperchromatic nuclei, often containing a prominent nucleolus. With P stain, the chromatin of the large nuclei is finely granular and visible mostly in the periphery of the nucleus, resulting in a perinucleolar halolike clear area (Fig. 16-7B). Differential diagnosis includes chondrosarcoma and mucus-producing adenocarcinoma.

Other Malignant Tumors of Bone

Other sarcomas of bone, such as fibrosarcoma, malignant fibrous histiocytoma, angiosarcoma, and leiomyosarcoma, show patterns resembling their counterparts in soft tissue (see Chap. 15).[3,13,14]

DISEASES OFTEN CLINICALLY RESEMBLING BONE NEOPLASMS

Osteomyelitis

Infection of bone is most often a hematogenous process. Patients diagnosed as having suppurative osteomyelitis are usually infants and young children, although it may occur in adults. In children its symptomatology can mimick Ewing's sarcoma. In adults osteomyelitis sometimes develops during the course of chemotherapy for cancer; and, not infrequently, such patients are referred with a suspicion of bone metastasis. In the presence of roentgenologic bone destruction, FNA may be resorted to for diagnosis. On such occasions an abscess may be emptied.

If no pus is aspirated, a representative cell sample may consist of fragments of necrotic bone or bone marrow, inflammatory cells, and evidence of bone resorption and production. The inflammatory infiltrate may include granulocytes, histiocytes, lymphocytes, and plasma cells occurring singly or in solid fragments of granulation tissue. Manifestations of bone renewal are sheets or clusters of osteoblasts together with osteoclasts. Osteoblasts (Fig. 16-8) are cuboidal to cylindric with dark basophilic cytoplasm and sharply delineated, somewhat bulging edges. They contain an oval nucleus that generally abuts one cell pole. Osteoblasts characteristically have a distinct, pale-staining subnuclear space that often is somewhat smaller than the nucleus. With P, nuclei show homogeneously granular dark chromatin with a visible small nucleolus.

When there is a clinical suspicion of osteomyelitis or when the macroscopic aspect of the cell sample suggests osteomyelitis, part of the aspirate may be processed for bacteriologic culture. If the causative agent for osteomyelitis is large enough, that is, is a yeast or a fungus, it can be demonstrated in smears from aspirates as in, for example, cryptococcal osteomyelitis in a cancer patient.[6] Such cases emphasize the value of FNA cytology in diagnosis and follow-up study of patients with suspected or proven malignancies.

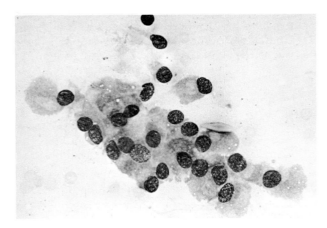

Fig. 16-8. Osteoblasts from an area of bone production from a case of osteomyelitis. Note the sharply demarcated dense cytoplasm and the regular nuclei with a paler paranuclear zone. (MGG stain, ×600)

Eosinophilic Granuloma

Eosinophilic granuloma is a benign, nonneoplastic, solitary lesion of bone, presumably arising from the reticuloendothelial system (RES), with a peak incidence between the ages of 5 and 10 years.[3,13,14] The skull is one of the most common sites, followed, in decreasing frequency, by the mandible, humerus, ribs, and femur. The lesion shows up roentgenographically as a discretely outlined defect in bone but occasionally may produce a shadow suggesting a malignant bone tumor.

Histologically, the lesion shows some similarity with the RES proliferations seen in Letterer–Siwe and Hand–Schüller–Christian disease, both disseminated, potentially lethal conditions. Microscopically eosinophilic granuloma contains proliferating histiocytic cells, usually with some multinucleated histiocytic giant cells. In addition, an inflammatory infiltrate, predominantly consisting of eosinophilic and neutrophilic granulocytes and some lymphocytes, is seen.

Franzén and Stenkvist described seven cases and Thommesen and colleagues described eight cases diagnosed cytologically from smeared needle aspirates.[5,17] Smears from eosinophilic granuloma are highly cellular and contain large numbers of histiocytes with moderate amounts of cytoplasm and pale-staining regular nuclei. A certain number of foam cells are often seen, as are multinucleated histiocytic giant cells. Usually, a prominent mixture of eosinophils and neutrophilic granulocytes is recognizable. As a rule, representative cell samples are obtained without difficulty and cytodiagnosis does not pose problems (Fig. 16-9).

In children, adolescents, and young adults, the condition may clinically and radiologically simulate malignant small cell tumors in bone, for example, Ewing's sarcoma and metastatic neuroblastoma; but it can cytologically readily be differentiated from those lesions.

Benign Lytic Lesions of Bone Showing Fibrogenesis, Osteoid Formation, and Osteoclastic Giant Cells

Fibrous dysplasia, aneurysmal bone cysts, giant cell reparative granulomas, and simple cysts, which generally occur in younger age groups, may require needle aspiration for cytologic diagnosis because of suspicion of a primary bone tumor.[3,13,14]

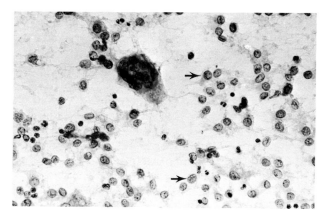

Fig. 16-9. Lytic lesion of rib, eosinophilic granuloma. Needle aspirate. This smear is from a fine needle puncture of a rib. There are multiple, scattered, benign histiocytes *(arrows)* intermixed with polymorphonuclear cells and occasional eosinophils. One atypical giant cell is present. (P stain, ×132)

Representative aspirates from the above lesions may be either hemorrhagic and poorly to moderately cellular or consist of cyst contents. More infrequently, rich amounts of cell material can be obtained, especially from aneurysmal bone cysts. In the latter smears, cohesive clusters of spindle cells with fibrillary stromal substance dominate, with, in the background, dispersed fibroblastlike and histiocytelike connective tissue cells. Multinucleated osteoclastic giant cells are often conspicuous, mostly lying free. Varying amounts of hemosiderin- and fat-laden macrophages occur in most cases. Sometimes osteoid can be recognized, as can scattered groups of osteoblasts.

Such lesions, in which the osteoclastic giant cell component is obvious, must be differentiated from true benign giant cell tumor of bone. The latter, in uncomplicated typical cases, shows large yields of more cellular clusters without detectable intercellular substance, built up of mononuclear cells that readily disperse into sheets or single free cells and of an abundant component of osteoclastic giant cells.

CLINICAL APPLICATION

Aspiration biopsy in skeletal lesions yields a variety of cell patterns, and only the most frequent conditions have been discussed in this chapter. Bone tumors often show a preferential age, sex, and topographic distribution and a reasonably typical radiologic picture.[3,13,14] Within this context, with the exception of those lesions inaccessible to a fine needle, ABC can provide valuable information of differential diagnostic significance. Stormby and Åkerman estimate that diagnostic cell material can be obtained in about 66% of patients who clinically and radiologically are believed to have primary bone lesions.[15]

Histopathologic recognition of bone tumors has its difficulties. Attempts at defining tumor type from cytologic smears should thus be undertaken with even more caution than usual. Since primary bone tumors are only infrequently encountered outside of large medical centers, fine needle aspirates should be interpreted by an experienced cytopathologist cognizant of diagnostic possibilities and limitations. Furthermore, clinicians should understand the scope of the method. Preferably, both cytopathologist and clinician should determine patient management and treatment.

In addition to its role in the diagnosis of primary bone tumors, ABC has a place in documenting recurrences or metastatic spread of bone tumors during follow-up study.

REFERENCES

1. Carvalho G, Humberto L: Chordoma of rhinopharynx. Acta Cytol 18:425, 1974
2. Coley BL, Sharp GS, Ellis EB: Diagnosis of bone tumors by aspiration. Am J Surg 13:215, 1931
3. Dahlin DC: Bone Tumors. General Aspects and Data on 6,221 Cases., 3rd ed. Springfield, IL, Charles C Thomas, 1977
4. Evans HL, Ayala AG, Romsdahl MM: Prognostic factors in chondrosarcoma of bone. A clinicopathologic analysis with emphasis on histologic grading. Cancer 40:818, 1977
5. Franzén S, Stenkvist B: Cytologic diagnosis of eosinophilic granuloma-reticulo-endotheliosis. Acta Pathol Microbiol Scand 72:385, 1968
6. Ganjei P, Evans DA, Fischer NL: Diagnosis of cryptococcal osteomyelitis by fine needle aspiration cytology. A case report. Acta Cytol 26:224, 1982
7. Hajdu SI, Hajdu EO: Cytopathology of Sarcomas and Other Nonepithelial Malignant Tumors, pp 257–289. Philadelphia, WB Saunders, 1976
8. Hajdu SI, Melamed MR: Needle biopsy of malignant bone tumors. Surg Gynecol Obstet 133:829, 1971

9. Kreicbergs A, Slezak E, Söderberg G: The prognostic significance of different histomorphologic features in chondrosarcoma. Virchows Arch Pathol Anat 390:1, 1981

10. Lefer LG, Rosier RP: The cytology of chordoma. Acta Cytol 22:51, 1978

11. Lopes–Cardozo P: Atlas of Clinical Cytology. Disorders of Bone and Cartilage, pp 175–196. Distributed by JB Lippincott, Philadelphia

12. Reif RM: Ewing's sarcoma cytology of large cell type. Acta Cytol 24:175, 1980

13. Schajowicz F, Ackerman LV, Sissons HA et al: Histological Typing of Bone Tumours, no. 6. International Histological Classification of Tumours. Geneva, World Health Organization, 1972

14. Spjut HJ, Dorfman HD, Fechner RE et al: Bone and cartilage. In Atlas of Tumor Pathology. Washington, D.C., Armed Forces Institute of Pathology, 1971

15. Stormby NG, Åkerman M: Cytodiagnosis of bone lesions by means of fine needle aspiration biopsy. Acta Cytol 17:166, 1973

16. Szyfelbein WM, Schiller AL: Cytologic diagnosis of giant cell tumor of bone metastatic to lung. A case report. Acta Cytol 23:460, 1979

17. Thommesen P, Frederiksen P, Löwhagen T et al: Needle aspiration biopsy in the diagnosis of lytic bone lesions in histiocytosis X, Ewing's sarcoma and neuroblastoma. Acta Radiol 17:145, 1978

Aspiration Biopsy Cytology of Tumors of the Central Nervous System and the Base of the Skull

JAN–SILVESTER WILLEMS

Patients with neurologic or endocrine symptoms related to an intracranial tumor generally are thoroughly investigated before they are subjected to fine needle aspiration (FNA) biopsy. FNA biopsy is one of the final diagnostic measures for intracranial tumor; this contrasts with its usual role as a first-choice office or bedside procedure for quick diagnosis. However, in a patient with a suspected brain tumor, the neurosurgeon must decide, often intra-operatively, whether to use FNA for cytologic diagnosis.

In the past, patients with neuroradiologically diagnosed inoperable brain tumors have often been treated without biopsy. However, even nonsurgical treatment of brain tumors should be preceded by a morphologic diagnosis so that the effect of radiation or drug treatment can be assessed. Therefore, at Karolinska Hospital, stereotaxic needle biopsy is part of the diagnostic approach in patients with intracranial space-occupying lesions. The precision of stereotaxic biopsy makes most brain areas, although small, accessible to biopsy with a variety of narrow needles.[6] Space-occupying lesions in anatomically well-defined areas, especially in the sellar, third ventricle, and pineal regions, can be easily investigated, as can larger intrahemispheric tumors.[4,10]

FNA biopsy can be performed through a burr hole or a craniotomy either before or after a spiral needle biopsy for histologic examination. The cell sample for cytologic examination is thus obtained along the same route as the histologic specimen. When histologic and cytologic examinations are performed simultaneously, they complement one another. Moreover, should the larger spiral needle biopsy encounter hard tissue with the risk of trauma, the surgeon may choose to perform only a FNA biopsy. Cytologic examination can also be relied on if insufficient material is obtained from the target lesion for histologic processing.

Stereotaxic spiral needle biopsy and FNA biopsy are not performed when neuroradiologic evidence is virtually pathognomonic for a tumor, as in, for example, meningioma. Neither is a biopsy made in tumors that are bound to be operated on, such as tumors of the posterior fossa with raised intracranial pressure, seen, for example, in children with cerebellar medulloblastomas.

Lesions near or in the base of the skull do not require transcerebral aspiration biopsy. Chordomas and paragangliomas can be biopsied by directing a needle under fluoroscopic monitoring transnasally or transpharyngeally through the foramen ovale or through the soft tissues of the neck (see Fig. 3-2).[3,8]

Procedure

To proceed the neurosurgeon introduces the aspirating thin needle through a needle guide with an outer diameter of 0.8 mm. Optimally, the material should be prepared immediately, that is, in the operating room. It should be carefully but readily smeared and immediately fixed in 90% ethyl alcohol. When sufficient material is available, make air-dried preparations. Wet alcohol fixation has proved to be the most satisfactory for two reasons. First, together with the Papanicolaou (P) stain, it displays the fibrillar structure of astrocytic tumors more distinctly than air-dried May–Grünwald–Giemsa (MGG) preparations. Second, air-dried cell samples are sometimes damaged by the mixture of fluid obtained at aspiration.

TUMORS OF THE CENTRAL NERVOUS SYSTEM

Astrocytomas

Astrocytomas of grade I to II constitute 25% to 30% of all cerebral gliomas, according to Kernohan.[9,12] They arise at any age, mostly in males. Depending on the patient's age, they are noted to occur preferentially in certain sites of the brain. In adults they grow in the cerebral hemispheres; in children and adolescents they involve the hypothalamus, cerebellum, and pons. Symptoms are determined by tumor location and generally slow growth.

Histologically, astrocytomas can be divided into fibrillar, protoplasmic, and gemistocytic types, but other varieties, such as those with piloid types of cells, are also seen.[13] There is a frequent tendency to anaplastic change, and when it occurs tumors correspond to grades III to IV of the Kernohan classification.[9] Glioblastoma multiforme is mostly seen in the frontal or temporal lobes. It is a rapidly growing tumor with a histologic picture varying considerably from highly pleomorphic to fairly uniform tumor cells.[18] The general histologic picture, which includes evidence of vascular endothelial proliferation and necrosis, allows the tumor to be classified as grade III to IV.

The smear patterns depend on tumor type. Fibrillar astrocytomas are difficult to smear, while high-grade gemistocytic and anaplastic astrocytomas tend to spread more easily. The tumor cells occur partly in clusters and partly as free cells and stripped nuclei.

Tumors of fibrillary and protoplasmic astrocytes consist of cells with cytoplasm concentrated around the nucleus and very long, slender, or almost threadlike cytoplasmic extensions. The nuclei are single and oval to round. Cellularity, nuclear pleomorphism, hyperchromasia, nucleoli, and mitotic figures can be judged in smears, as can vascular endothelial proliferation and necrosis. These characteristics can be used to grade the tumors into low-grade (I, II) and high-grade (III, IV) varieties according to the system proposed by Kernohan and adapted to cytologic material.[9] Be aware, however, that the spectrum of variation is gradual, and separating low-grade from high-grade tumors is not always easy.

Well-differentiated astrocytomas of grade I are rarely seen. The cells are more closely arranged than in normal brain tissue, the nuclei being slightly enlarged but monomorphous, and no mitoses are seen. This tumor type is very difficult to differentiate from gliosis, and cytologic reporting is apt to be inconclusive.

Grade II astrocytic tumors show obvious neoplastic characteristics in smears (Fig. 17-1). Cellularity is definitively increased. With MGG stain, the fibrillar character of the tissue is more coarse and the cell bodies are more prominent. The nuclei are larger than normal astrocytes and grade I tumors, although only slightly pleomorphic. A few mitotic figures may be seen, but no necrosis and, as a rule, no evidence of vascular endothelial proliferation are present.

In *grade III astrocytic tumors* nuclei are pleomorphic (Fig. 17-2) and show a coarser chromatin pattern and often multiple nucleoli. Fragments of ramifying hypertrophic blood vessels showing several layers of endothelial cells can be seen, some of which show mitosis (Fig. 17-3). Necrotic tissue in the smears is seen very easily with MGG. Sometimes, as can also be the case in grade IV tumors, the entire aspirate may consist of necrotic material, thus making for an inconclusive report. In the absence of necrosis or blood vessel fragments, a confident cytologic distinction between grade II and grade III tumors cannot always be made on cellular features alone.

Grade IV astrocytic tumors including glioblastoma multiforme show often frank cellular pleomorphism with spindle, stellate, and round cells (see Color Plate XXXVIII C–E). The nuclei show varying degrees of hyperchromasia (Fig. 17-4), and they can be very bizarre, sometimes with intranuclear cytoplasmic inclusions. Multinucleated cells are common. However, tumors with a relatively monotonous cell and

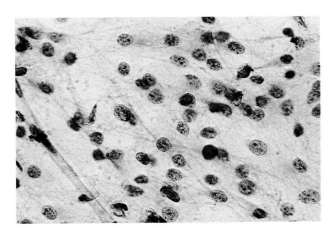

Fig. 17-1. Astrocytoma grade II. Fibrillar astrocytic tumor cells with large, rather monomorphous nuclei show visible nucleoli and chromocenters. Note thin fibrillary extensions and capillary *(bottom left)*. (P stain, ×600)

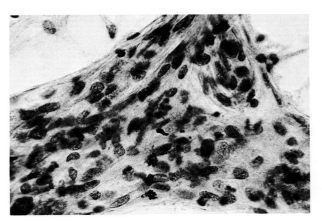

Fig. 17-3. Astrocytoma grade III. Ramifying hypertrophic blood vessel shows endothelial proliferation. Note mitosis *(bottom)*. (P stain, ×600)

nuclear picture occur. In many tumors, fibrillary cytoplasmic extensions are still present. However, in some they can be hard to detect, and small, optically empty cytoplasmic vacuoles may be prominent. In anaplastic tumors without demonstrable cytoplasmic fibrillary extensions, differential diagnosis with poorly differentiated carcinomas can be problematic.[11] Conversely, metastatic carcinomas may be misdiagnosed as grade IV gliomas. Immunocytochemical staining for glial fibrillary acidic protein (GFAP) and other tumor markers on aspirates are

useful in the differential diagnosis between glioblastoma and nongliomatous neoplasms.[5] Some glioblastomas consist, in part, of rather monomorphous astrocytic cells; and in aspirates their grade may be underestimated if no frankly pleomorphic areas, necrotic tissue, or vascular endothelial proliferation can be demonstrated. Occasionally, giant cell glioblastoma is aspirated, characterized by a cell flora of extremely bizarre pleomorphic giant cells.

Gemistocytic astrocytomas are composed of tumor cells with abundant dense cytoplasm, often with a polygonal appearance (Fig. 17-5) and only occasionally showing cytoplasmic extensions. The

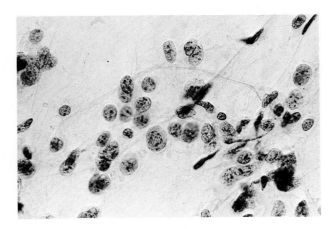

Fig. 17-2. Astrocytoma grade III. Fibrillar astrocytic cells with several large pleomorphic nuclei, each with multiple nucleoli, are evident. (P stain, ×600)

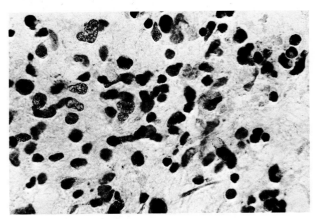

Fig. 17-4. Astrocytoma grade IV. Conspicuous nuclear pleomorphism and hyperchromasia are present. (P stain, ×600)

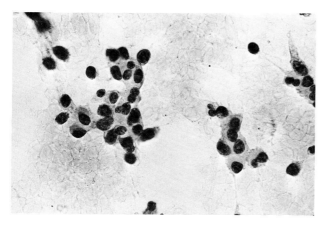

Fig. 17-5. Recurrent astrocytoma of gemistocytic type. Well-delineated cell bodies and hyperchromatic nuclei are present. Note the sparse, thin fibrillar extensions. (May–Grünwald–Giemsa [MGG] stain, ×600)

latter can be difficult to see with MGG stain. Gemistocytic cells have one or more eccentrically placed nuclei showing varying degrees of pleomorphism. In difficult cases positive stainability for GFAP is of decisive diagnostic importance. These astrocytomas are usually not graded for malignancy, but they tend to dedifferentiate into anaplastic forms.[13]

Oligodendroglioma

Oligodendrogliomas are a relatively rare type of intracranial glioma generally arising in the cerebral cortex and subcortical white matter. The tumor is a well-defined, solid, soft mass, sometimes with central foci of necrosis or cystic or mucinous changes. Histologically, oligodendroglioma is made up of compact masses of swollen cells. The nuclei show more variation in the central parts of the tumor, being almost indistinguishable from normal tissue in the peripheral areas. Mitoses are rare. The cells are intersected by a scanty, delicate stroma of blood vessels and collagen. Some of the blood vessels show fine lamellar or coarse clumps of calcium salts in the walls.

Many oligodendrogliomas grow slowly, but their behavior is quite unpredictable. A certain proportion undergo malignant change, heralded by prominent mitotic figures. Massive calcification, which is roentgenologically and histologically demonstrable, does not always imply a good prognosis.

Oligodendrogliomas yield abundant material at aspiration, which spreads rather easily. Tumor cells display a monotonous picture and are seen against a finely granular background substance containing small empty spaces (Fig. 17-6). Their cytoplasm stains palely or is clear in certain cells, in which a perinuclear halo can be seen. However, cell borders are generally not as well defined as in histologically processed tissue. The nuclei are mostly oval and have a finely and evenly distributed chromatin with some chromocenters and usually one small nucleolus. Some of the nuclei are larger than the majority. Thin, delicate, ramifying capillaries are often seen. Some purple-staining amorphous or laminated calcified bodies of diverse shapes can be observed with P, sometimes very abundantly. The above picture is, in most instances, diagnostic for pure oligodendroglioma, but poorly differentiated ependymoma and chromophobe adenoma of the pituitary gland are a differential diagnostic possibility, especially when smearing artifacts produce predominantly naked nuclei. When a cell component with fibrillar extensions is observed, assume that an astrocytic component is present.

In some instances oligodendroglioma recurs as a dedifferentiated tumor. The smears in such cases resemble pleomorphic glioblastoma multiforme.

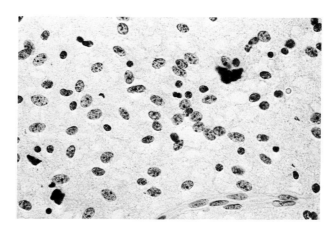

Fig. 17-6. Oligodendroglioma. A monomorphous cell population exists against a thin granular background. Note finely granular chromatin with one or two nucleoli. Two dark calcified bodies (*corners*) and a narrow capillary are seen. (P stain, ×600)

Ependymoma

Ependymomas are generally slow-growing tumors arising from cells lining the ventricular system, the most common site being the fourth ventricle, and the lumbosacral segments of the spinal cord. The tumors are usually well defined and homogeneous, the larger ones showing areas of cystic degeneration.[12] Ependymomas can spread along the neural canal but rarely, they grow rapidly and cause remote metastases. Histologically, the tumors are cellular. A diagnostic feature is the presence of ependymal rosettes when tumor cells are grouped around small spaces. Perivascular pseudorosettes are another characteristic useful for diagnosis.

Ependymomas yield abundant material at aspiration and can be smeared easily. In well-differentiated forms there is a repetitive pattern throughout the smear consisting of numerous pseudorosette cell clusters. In addition, numerous free-lying cells are seen, either in small groups or singly. There are also many fragments of magenta-staining intercellular substance scattered over the smear. The pseudorosettes (Fig. 17-7) are composed of many elongated cells concentrically arranged around a central blood vessel that sometimes branches.[2] The tumor cells are elongated and spindle shaped, with one end tapering off toward the capillary wall. They have one round or oval uniform, consistently eccentric nucleus with granular chromatin. The smaller ependymal rosettes, consisting of concentrically arranged cells with tapering ends radiating toward a center, are more frag-

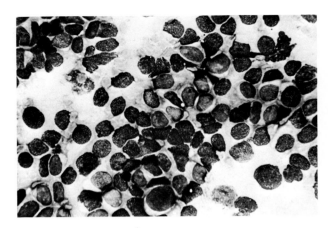

Fig. 17-8. Medulloblastoma. Immature round, sometimes molded nuclei with scanty cytoplasm are present. Variation in nuclear size is recognizable. (MGG stain, ×600)

ile. They are more rarely seen and may not always be demonstrable. The free-lying cells are elongated, but a large number of plump, rounded cells are also seen. The long cytoplasmic processes extending toward the center of the ependymal rosettes are useful in the differential diagnosis against adenocarcinoma.[16] In less-differentiated variants ependymal rosettes and pseudorosettes may be absent. The smears are then characterized by a monomorphous tumor cell population generally occurring as single cells with oval nuclei stripped of their cytoplasm. (See Color Plate XXXIX A–B.)

Medulloblastoma

Medulloblastomas are malignant cerebellar tumors occurring in childhood, adolescence, and early adulthood and causing symptoms similar to posterior fossa tumors. It is a very cellular solid tumor consisting of primitive small cells with hyperchromatic nuclei. Frequent metastases are observed along the cerebrospinal pathway.

Needle aspirates from medulloblastomas yield highly cellular smears consisting of very easily dispersable immature tumor cells. They occur in thin, almost monolayered large and small sheets and singly. The tumor cells, although small, have a proportionally large, oval, and slightly pleomorphic nucleus with a small amount of unevenly distributed cytoplasm (Fig. 17-8). Some of the nuclei show molding. Rosettes may occasionally be recognized. Mitotic fig-

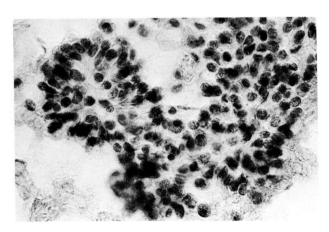

Fig. 17-7. Ependymal pseudorosette. Cellular clusters of elongated tumor cells are arranged around a central blood vessel. (P stain, ×600)

Fig. 17-9. Meningioma. Note typical onion-peel-like arrangement of cells with regular nuclei. (P stain, ×600)

ures are common. The cytologic picture of medulloblastoma is not pathognomonic, and circumstantial clinical evidence is necessary for a conclusive diagnosis. The cytologic differential diagnosis consists of other undifferentiated small cell malignant tumors.

Meningioma

Meningiomas are slow-growing, well-circumscribed, and generally benign tumors of arachnoidal cells occurring at any age, but mainly in adults. They are most frequently located in the parasagittal region and the lateral cerebral convexity. Histologically, several variants can be discerned without histogenetic or prognostic significance.[12] Psammoma bodies appear in some tumors. Meningiomas are rarely malignant. In paraffin sections malignant forms show an increased cellularity, a slight nuclear pleomorphism, and mitotic figures. Infiltration in brain tissue, the single most powerful diagnostic feature of malignancy cannot be appreciated in smears.

Meningiomas are not often aspirated, since they have a pathognomonic radiologic picture in most instances. However, some meningiomas erode the skull or occur intra-orbitally (see Fig. 17-9) and are thus a target for FNA. Biopsy usually yields richly cellular cell samples from the various tumor types. Typically, the smeared cells have a macroscopic granular appearance. Microscopically, they show lobulated, budding, thick cell clusters with rounded edges. In some clusters the cells lie concentrically and are flattened like onion peels or whorls (Fig. 17-9).

Only few cells lie free. The cells have pale-staining cytoplasm—green with P and gray-blue with MGG—with ill-defined cell boundaries. They have oval nuclei with pale-staining, evenly distributed chromatin surrounded by a delicate nuclear membrane and containing a small nucleolus. Intranuclear cytoplasmic inclusion bodies are seen. In MGG fibrillar connective tissue can be seen between the cell aggregates. The cytologic picture of meningioma is, as a rule, pathognomonic. However conclusive distinction between borderline and malignant forms cannot be done.

TUMORS OF THE PINEAL REGION

Germinoma

Germinoma is the most common tumor, accounting for at least 50% of all neoplasms in the pineal region.[12] The great majority appear in persons aged 10 to 20 years, predominantly in males. Clinical symptoms are often nonspecific on tumor location or nature and often due to visual or hypothalamic disturbances as the tumors infiltrate neighboring structures. Histologic sections show two distinct cell types—large spheroidal cells and lymphocytes—as in testicular seminoma and ovarian dysgerminoma.

Aspirates contain very abundant material, are easy to smear, and show a typical picture. Smears contain three-dimensional tumor cell clusters, single

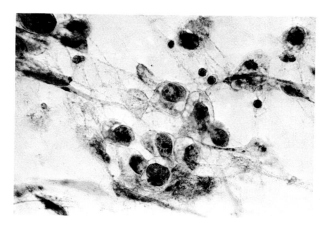

Fig. 17-10. Germinoma of the pineal gland. A group of large pleomorphic tumor cells with a few interspersed small lymphocytes is present. (P stain, ×600)

large tumor cells with pleomorphic nucleolated nuclei, and very fragile cytoplasm that often produces a "tigroid" pattern as background substance (with MGG) and lymphocytes. The cytologic picture (Fig. 17-10) is identical to that seen in seminoma.

Pineocytoma and Pineoblastoma

Pineocytomas and pineoblastomas, which are derived from pineal parenchymal cells, are rare. Pineocytoma is slowly growing and in aspirates resembles neuroblastoma very closely. Cytologically, the tumors cannot be differentiated confidently. Pineoblastoma is a more aggressive, highly malignant tumor that in smears is indistinguishable from medulloblastoma.

Other Pineal Gland Tumors

Gliomas can arise as primary tumors in the pineal gland or infiltrate it from neighboring structures, such as the corpora quadrigemina, or from the wall of the third ventricle (see the section on astrocytic tumors).

Teratoma, choriocarcinoma, and embryonal carcinoma show a cytologic picture fully comparable to similar tumors in the ovaries and testes (see Chaps. 10 and 12).

ACOUSTIC NERVE TUMORS

Acoustic nerve tumors generally occur in middle-aged patients as a pontocerebellar lesion. Patients exhibit cerebellar dysfunction, deafness, and other cranial nerve disturbances. These lesions can be reached by stereotaxic aspiration biopsy and show a cytologic picture similar to peripheral neurilemoma (see Chap. 16).

TUMORS OF THE BASE OF THE SKULL

Chordoma

Speno-occipital chordoma may cause symptoms referrable to any of the cranial nerves, but those resulting from involvement of the nerve to the eye are by far the most common. Cranial chordoma may also

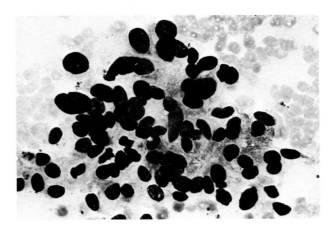

Fig. 17-11. Paraganglioma. This is a cell cluster with concentrically arranged tumor cells, some of which have small oval nuclei while others have distinctly larger nuclei. (MGG stain, ×600)

cause symptoms related to infiltration of the pituitary gland or symptoms suggesting cerebellopontine angle tumors. A roentgenologic lesion is almost always present, but ventriculography and cerebral angiography may help locate the lesion so that it can be aspirated along the most suitable route.

For the cytologic picture, see Chapter 16.

Paraganglioma

Paragangliomas, when located in the base of the skull, usually originate in the vagal or jugular paraganglion, mostly in women. Their histologic structure is almost identical to carotid body paragangliomas, with reticulin strands isolating cell nests. They are well vascularized, and nuclear pleomorphism is frequent.

Paragangliomas tend to bleed when operated on and should be aspirated with caution because of the risk of hemorrhage and thrombosis. Aspirates are usually bloody and contain clusters of tumor cells that, in 50% of patients, show a follicular arrangement.[7] The tumor cells have oval or round but sometimes spindle shaped nuclei. Frank anisokaryosis (Fig. 17-11) and giant nuclei are present. They have ill-defined cell borders and abundant cytoplasm that often contains fine reddish granules best seen with MGG. The differential diagnosis includes metastatic thyroid gland carcinoma and other well-differentiated adenocarcinomas, which usually show more

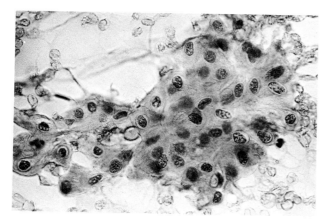

Fig. 17-12. Craniopharyngioma. A large sheet of squamous epithelium with regular nuclei is evident. (P stain, ×600)

nuclear monomorphism and lack scattered giant nuclei. When spindle-shaped nuclei occur, a neurofibromatous tumor should be excluded (see Chap. 3).

Craniopharyngioma and Epidermoid Cysts

Craniopharyngiomas are mostly suprasellar (sometimes partly intrasellar) cystic or partly cystic and partly solid lesions. The tumor is slow growing and, in 50% of patients, appears before adulthood. Clinical features depend on size and site and result from increased intracranial or local pressure on the pituitary gland, hypothalamus, or optic chiasm. Cystic tumors are lined by nonkeratinizing stratified squamous epithelium and contain an oily fluid in which cholesterol crystals are suspended. The solid varieties consist of trabeculae of squamous cells with foci of calcification.

Craniopharyngiomas most often yield oily cystic contents. Smears show abundant amorphous granular material mixed with easily recognizable cholesterol crystals. A few macrophages are usually present, and isolated squamous cells may be seen.[17] If material from the cyst wall or a more solid portion of the tumor is obtained, clusters of squamous epithelium and immature polygonal squamous cells are evident (Fig. 17-12). The cells show evidence of keratinization; some may show nuclear pyknosis, but no atypias are seen. (See Color Plate XXXIX C and D.)

Epidermoid cysts generally occur in middle age and are situated in the midline or at the cerebellopontine angle or the parapituitary region. The cyst is usually filled with soft waxy or flaky material and is lined by simple, stratified, squamous epithelium showing a variable degree of keratinization. Aspirates consist of large amounts of easily smeared gritty material that, in smears, is made up of anucleated squames and mature squamous cells without atypia (Fig. 17-13). However, there are instances in which confident differentiation between craniopharyngioma and epidermoid cyst cannot be made cytologically. Those conditions can best be demonstrated with P stain.

Chromophobe Adenoma

Chromophobe adenomas become apparent mainly in young and middle-aged adults as a slowly growing tumor. Symptoms are due to endocrine disturbances and the effects of the suprasellar extension of the growth, which impinges on optical pathways, hypothalamus, and mid-brain. Around 10% of lesions show cystic change. Histologically, the tumor is built of small cells with scanty, poorly stainable cytoplasm and compact hyperchromatic nuclei. In other cases or focally in the same neoplasm, the cells are larger with more abundant cytoplasm and larger nuclei and a less compact chromatin texture.

Smears from noncystic portions show abundant cell material evenly smeared on the glass surface and consisting of sheets of cells, very small clusters, and single cells, sometimes seen as stripped nuclei. Cytologically, both small and larger cells can be dis-

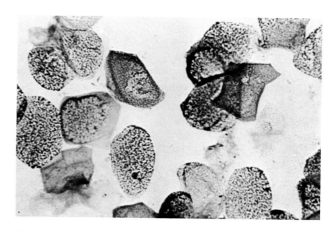

Fig. 17-13. Epidermoid cyst. Anucleated squames and one mature squamous cell with a small pyknotic nucleus are present. (MGG stain, ×600)

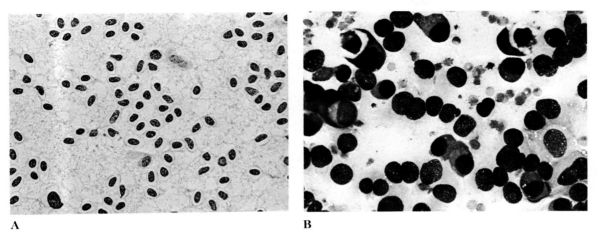

A   B

Fig. 17-14. Chromophobe adenoma. **A.** Small cells with nonstaining cytoplasm but distinct cell boundaries. A few cells have a large amount of pale-staining cytoplasm and larger pale nuclei. Note one hyperchromatic bizarre nucleus. (P stain, ×600) **B.** Predominantly small cells with dark nuclei, partly as stripped nuclei. Two vacuolated cells, each with an engulfing cell, are evident *(top)*. Note pseudonucleolus *(bottom right)*. (MGG stain, ×600)

cerned. With MGG, the smaller free-lying cells have dark gray–staining cytoplasm with sharply delineated cell borders (Fig. 17-14B). This cell population displays a variety of cell shapes and is best seen with MGG stain. They are round, oval, elliptic, dome shaped, or tadpolelike with short or longer cytoplasmic extensions or rhomboid to triangular with rounded edges. Some cells are spindle shaped and located, like a crescent, around an adjacent circular cell. The latter often show several small peripheral vacuoles. Sometimes the vacuoles are confluent, forming an optically empty ring, and this cell-in-cell arrangement gives the structure the appearance of a bird's eye.[14] With P, these cells have nonstaining cytoplasm with distinctly staining cell borders that stand out as sharp lines (Fig. 17-14A). The small cells have oval or round hyperchromatic nuclei and somewhat coarsely granular nucleoplasm that, with P, show the nucleolus. Pseudonucleoli (intranuclear cytoplasmic inclusions) can be seen, although with some difficulty.

The larger cell type has relatively more cytoplasm, which stains pale gray-blue with MGG, and, in air-dried smears, cell boundaries are rather hazy. However, with P stain the cytoplasm appears well defined and shows a slight granularity. The nuclei are large, oval, finely granular, and normochromatic.

They show a small nucleolus, visible with both MGG and P stain, and often contain clearly visible pseudonucleoli. Sometimes an occasional cell with a distinctly large nucleus is scattered in the smear. The background substance is finely granular. Occasionally, calcifications can be found. The general cell picture is quite characteristic. Differential diagnosis with the uncommon pituitary gland carcinoma is on purely cytologic grounds impossible. In some instances where the cell picture is dominated by naked nuclei, pituitary gland carcinoma and oligodendroglioma should be excluded. (See Color Plate XXXIX E and F.)

CLINICAL APPLICATION

In the opinion of neurosurgeons at Karolinska Hospital, who have practiced stereotaxic needle biopsy since the mid-1960s,[1] the reliability of cytologic diagnosis of brain tumors in smear preparations is comparable with that of histologic examination. Reported diagnostic accuracy from other authors varies from 80% to 94%.[2,11] Willems reports a cytodiagnostic accuracy of 87% in a consecutive series of 95 histologically investigated malignant intracranial neoplasms of the central nervous system.[15] In many of the space-occupying lesions investigated, FNA biopsy was chosen instead of histologic biopsy; and in many tumor patients treatment was based on the cytologic report alone, as, for example, in pinealoma.

Complications of stereotaxic needle biopsy in general are reported to be very low. Edner observed deterioration in 11 of 345 patients; 3 died, constituting a mortality of under 1%.

The ever-increasing precision of the instruments used in FNA can be expected to widen its applicability and implies that cytologic specimens will be sent to cytopathologists for analysis more frequently.

REFERENCES

1. Backlund EO: A new instrument for stereotaxic brain tumour biopsy. Acta Chir Scand 137:825, 1971
2. Barnard RO: Smear preparations in the diagnosis of malignant lesions of the central nervous system. In Koss LG, Coleman DV (eds): Advances in Clinical Cytology. Scarborough, Ont, Butterworth & Co, 1981
3. Bergström B, Drettner B, Stenkvist B: Transnasal aspiration biopsy of central skull base destructions. Acta Cytol 17:425, 1973
4. Boëthius J, Collins VP, Edner G et al: Correlation between histology and computer tomography in gliomas assessed by multiple needle biopsies. Acta Neurochir 37:304, 1977
5. Collins VP: Monoclonal antibodies to glial fibrillary acidic protein in the cytologic diagnosis of brain tumors. Acta Cytol (Baltimore) 28:401–406, 1984
6. Edner G: Stereotactic biopsy of intracranial space occupying lesions. Acta Neurochir 57:213, 1981
7. Engzell U, Franzén S, Zajicek J: Aspiration biopsy of tumours of the neck. II. Cytological findings in 13 cases of carotid body tumour. Acta Cytol 15:25, 1971
8. Franzén S: Needle biopsy in skull base tumours. In Disorders of the Skull Base Region, pp 285–287. Stockholm, Proceedings of the Tenth Nobel Symposium, 1968
9. Kernohan J, Mabou R, Svien H et al: A simplified classification of the gliomas. Proc Mayo Clin 24:71, 1949
10. Lewander R, Bergström M, Boëthius J et al: Stereotactic computer tomography for biopsy of gliomas. Acta Radiol 19:867, 1978
11. Marshall LF, Adams H, Doyle D et al: The histological accuracy of the smear technique for neurosurgical biopsies. J. Neurosurg 39:82, 1973
12. Rubinstein LJ: Tumours of the central nervous system. In Atlas of Tumor Pathology, sec. series, fasc. 6. Washington, D.C., Armed Forces Institute of Pathology, 1972
13. Russell DS, Rubinstein LJ: Pathology of Tumours of the Nervous System, 4th ed. London, Edward Arnold & Co, 1977
14. Watson CW, Hajdu SI: Cytology of primary neoplasms of the central nervous system. Acta Cytol 21:40, 1977
15. Willems JGMS, Alva–Willems JM: Accuracy of cytologic diagnosis of central nervous system neoplasms in stereotactic biopsies. Acta Cytol (Baltimore) 28:243–249, 1984
16. Woyke S, Czerniak B: Fine needle aspiration cytology of metastatic myxopapillary ependymoma. Acta Cytol 22:312, 1978
17. Zaharopoulos P, Wong JY: Cytology of common primary midline brain tumours. Acta Cytol 24:384, 1980
18. Zülch KJ: International histological classification of tumours, no. 21. Histological Typing of Tumours of the Central Nervous System. Geneva, World Health Organization, 1979

Advances in Cytology

MIGUEL A. SANCHEZ

GERT AUER

ANDERS ZETTERBERG

During the many years of clinical usage of aspiration biopsy cytology (ABC), the main goal has been to reach an accurate diagnosis using relatively simple stains. Recently the scope of usage of the aspirate has been expanded thanks to the application of new techniques to improve the diagnostic specificity (immunocytochemistry, electron microscopy) and to add a new dimension with measurement of quantitative parameters of importance in helping in the prognosis or the understanding of the biological behavior of tumor. (*e.g.,* DNA [deoxyribonucleic acid] analysis, estrogen receptors, morphometry). In the next few pages we will focus on some aspects of this relatively new usage of fine needle aspiration (FNA), mainly analysis of nuclear DNA contents, transmission electron microscopy, and immunocytochemistry. Cells obtained by the means of aspiration are particularly suitable for the performance of these special studies, and a great deal of experience has accumulated at the Karolinska Hospital and many other institutions.

NUCLEAR DNA ANALYSIS

The measure of nuclear DNA content in individual cells has been demonstrated to contribute diagnostic and prognostic information complementary to that

Acknowledgments: Dr. Sangho Cho, Assistant Professor of Pathology, Albert Einstein School of Medicine, provided the electron microscopy studies. Mrs. Deborah Whittaker is responsible for secretarial assistance.

obtained by clinical and morphological parameters. Two different methods for DNA analysis are employed: (1) DNA measurement in cells or tissues mounted on a slide (static photometry) and (2) DNA determination in isolated cells or cell nuclei in suspension (flow cytometry). The most important advantage of static photometry is that DNA analysis is carried out on morphologically identified cells—at present the only way to determine the cell type specific DNA content. DNA measurement employing flow cytometry makes it possible to examine large numbers of cells in a relatively short time. In pure tumor cell population, as in cultured cell lines, the DNA values can be measured with high precision, which enables small differences to be quantified and even kinetic cell data to be obtained (*e.g.,* the fraction of the cell population being in S-phase). In contrast, clinical specimens usually present heterogeneous cell populations that lead to interpretation problems when using flow cytometry, thus limiting the value of DNA determination using this method.

Based on the supposition that benign tissues are euploid and tumors of high-grade malignancy usually aneuploid, DNA measurements have also served as a complement to cytologic and histologic diagnosis in cases where difficulty has been encountered in deciding if the cells were benign or malignant.[8] Thus, in squamous bronchial carcinoma, esophageal carcinoma, gastric carcinoma, pancreatic carcinoma, and osteosarcoma, measurement of tumor cell DNA content can be of decisive diagnostic help. Since the genetic disorder reflected by DNA-aneuploidy precedes final malignant transformation,[34] nuclear DNA measurement can be used as a highly sensitive diagnostic method discriminating nonspecific cellular alterations from true premalignant cells.[45]

The main reason why measurements of tumor cell DNA content have spawned so many reports on various neoplastic diseases is the lack of good morphological parameters to assess the grade of malignancy. Experience from both Sweden and elsewhere confirms that tumor cell DNA content is one of the best prognostic indices of the malignant potential in a wide range of cancers. It has been shown that DNA measurements give additional prognostic information compared to conventional clinical and histopathological examinations alone.[7,21] In combination with FNA, DNA analysis makes it possible to obtain this additional prognostic information pre-operatively, which is of utmost importance for planning therapy.

Preparation of Specimens

In the tumors analyzed in our studies, cytologic material was obtained by means of FNA biopsy technique.[27] For sampling of breast and thyroid tumors, a thin needle (0.6 mm to 0.7 mm in outer diameter) was inserted into the mass and cells were aspirated into the lumen of the needle by a syringe as described below. Material from prostatic carcinomas was collected by using the Franzen technique.[26] The biopsy instrument consisted of a 20-cm-long 22-gauge needle, a needle guide, and a syringe of Luer-Lok type fitted with a special handle that permitted a single-hand grip during aspiration. The needle was advanced transrectally into the prostatic lesion, and the material was aspirated by quickly retracting the plunger of the syringe. After the needle had been moved back and forth three or four times within the target, the negative pressure in the syringe was equalized by allowing the plunger to return to its neutral position. The needle was then withdrawn from the prostate and returned into the guide. The material contained in the needle was expressed onto a glass slide. The drop of the aspirated material was divided in two halves by means of a coverslip. One half was rapidly smeared on a slide and fixed in 10% neutral formalin for cytophotometric analysis. The remaining material, again divided into two halves, was then processed for conventional morphologic analysis. One smear was air-dried and stained with May–

Grünwald–Giemsa (MGG); the other was fixed with methanol and stained according to Papanicolaou (P). Papanicolaou-stained cells show good nuclear details that permit comparison with tissue sections, while MGG gives better cytoplasmic details. The two methods thus complement each other.

FNA yields tumor cell material that is representative for the entire tumor, probably due to the fact that cells are sampled from multiple spots throughout the tumor.[6] In the absorption cytometry method using aspirates, it is usually easy to discriminate between inflammatory mesenchymal and epithelial cells. However, there might be difficulties in separating admixed benign epithelial cells from epithelial tumor cells, especially in histologically highly differentiated tumors. Slide DNA analysis in aspirates may thus result in normal cells in the benign diploid region erroneously judged as tumor cells. In the flow technique, distinctions between tumor cells and nontumor cells are at present impossible to make. In flow DNA histograms, overrepresentation of nontumor cells might result in a profile totally dominated by diploid, nontumor cells, where the relatively few nondiploid tumor cells disappear in the background "noise."

DNA Staining

During the last decades a number of nucleic acid or DNA specific staining procedures have been described. However, the DNA staining method described by Feulgen in 1924[22] still serves as a standard reaction to which the other methods can be compared. The Feulgen staining technique includes acid hydrolysis of DNA, which removes purine bases and thereby unmasks the aldehyde groups of deoxyribopentose molecules. Decolorized or leuco-Schiff reagent is then allowed to react with the aldehyde groups. The leuco-Schiff reagent is thereby converted into its colored form and covalently bound to DNA. RNA (ribonucleic acid) and proteins do not react with the Schiff's reagent, which means that the Feulgen staining procedure does not demand removal of these cell components. One part of the Feulgen staining procedure is the acid hydrolysis step, which is of varying efficiency in different cell types and in cells with different chromatin compactness. The task is to efficiently remove the purine bases without overdenaturing and destroying DNA.

The conventional Feulgen hydrolysis is performed in 1 M HCl at 60° C for 8 minutes. We use hydrolysis in 5M HCl at 22° C (room temperature) for 60 minutes.[28] The Feulgen staining procedure used in all studies reported herein is described in detail below.

A. Preparation of the Schiff reagent
  1. Dissolve 5 g pararosaniline in 150 ml 1 M HCl, room temperature.
  2. Add a solution containing 5 g $K_2S_2O_5$ in 850 ml distilled water (keep dark over night).
  3. Add 3 g activated carbon, shake for 2 minutes, and filter through filter paper.
B. Preparation of natriumbisulfite
  180 ml distilled water
  10 ml 1 M HCl
  10 ml 10% $Na_2S_2O_5$
C. Procedure
  1. The tissue sections/cytological slides are fixed in 10% neutral formalin overnight.
  2. Rinse carefully in water.
  3. Acid hydrolysis obtained by incubation in 5 M HCl at 22° C (room temperature) for 1 hour.
  4. Rinse carefully in water.
  5. Stain in Schiff reagent for 2 hours at room temperature (keep dark).
  6. Rinse carefully in water.
  7. Transfer the slides into three changes of sulfurous acid and bleach for 30 minutes.
  8. Rinse in tap water followed by distilled water; dehydrate through a graded series of alcohols to xylene and mount.

DNA Analysis

Normal cells usually exhibit diploid DNA values when in G0/G1 phases and DNA values up to twice the diploid amount when passing through the S and G2 phases of the cell growth cycle. Malignant tumor cells may contain DNA amounts comparable with those found in normal cells or may clearly deviate from normal, with increased and highly variable DNA amounts (Fig. 18-1).

Two main cytochemical procedures for quantitative analysis of nuclear DNA in individual cells can be distinguished. The first is image cytometry, which

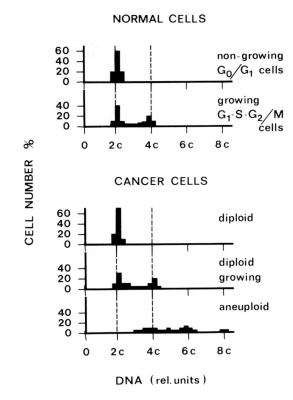

Fig. 18-1. Normal and tumor cells may share a similar quantity and distribution of DNA. Aneuploid cells, otherwise, represent the widest deviation from the norm and usually are associated with the worst biological behavior.

can be subdivided into static fluorometry and absorption cytometry, and the second is flow cytometry (Fig. 18-2).

Absorption cytometry allows the performance of DNA measurement in cytologically identified cells without the problem of fading as is the case with image fluorometry and with an in general much higher tumor cell specificity than with the use of flow cytometry (Fig. 18-3). In this chapter we will only describe the absorption cytometry procedure. For detailed information concerning static fluorometry and flow cytometry techniques, we refer you to other literature.[9,25,29,35,37,39,44,49]

The quantitative microscopical methods of cytochemistry were introduced by Caspersson in the early 1930s[11] and have been continually developed since then. Special emphasis has been put on refining instruments, allowing cytochemical analysis of mor-

CYTOPHOTOMETRY

SLIDE

FLUORESCENCE   ABSORPTION

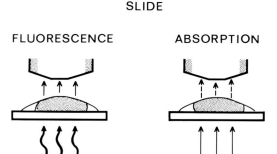

FLOW

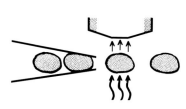

Fig. 18-2. Fluorescence, absorption, and flow cytometry represent the three most commonly used techniques of DNA measurement.

phologically identified cells, and being rapid enough to perform large scale population work. Image cytometric determination of the amount of absorbing substances within cells and cell organelles (*e.g.*, nuclear DNA) involves measuring absorbance in a large number of small measuring spots distributed over the object. This can be achieved by means of the so-called object scanning method[12] or two-dimensional (CCD) sensor-based measuring devices.[10] In the work reported in this chapter cytometric measurements of Feulgen-stained cells were performed in a rapid scanning and integrating microspectrophotometer developed by Caspersson.[12] The absorption was measured at 546 nm and used as a measure of the total amount of DNA in the cell nuclei. Generally, 100 morphologically identified tumor cells per case were measured randomly. Admixed normal cells (epithelial cells or granulocytes) were used as an internal diploid DNA standard to determine the modal value of the tumor cells.

Prognostic Value of DNA Measurements

The assessment of prognosis in patients with malignant tumors is, in general, based on clinical and morphological criteria. Morphologically, malignant tumors are usually graded or classified into well, moderately, and poorly differentiated forms. Although well-differentiated tumors on the whole are generally less malignant than poorly differentiated tumors, at least for certain tumor types, morphological features are unfortunately not reliable indicators of prognosis in most cases. This is especially evident in cytologically diagnosed tumors. This is not surprising since morphological features cannot easily be determined in an objective and quantitative fashion. Furthermore, many tumors are morphologically heterogeneous (*i.e.*, different morphological features are seen in different regions of the same tumors), which makes a morphological analysis even more difficult. Maybe even more important in this context is the possibility that many of the morphological features expressed by the tumor cells are not necessarily related to the biological properties that determine the malignant behavior of the tumor.

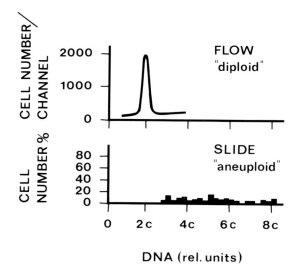

Fig. 18-3. Easily identifiable tumor cells are seen among abundant numbers of non-tumor cells resulting in a "diploid" DNA profile using flow cytometry. The few tumor cells are lost in the background noise. By measuring of absorption cytometry it is possible to selectively measure the nuclear DNA content of true tumor cells resulting in an "aneuploid" DNA profile.

Quantitative photometric measurements in malignancies including prostate, thyroid, breast, ovary, colon-rectum, endometrium, and cartilage[4,5,7,19,24,36,42,54,55] have proved that there is a marked correlation between tumor nuclear DNA content and clinical course. Since the data obtained in these malignancies are quite similar, we will use the findings in one of these tumor types, namely breast adenocarcinoma, as an example for the prognostic values of DNA measurements in fine needle aspirates.

The results of retrospective studies demonstrate a strong relationship between nuclear DNA content of breast cancer cells and prognosis. Low-grade malignant tumors were characterized by euploid DNA content, DNA content predominantly in the diploid or tetraploid region. In contrast, high-grade malignant breast carcinomas were characterized by aneuploid DNA content, frequently increased and highly variable DNA values deviating from normal (Fig. 18-4). Since precise data is lacking as to which aspects

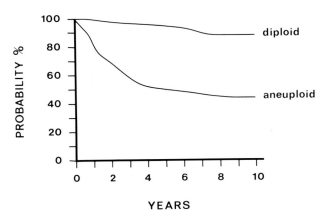

Fig. 18-5. Life table probability of distant recurrence-free survival in cancer patients according to DNA distribution.

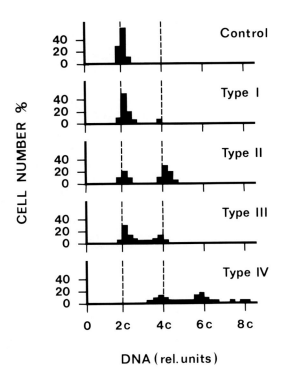

Fig. 18-4. DNA profiles from non-proliferating normal mammary epithelial cells (control) and from four groups, characterized by their DNA patterns (Types I–IV).

of the DNA histogram best reflect the inherent malignancy potential of the tumor, various criteria were used to characterize the DNA histogram patterns. By using the log-rank method, the significance of the relationship between a particular DNA variable and distant recurrence-free survival was tested. Our results showed that regardless of the percentage of cells above 2.5c or 5c (2c = diploid, or the normal DNA content), DNA index (DI), modal value (MV), or the histogram typing (HT)[5] were utilized to discriminate low-grade from high-grade malignant cases. A significant correlation between nuclear DNA content and prognosis were found (Fig. 18-5). It is of great interest that even a simple DNA measure (*i.e.*, the percentage of cells above 2.5c) shows a significant correlation to prognosis.

The DNA variables tested in our studies were also shown to be significantly correlated to each other. With the aid of the Cox regression method,[16] the additional prognostic value of any given variable was tested against the others. The results from the Cox analysis show that DNA gives significant prognostic information in addition to that obtained by any other variable. Thus, it is clear from our studies that DI, frequently used in flow cytometry, or MV do not optimally reflect the malignancy potential of breast carcinomas. In comparison with DI/MV, HT was shown to be a considerably more powerful tool in discriminating low-grade malignant from high-grade malignant cases.

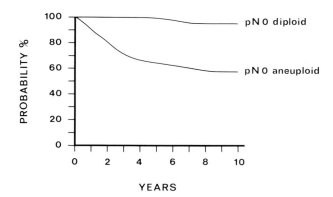

Fig. 18-6. Distant recurrence-free survival probabilities in axillary negative nodes with different DNA distribution.

The predictive power of DNA content in relation to well-known established predictors was also statistically evaluated. The main result is that, by the use of Cox multivariate analysis, nuclear DNA content provides significant prognostic information in addition to that yielded by all other clinical and histomorphologic variables. This fact indicates that the DNA content of breast cancer cells reflects biological properties associated with the malignant behavior of the tumor other than those determining the stage of the disease.

Previous studies have shown conflicting results concerning the relationship between axillary node status and the ploidy level of breast tumors.[20,43] The relation between node status and ploidy was of par-

ticular interest to study, since the prime importance of axillary node status in predicting recurrence and survival is generally recognized.[23,48] Our results, showing no clear correlation between nuclear DNA content and axillary node status, are therefore noteworthy and indicate that these two factors are independent prognostic variables. It is also clear from the present data that determination of nuclear DNA content provides additional prognostic information within both the node negative (Fig. 18-6) and node positive (Fig. 18-7) patient groups. Furthermore, it seems that the prognostic information contributed by DNA is equal for both node categories. By the use of DNA analysis, it is thus possible to separate both node positive and node negative patients into different prognostic subgroups.

Patients with diploid, node negative tumors were found to have an excellent prognosis of 95% probability of 10-year survival. In contrast, patients with aneuploid, node positive tumors were shown to have an extremely bad prognosis with only a 31% probability of 10-year survival (Fig. 18-8). These two groups comprised 41% of the entire patient material, thus indicating that DNA analysis, in combination with lymph node assessment can provide a highly accurate prediction of prognosis for large patient categories. While staging is a static rather than a dynamic concept of malignancy grading, determination of nuclear DNA content gives information about the intrinsic malignancy of the mammary carcinomas and is therefore a powerful tool for a more accurate prognostic evaluation of the expected tumor aggres-

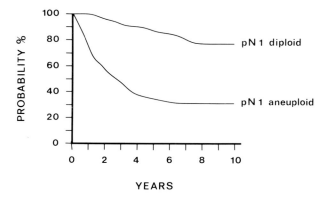

Fig. 18-7. Distant recurrent-free survival probability in patients with positive axillary nodes and different DNA distribution.

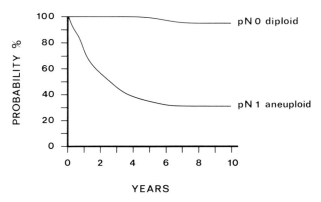

Fig. 18-8. Distant recurrent-free survival probabilities by axillary node status and DNA histograms for patients with favorable and poor prognosis.

siveness in each individual case. It must, however, be emphasized that ploidy determination is not supposed to replace the traditional staging of breast carcinomas. Even though the results reported here clearly demonstrate that nuclear DNA analysis can provide prognostic information additional to that of all other clinical predictors, our investigation shows that established prognostic factors, such as axillary node status and tumor size, also provide prognostic information supplementary to DNA.

In summary, our results indicate that nuclear DNA content, as an objective biological marker of tumor aggressiveness, can significantly improve our prognostic capabilities within the currently designated stages. The number of patients with diploid, slowly growing tumors being overtreated today would be reduced. Furthermore, nuclear DNA analysis of breast cancer cells can become of utmost importance in clinical practice for defining subsets of patients with a high probability of short-term relapse, thus making them candidates for cytotoxic or some other form of adjuvant therapy. In combination with FNA biopsy, this method also allows prognostication prior to making therapeutic decisions both an important step toward individualized therapy and a basis for selective and appropriate stratification of clinical trials.

Biologic Aspects of Aneuploidy

The reason why aneuploid tumors are more malignant than diploid tumors remains to be clarified. The clearly aneuploid and variable cytophotometric DNA pattern in the aneuploid tumors may reflect a high degree of genomic instability in such tumor cell populations, an instability that in itself may lead to a rapid generation of new phenotypes that may be a prerequisite for a rapid progression of the tumor disease. A genetic instability of this kind could be the result of an increased instability at the karyotypic level. The tumor cells could for example have acquired the ability to gain and lose chromosomes more readily. This would favor the development of the specific karyotypes (or the specific gene dosage imbalances) required for overcoming the various stages in the progression of the tumors. Alternative hypothesis including increased mutability, high frequency of recombination events, gene amplification, or other DNA rearrangements have to be considered.

Although the genetic background for diploid and aneuploid tumors is not understood at present, cytophotometric determinations of nuclear DNA content undoubtedly provide us with a powerful diagnostic method for discriminating between low-grade and high-grade malignant tumors.

ELECTRON MICROSCOPY AND IMMUNOCYTOCHEMISTRY

Electron Microscopy

The use of electron microscopy (EM) in morphologic diagnosis has been studied and reported in depth for many years.[18,53] The ability to utilize material obtained by FNA has also found its way into some excellent publications.[1,2]

From the practical point of view, the rapid development of immunomorphology in the last few years has decreased utilization of electron microscopy in diagnostic histopathology and cytopathology. Nevertheless a logical approach embracing both cytochemical and ultrastructural studies will show how complementary these technologies are, providing the most efficient way of arriving at a diagnosis. The advantages of aspiration over exfoliative cytology samples include frequently good preservation of intercellular connections, major value in ultrasound diagnosis, and the accessibility of tumors for easily securing repeated samples for complementary studies.

Preparation of the Specimens

Adequate preservation of the material to be utilized for EM cannot be overemphasized. Not infrequently major efforts to reach diagnosis will be jeopardized by inadequate fixation or prolonged storage of tissue samples. Several methods have been described, and the enthusiasm of the authors is probably a reflection of the familiarity with the different approaches. In a cooperative book that combined experiences of both sides of the Atlantic, a brief description of the methods for processing FNA material is appropriate. Table 18-1 is a modification of Collins and Eversson[15] as used at the Karolinska Hospital. A modified Karnovsky fixation is favored by some authors in the United States.[40] Recently, Akhtar and colleagues have described an ingenious device using a nylon

Table 18-1. EM Processing

Prim fix 2% glutaraldehyde	30 min
Centrifugation + washing	10 min
Agar centrifugation	20 min
On ice	10 min
OsO$_4$	90 min
70% alcohol	15 min
95% alcohol	15 min
Abs. alc. ×4	60 min
Prop. oxide ×2	30 min
50/50 prop. oxide + Epon	60 min
Epon	30 min
Polymerization	16 h
Total	Approximately 22 h

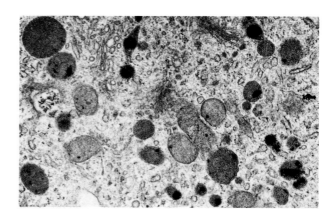

Fig. 18-10. High-power shows solitary melanosomes within vesicles. (×52,500)

sieve to free the aspirate from the not uncommon abundant red cells mixed with the diagnostic elements.[3] Of major importance, as previously mentioned, is the ability to promptly decide if electron microscopy or immunocytochemistry is going to be necessary for diagnostic purposes. The ideal, as has been repeatedly shown, is that the physician performing the aspirate is the same person interpreting it. If, as is common in the United States, performance and interpretation are carried out by different individuals, the presence of the cytopathologist or the immediate evaluation of an aliquot of the aspirate is a must. A simple procedure that we found extremely useful consists of preparing and immediately staining a slide to evaluate the adequacy of the specimen. At

this time the rest of the specimen is suspended in Plasmalyte. Using a rapid stain and with some experience and common sense, in almost 100% of the cases, it can be predicted if special studies are going to be needed. Repeated punctures can then be performed and a sample fixed in electron microscope fixative. For immunoperoxidase the material obtained is centrifuged and the pellet mixed with a small amount of thrombin and plasma.[17] An easily manageable "clot" can then be fixed and submitted for routine histological sectioning. An alternative[30] is to encase the pellet in melted agar after at least 1 hour of formalin fixation and embedding in paraffin.

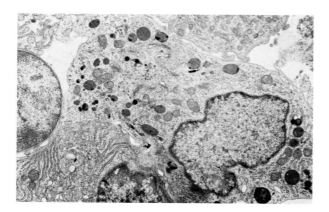

Fig. 18-9. Low-power view of cells containing many cytoplasmic organelles including mitochondria, rough and smooth endoplasmic reticulum, and free polysomes. Many solitary melanosomes are present. (×18,000)

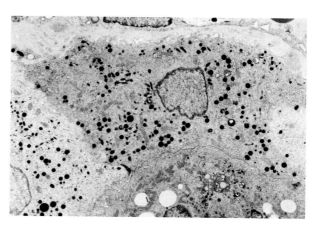

Fig. 18-11. Thyroid cells containing many exocrine granules. (×12,000)

Transmission Electron Microscopy

The application of electron microscopy in the diagnosis of tumors is identical to that used in surgical pathology. The morphological bases are as follows:

1. Recognition of granules (*i.e.,* carcinoids, melanomas), filaments, tubules, organelles, and other cytoplasmic structures
2. Configuration of the nucleus and nucleolus
3. Membranous structures and relations with adjacent cells (junctions, microvilli, desmosomes, pinocytolic vescicles)

The clinicopathologic problems resolved or helped by electron microscopy in FNA are also identical and as limited as in surgical pathology and include anaplastic round cell tumors, spindle cell tumors, and small cell tumors. A few case examples follow.

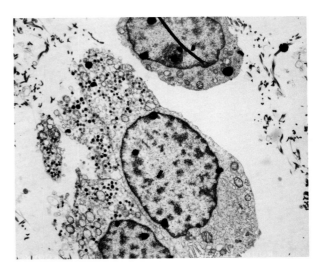

Fig. 18-13. Ovoid nuclei with smooth borders and coarse chromatin pattern. Many cytoplasmic granules. (×4400)

■ *Case 18-1*

A 57-year-old man presented with a recent history of a large right axillary mass, clinically suspected to be a lymphoma. FNA immediately evaluated showed a poorly differentiated tumor. At that time a second aspirate was fixed and submitted for transmission electron microscopy (Figs. 18-9 and 18-10). Numerous cells contained solitary melanosomes. Many cytoplasmic organelles, including mitochrondria, rough and smooth endoplasmic reticulum, and free polysomes, were also present. A pigmented lesion re-moved a few years earlier from the back proved to be malignant melanoma. The diagnosis was metastatic malignant melanoma.

■ *Case 18-2*

A 33-year-old woman was found to have a 3-cm nodule in the right lower lobe of the thyroid. FNA suspected carcinoma, probably medullary. Transmission electron microscopy of the aspirate showed many secretory granules (Figs. 18-11 and 18-12). A thyroidectomy confirmed the diagnosis of follicular carcinoma of the thyroid. Tissue stains for calcitonin were negative. The diagnosis was follicular thyroid carcinoma.

■ *Case 18-3**

A 53-year-old man presents with a retrorectal mass. FNA showed round malignant cells. Electron microscopy showed cells devoid of junctional complexes and microvilli. Numerous dense core osmophilic granules measuring 575 mm to 640 mm were present in the cytoplasm (Figs. 18-13 and 18-14). The diagnosis was neoplastic cells with endocrine differentiation.

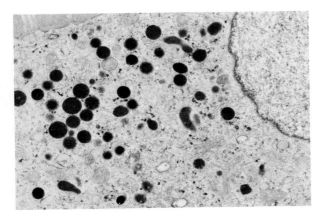

Fig. 18-12. Large electron dense granules probably representing colloid. (×30,000)

* This case was provided by Drs. C. Sian and A. Avitabile from the St. Lukes–Roosevelt Hospital Center of New York.

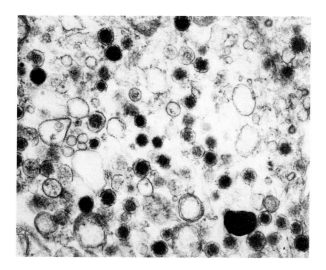

Fig. 18-14. Numerous dense spherical large endosecretory granules. (×30,000)

Compare the exocrine and the endocrine granules from cases 18-2 and 18-3.

■ *Case 18-4*

A 65-year-old man presented with a submental mass. FNA showed a poorly differentiated tumor without further classification possible. Transmission electron microscopy revealed cells that were oval in shape, having large heterochromatic

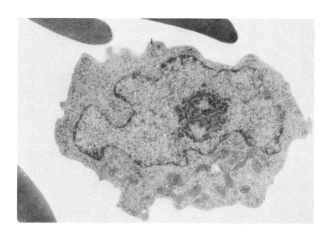

Fig. 18-16. Cells showing markedly indented nuclei. (×24,000)

nuclei and scant amount of cytoplasm. Many nuclei showed marked indentations and had prominent nucleoli. The cytoplasm contained a few mitochondria and many polysomes. There were no base membranes, cell junctions, or microvilli (Figs. 18-15 and 18-16). The diagnosis of malignant lymphoma was strongly suggested. An excision biopsy was studied for lymphoid markers revealing a pure population of B-cells. The diagnosis was poorly differentiated malignant lymphoma.

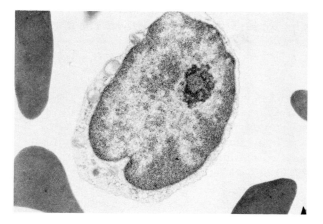

Fig. 18-15. Round cells showing few organelles, lack of epithelial junctions and small indentation of the nucleus. (×24,000)

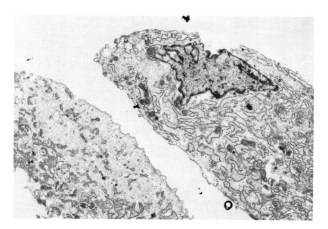

Fig. 18-17. Spindle cells showing branching rough endoplasmic reticulum characteristic of fibroblast. (×15,000)

■ *Case 18-5*

An 80-year-old woman presented with a 5 cm mass in the back fixed to the skin and subcutaneous tissue. FNA showed a scanty specimen with spindle cells. Transmission electron microscopy revealed two kinds of cells (Figs. 18-17 and 18-18); fibroblasts with elongated spindle-shaped cells with indented irregular nuclei; no nucleoli; abundant rough endoplasmic reticulum; many polyribosomes; cytoplasmic intermediate filaments; no basal lamina, and also myofibroblasts with more elongated indented nuclei, myofilaments with focal densities. The mass was excised with the knowledge that this tumor probably represented fibromatosis or pseudosarcomatous fasciitis, so an adequate margin was obtained to prevent local recurrence. The diagnosis was subcutaneous fibromatosis.

■ *Case 18-6*

A 59-year-old man known to have multiple myeloma developed tenderness and swelling in both palmar areas. One of them was relieved through surgery, and tissue failed to reveal any deposits of amyloid. The other side was aspirated using a 22-gauge needle, and the specimen obtained was submitted for electron microscopy. The electron microscopy showed amorphous fibrils of variable lengths, nonbranching and slender (Figs. 18-19 and 18-20) consistent with amyloid. The diagnosis was amyloid deposits in a patient known to suffer from multiple myeloma.

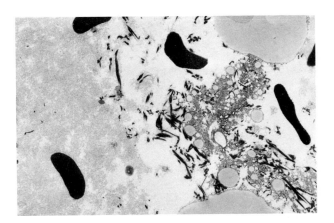

Fig. 18-19. Amorphous fibrillar material. (×7200)

Immunocytochemistry

The rapid development of commercial antibodies to different cellular components has made immunoperoxidase and related techniques routine in most sophisticated pathology laboratories. Most immunohistological methods can easily be applied to cytological specimens and material obtained by FNA. Interpretation of the results must be carried out with caution.[5,13] Neither FNA nor immunocytochemistry are panaceas that can replace common sense and good judgment.

In our experience, cell blocks prepared with aspirated specimens are preferable to direct smears.

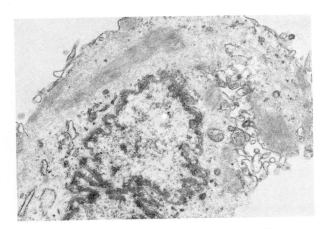

Fig. 18-18. Myofibroblast showing myofilaments. (×36,000)

Fig. 18-20. Variable lengths slender fibrils. Metachromatic stains confirmed the identity of amyloid deposits. (×48,000)

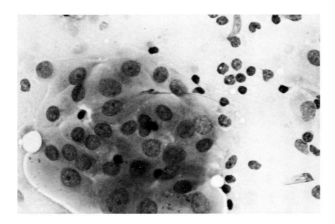

Fig. 18-21. Cluster of malignant cells aspirated from submental area.

The interpretation is easier and the artifacts less. Different laboratories have reported good results with Papanicolaou destained smears[52] to which immunoperoxidase is secondarily applied.[31] The Pap stains serve as a counter stain and that step can be avoided in the immunoperoxidase stain procedure.

Perhaps the most valuable use of immunocytochemistry in FNAs is found in the common problem of determining the primary site of an anaplastic tumor. An excellent review has been published addressing this issue for histological sections.[51] In our experience, the most common problem is separating lymphomas, anaplastic carcinomas, and sarcomas. A practical approach to efficiently utilize the laboratory would be to select the most likely positive antibody among the following: common leukocyte antigen (CLA), carcinoembryonic antigen (CEA), low and high molecular weight cytokeratins, vimetin, and S100 protein. The aforementioned antibodies will provide excellent focusing on lesions outside the central nervous system and may be easily performed sequentially. An illustrative case may serve as a simple example.

■ *Case 18-7*
A 23-year-old man presented with a 1 cm nodule in the soft tissues of the submental area. FNA revealed malignant cells (Fig. 18-21). Thyroglob-

ulin staining was positive (Fig. 18-22), clarifying the presence of an occult primary thyroid carcinoma. (See Color Plate XL.)

Hormone Receptor Assay

The measurement of estrogen and progesterone receptors are routinely performed in breast tumors and used for the management of patients with breast cancers.[14,38] Early results in other tumors also promise that the recognition of estrogen receptors may be of prognostic significance.[46,47] The standard technique for measuring estrogen receptors requires a substantial amount of tissue unobtainable by FNA. Techniques developed in Europe that permit measurement of estrogen receptors with minute amounts of cellular sampling have not found their way into the routine laboratory practice of the United States.[50] Recently, the detection of estrogen receptors using monoclonal antibodies[32,33] has replaced in many laboratories the charcoal dextran technique.

Two assays are now available on the market; one (ER-EIA Abbott), a quantitative method, uses immune methodology and another (ERICA Abbott) is an immunocytochemical procedure for the visual localization of estrogen receptors in the nucleus of cells. Both can be easily modified[41] to utilize FNA specimens and will without a doubt replace other more cumbersome methodology.

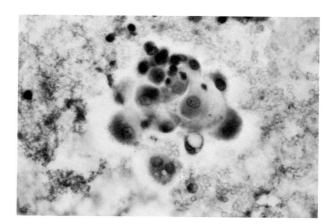

Fig. 18-22. Positive thyroglobulin staining identifies thyroid primary.

REFERENCES

1. Akhtar M, Ali MA, Owen E: Application of electron microscopy in interpretation of fine-needle aspiration biopsies. Cancer 48:2458–2463, 1981
2. Akhtar M, Ali MA, Sabbah R et al: Fine-needle aspiration biopsy in diagnosis of round cell malignant tumors of childhood: A combined light and electron microscopic approach. Cancer 55:1805–1817, 1985
3. Akhtar M, Bakry M, Nasah E: An improved technic for processing aspiration biopsy for electron microscopy. Am J Clin Pathol 85:57–60, 1986
4. Atkin NB: Modal deoxyribonucleic acid value and survival in carcinoma of the breast. Br Med J 86:271–272, 1972
5. Auer G, Caspersson T, Wallgren A: DNA content and survival in mammary carcinoma. Anal Quant Cytol 3:161–165, 1980
6. Azavedo E, Tribukait B, Konaka C et al: Reproducibility of the cellular DNA distribution patterns in multiple fine needle aspirates from human malignant tumors. Acta Pathol Microbiol Immunol Scand 90:79–83, 1982
7. Backdahl M, Carstensen J, Auer G et al: Statistical evaluation of the prognostic value of nuclear DNA content in papillary, follicular and medullary thyroid tumors. World J Surg 10:974–980, 1986
8. Bohm N, Sandritter W: DNA in human tumors: A cytophotometric study. Curr Top Pathol 60:151–219, 1975
9. Braylan RC: Flow cytometry. Arch Pathol Lab Med 107:1–6, 1983
10. Carlson L, Auer G, Kudynowski J et al: Instruments for rapid cytophotometric analysis of morphologically identified clinical cell material. Cytometry, 5:319–326, 1984
11. Caspersson T, Kudynowski J: Cytochemical instrumentation for pathological work. Int Rev Exp Pathol 21:1–54, 1980
12. Caspersson T, Lomakka G: Recent progress in quantitative cytochemistry: Instrumentation and results. In Wied GL, Bahr GF (eds): Introduction to Quantitative Cytochemistry II, pp 27–56. New York, Academic Press, 1970
13. Chess Q, Hajdu SI: The role of immunoperoxidase staining in diagnostic cytology. Acta Cytol 30(1):1–7, 1986
14. Clark GM, McGuire WL, Hubay CA et al: Progesterone receptors as a prognostic factor in stage II breast cancer. N Engl J Med 309:1343–1347, 1983
15. Collins VP, Eversson B: Tumor classification by electron microscoy of fine needle aspiration biopsy material Acta Path Microbiol Scand (Sect A) 89:103–105, 1981
16. Cox DR: Regression models and life tables. J R Statist Soc (B) 34:187–220, 1972
17. DeGiromali E: Application of plasma thrombin cell block in diagnostic cytology. Path Ann 12:91–100, 1977
18. Earlandson RA: Diagnostic transmission electron microscopy of human tumors. New York, Masson Publishing USA, 1981
19. Erhardt K, Auer G, Bjorkholm E et al: Prognostic significance of nuclear DNA content in serious ovarian tumors. Cancer Res 44:2198–2202, 1984
20. Ewers SB, Langstrom E, Baldetorp B et al: Flow-cytometric DNA analysis in primary breast carcinomas and clinicopathological correlations. Cytometry 5:408–419, 1984
21. Fallenius A, Franzen S, Auer G: Predictive value of nuclear DNA content in breast cancer in relation to clinical and morphologic factors. A retrospective study of 409 consecutive cases. Cancer (in press)
22. Feulgen R, Rossenbeck H: Mikroskopisch-chemischer Nachweis einer Nukleinsäure vom Typus der Thymonucleinsäure und die darauf beruhende elektive Färbung von Zellkernen in mikroskopischen Präparaten. Z Physiol Chem 135:203, 1924
23. Fisher B, Slack NH: Number of lymph nodes examined and the prognosis of breast carcinoma. Surg Gynecol Obstet 131:79–88, 1970
24. Forsslund G, Cedermark B, Ohman U et al: The significance of DNA distribution pattern in rectal carcinoma: A preliminary study. Dis Colon Rectum 27:579–584, 1984
25. Frankfurt OS, Arbuck SG, Chin JL et al: Prognostic applications of DNA flow cytometry for human solid tumors. In Andreef M (ed): Clinical Cytometry, pp 276–290. New York, Academy of Sciences 1986
26. Franzen S, Giertz G, Zajicek J: Cytological diagnosis of prostatic tumours by transrectal aspiration biopsy: A preliminary report. Br J Urol 32:193, 1960
27. Franzen S, Zajicek J: Aspiration biopsy in diagnosis of palpable lesions of the breast. Acta Radiol (Stockholm) 7:241–262, 1968
28. Gaub J, Auer G, Zetterberg A: Quantitative cytochemical aspects of a combined Feulgen naphtol yellow S staining procedure for the simultaneous determination of nuclear and cytoplasmic pro-

teins and DNA in mammalian cells. Exp Cell Res 92:323–332, 1975

29. Herman CJ, Mardar RJ: Recent progress in clinical quantitative cytology. Arch Pathol Lab Med 111:505–512, 1987

30. Johnston W, Szpak CA, Lottich SC et al: Use of a monoclonal antibody (B72.3) as a novel immunohistochemical adjunct for the diagnosis of carcinomas in fine needle aspiration biopsy specimens. Hum Pathol 17:501–513, 1986

31. Keshgegian AA, Kline TS: Immunoperoxidase demonstration of prostatic acid phosphatase in aspiration biopsy cytology (ABC). Am J Clin Pathol 82:586–589, 1984

32. King WJ, DeSombre ER, Jensen EV et al: Comparison of immunocytochemical and steroid-binding assays for estrogen receptor in human breast tumors. Cancer Res 45:293–304, 1985

33. King WJ, Greene GL: Monoclonal antibodies localize oestrogen receptor in the nuclei of target cells. Nature 307:745–747, 1984

34. Konaka C, Auer G, Nasiell M et al: Pathogenesis of squamous bronchial carcinoma in 20-methylcholantrene-treated dogs. Anal Quant Cytol 4:61–71, 1982

35. Koss LG, Tribukait B: Flow cytometry in surgical pathology and cytology of tumors of the genitourinary tract. In Advances in Clinical Cytology. New York, Masson Publishing USA, 1984

36. Kreicbergs A, Soderberg G, Zetterberg A: The prognostic significance of nuclear DNA content in chondrosarcoma. Anal Quant Cytol 4:271–278, 1980

37. Laerum OD, Farsund T: Clinical application of flow cytometry: A review. Cytometry 2:1–13, 1981

38. Lippman ME, Allegra JC: Quantitative estrogen receptor analyses: The response to endocrine and cytotoxic chemotherapy in human breast cancer and the disease-free interval. Cancer 46:2829–2834, 1980

39. Loebl J: Image analysis: Principles and practice. England, Marquisway Team Valley Gateshead, Tyne & Wear NE11 0QW, 1985

40. Mackey B: Diagnostic electron microscopy of tumors. Clinics in Laboratory Medicine, Vol 7, No 1, Philadelphia, WB Saunders, 1987

41. Magdalenat H, Merle S, Zajdela A: Enzyme immunoassay of estrogen receptors in fine needle aspirates of breast tumors. Cancer Res 46:4265s–4267s, 1986

42. Moberger B, Auer G, Forsslund G: The prognostic significance of DNA measurements in endometrial carcinoma. Cytometry 5:430–436, 1984

43. Moran RE, Black MM, Albert L et al: Correlation of cell-cycle kinetics, hormone receptors, histopathology, and nodal status in human breast cancer. Cancer 54:1586–1590, 1984

44. O'Hara MF, Bedrossian CW, Johnson T et al: Flow cytometry in cancer diagnosis. Stefanini's Progress in Clin Path, pp 135–153. New York, Grune & Stratton, 1983

45. Ono J, Auer G: The significance of DNA measurements for the early detection of bronchial cell atypia. Cytometry 5(3):140–344, 1983

46. Pertschuck LP, Tobin EM, Tanapat P et al: Histochemical analysis of steroid hormone receptor in breast and prostate carcinoma. J Histochem Cytochem 28:799, 1980

47. Press MD, Green GL: An immunocytochemical method for demonstrating estrogen receptors in human uterus. Lab Invest 50:480, 1984

48. Schoidt T: Breast carcinoma. A Histologic and Prognostic Study of 650 Followed-up Cases. Copenhagen, Munksgaard, 1966

49. Shapiro HM: Practical Flow Cytometry. New York, Alan R Liss, 1985

50. Silfversward C, Gustafsson JA, Gustafsson SA et al: Estrogen receptor concentrations in 269 cases of histologically classified human breast cancer. Cancer 45:2001–2005, 1980

51. Taylor CR: Immunomicroscopy: A diagnostic tool for the surgical pathologist. Philadelphia, WB Saunders, 1986

52. Travis WD, Wold LE: Immunoperoxidase staining of fine needle aspiration specimens previously stained by the Papanicolaou technique. Acta Cytol 517, 1987

53. Trump BF, Jesudason ML, Jones RT: Ultrastructural Features of Disease Cells. Diagnostic Electron Microscopy, Vol 1, New York, John Wiley & Sons, 1978

54. Wolley RC, Schreiber K, Koss LG et al: DNA distribution in human colon carcinomas and its relationship to clinical behavior. JNCI 69:15–22, 1982

55. Zetterberg A, Esposti LP: Prognostic significance of nuclear DNA levels in prostatic carcinoma. Scand J Urol Nephrol 55:53–58, 1980

Appendix

Response to Fine Needle Aspiration and a Word of Caution

Pathologists and clinicians who have expressed interest in fine needle aspiration have, in most instances, responded enthusiastically. This is a relatively new direct and dynamic technique that carries the potential for rapid problem solving. A word of caution must be introduced, however. The procedure for obtaining material varies according to body region. The cytologist must work with the neurosurgeon, radiologist, gynecologist, and selected other specialists. Experience in puncturing surface lesions may be insufficient for approaching deeper targets. In short, there is more than one aspiration biopsy cytology (ABC). Written descriptions are clearly insufficient and must remain secondary to actual experience.

In addition, the method of making smears cannot be overstressed. Cells that are fragile and, in particular, cell groupings that can be easily obscured or disrupted must be properly preserved by careful technique. Cellularity should be florid, and scanty smears are too often nondiagnostic unless they clearly are extracted from scirrhous or fibrotic lesions (see Fig. 6-36). The cytologist must evaluate smears critically and repeat puncture when indicated.

Interpretation of cytologic specimens is immediately compromised by poor smears. Since ABC lends itself to regionalization and referral of slides for diagnosis, poor material not only strains the interpreter but is hazardous to the patient, since many therapies follow cytologic diagnosis, including radical surgery and radiation.

The ability to interpret cytomorphologic material provided by fine needle aspiration clearly does not automatically emerge from training in pathology, exfoliative cytology, and hematology. Each organ and region has a set of morphologic parameters with many variables. Individual cell morphology may be dramatic in one clinical disorder; cell grouping may be dramatic in another. Fairly intensive and prolonged experience with large numbers of slides in each category of review is necessary to achieve a safe level of competence. Nevertheless, a beginning must be made, and the organization of training programs is necessary.

Organizing a Training Program

The pioneer program at Karolinska Hospital in Stockholm, Sweden, can serve as a model, although many variations can be introduced.

The aspiration cytology section is part of the division of pathology. At Karolinska, this department and the puncture clinic are separate from the section of exfoliative cytology. Although the two cytology sections share many common features, the clinical orientation of the aspiration cytology section sets it apart. In our opinion, such a department should be separate. Histopathology, however, is the diagnostic base on which cytopathology is erected, and the two cannot be separated.

A regionalized puncture clinic that will receive referred patients is essential. It should serve 20 to 30 patients a day, providing adequate and varied puncture experience in palpable disease. Specialized punctures, usually carried out in radiographic suites within the hospital complex, should be available to trainees. Archives of slides, stored and collated with clinical data, should be available. A residency or fellowship program of not less than 1 year, and preferably 2, is required. Training consists of full puncture experience, initially supervised. Patient contact and pertinent review of historical and physical findings are implicit. The trainee reads his own slides, initially with a supervising physician and two-head microscope.

Systematic review of all lesion types in the slide archives, together with study of collections of unknown smears, is essential. The trainee participates in clinical pathologic conferences, particularly with reference to current patients, and maintains contact with histopathologic, clinical, and radiographic aspects of each case. Ideally, he becomes part of the decision-making team that plans future diagnosis and therapy.

An outline of such a program is presented below:

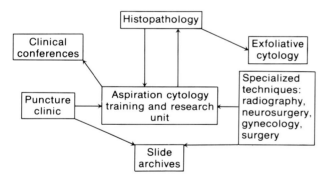

Such a program cannot spring fully developed from an existing pathology department. Someone, usually a member of the pathology department, must express interest, visit an ongoing program, and attend workshops, courses, or tutorials. The concept must then be presented to the clinicians. Collaboration of clinician and cytopathologist must be determined, and this will vary among institutions. It is the thesis of this book that best results are obtained when the operator also reads his slides. However, intimate cooperation between clinician and cytopathologist can produce comparable results.

Study sets and slide collections are gathered from patient material and by aspiration of surgical specimens at the bench. Tissue sections of the same specimen are included to complete the study set. This approach will lead to the development of a full training program.

· · ·

ABC is a fully developed discipline with an impressive literature and several decades of growing use and achievement. Current failure to reach into all the major clinical and pathologic institutes reflects some degree of inertia. A movement of clinicians and pathologists beyond traditional methods of diagnosis is necessary. We hope that this volume will provide an appropriate stimulus for that move.

Index

ISBN 0-397-50826-3